REVIEWS IN ENGINEERING GEOLOGY
VOLUME XIII

MILITARY GEOLOGY IN WAR AND PEACE

Edited by

James R. Underwood, Jr.*
Department of Geology
Kansas State University
Manhattan, Kansas 66506-3201

and

Peter L. Guth
Department of Oceanography
U.S. Naval Academy
572 Holloway Road
Annapolis, Maryland 21402-5026

The Geological Society of America, Inc.
3300 Penrose Place, P.O. Box 9140
Boulder, Colorado 80301
1998

*Present address: 9518 Topridge #3,
Balcones Place, Austin, Texas 78750.

The Reviews in Engineering Geology series was expanded in 1997 to include Engineering Geology Case Histories, 11 volumes of which were published by the Geological Society of America from 1957 to 1978 with ISBNs from 0-8137-4001-0 to 0-8137-4011-8. Beginning with Volume XI, Reviews in Engineering Geology may include both reviews and case histories, under the ISBN 0-8137-4111-4 and subsequent numbers.

Published by The Geological Society of America, Inc.
3300 Penrose Place, P.O. Box 9140, Boulder, Colorado 80301

Printed in U.S.A.

GSA Books Science Editor Abhijit Basu

Library of Congress Cataloging-in-Publication Data

Military geology in war and peace / edited by James R. Underwood, Jr. and Peter L. Guth.
p. cm. -- (Reviews in engineering geology ; v. 13)
Includes bibliographical references and index.
ISBN 0-8137-4113-0
1. Military geology--History. 2. Military geology.
I. Underwood, James R., 1927- . II. Guth, Peter L. III. Series.
TA705.R4 vol. 13
[UG645] 98-7163
[623'.01'55]--dc21 CIP

10 9 8 7 6 5 4 3 2 1

Contents

Foreword

The history of geology is bound to practical matters. From the works of Agricola on mining, to those of civil engineer William Smith, practical considerations have given rise to theories, which in turn have been applied to new practical problems.

As shown in the chapters in this book, military geology followed this classic pattern. Terrain analysis was its earliest manifestation, as practiced by French military engineers. Their development of the hachured map was a form of geomorphic expression that allowed freedom of interpretation and emphasis on critical features that were not possible within the literal strictures of the contour maps. Such cartographic emphasis, designed to guide the user, is a fundamental tenet of military geology. The "go–no go" system, described or implied in several chapters, is an example of this emphasis, as is the important cartographic system of preparing basic-data geologic maps founded entirely on lithology rather than on geologic age as expressed in formally named formations.

Scale, brevity, and emphasis are key concerns in military geology. Fortunately, the traditional military chain of command furnishes criteria that can be used by the military geologist in designing reports and recommendations. Strategic and tactical needs at all levels of command dictate the format and context of the military geology report as military organizations evolve and as military requirements change.

The decision of what to present and how to present it requires, of course, knowledge of both earth science and military science. The procurement, training, and organizational position of military geologists became a subject of debate immediately after World War II, and the debate continues. Should the military geology advisors be highly trained civilian scientists who, when needed, could be attached to headquarters and sometimes to tactical troops in a theater of operation? This was the system used during World War II by the Military Geology Unit (MGU) of the U.S. Geological Survey.

The MGU, located in Washington, D.C., was a self-contained organization of civilian specialists that, in addition to its primary duties of intelligence production, served as a cadre to supply experts for integration into military units. Such use of civilian specialists, which was refined during and immediately after World War II, was from the beginning greatly influenced by the thoughts of Charles B. Hunt, who was instrumental in the founding of the MGU and in establishing its relations with the U.S. Army Corps of Engineers. It was under this arrangement that I was attached to general headquarters of the supreme commander of the Allied powers in Tokyo where, although a civilian, I functioned as a staff officer. In 1946 I was assigned to the headquarters of XXIV Corps in Korea where I carried out terrain analyses essentially on my own.

Should military geologists be fully trained as soldiers and then assigned to special advisory organizations, as is now practiced in the British army reserve? As pointed out in this volume, W. B. R. King, as a British army officer in both world wars, was influential in the use of geologists who also were members of the military. It would appear that such integration of scientists into the services would disseminate them more widely; on the other hand, use of a civilian organization may encourage innovation and also may reduce the likelihood that, once a specialist is in uniform, diversion to nongeologic duties would occur. These are questions well worth discussing.

Given a shortage of trained geologists in military field units, what is the utility and what are the pitfalls of the use of terrain-analysis and remote-sensing keys by technicians who do not have earth-science training? What are the career opportunities for military geologists in a science, geology, that bestows recognition on the basis of published research? These questions are among those explicitly or implicitly addressed in the chapters that follow, and it is good that this discussion continues.

From the historic point of view, it is fascinating to examine the development of military geology techniques. The go–no go concept was used by the MGU in World War II, and it is implied in this book that it also was used by the British army. Charles B. Hunt of MGU visited British intelligence organizations in 1944, and shortly thereafter, Major John Farrington of the British Interservice Topographic Detachment made two visits to MGU. The go–no concept already was in use at MGU at that time, but I do not know whether it was in use in Britain. It is my opinion, based also on examining the products of the German Wehrgeologie organization, that earth scientists, faced with military problems, tended to come up independently with similar solutions. An exception in World War II was Japanese Intelligence, whose terrain reports were devoted mainly to living off the invaded country.

In discussing British-American military geology relations, I must digress to recount an example of how even allies cannot be fully open with each other. At the time of one of John Farrington's visits to MGU, we were doing terrain studies of China using 1:50,000-scale topographic maps prepared by the Chinese, on condition that we not inform the British. So, when notified that Farrington was coming, we moved the China project to the top floor of the building and set up a dummy project for Farrington to inspect.

Much is recorded in this collection about the applications of earth-science principles to military problems, but there also is indication of the impetus provided by military geology to the expansion of geologic knowledge. A prime example of this is the study of permafrost that was, until World War II, almost completely a Russian domain of understanding and research. Simon W. Muller of MGU, a Russian-born paleontologist, translated the major Russian book on the subject, giving rise to the postwar program of the Alaska Terrain and Permafrost Program of the Military Geology Branch, which made detailed studies in areas of perennial and sporadic permafrost in Alaska and elsewhere. The new concern with permafrost also led to the establishment of the cold regions research laboratory of the Corps of Engineers and of the Air Force Laboratories at L. G. Hansscom Field, described in this book.

The Pacific Island Mapping Program, also described here, has given us new awareness of regional geology—a useful contribution to plate-tectonic studies—and expanded our knowledge of the engineering properties of tropical soils and the cyclic phenomena of coral reefs, which are most useful in evaluating environmental change.

Perhaps the most important, although not the most obvious, contribution of military geology is its development of ways of thinking and presenting data for the use of nonscientists, especially policymakers. The fields of engineering geology and environmental sciences, in their development during the past 50 years, have become much more effective because of the acceptance by our profession of these approaches.

These chapters reflect a lively branch of science in all of its ramifications. Much historic information is presented, and this should stimulate further research into the history of military geology. It is clear, however, that military geology is not by any means an endeavor only of the past. This book provides the groundwork for consideration of not only the history but also the present state and future development of this challenging and important discipline.

Frank C. Whitmore
Washington, D.C.
August 1997

Geological Society of America
Reviews in Engineering Geology, Volume XIII
1998

Military geology in war and peace: An introduction

Peter L. Guth
Department of Oceanography, U.S. Naval Academy, 572 Holloway Road, Annapolis, Maryland 21402-5026

Military Geology: Those branches of the earth sciences, especially geomorphology, soil science, and climatology, that are applied to such military concerns as terrain analysis, water supply, cross-country movement, location of construction materials, and the building of roads and airfields (Bates and Jackson, 1987).

ABSTRACT

In warfare military geologists pursue five main categories of work: tactical and strategic terrain analysis, fortifications and tunneling, resource acquisition, defense installations, and field construction and logistics. In peace they train for wartime operations and may be involved in peace-keeping and nation-building exercises. Although many geologists view military geology as a branch of engineering geology, the U.S. military does not include geologists in its force structure and gets geological assistance on an ad hoc basis. The army does, however, include organic terrain teams at division and higher levels to provide routine information for mission planning and execution. The classic dilemma for military geology has been whether support can best be provided by civilian technical-matter experts or by uniformed soldiers who routinely work with the combat units.

INTRODUCTION

Geology has shaped the art of warfare since its inception. Cavemen moved into caves for protection from the elements, wild animals, and other humans; they must have used their knowledge of the earth to help find suitable locations. The Greeks and Trojans fought the first war celebrated in literature at a site where geology dictated the strategic importance of a hilltop dominating the coastal plain near a strategic strait. But geology did not emerge as an independent, defined science until the end of the 18th century, and military geology did not develop as a separate subdiscipline until the middle of the 19th century (Kiersch and Underwood, this volume).

In warfare military geologists pursue five main categories of work: tactical and strategic terrain analysis, fortifications and tunneling, resource acquisition, defense installations, and field construction and logistics. Engineer, intelligence, and logistics officers handle these functions on military staffs. In the current international climate, military geologists also support nation-building and peace-keeping efforts, environmental cleanup, and implementation of the eternal dream of converting swords into plowshares. Finally, in the United States, through interesting quirks of history, the Army Corps of Engineers has a major role in civil works such as river and beach erosion and stabilization. To support this mission the Corps has major research laboratories that employ many geologists (Waterways Experiment Station, Topographic Engineering Center, and Cold Regions Research and Engineering Laboratory), and Corps of Engineers work often combines civil works with support of the active army.

To get a sense of the state of military geology and its history, a search was conducted in GEOREF on the terms "military geology" and "military geography." As will be clear from the results, GEOREF does not provide complete coverage of military geology, but it does point to both the successes and the challenges facing military geology. Through mid-1996, GEOREF contains 526 references to military geology, which might be approximately half of the true total (A. W. Hatheway, pers. comm., 1996). Of those with a language indicated, almost all (424) are in English. The German heritage of military geology appears in 41 references, but this number probably grossly underestimates the true German and Austrian contributions (Hatheway, 1996). Coverage drops dramatically to 4 or 5 each in Russian, French, and Finnish, and 1 or 2 each in Polish, Spanish, Hungarian, Dutch, and Japanese.

Figure 1 shows the references to military geology plotted by decade. The general increase in knowledge, or at least the frenzy to publish, is obvious over time, but there are distinct peaks after both major world wars. From the historical perspective, three works

Guth, P. L., 1998, Military geology in war and peace: An introduction, *in* Underwood, J. R., Jr., and Guth, P. L., eds., Military Geology in War and Peace: Boulder, Colorado, Geological Society of America Reviews in Engineering Geology, v. XIII.

stand out for military geology in the United States. Vogdes (1884) marks the earliest work on military geology contained in GEOREF, and Vogdes would make an appropriate father for American military geology. A drummer boy in the Civil War who rose through the ranks to retire as a brigadier general, Vogdes was an internationally known trilobite paleontologist and a founding member of the Geological Society of America (Dumble, 1923). Among the flurry of publications marking World War I, the paper by Brooks (1920) serves as the definitive statement on the role of geology in that conflict; the paper by Hunt (1950) has a similar role for the Second World War. Major papers also appeared in the Bulletin of the Geological Society of America for each world war (Cross, 1919; Erdmann, 1943), one of which was a presidential address by Whitman Cross. No similar comprehensive reviews of military geology appeared after the Korean, Vietnam, or Gulf War conflicts.

GEOREF indexes references with key terms, and the 526 references to military geology have a total of 5,874 index terms, an average of 11.2 terms per paper. A total of 1,804 different descriptors appear, although the database contains synonyms that may not have been consistently applied, e.g., "terrain analysis," "terrains," "terrain intelligence," "terrain evaluation," "terrain studies," and "terrian [sic] analysis" all appear. Only three descriptors appear in more than 100 of the papers (Military geology, United States, and Engineering geology), 30 terms appear 26 or more times, and 638 terms appear at least twice. Thus almost two-thirds of the descriptors used apply to only one publication.

Table 1 shows the geographic coverage of military geology as reported in GEOREF; there is a clear North American bias. Table 2 shows the breakdown by topics; engineering geology and hydrogeology are the most common disciplines. Terrain analysis, the field recognized by the U.S. Army as important to its needs in creating teams for each division and corps, shows up much less often in the geologic literature. This could be a reflection of the cross-disciplinary nature of terrain analysis, covering geology, photogrammetry, remote sensing, geography, and other fields. Table 3 shows the publications and series in which the papers have appeared; the U.S. Geological Survey clearly dominates. A search for primary authors found only 10 authors with multiple publications and only one (C. V. Theis with 42) with more than two.

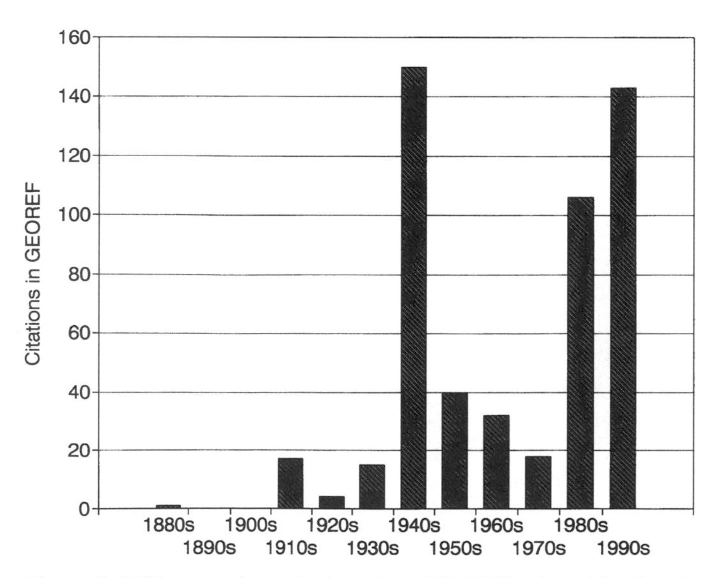

Figure 1. Military geology citations from GEOREF, plotted by decade. Note the peaks in the 1910s and 1940s marking the two world wars.

TABLE 1. GEOGRAPHIC COVERAGE OF MILITARY GEOLOGY FROM GEOREF

Papers	Region
144	United States
67	Europe
52	Asia
39	Far East
34	Western Europe
30	New Mexico
21	Canada
21	Central Europe
19	Alaska Highway
16	Germany
16	U.S.S.R.
14	France
13	Alaska
13	Burma
11	United Kingdom
10	Africa
10	Great Britain
10	Micronesia
10	Oceania

The relationship between the U.S. military and the geologic community remains enigmatic. In a pattern that repeats over the last century, between wars geologists and the military have little in common. The U.S. Geological Survey or military laboratories supply geologic advice and data when needed, but staffs and units do not have geologists assigned. A normal, automatic working relationship never develops between military leaders and civilian scientists brought in for specific missions. During wars the military enlists geologists into ad hoc organizations to supply field units with geologic assistance. After the war the geologists involved suggest a permanent, integrated corps of geologists in the military so that all staff planning incorporates geologic analysis. After World War I, Brooks (1920, p. 124) suggested "a staff of geologic engineer reserve officers should be organized." After World War II Hunt (1950) suggested expansion of the model set by the Military Geology Unit during that war, and the assignment of geologists down to at least the level of Corps. Whitmore (1953, 1955) discussed the problems in getting trained earth scientists to solve military needs for aerial photography analysis.

In the 1980s the U.S. military finally recognized the need for routine input of terrain analysis for staff planning. Each army division and Corps has an organic engineer terrain-analysis team assigned. This team, led by a warrant officer, provides terrain analysis, usually to the intelligence (G2) staff section. Although these soldiers receive some training in geology at military schools, they are not geologists. Their considerable expertise comes from military schools and practical experience and not a standard geology education.

Many geologists consider military geology as a branch of engineering geology; the topic coverage of the military geology papers indexed in GEOREF in Table 2 demonstrates this orientation. This does not coincide with the definition quoted earlier in this chapter (Bates and Jackson, 1987) or with the U.S. Army's current organization. Terrain-analysis support comes from organic teams assigned to each division, whereas engineering geology expertise is supplied by two Engineer commands in the Army Reserve, the Corps of Engineers research laboratories staffed with civilian scientists, or the U.S. Geological Survey. Assignment of geologists to support particular missions may accomplish the particular task, but geologic input does not become a normal part of staff action considered by commanders in the planning and execution phase of operations. Because it is an organic asset, the terrain team routinely provides products and input, and commanders develop a relationship with it during training that translates into reliance on its advice during operations.

Military geology uses the expertise of all fields. During World War II micropaleontologists and mineralogists helped trace the source of balloon-delivered bombs that followed the jet stream from Japan (Mikesh, 1973; McPhee, 1996). A military geologist must be a generalist who provides a rapid assessment that can be incorporated into military plans.

TABLE 2. TOPIC COVERAGE OF MILITARY GEOLOGY FROM GEOREF

Papers	Topic
101	Engineering Geology
88	Ground water
57	General
57	Maps
53	History
46	Surveys
42	Hydrogeology
41	Geology
40	Water supply
34	Geologic maps
34	Site exploration
32	Aquifers
31	Government agencies
30	Resources
30	Waste disposal
29	Pollution
28	Soils
27	Catrography
27	Data processing
27	Explosions
26	Applications
26	Geomorphology
24	Remote sensing
23	Economic geology
23	Environmental geology
23	Nuclear explosions
23	Physiographic geology
23	Wells
21	International cooperation
21	Seismology
21	Water wells
20	Highways
20	Mineral resources

TABLE 3. PERIODICALS WITH MILITARY GEOLOGY PAPERS FROM GEOREF

Papers	Publication
62	Open-File Report U.S. Geological Survey
37	Abstracts with Programs Geological Society of America
27	Geological Society of America Bulletin (17 abstracts only)
22	EOS, Transactions, American Geophysical Union (all in two volumes)
16	Water-Resources Investigations U.S. Geological Survey
13	Science
12	U.S. Geological Survey Circular
10	India, Geological Survey, Strategic Branch, Technical Note
9	Economic Geology and Bulletin of the Society of Economic Geologists
7	The American Association of Petroleum Geologists Bulletin
7	The Military Engineer
6	U.S. Geological Survey Professional Paper
5	The Royal Engineers Journal
5	Geotimes

MILITARY GEOLOGY IN WAR AND PEACE

This volume grew out of a symposium organized for the 1994 annual meeting of the Geological Society of America in Seattle, Washington. The symposium had two half-day sessions, with 16 presentations in the morning and 14 in the afternoon. Abstracts for all the papers presented in the symposium, except for the informal introduction, appeared in the abstracts volume for the meeting (Geological Society of America, 1994), and Hatheway (1996) provided a discussion of 27 of the 30 papers presented.

In addition to the introduction, this volume includes 24 papers that grew out of the symposium and have been organized into six sections. The first group of 11 papers covers selected aspects of the history of military geology. The next three papers deal with the current status of geology in the military. Three papers deal with terrain analysis, three with engineering geology in the military, and three with the use of military geology for diplomacy and peace keeping. The last paper deals with future relations of military geology and the U.S. Army.

The section on the history of military geology contains 11 papers arranged in chronological order. The overview paper by *Kiersch and Underwood* traces the use of geology in military operations from 1800 to 1960. The paper by *Rose and Rosenbaum* discusses military geology in Britain during the 19th and 20th centuries. *Pittman* covers the use of geology and the role of American geologists during World War I—the last great trench warfare with mining and tunneling playing key roles in huge bloody conflicts fought for minor advances of the front lines. In the first of two papers, *Terman* covers the role of the U.S. Geological Survey Military Geology Unit in World War II. *Rose and Pareyn* relate geology to the Normandy landings, the largest amphibious operation on the western front in Europe during World War II. A posthumous paper by *Corwin* chronicles the post-World War II mapping program of Pacific Islands by the

U.S. Geological Survey for the U.S. Army Corps of Engineers. This mapping represented part of the larger program of the Military Geology Branch after the war and is described by *Terman* in his second paper. *Cameron* relates major offensives in Korea to secure limited objectives prior to the peace talks as fundamentally related to the geologic structure of the ridges and an appreciation from terrain analysis of the importance of the ridges in securing a defensible truce line. A second paper by *Cameron* describes the underground tunnels used to get under that defensible truce line. *Neal* describes how military geology during the cold war produced scientific advances, an example of the Biblical injunction to "turn swords into plowshares." *Knowles and Wedge* describe the use of military geology in the Gulf War.

The section on the current status of military geology covers the situation in the British and U.S. military. *Rosenbaum* discusses recent applications of military geology in the British armed forces. *Jens and Stevens* highlight the program of geologic education in the U.S. Army; the education program reveals the army's vision of what it needs in the way of geologic support and how the Army supplies that support in the form of soldiers assigned to field units. *Leith and Matzko* discuss current support of the military from the U.S. Geological Survey.

The U.S. Army currently recognizes the importance of terrain analysis for military operations with the assignment of a terrain analysis team to each division and corps. *Rinker* discusses the relationship of terrain analysis and remote sensing to military examples. *Ehlen* discusses the use of remote sensing to characterize fracture patterns in politically or geographically inaccessible areas as a precursor to more complete terrain analysis. *Neal* discusses significance of playas in military operations.

Three examples cover the variety of engineering applications of military geology. *Eastler, Percious, and Fisher* discuss geologic aspects for the siting of underground facilities, which range from the foxholes of individual soldiers to national command and control facilities. *Krinsley* recounts the story of locating and designing airfields in Greenland during the early days of the cold war. *DeGoes and Neal* provide a history of the construction of Distant Early Warning (DEW) Line.

As the threat of another major world war diminishes, many nations are considering other uses for their military forces. *Nathanail* served as a reserve engineer officer with British peace keeping forces in Bosnia, and he describes the engineering geological support rendered there. *Knowles* discusses the military geology involved in supporting engineer construction projects in Latin America, efforts that provide realistic training for soldiers and generate good will in support of foreign policy. *Baehr* provides details of the importance of good geologic background work to ensure the success of well drilling projects.

The final paper by *Hatheway and Stevens* provides a look to the future and how the U.S. Army might increase its geologic expertise even as the army downsizes with the end of the cold war. This represents the classic dilemma for military geology: how to get critical geologic information into the decision making process for commanders, and whether that can best be done through uniformed geologists who train and fight with the total team, or by experts—probably most of whom would be civilians called on only as necessary.

ACKNOWLEDGMENTS

The impetus for this symposium grew out of our associations with the First Infantry Division (Big Red One) at Fort Riley, Kansas. Jim Neal had the inspiration for the title. The original symposium and this volume would not have been possible without the help and enthusiasm of the many individuals who (1) reviewed the abstracts of papers proposed for the symposium, (2) prepared and presented the papers at the symposium, (3) provided critical reviews of the papers submitted for the volume, and (4) prepared the final manuscripts included here. We express our appreciation to each of them, to Ms. Linda Clough and Katherine L. Bier of Kansas State University for their assistance with many of the manuscripts, and to the Geological Engineering Division of the Geological Society of America for sponsoring the symposium at the 1994 annual meeting of the Society.

Our reviewers include: Allen F. Agnew, Ernest Angino, Todd Bacastow, Lawrence D. Bonham, John Brockhaus, Christopher P. Cameron, George Clark, Raymond E. Coveny, Jr., Bo Dunaway, Thomas E. Eastler, Paul Fisher, Joe Fucella, Allen Hatheway, J. Ponder Henley, John Jens, Marie Johnson, George A. Kiersch, Donald W. Klick, Bob Knowles, Daniel B. Krinsley, William Leith, John R. Matzko, Ward S. Motts, C. P. Nathanail, Jim Neal, Cliff Nelson, Walter E. Pittman, Jack Rachlin, John Robertson, Edward P. F. Rose, Michael S. Rosenbaum, Joe Schweitzer, Carl J. Smith, Merrill Stevens, George E. Stoertz, Al Strong, Joshua Tracey, Ed Tremba, Kenneth Verosub, and Frank Whitmore.

REFERENCES CITED

Bates, R. L., and Jackson, J. A., 1987, Glossary of geology: American Geological Institute, 3d ed., 788 p.

Brooks, A. H., 1920, The use of geology on the western front: U.S. Geological Survey Professional Paper 128-D, p. 85–124.

Cross, W., 1919, Geology in the World War and after: Geological Society of America Bulletin, v. 30, p. 165–188.

Dumble, E. T., 1923, Memorial of Anthony Wayne Vogdes: Geological Society of America Bulletin, v. 35, p. 37–42.

Erdmann, C. E., 1943, Application of geology to the principles of war: Geological Society of America Bulletin, v. 54, p. 1169–1194.

Geological Society of America, 1994, Abstracts with programs, 1994 Annual Meeting, v. 26, no. 7, p. 275–278, 345–348.

Hatheway, A. W., 1996, Conference report: 1994 Seattle symposium reveals complexity and value of military geology: Engineering Geology, v. 44, p. 245–253.

Hunt, C. B., 1950, Military geology, *in* Paige, S., ed., Application of geology to engineering practice: Geological Society of America, Berkey Volume, p. 295–327.

McPhee, J., 1996, Balloons of war: New Yorker, v. 71, no. 46, p. 52–60.

Mikesh, R. C., 1973, Japan's World War II balloon attacks on North America: Smithsonian Annals of Flight, no. 9, 85 p.

Vogdes, A. W., 1884, Course of science applied to military art; Part I, geology and military geography: Fort Monroe, Virginia, 176 p.

Whitmore, F. C., Jr., 1953, The dilemma of military photo interpretation: Photogrammetric Engineering, v. 20, p. 425–427.

Whitmore, F. C., Jr., 1955, Manpower for military photo interpretation of terrain: Photogrammetric Engineering, v. 22 , p. 717–719.

Manuscript Accepted by the Society October 29, 1997

Printed in U.S.A.

Geological Society of America
Reviews in Engineering Geology, Volume XIII
1998

Geology and military operations, 1800–1960: An overview

George A. Kiersch
Professor Emeritus, Cornell University, 4750 North Camino Luz, Tucson, Arizona 85718
James R. Underwood, Jr.
Department of Geology, Kansas State University, Manhattan, Kansas 66506

ABSTRACT

The first recorded use of terrain analysis was in 1813 during the Napoleonic Wars, and in most major military operations since that time, geologic counsel and assessment have played important roles. Intelligent use of the terrain of the battlefield, movement of supplies and personnel, and the procurement of adequate supplies of water and of construction materials all have relied on an understanding and application of geologic principles.

During the 19th century, as the value of geologic insight came to be recognized, books on military geology appeared as did basic courses in geology at military academies in the United States and abroad. Beginning in World War I, vital geologic data were placed on increasingly sophisticated specialized terrain maps and used both tactically and strategically. Successful military mining beneath enemy fortifications in World War I required an understanding of subsurface geology, including hydrogeology. And in the 1940s and 1950s, geologic principles were applied on an unprecedented scale to the construction of massive underground installations. Moreover, in the 1950s, these principles, applied in a massive research effort, resulted in the ability to distinguish the release of energy by an underground nuclear test from that produced by a natural seismic event.

As weapons and defenses against them continue to evolve, geoscience and geoscientists will play an increasingly important role in military planning and operations in diverse and challenging environments worldwide.

INTRODUCTION

The geological characteristics of contested terrain have influenced the outcome of battles and wars since individuals, tribal groups, and nations began vying for power and control. But only when the relatively new science and profession of geology evolved in the late 18th–early 19th centuries did those with some understanding of geology begin to have a direct influence on military planning and operations.

From that period to the present, the ways in which geology and geologists have influenced military affairs have been numerous and varied and, in some instances, critical. This influence has ranged from analysis of the best use of terrain to the World War II production of sophisticated preinvasion geologic folios and the construction of underground factories and shelters to the postwar siting of huge underground military command and control centers.

EARLY GEOLOGIC CONCEPTS OF WARFARE

Although early reports mention that two "geologists" were attached to Napoleon's expeditionary forces when they invaded Egypt in 1798, the first documented military operation using geologic guidance was in 1813. Professor K. A. von Raumer analyzed the terrain of Silesia for General von Blucher, who defeated Napoleon's forces there under General Jacques Macdonald in the battle of the Katzback River, August 26, 1813 (Betz, 1984, p. 238–241). A similar geologic analysis of the terrain of Luxembourg was made in 1843 for the French military (Marga, 1885;

Kiersch, G. A., and Underwood, J. R., Jr., 1998, Geology and military operations, 1800–1960: An overview, *in* Underwood, J. R., Jr., and Guth, P. L., eds., Military Geology in War and Peace: Boulder, Colorado, Geological Society of America Reviews in Engineering Geology, v. XIII.

Barré, 1897–1902). In 1826, Johann Samuel of Freiberg Academy, Saxony, presented "Geology and Military Science," believed to be the first paper on military geology (Betz, 1975, p. 95).

The Revolutionary War campaigns in New York and New Jersey in 1776–1777 were influenced significantly by terrain features, including relict glacial deposits of the region (Fig. 1). For example, the drift of the Harbor Hill terminal moraine (late Wisconsinan) was distributed across central Long Island, the Narrows of Hudson River valley, and Staten Island. This terminal moraine is the topographic high in the area and thus was the ideal location for fortifications to protect New York City harbor, and Forts Hamilton and Wadsworth (Fig. 1). Perhaps the terrain lessons learned during the Revolutionary War campaigns influenced the U.S. Military Academy, in 1823, to be among the first institutions to introduce formal instruction in geology, even though geology was an embryonic science (Smith, 1964, p. 312).

In America, the historic Battle of Gettysburg in 1863 was critically and decisively affected by the terrain (Brown, 1961).

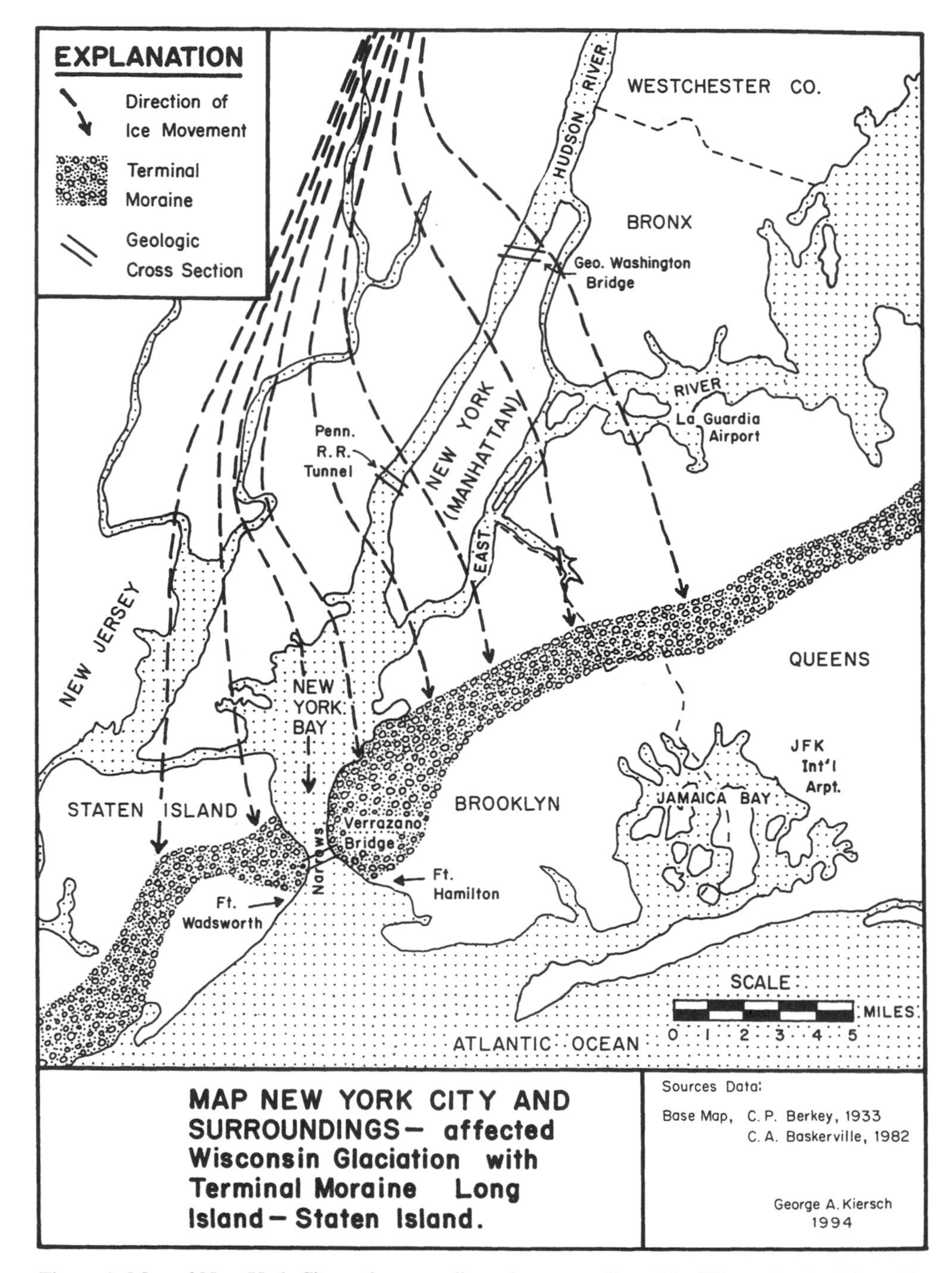

Figure 1. Map of New York City and surroundings that were affected by Wisconsin glaciation. The major feature, Harbor Hill terminal moraine, extends across Long Island and Staten Island.

For example, several strategic heights and commanding hilltop areas such as Little Round Top and Cemetery Ridge (Fig. 2) are underlain by a diabase sill that occurs within the softer shale and sandstone of the Gettysburg Triassic basin (Fig. 2, index map). Consequently, the Confederate troops attacking from the west were unable to dislodge the Union defenders who were dug in along the diabase outcrops (Fig. 2) and boulder fields at such localities as Devil's Den, Plum Run, Round Top, and Little Round Top. The Confederate forces never fully recovered from the Gettysburg defeat.

During the Union army's siege of Vicksburg during May–July 1863, one of General Grant's options was to stop all Mississippi River traffic with the city and force the capitulation of the Confederate forces. This effort included construction of the Williams-Grant Canal at the Tuscumbia Bend of the main Mississippi River, situated at the site of an earlier canal south of Vicksburg that bypassed the city. The canal would divert the main river flow away from the large meander-loop channel that served the Vicksburg waterfront (Fig. 3). The 1863 diversion effort was not fully implemented nor successful. Within a few years, how-

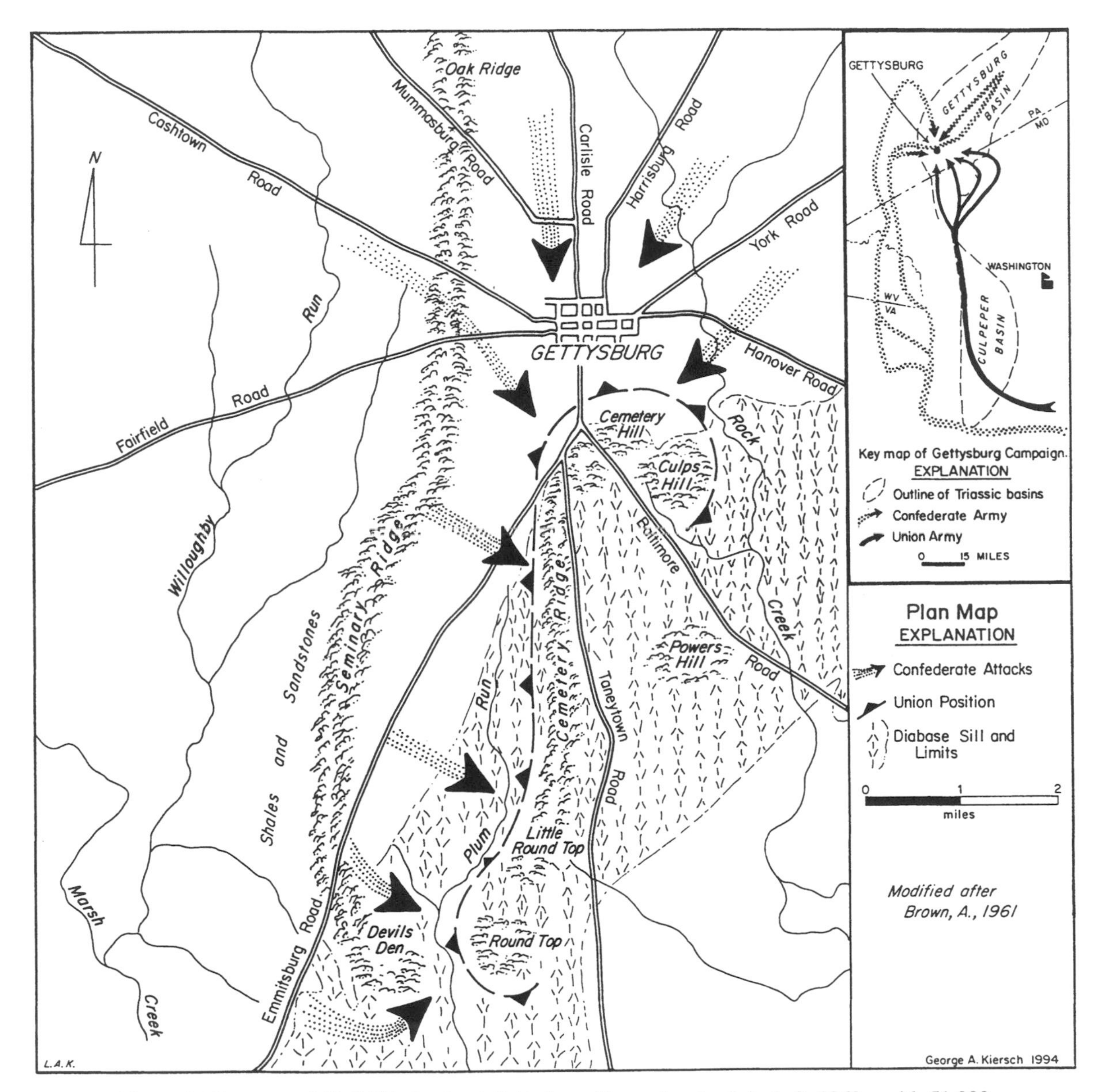

Figure 2. Plan map of Civil War battle at Gettysburg, Pennsylvania, July 1–3, 1863—with 51,000 casualties the most costly battle in U.S. history. The diabase outcrop, a sill within the sequence of Triassic sediments, was strategic to the battle's outcome. The boulder-strewn terrain of Devil's Den, Cemetery Ridge, Little Round Top, and Plum Run sectors were used by Union forces as protective fortifications against Confederate forces that advanced over open fields; they could not dislodge the dug-in Union troops.

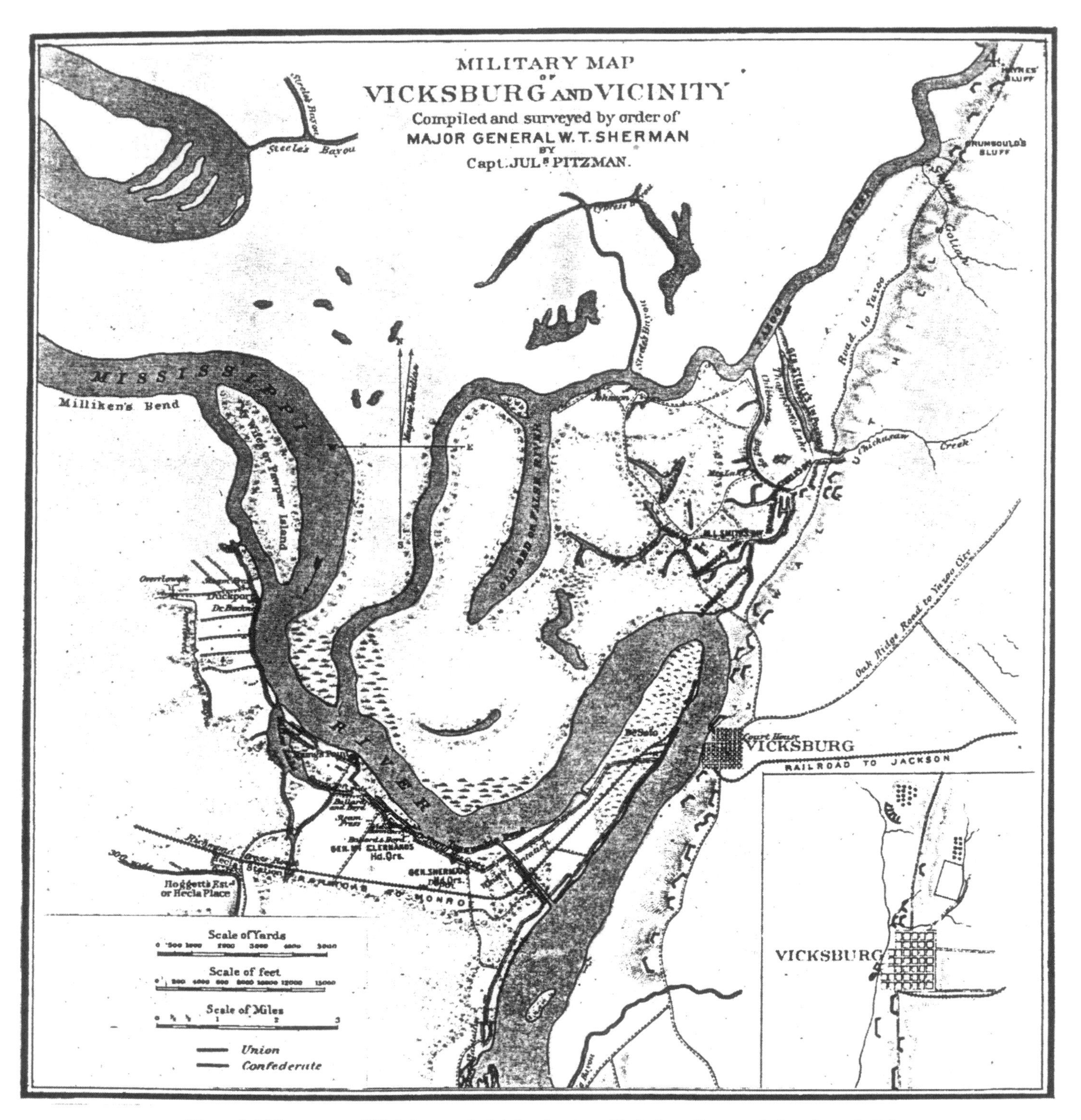

Figure 3. Military map of Vicksburg and vicinity showing: (1) the broadly meandering channel of the Mississippi River upstream from the city and (2) the location of the Williams-Grant Canal project of 1863 at the Tuscumbia Bend, downstream from the city. Note the changes in the channel of the Mississippi shown in Figure 4 soon after Grant's attempt to divert the river's flow from the Vicksburg waterfront. (From the Official Union and Confederate Army Atlas; reproduced with the permission of the Heritage Publishing Co., 1995.)

ever, the Mississippi River diverted naturally to a new channel close by the Williams-Grant alignment, which isolated Vicksburg from the main river and its traffic. More recently, the U.S. Army Corps of Engineers dredged a connection between the Yazoo River and the old 1863 channel so that today river water again flows past Vicksburg (Fig. 4).

On July 4, 1863, another Union victory was the climax of the Tullahoma campaign under General William Rosecrans during which his forces maneuvered the Confederate army of Tennessee out of strong regional positions and into the Confederate base in Chattanooga. Much of Rosecrans' success was due to his knowledge of geology and the rugged terrain over which troop movements were conducted (Brown, 1963a, b). Using an 1856 geologic sketch map of Tennessee, Rosecrans maneuvered within the flat stream valleys of the dissected Highland Rim country and its spur features. Suitable topographic maps were nonexistent for most areas and Rosecrans often utilized an ingenious tactic to supply critical terrain information. Union cavalry forces would fan out in advance to reconnoiter the unknown terrain, a tactic many criticized as "useless" raids; they were not aware of the underlying purpose, which was to obtain badly needed topographic information in advance of infantry movements (Brown,

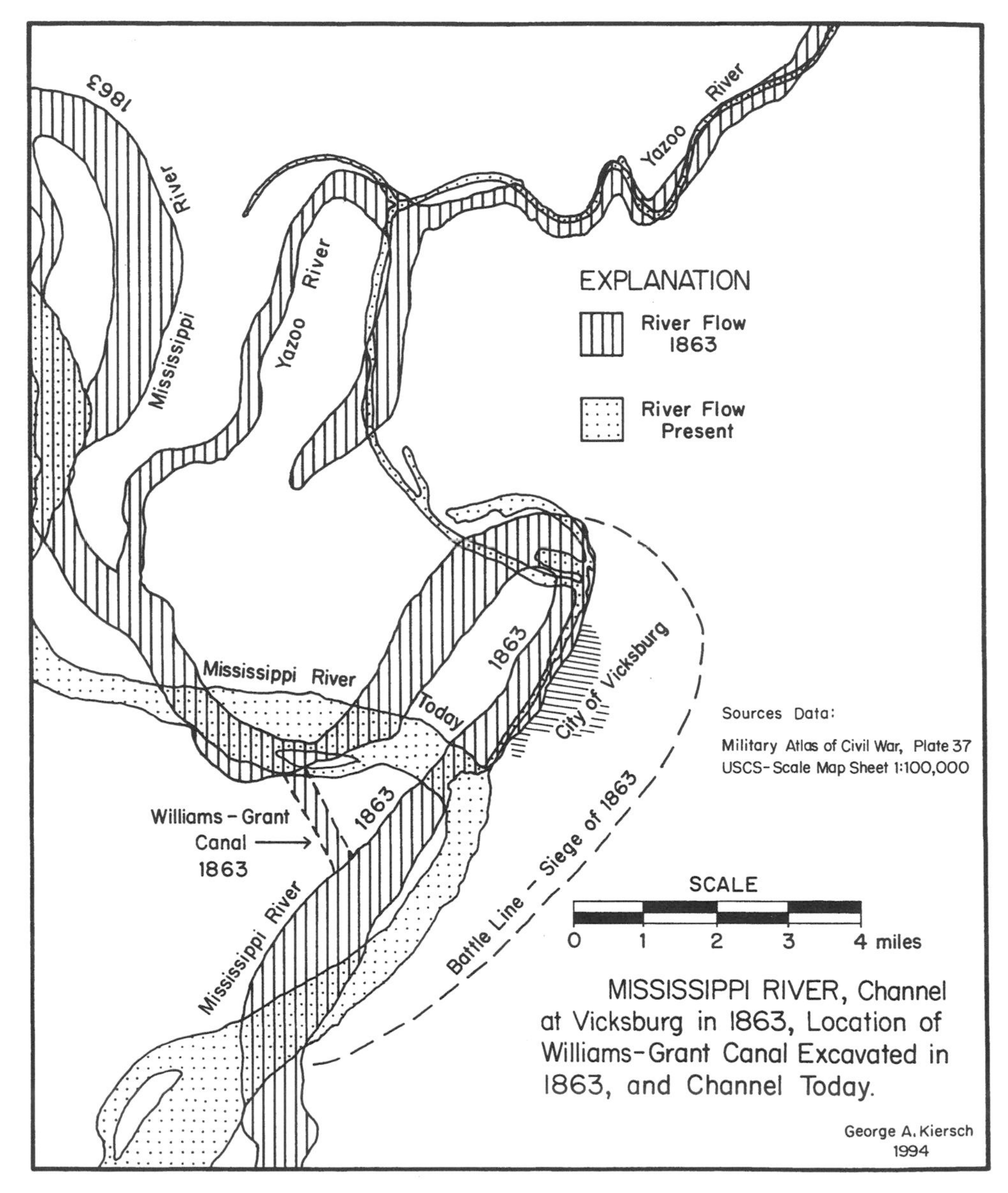

Figure 4. Map of the Mississippi River meandering channel at Vicksburg in 1863 and the Williams-Grant Canal, dug that year by the Union army in an effort to divert the river's flow and defeat the Confederates under siege in the city (May–July, 1863). During high-flood stage of the river a few years later, the main channel diverted itself to its present location close by the Williams-Grant canal site, which bypasses Vicksburg.

1963a,b). Rosecrans' activities were milestones in the history of the use of prebattle maps by U.S. Army Topographic Engineers. The subsequent Chickamauga campaign forced the Confederate forces to evacuate their Chattanooga base, a result of the skillful and daring use of the geology and topography of rugged terrain by the Union forces (Brown, 1964, p. 1).

In June 1864, Union troops were engaged with well-entrenched Confederate forces at Petersburg, Virginia. To advance, Colonel Henry Pleasants, a mining engineer, along with a group of volunteer Pennsylvania coal miners, proposed the construction of a tunnel system in the clay and sand-clay beds beneath the Confederate fortifications. The tunnel consisted of a main gallery ~155 m long with lateral openings for the explosive charges beneath the Confederate works (Fig. 5). By July 23 the main gallery with a ventilation shaft was ready for placing explosive charges (Fig. 5). Approximately 3,630 kg of gunpowder were detonated at 4:45 A.M. on July 30, causing the earth to erupt catastrophically with a heavy loss of enemy troops and equipment. The blast formed a crater ~50 m long, almost 20 m wide, and some 10 m deep (Fig. 5). Although the explosion created a major break in the enemy lines, the Union forces were not equipped nor prepared to overrun the steep-walled crater immediately, and within a short time Confederate forces regrouped and held on (Powell, 1989). During World War I, a similar use of mining and tunneling by British forces beneath German lines on the western front in 1917 was very successful (Figs. 6 and 7).

A British soldier and geologist, General J. R. Portlock of the Royal Engineers, recognized in his *Treatise on Geology* (1859) that a knowledge of geologic principles would be an aid in waging war. French Commandant A. Marga (1885), in his two volumes on *Geographie Militaire*, recognized the influence of the physical character of surficial material, (e.g., free-draining granite vs. relatively soft, water-absorbent clay and argillite) on troop movements. From 1897 to 1902, French Commandant O. Barré,

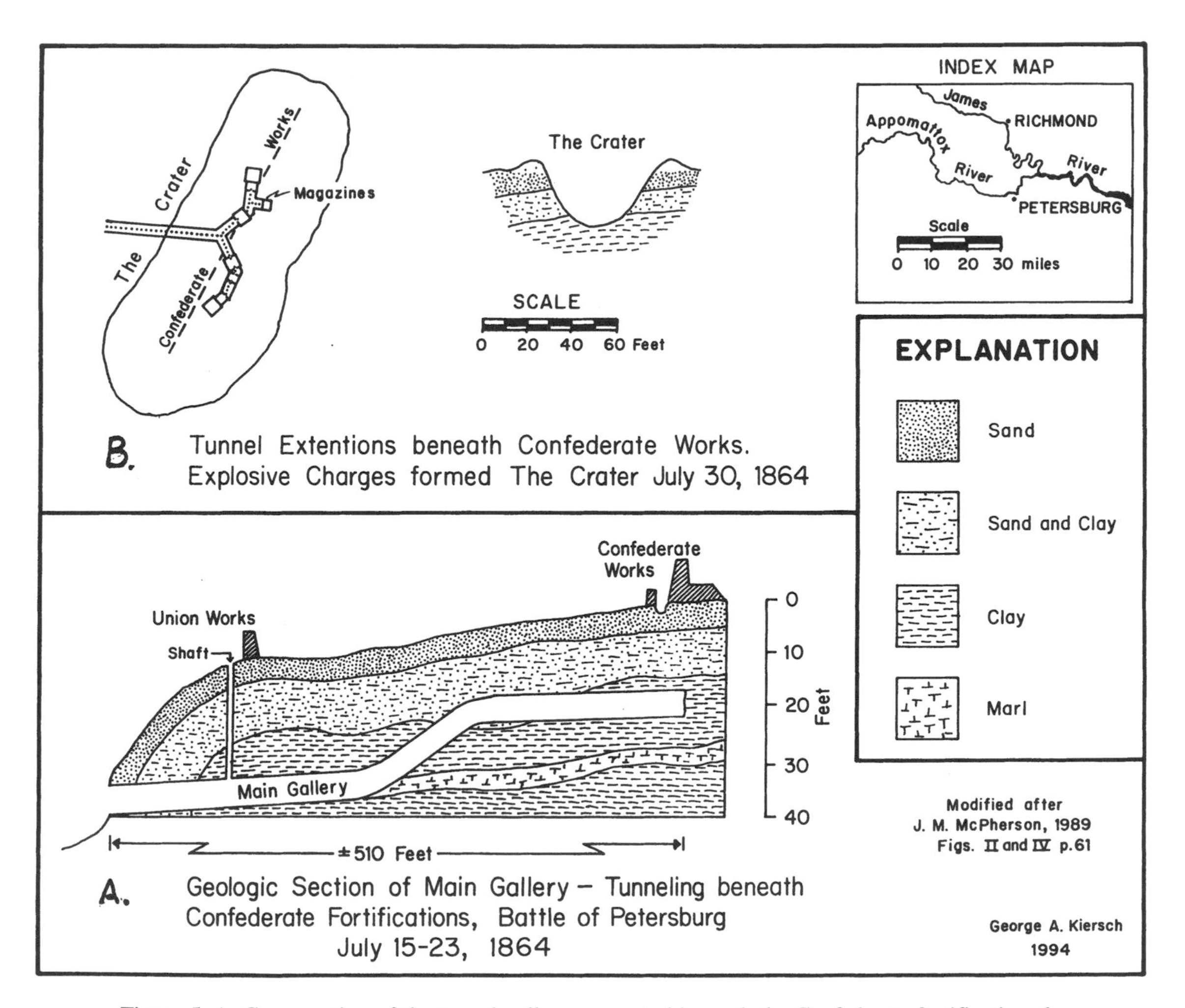

Figure 5. A. Cross-section of the tunnel gallery excavated beneath the Confederate fortifications by Union forces at the Battle of Petersburg, Virginia, in July 1864. B. Map view of the mine tunnels and the crater limits with a cross section of the crater formed. Detonation of explosive charges destroyed enemy forces and equipment and formed a large-scale crater at the surface, but the Union forces did not capitalize properly on their advantage. This type of warfare was very common during the World War I years of 1916–1917.

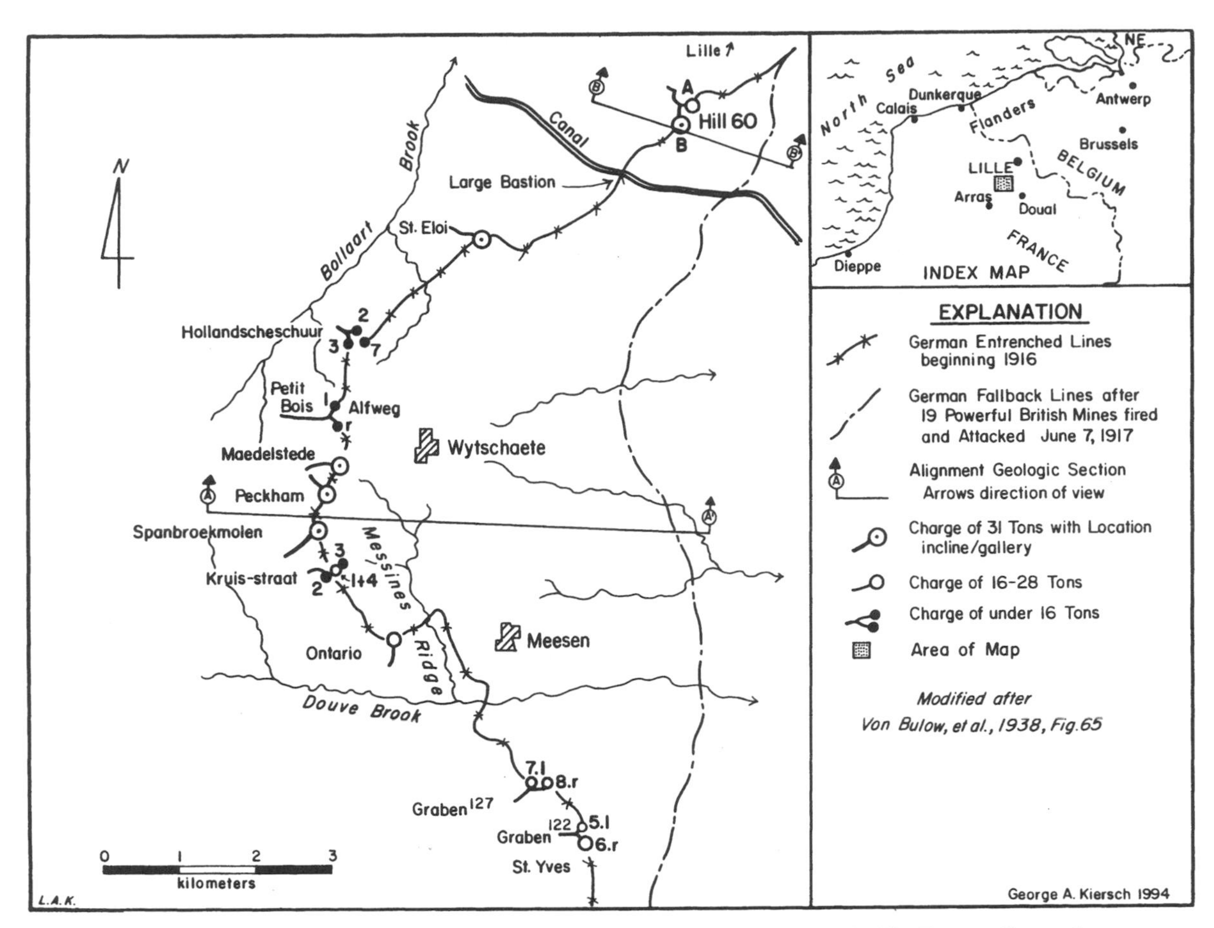

Figure 6. Plan map of the entrenched German lines on June 7, 1917, south of Lille, France. Shows the location of British lines and the location of the 19 separate underground galleries in which charges were placed and fired that demolished the fortifications of the Ontario Ridge, St. Eloi, Hill 60, and Peckham areas. Note the German fallback line. In places it is as far as 5 km east of its position prior to the attack on June 7.

eminent geologist and soldier, lectured at Ecole d'Application de l'Artillerie et du Genie, Fontainbleau, on the military importance of geology, physiography, and the impact of tectonic history on landforms (Barré, 1897–1902). In Britain, Colonel Charles Cooper King, geologist and artillery officer, lectured at the British Staff College from 1886–1898; he was among the first to recognize the wide application of geology, other than merely topography, to military problems and operations (Brooks, 1920, p. 90).

Following the Spanish-American War of 1898, G. F. Becker of the U.S. Geological Survey was sent to the Philippine Islands to make a geological study of resources (M. F. W., 1899). The Philippine Insurrections in 1899 interrupted this work, and he became attached to the Bureau of Military Information of the U.S. Army. Becker undertook more than 14 major reconnaissance missions that included repeated acts of gallantry; for this soldierly usefulness, Becker has been called the "First American military geologist" (Erdmann, 1943, p. 1177). During those years, Professor William O. Crosby of Massachusetts Institute of Technology was gaining recognition as the "Father of Engineering Geology in North America" for his many projects in which geology was utilized for engineered works (Kiersch, 1991, p. 44).

The first large-scale use of geology for military operations apparently dates from the Russo-Japanese War of 1904–1905. The Russians used a number of geologists as advisors, particularly in constructing fortifications (Whitmore, 1954, p. 212). The Japanese took the war as an opportunity to perform a regional geologic survey of Korea (Chosen), and it is likely that they utilized some of the data for military purposes. Similarly, during the pre-World War I years, selected geologic maps of Germany and Austria appear to have been made by order of the military authorities to develop a database on water supply (Brooks, 1920, p. 91), including ground-water resources for military posts.

WORLD WAR I

Little further attention was given to military geology in Europe or America until shortly before the outbreak of World War I. In 1913, Captain Walter Kranz, a geologist and artillery officer of the German army, called attention to the importance of

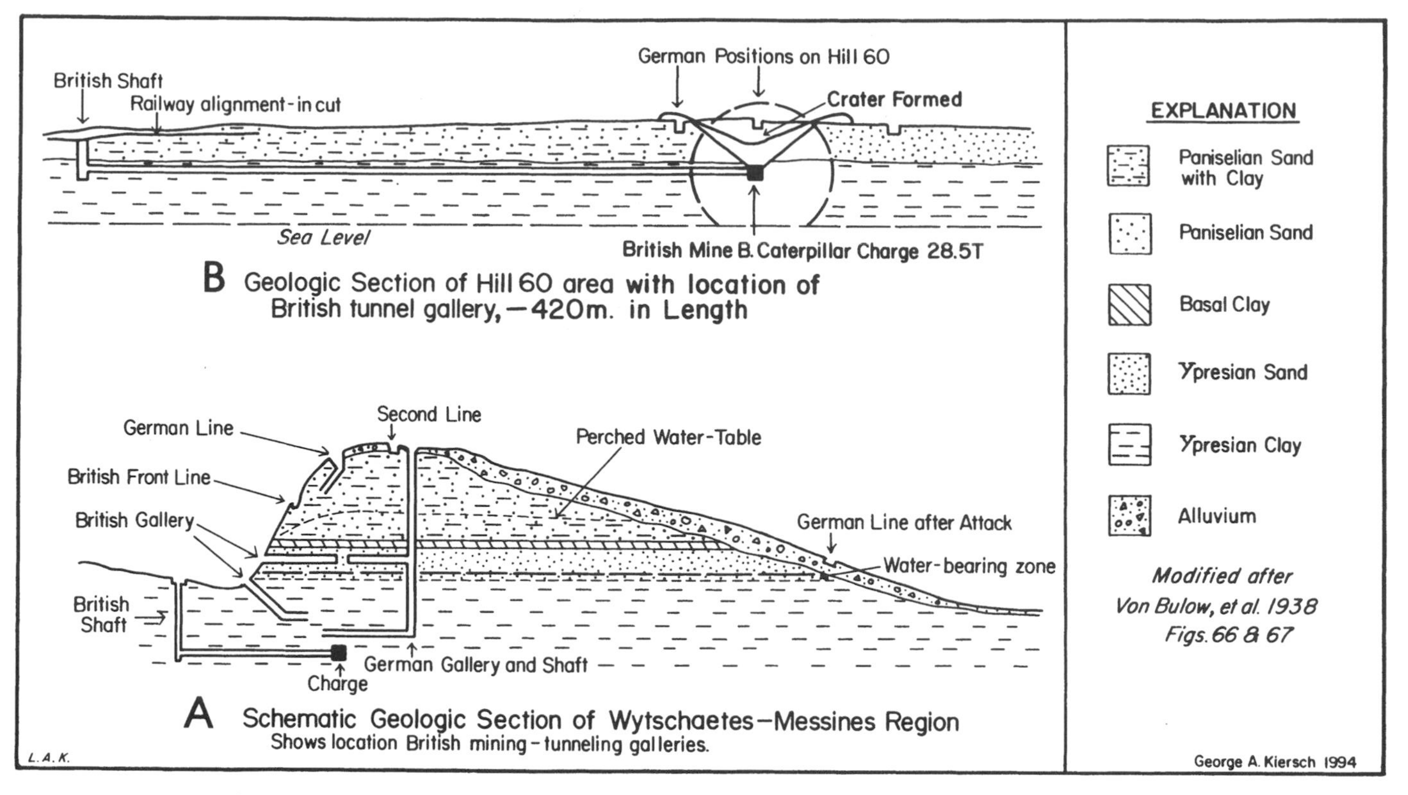

Figure 7. Schematic geologic sections of Wytschaetes-Messines region (Fig. 6) showing location of British mining-tunneling galleries at depth beneath the German lines and unknown to them. A. A schematic geologic section along one of the 19 galleries loaded with explosive charges (Fig. 6). B. A schematic geologic section across Hill 60, showing the location of the tunnel gallery that was 420 m long. Note the large surface crater that formed and demolished German positions and collapsed open cuts of the railroad.

geology in war in his text *Militargeologie* (Kranz, 1913), in which he advocated the recognition of geology as a special profession. Few among the Germans, French, or British heeded or acknowledged Kranz's warnings (Brooks, 1920, p. 91).

The German army began to develop the use of geology in warfare, "Wehrgeologie," during 1915. Initially concerned with identifying and utilizing the underground water supplies of the Moselle Valley, geological activities broadened in 1916 with the help of 20 geologists who were also occupied with terrain classification for troops and vehicles, advising on trench and underground fortifications, and locating construction materials (Brooks, 1920, p. 95). Professor Albrecht Penck, a prominent geomorphologist, was an advisor to the German General Staff on matters relating to geology and geography. By 1918, some 100 German geologists were contributing to the solution of military problems on the western front (Brooks, 1920, p. 96) that included terrain classification for troop and vehicle mobility, sitings for well borings and underground factories, and mining-tunneling for underground warfare.

By the outbreak of World War I, the ancient study of military geography had been refined by the French. They recognized that regional geologic history and landforms were considered basic to any terrain analysis concerned with military operations (Brooks, 1920, p. 89). Furthermore, the French countryside had been mapped geologically, and French military engineers had been educated in general geologic principles, including the analysis of maps for water sources and the siting of trench dugouts that recognized the crucial control of ground-water level over the performance of fieldworks (Fig. 8). Moreover, the French were leaders in the military application of hydrologic principles, and they actively drilled and outfitted thousands of water wells in areas lacking surface supplies. By 1918, several prominent French and Belgian geologists were attached to the Service Geographique to prepare "tank maps" that delineated the physical conditions produced by the surface and near-surface soils and materials of future battle grounds (Brooks, 1920, p. 93).

In 1915, the British Army attached Captain W. B. R. King of the British Geological Survey to the Chief of Engineer's Staff; King's roles were to serve as an advisor and to develop water supplies. In 1916, Lieutenant Colonel T. E. David was assigned as geologic advisor on underground mining and tunneling operations, which by 1917 became a major battle-related activity (Brooks, 1920, p. 93–94, 105). The British geologists performed their most notable service when they determined the areal and site-related conditions that affected underground warfare, e.g., those related to the German fortifications throughout the Messines-

Wytschaetes region of the western front (Figs. 6 and 7). Many cite the action on June 7, 1917, as the most destructive military mine explosion of World War I. Charges placed in 19 separate mining operations beneath enemy lines (Fig. 6) were exploded simultaneously. The shaking ground resembled an earthquake, and the city of Lille about 20 km behind German lines experienced sharp, earthquake-like shocks. The hitherto impenetrable German lines were destroyed (Fig. 7); large craters developed, the surface collapsed and the depressions quickly filled with water, and "quicksand" flowed into railway cuts as at Hill 60 (Fig. 7). A British artillery barrage followed, and troops quickly captured the fortresses and hills of Messines and Wytschaetes with comparatively light casualties (King, 1919, p. 208).

The British geologists Colonel David and Captain King had made a detailed investigation of subsurface-rock units throughout the Messines sector of the German lines (Fig. 6). They identified two sandy clay units: the greenish Paniselian Formation and the blue-gray Ypresian Clay (Fig. 7). Both horizons were suitable for mining and tunneling (King, 1919, p. 207–208) and are overlain near the surface by a water-bearing sand (Fig. 7). The areal distribution of clay units was delimited by correlating existing water-well logs and by drilling exploratory wells in the lesser-known areas. In addition to determining the location of the Ypresian Clay bed, the British geological study determined that thicknesses varied because of "erosion" in ancient time, and that safe tunneling and mining required knowing what was immediately ahead of the progressing tunnel face (Fig. 7). The German geologists apparently were unaware that a thick horizon of Ypresian Clay suitable for tunneling occurred beneath the Messines-Wytschaetes fortification system (Figs. 6 and 7).

When British forces later occupied the city of Lille near the Belgium border (Fig. 6, index map), Professor M. Ch. Barrois, eminent geologist of the University of Lille, related that the Germans, beginning with Captain W. Kranz in 1915, had used his lecture halls throughout the war for geological studies; 30–60 geologists under Captain Karl Regelman were active on the western front. The morning after British General Harvey blew up Messines ridge and captured the system of fortifications, a German general and his staff appeared at the lecture hall. The contingent of geologists was assembled, and the general proceeded to curse them vigorously because they had not known about, nor informed him of, the British mining-tunneling beneath the German lines and fortifications (Fig. 7). He then ordered all of the geologists older than 40 to Berlin; all younger geologists were assigned to the front. The moral seems to have been that the two experienced geologists with the British forces were superior to more than 20 German geologists with an incomplete knowledge of the areal geology.

The earlier lesson learned by the French in the Battle of Verdun (1916) became widely known; troops ordered to "dig in" on the high plateau of Cotes de Meuse met with disaster. The plateau area is underlain by hard limestone with only a foot of soil cover and cannot be excavated with hand entrenching tools. A brief study of the existing geologic map would have alerted the French forces (Brooks, 1920, p. 87).

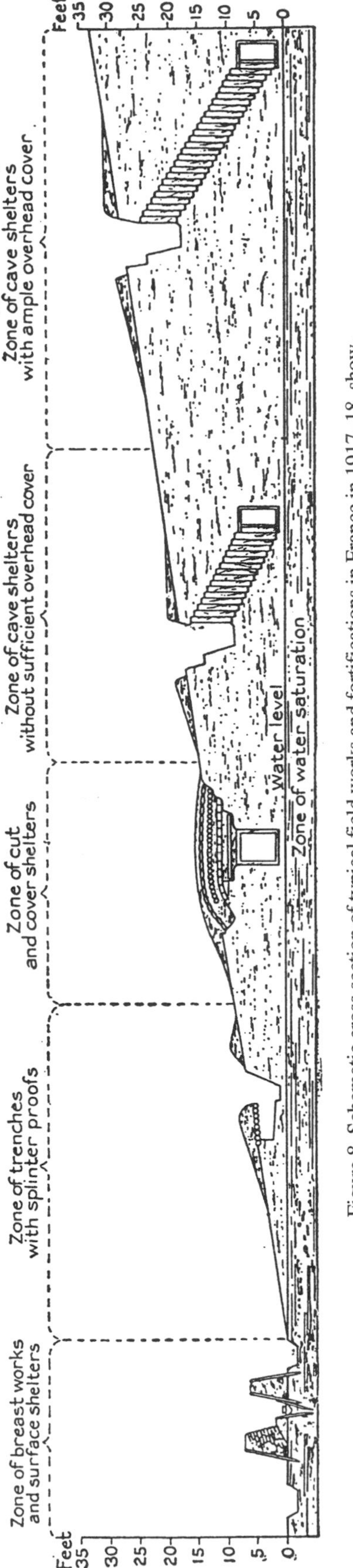

Figure 8. Schematic cross section of typical field works and fortifications in France in 1917–18, showing how the depth to the ground-water table can influence or control their construction. (From Brooks, 1920, fig. 9, p. 101.)

The U.S. Expeditionary Forces had two experienced geologists attached to the staff of general headquarters in France in September 1917: Lieutenant Colonel A. H. Brooks of the U.S. Geological Survey and Major E. C. Eckel, a well-known engineer-geologist. Additional geologists who arrived in mid- to late 1918 included M. F. La Croix, C. H. Lee, R. S. Knappen, T. M. Smithers, H. F. Crooks, Kirk Bryan, Wallace Lee, and A. W. Diston. The geologic maps for terrain analysis prepared by the U.S. forces differed from those of the other armies—German, British, or French. Termed "engineering geologic maps" (scale 1:50,000), the areas covered extended far within the enemy lines and provided the physical conditions and constraints for construction of field fortifications (Figs. 8 and 9), a distinct value even though they were based on earlier French hachured base maps and only approximate (Brooks, 1920, p. 110). By the fall of 1918, locating the most suitable locations for river and stream crossings became important geological tasks when armies were advancing rapidly (Brooks, 1920, p. 114–115). A related effort involved the training by Professor Warren J. Mead at the University of Wisconsin of more than 1,500 students in military mapping and principles of geology relevant to military activities (Kiersch, 1991, p. 33).

By the end of World War I, the participants and industrialized nations had come to recognize the importance of geologists in wartime (Smith, 1918; Cross, 1919) and their contribution in three major areas of activities: (1) economic natural resources and war materials; (2) geographic applications related to strategy and logistics; and (3) disposition of armies involving battle areas, zones of combat, maneuvers, and the extent to which geologic conditions can control the physical features of the battleground and, thus, tactics (Erdmann, 1943, p. 1175). Activities reviewed herein largely are those of the third area of activity, i.e., military engineering geology.

WORLD WAR II

German forces

Following World War I, the German army continued an interest in geology; in 1935 they reestablished the Wehrgeologie, a German army geology group (Whitmore, 1954, p. 213) and released several publications on military engineering geology. Two of the most significant were *Wehrgeologie* by Wasmund (1937) and *Wehrgeologie* by Von Bulow et al. (1938). The latter was a widely studied field manual that included a quantitative classification of earth materials (Fig. 10) and a summary of the basic principles of geology useful to troops in the field. From these studies and others, the Germans became aware of many new and unrecognized possibilities for using geology in warfare; some hitherto untried applications helped promote the type of war they were planning (Erdmann, 1943, p. 1180). The American military was impressed by the manual of Von Bulow et al. (1938), and in 1943 geologist Kurt E. Lowe, City College New York, translated the text for the Intelligence Branch, Office of the Chief of Engineers, U.S. Army.

The German Armed Forces had three top-level, geology-related agencies (Smith and Black, 1946). The Mil-Geo of the Army, established prior to 1939, was the largest and best-known group. This unit initially prepared Mil-Geo handbooks on the countries in Europe, Asia, North Africa, and ultimately on the countries in the Indian Ocean region. Maps were prepared (scale 1:500,000) with a descriptive text on the natural features of landscape and vegetation; special maps of coastlines (scale 1:200,000) were prepared after 1942.

The Forschungsstaffel of Supreme Command (OKW) developed into one of the most focused and successful of German military geology units. The concept was established by a small expedition to the Amazon in 1935–1937; the wartime group was formed in 1942. Its first assignment was to investigate the Egyptian-Libyan Desert relative to troops moving from the south or west; a three-volume atlas of maps (scale 1:200,000) and text was completed in 1943 by field-experienced geologists and other specialists. Details were provided on major physical features, water supply, terrain characteristics, and vehicle trafficability. For example, the location and extent of such topographic features as wadis were mapped and each wadi described relevant to its underground-water potential and as a natural, steep-walled canyon-like feature (10–30 m deep) that would be a barrier to large-scale mechanized movements. This knowledge of the terrain was critical to the early successes of the German North African campaign in the 1940s. Reportedly, tens of geologists were attached to all levels of the field forces from the army command to battalion- and company-sized units (Professor Georg Knetch, University of Würzburg, a senior geologist with the Afrika Korps, personal commun., 1962).

The Mar-Geo group was established by Naval High Command in late 1942 to prepare special nautical maps of selected coastal areas showing landforms, soil and vegetation patterns, and the underwater conditions, e.g., the rock platform and related swampy areas along the Normandy coast. The Mar-Geo group worked closely with the Mil-Geo group and Forschungsstaffel (research) units when concerned with coastal works, construction, fortifications, or flooding potential, e.g., when Royal Air Force bombing breached the Walcheren Island dikes in October 1944, resulting in the flooding of much of the island.

The German army and navy became noted for utilizing geological guidance when planning and constructing major field works, whether fortifications, underground installations and factories, airfields, military bases, or secret command facilities. Two separate projects that were confronted with adverse geological conditions, yet were successfully constructed and operated, illustrate the geologic insight and on-site guidance of the German geologists-engineers.

The German navy located U-Boat bases on the northern coast of Norway and decided early on to construct bomb-proof submarine pens at selected locations. Two massive 1-m-thick, above-ground concrete-block pens (Dora 1 and 2) were constructed on the bay shore in downtown Trondheim, Norway. Beach sands and weak, sensitive clay—treacherous materials for

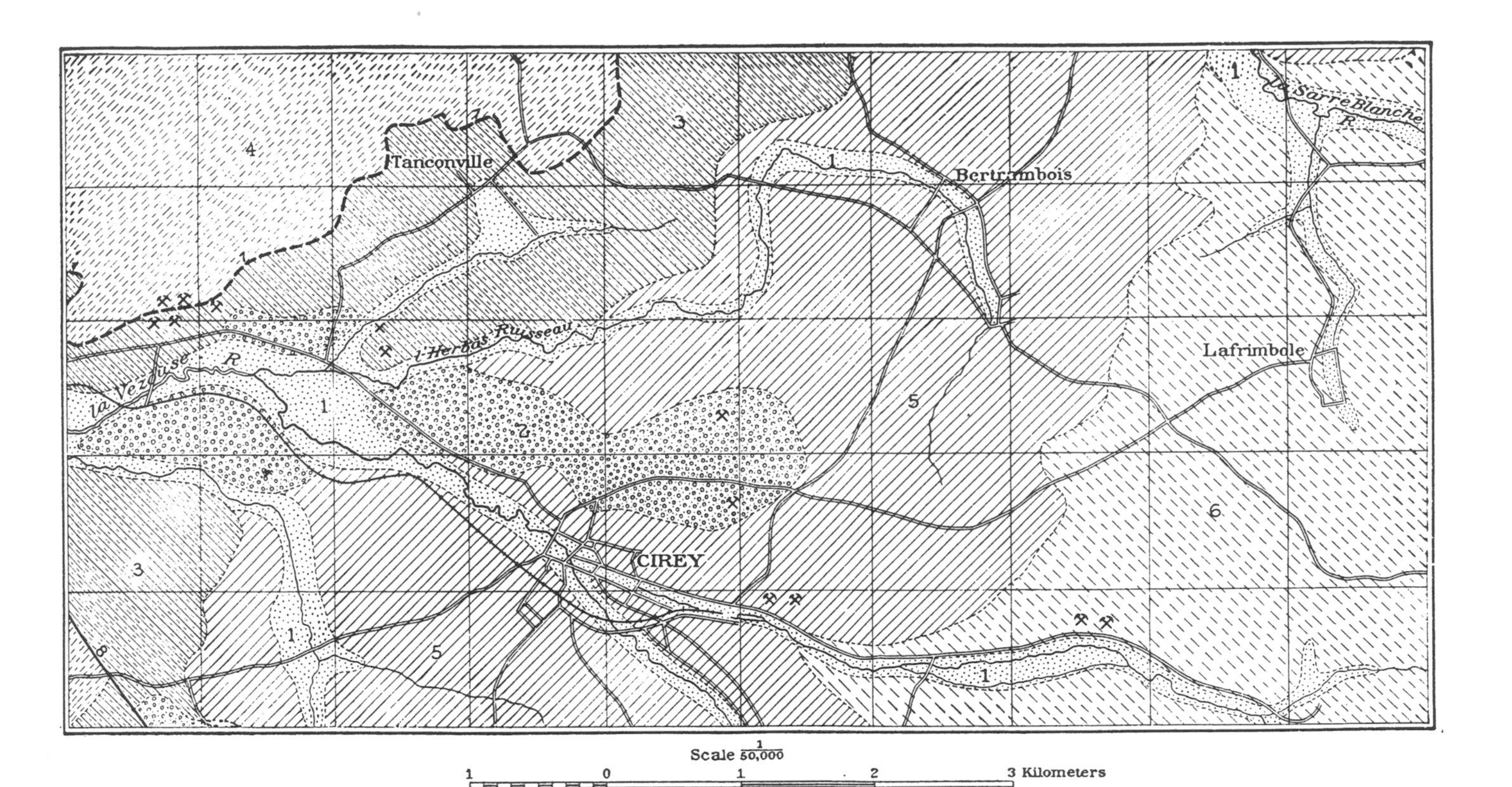

Scale $\frac{1}{50,000}$

1 0 1 2 3 Kilometers

DESCRIPTION OF FORMATIONS

1

Silt, mud, sand, and gravel, with some clay

The valleys of the Meurthe and parts of the Sanon (Parroy to Mouzy) and the Vezouze up to Blamont, as well as those of the small streams within the sandstone area, are chiefly floored with sand and gravel. In the other valleys the floor is chiefly silt and mud. The thickness of these deposits ranges from less than 1 meter in small valleys to 2 or 3 meters along Sanon and Vezouse rivers.

They are usually saturated, except during the dry season (June to September); in part subject to flood and unfavorable to trenches and dugouts.

The silt and mud floored valleys usually afford soft footing, even during the dry season. They are for the most part impassable during the wet season (October to May), except when frozen. The gravel and sand floored valleys are better drained and are passable except during floods. Floods are likely to occur in any month but are most frequent and highest between November and April.

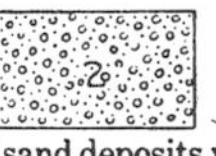

High gravel and sand deposits with some clay (½ to 20 meters thick)

Well drained near surface but contain much water at bottom. Trench construction easy but requires complete revetment. Cave shelters possible where deposits are thick enough but require heavy timbering to maintain roof.

The surface is usually firm, even during wet weather.

3

Clay 1 to 3 meters thick resting on limy shale

The clay is thinnest at the crests of deep slopes, thickening toward the summits of divides and toward the bases of slopes, where it may exceed the thickness given above by accumulation of slide material. During the wet season the surface zone is more or less saturated, and the ground is muddy to a depth of 1 meter or more. Ground water usually within 3 or 4 meters of surface, except at crests of slopes, where it is deeper. The formation also contains some thin beds of limestone, usually water bearing. The presence or absence of underground water should always be determined in advance of cave-shelter construction by test shafts or bore holes.

Trench construction easy but requires complete revetment and ample provision for surface drainage. Trenches located on clay slopes are liable to be destroyed by slides and need strong revetment. Cave shelters can be built, only where the terrain permits drainage tunnels.

The surface of this formation is usually soft and except where well drained on crests and slopes is likely to be muddy during the wet season (October to March); except when frozen, slopes are slippery during the wet season.

4

Clay with weathered limestone ½ to 1½ meters; compact limestone below, with some shale beds

Surface usually well drained, but some water may occur along shale beds.

Trenches difficult to excavate but require little revetment. Formation is favorable to cave shelters but requires hard-rock excavation and where rock is fractured heavy timbering.

Surface of ground usually firm, but during wet weather (October to April) some mud found on limestone. Limestone forms steep cliffs where capping hills.

5

Sandy clay soil with weathered sandstone 1 to 3 meters; medium-hard sandstone containing some clay below

Usually well drained; trench construction in weathered zone easy but requires heavy revetment on account of swelling. Cave shelters in sandstone require heavy rock work; some seepage water, and drainage tunnels usually necessary.

Surface of this formation usually well drained and firm.

6

Sandstone and conglomerate, in most places weathered to a depth of only ½ meter or less

Rock below soil very hard and requires blasting for trench construction. Cave-shelter construction requires use of power drills and blasting. In most places not much seepage water.

Surface ground firm and in places hard throughout the year.

7

Locus of springs and seepages

These should be avoided as far as possible in the location of fieldworks, especially of dugouts. Fieldworks should be placed above the lines of springs.

8

Line of dislocation (fault)

Water is likely to occur along these lines, due to the zones of fracture.

⚒ Quarry (in part abandoned)

× Sand and gravel pit

MILITARY GEOLOGIC MAP OF CIREY, FRANCE, AND VICINITY.

Figure 9. Typical engineering geology military map, scale 1:50,000, prepared for the U.S. Expeditionary Field Forces in 1918. The map text describes the characteristics of surficial sediments and near-surface rocks and the location of springs and seepages in the vicinity of Cirey, France, and describes terrain features, suitable locations for trenches and fieldworks, and sources of construction materials and water. (From Brooks, 1920, plate XVI.)

THE MOST IMPORTANT ROCKS AND THEIR PROPERTIES

[After von Bülow (1938); translated by Kurt Lowe]

Rocks	(a) Occurrence and Land Forms (b) Jointing	Condition of Weathered Rock Mantle in Horizontal Position[3]	Water Permeability[2]	Filtering ability[2]	Loosing Facility, Workability[1]	Stability[2]	Structural Roof Strength[3]	Loading Capacity[2]	Suitability for: Construction site[2]	Compacted subgrade[3]	Pavement[2]	Road metal[2]	Crushed rock[2]	Concrete aggregate[2]	Fill[2]	Remarks
Sedimentary Rocks Sandstone	(a) Knobs, hills, plateaus (terrace topography) (b) Thick-bedded, platy, jointed	Sandy or loamy, depending on cement of sandstone	II-III	II-III	4-5	I-III(g)	I-III	II-III	II	IV	II-IV	V	V	III-IV	If mixed with sufficiently fine-grained material: II-III	(g) Sometimes danger of sliding when dipping
Quartzite	(a) Steeper slopes, otherwise as above (b) Like sandstone	Stony, not very thick, only seldom actually loamy	IV-V	Mostly IV-V Percolation in joints	4-5	I-III	I-III	I-II	I-II	II-III	I-III	II-III	I-III	I-II		
Graywacke	Like sandstone	Sandy, loamy, often quite thick	III-V	II-III; reason same as above	4-5	I-III(g)	I-III	I-II	I	I	I-II	II-III	II-III	II-III		(g) Same as sandstone
Conglomerate, breccia	(a) Like quartzite (b) Thick-bedded to irregular, depending on geologic structure	Stony	Very variable, mostly III-I	Very variable, mostly IV-II	Very variable, II-V(h)	I-III	II-IV	II-III	I-II	II-V	Conglomerate III-IV; breccia mostly V	Conglomerate III-IV; Breccia II-III(h)	Conglomerate IV-V; Breccia I-III(h)	III-IV II-III(h)		(h) Loosely cemented breccia, easily loosed; after screening, suitable for road metal, crushed rock, concrete aggregate
Slate	(a) Hilly topography; upland surfaces with steep valleys (b) Cleavable in one or more directions	Loamy, partly sandy, mostly quite thick	IV-V	IV-III	3-5	I-IV(g)	II-IV	II-III	II	III-IV(i)	V	V	V	V		(g) Sometimes danger of sliding when dipping (i) Fragments should be placed on edge
Marl, shale, shaly marl, indurated marl	(a) Like sandstone but more subdued; depressions in the mountains, etc. (b) Shaly or laminated	Loamy or marly-clayey, also of plastic consistency	IV-V	IV-III	3	IV-V(k)	IV-III	III	III	V	V	V	V	V		(k) Sometimes danger of sliding when dipping and after rainfall
Limestone, dolomitic limestone, marble	(a) Knobs, hills, ridges, caves, sinkholes, plateaus (b) Thick-bedded, platy, jointed	Calcareous, loamy; often very thick and then poor in lime	II-IV	Owing to the numerous joints, mostly V-IV	4-5	I-III(l)	I-III	II	I-II	III	V	II-V(m)	V	II-III, loamy dolomitic limestone		(l) When dipping, IV; sometimes danger of sliding (m) Dolomitic limestone not suitable for railroad construction
Gypsum-rock	(a) Usually barren hills (b) Jointed, honeycombed with cavities	Usually in very thin beds, calcareous, loamy	IV-II	V-IV	3-4	III-IV	III-V	III-IV	III-V	V	V	V	V	V		
Volcanic tuffs	(a) Very varied, massive or stratified, porous	Mostly loamy	II-IV	II-III	3-5	III-IV	III-IV	III-IV	III	V	V	V	V	V		

Rocks	(a) Occurrence and Land Forms (b) Jointing	Condition of Weathered Rock Mantle in Horizontal Position [3]	Water Permeability [2]	Filtering ability [2]	Loosing Facility, Workability [1]	Stability [2]	Structural Roof Strength [2]	Loading Capacity [2]	Suitability for: Construction site [2]	Compacted subgrade [2]	Pavement [2]	Road metal [2]	Crushed rock [2]	Concrete aggregate [2]	Fill [2]	Remarks
Sand	(a) River banks, plains, often in form of knobs rising from level topography	Loamy sand, sometimes hard residual soil	I–IV(n)	III–I(o)	1–2	III–IV	IV	II(q)	II(q)	.....	.....	.....	.....	I(p)	I–IV	(q) When lateral displacement is impossible (n) Rises with increasing grain size (o) Increases with decreasing grain size (p) If free from loam and clay (requires washing)
Gravel, cobbles, and boulders	(a) Like sand	Stony soil	I	III–V	2–3	IV–V	IV	II(q)	II(q)	.....	.....	II–III	II–III	I(r)	I–IV	(n) Rises with increasing grain size (r) If free from loam and clay (washing) and if containing only few shaly fragments
Compacted moraine material, erratics	(a) In level topography: terminal moraine ridges, terrain with small knobs	Stony loam or sand soil	I	IV–V	2	II–V	IV	II–IV	II–IV	II	av.: III	II–III	II–III	II–III	av.: III	
Loam	(a) Plains, subdued hill topography, flood plains, lower slopes	Loam soil	III–IV, also V(s)	II–I	2	II–III; when wet: IV–V	IV	II–III	II–III	.....	.....	.....	.....	.....	III–IV(t)	(s) Rises with increasing sand content (t) Cavities form easily
Boulder clay (u) (Boulder marl)	As above, but not on flood plains	As above, stony	III–V(t)	II	2–3	II–III; when wet: IV–V	IV	II–IV(t)	II–IV(t)	.....	.....	.....	.....	.....	III–IV(t)	(t) Cavities form easily (u) Waterproofing material
Loess-loam	Slopes, margins of flood plains	Thick loam soil, free from stones	II–IV	I–II	2	III; when wet: V	IV	III; when dipping: IV–V	III–IV	.....	.....	.....	.....	.....	III–V(v)	(v) Cannot be used for hydraulic construction
Loess	Plains, plateaus, slopes	As above	I–II	I–II	2	III; when wet: V	IV	III; when dipping: IV–V	III–IV	.....	.....	.....	.....	.....	III–V(v)	(v) Cannot be used for hydraulic construction
Marl	Like loam	Loam soil, partially containing lime	II–III(w)	II–III	2–3	II–V	IV	II–III	II–III	.....	.....	.....	.....	.....	III–IV	(w) Decreases with rise in clay content
Calcareous and marly weathering crusts (caliche); bog lime and clay; ooses	Shore lines, swampy meadows	Calcareous humus	IV–V	V	2–3	III–V	V	IV–V	IV–V	.....	.....	.....	.....	.....	V	
Moor and peat	Lowlands	Wet, raw humus (peat)	III–V	I–II	1	V	V	V	V	.....	.....	.....	....	.....	V	
Clay (u)	Like loam	Tough, impervious soil	V	I–II	2–3 [4]	IV–V	IV	III; when wet: IV–V	III; when wet: IV–V	.....	.....	.....	.....	.....	I–V(z)	(z) Can be used for clay cores (u) Waterproofing material

The data above refer to fresh, unweathered rock.

[1] Classes of workability: 1, with shovel and spade; 2, pick and shovel; 3, mattock, crowbar, and iron wedge; 4, compressed-air drill and repeated blasting; 5, compressed-air drill and continuous blasting, together with crowbar, iron wedge, chisel, spade, and pick.

[2] Scale applicable to all other properties: I, excellent; II, good; III, adequate or fair; IV, poor, or usable only in emergency; V, inadequate, unsuitable, or absent.

[3] The rock mantle becomes thin along the slope and disappears altogether at scarps.

[4] When wet, best worked with hoe or drainage spade.

Figure 10. Construction materials, Von Bulow et al., 1938; reprinted from Whitmore, 1950. Reprinted with permission from *Applied Sedimentation*. Copyright 1950 by the National Academy of Sciences. Courtesy of the National Academy Press, Washington, D.C.

construction—occur at both sites. German army geologists and engineers, after a detailed investigation of the geological conditions, successfully utilized an electro-osmotic technique and sheet-walls to harden and stabilize the highly mobile clay and sand-silt at Dora 2; this provided stability to the open cuts and access to adequate foundation at depth for the massive concrete structures. Stabilization depended on a natural "ideal" clay-silt ratio and water content of the clayey sediments (Casagrande, 1952). Although Germans experienced difficulties with a layer of sensitive clay at Dora 1 site, electro-osmosis was not utilized there (L. Grande, personal commun. to G. A. Kiersch, 1995).

The Norwegian military was unable to destroy and remove the massive structures after the war, and they were in use as a small-ship military base in 1959 when G. A. Kiersch was allowed to enter and study the interior of Dora 1. One borehole inside the pen extends 52 m below sea level (L. Grande, personal commun. to G. A. Kiersch, 1995), probably into bedrock. The structures are intact today, and one has been converted into a downtown parking area (roof) and office building. Numerous attempts to duplicate the Trondheim stabilization success at sites in North America generally failed, reportedly because the clay-silt and the water-content ratios were incorrect. A second electro-osmosis project by the occupying Germans failed when they were unable to stabilize sensitive clay at the southwest entrance of a railway tunnel under construction to decrease the vulnerability of the junction of the Norwegian north-south rail systems at Trondheim (L. Grande, personal commun. to G. A. Kiersch, 1995).

A second project, also for submarine pens, involved the excavation and stability of large-scale underground chambers in granitic complex of the coastal area at Narvik, Norway. There, underground cavities were positioned and aligned so that the openings compensated for the active residual stresses and associated rebound-relief structures that formed because of the regional relaxation phenomenon. The glacial rebound and tectonic history of the Narvik region were additional factors in forming the prominent joint and sheet structure pattern, according to Kieslinger (1958, 1960), geologist for the planning and the construction of the submarine pens. The alignment of the pen chambers, relative to the pattern of rebound-relief structures that formed parallel to the steep granitic outcrops along the seacoast, was critical to retaining the stability of these large underground openings.

Earlier, German geologists had encountered an unexpected, strong pattern of rebound-relief structures parallel to the ancient canyon walls of the incised Salzach River in the late 1930s during construction of the underground Festival Hall at Salzburg, Austria (Fig. 11). This earlier experience became invaluable when designing the Narvik submarine pens (Kieslinger, personal commun., 1960, 1963) after the patterns of relaxation structures in granite were mapped and their origins recognized.

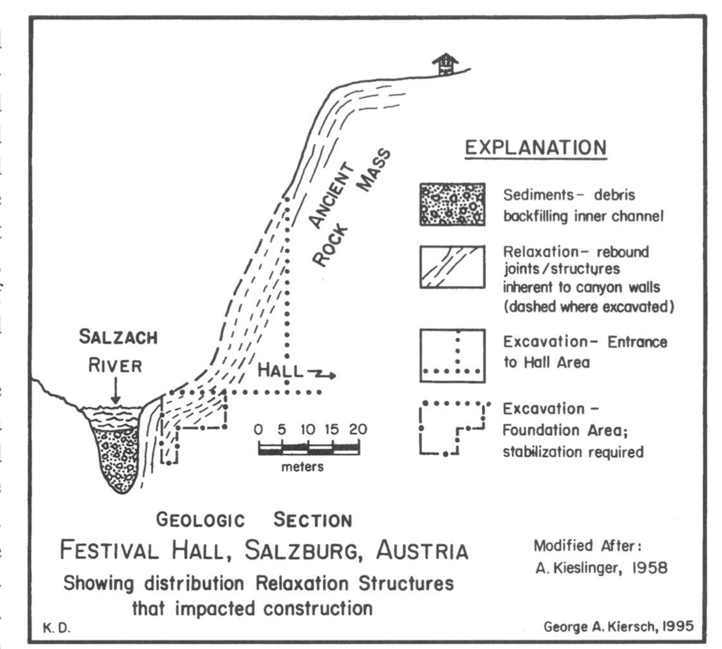

Figure 11. Site of the Festival Hall, Salzburg, Austria, showing the pattern and distribution of relaxation, relief-rebound structures in the ancestral inner and outer canyon walls of the Salzach River. This was the site of the music festival shown in the film *Sound of Music*. (After A. Kieslinger, 1960, fig. 5). The principles learned during the large-scale excavation project in the mid-1930s were used effectively to site and to design underground submarine pens in a granite complex at Narvik, Norway, in 1941–1942.

U.S. Armed Forces

The U.S. Armed Forces entered World War II in 1941 with only a sketchy insight into the many diverse geologic environs that would have an impact on the war's activities despite Mead's (1941) review of formal engineering geology practice covering the period 1888 to 1938. In his review Mead emphasized the value and importance of geology to military operations and as an area of growing professional practice. Most important were the many hazardous physical processes inherent to the dramatically different worldwide regions, from the frozen waters of the Arctic to the equatorial jungles, that were destined to become battlegrounds in 1942–1945. For instance, the permafrost phenomena of Alaska and the Aleutian Islands were a major hazard and adversely affected the construction and operation of the Alaskan Highway. This vital artery was built in eight months in 1942 by 10,000 U.S. troops following the Japanese landing on the Aleutians; the 29th Engineer Topographic Battalion established the right-of-way just ahead of the fast-moving constructors. Typical problems that were encountered are illustrated by a roadhouse that collapsed (Fig. 12) owing to the melting of ice-rich foundation materials and by consequent subsidence and lateral dislocation of such related engineered works as airfields and railroads (Muller, 1943).

Equally important throughout the island region of the South Pacific was the physical variability and properties of the terraced coral-reefs and associated soft-to-weathered, clayey mud and sand. A major difficulty experienced with coral deposits used at bases in 1943–1945 was its strength. Fresh coral is a suitable foundation for roads and airfields and serves as a self-cementing crushed rock. The weathered, karstic deposits with a high clay content are usually unsatisfactory, particularly during the rainy season (Aberdeen, 1945, p. 586; Whitmore, 1950, p. 648). The uplifted and terraced coral reefs, such as those along the northern coast of New Guinea, commonly have undergone adverse weathering and substantial physical changes, even though they appear usable when first reconnoitered.

By early 1942, along the U.S. East Coast German U-boats were destroying a substantial share of the coastal shipping, especially off North and South Carolina. The Germans had earlier investigated the drowned river valleys of Carolina coastal waters and were using the inland channels at night as fresh-air rest stops to recharge their batteries. Because adequate maps and ground control for precise location and military response were lacking in the region, the 30th Engineer Topographic Battalion established a first-order geodetic network in coastal areas in 1942–1943 to assist in monitoring U-boat positions. G. A. Kiersch served as detachment officer of the Survey crews.

To assist allied forces worldwide with information on all types of terrain, the Military Geology Unit (MGU) was organized in 1942, largely with U.S. Geological Survey personnel (Terman, Chapter 5, this volume). The unit acted under the U.S. Army Corps of Engineers providing geologic guidance at the strategic level of military planning as well as preparing folios of planned and possible invasion sites. Each of the folios and reports had maps covering terrain characteristics and their effect on trafficability, movement, concealment of troops and supplies, possible beachhead and battlegrounds areas, water supplies, roads and construction materials, sites for airfields and military works at forward bases, and possible geologic problems to be expected in an area or location (Hunt, 1950).

By 1943, the military realized that geology could be important and often critical or indispensable in both the planning and field operations for the army, navy, and air force. The unit's folio on eastern Sicily of 15 separate 1:100,000-scale quadrangle maps and accompanying tables on all types of terrain and construction information established confidence in MGU by the military and launched a broader phase of activity (Hunt, 1982, p. 13). So many new requests for geologic guidance followed that the number of MGU personnel doubled to 50 and later to more than 100 members (Hatheway, 1993; Terman, Chapter 5, this volume). The success of military geology folios, emphasizing the likely terrain problems to be encountered, was in direct proportion to the amount of basic mapping data available or completed in advance (Betz, 1984a; Hunt, 1950; Russell, 1950; Whitmore, 1950). This is demonstrated by Whitmore (1950) in a set of maps containing basic data (Fig. 13A–D). Further insight into warfare and its intimate relationship to geography, geology, and landscape as represented on maps was provided by Putnam (1943) in his text for training military students.

By 1945 the MGU had completed 140 folios and reports (each 25–200 pages) both on possible and on major combat areas (southern Europe and Pacific); personnel had increased to more than 100 geologists, hydrologists, soil scientists, and other specialists (Dryden, 1945, p. 589). Furthermore, some 75 unit members were sent to the Pacific and to Europe in 1944–1945 for short assignments to assist with tactical planning, base construction, and reconstruction problems. The next phase of MGU's activities was directed to the Pacific Theater of Operations where three members participated in the Philippine landings at Leyte and did field work under fire (Hunt, 1982, p. 14). Moreover, in 1943–1944, trained geologists serving with field troops often were included as members of three- to five-man advance, prelanding parties that reconnoitered a proposed beachhead, airhead, supply route, or proposed areas for military facilities. This behind-the-lines reconnaissance provided a ground-truth evaluation of the geologic environs and physical features that sometimes changed the proposed plans for an invasion or an installation; frequently the party, or one or more of its members, did not return from such a mission (G. A. Kiersch, personal observation, 1943–1944).

Figure 12. Destruction of a roadhouse at kilometer 443.6 along the Alaskan Highway, resulting from ground subsidence accompanying thawing of ice-rich permafrost at depth. Heat from a furnace in a corner of the building caused the structure to subside. (Photograph courtesy of Troy L. Péwé.)

Such activities were critical to the June 1944 Normandy beach landings. The Beach Intelligence Service of the Beach Erosion Board began studies in 1942 of the coastal beaches of France from Cherbourg to Dunkirk and, later, of numerous Pacific beachhead areas. Operational reports with emphasis on trafficability characteristics of beaches were prepared in coordination with the Military Geology Unit (Edmunds, 1945, p. 590); the reports served a joint army and navy intelligence group.

The U.S. Military Geology Unit was not responsible for preparing folios inland from the beaches of Normandy and in western Europe. The British had established a military geology unit at Oxford, England, in 1942 patterned after the MGU; reportedly they were impressed by the MGU's trial-run report on Madagascar, which demonstrated the breadth and substance of

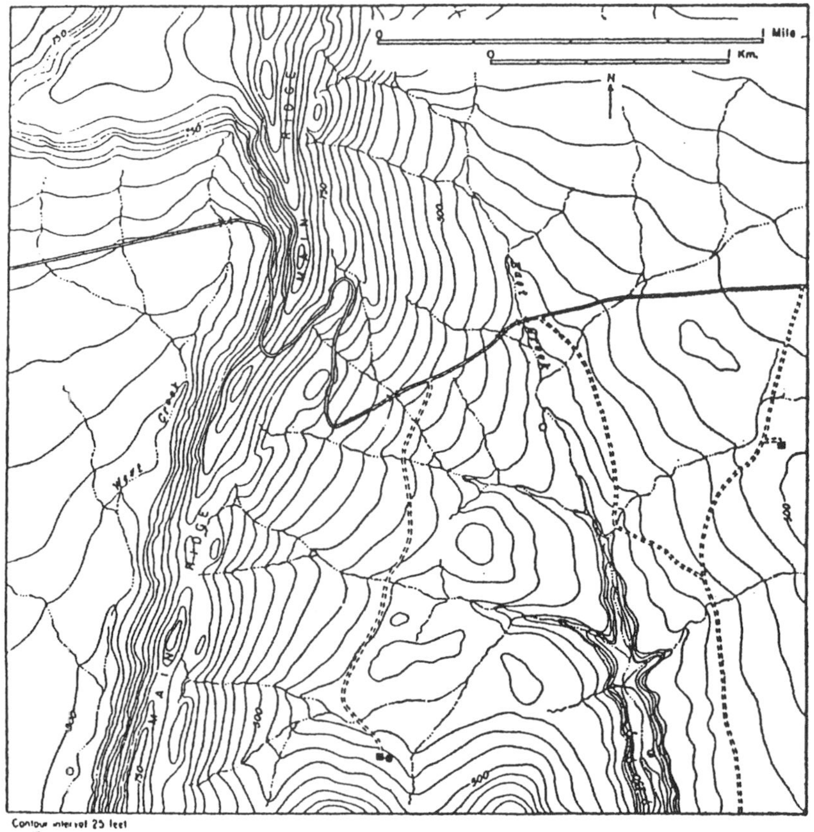

Contour interval 25 feet

A

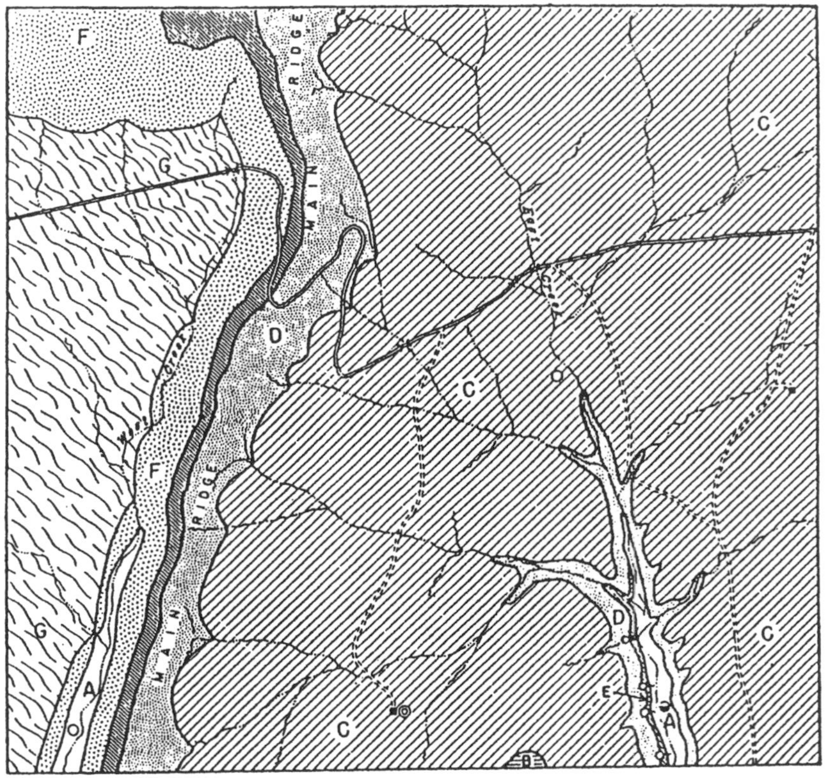

B

EXPLANATION OF GEOLOGIC MAP

SYMBOL	KIND OF MATERIAL	THICKNESS IN FEET	TOPOGRAPHIC EXPRESSION	DISTINCTIVE FEATURES	WATER-BEARING PROPERTIES
A	Alluvium; gravel, sand, silt, some clay.	< 50	Flood plain and valley bottoms.	Unconsolidated.	Permeable, small underflow; perennial yield only where underlain by impervious rocks.
B	Limestone.	max. 45	Caps flat-topped buttes.	Thick massive beds; hard; resistant.	Impervious.
C	Shale; kaolinitic clay shale; lower 50' is sandy shale interbedded with sandstone in beds up to 20'.	200	Low hills; lower sandy beds form minor ridges on east slope of Main Ridge.	Clay shale is finely laminated, easily eroded; sandstone is hard. Sharp contact with limestone above.	Shale impervious; basal sandy beds slightly permeable, very small yield.
D	Sandstone; quartz-sandstone with lime cement.	125	Forms crest of Main Ridge.	Fine-grained; thick massive beds; resistant.	Coefficient of permeability-45; small yield; water is high in carbonates.
E	Shale; gypsiferous shale containing shaly sandstone lenses and pure clay layers.	200	Forms broken slopes along East Creek and west side of Main Ridge.	Sandstone lenses are fine to coarse-grained; poorly cemented.	Shale impervious; sandy beds slightly permeable; water is brackish.
F	Sandstone.	250 to 300	Moderately resistant; hard beds form ridge along West Creek valley and cap northwest upland.	Coarse-grained (in part containing pebbles up to 1" diameter); poorly cemented; some hard, resistant beds.	Coefficient of permeability 350; large yields; water is low in dissolved minerals.
G	Schist; micaceous rock cut by irregular quartz views.	900 + M	Forms western lowland.	Finely laminated in general; a few massive granitic layers.	Impervious; no water except in occasional open fissures within 200 ft. of surface.

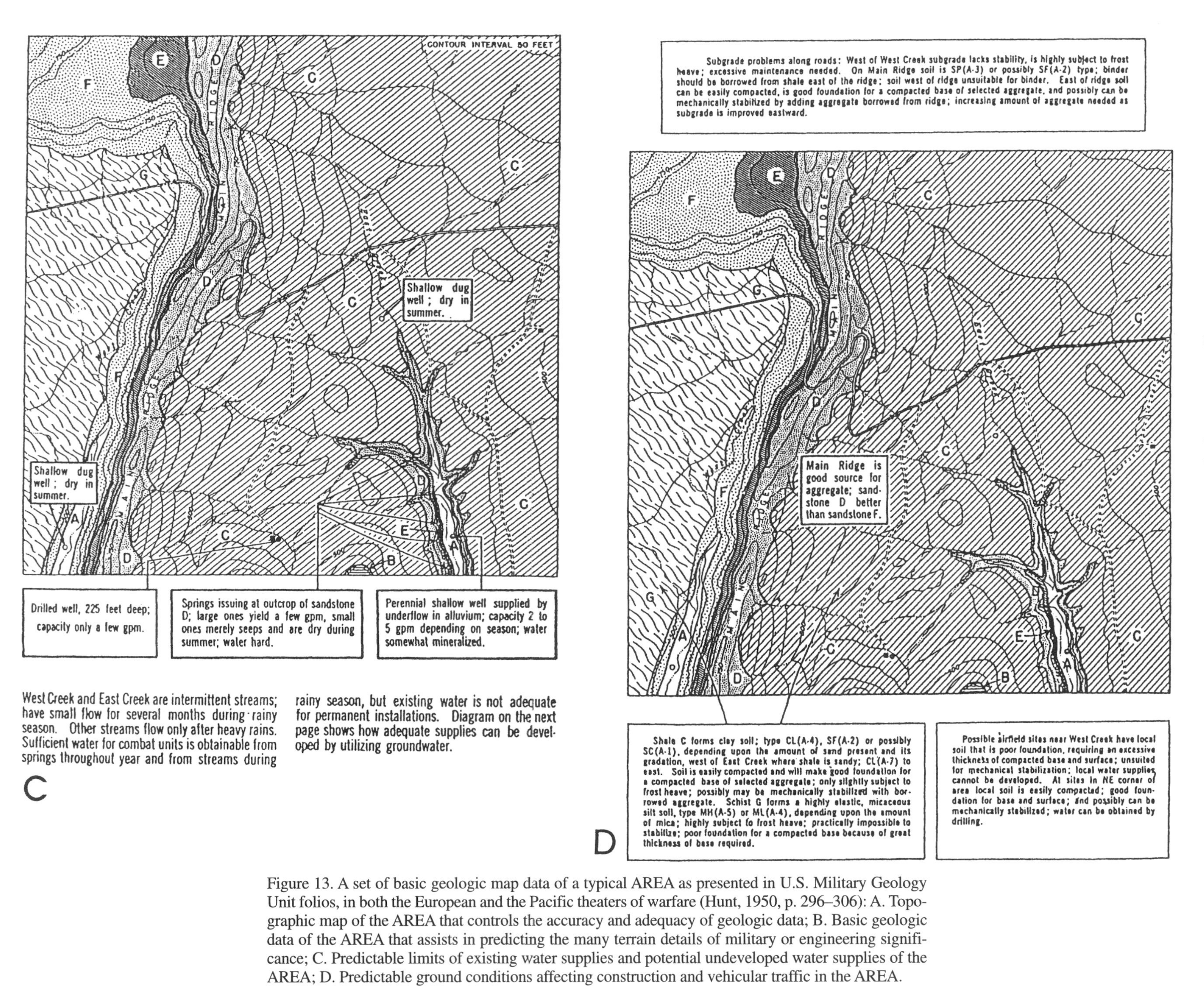

Figure 13. A set of basic geologic map data of a typical AREA as presented in U.S. Military Geology Unit folios, in both the European and the Pacific theaters of warfare (Hunt, 1950, p. 296–306): A. Topographic map of the AREA that controls the accuracy and adequacy of geologic data; B. Basic geologic data of the AREA that assists in predicting the many terrain details of military or engineering significance; C. Predictable limits of existing water supplies and potential undeveloped water supplies of the AREA; D. Predictable ground conditions affecting construction and vehicular traffic in the AREA.

information that could be provided in advance of operations (Hunt, 1982). The coordinated American-British geologic activities included those of the U.S. Navy Hydrographic Office for anchorages and the British Beach Erosion Board on landing areas and beaches. Only near the end of the war did U.S. geologists participate in reviewing European resources, underground factory installations, and reconstruction problems.

Military installations and many critical manufacturing plants were relocated underground throughout Europe during and after the war. Among them, one German war-time project cleverly incorporated the physical characteristics of a thick loess deposit in the Danube River valley a short distance north of Linz, Austria (Fig. 14). There, an underground facility was constructed with a modified room-and-pillar system (Kiersch, 1949, Fig. 3) with large interconnected rooms as high as 5 m; the facility housed a German aircraft-parts assembly plant. The plant site withstood aerial bombing attacks because the soft-to-spongy loess deposit effectively dissipated the shock waves of conventional explosions (Kieslinger, oral commun., 1963). Edwin B. Eckel of the U.S. Geological Survey and others prepared reports on many of the German factory sites in late 1945–1946 (Eckel, 1945, reviewed four major types of facilities).

British army

A review of British military geology activities (1939–1945) is provided by Rose and Pareyn (this volume). Features inherent to the geological setting contributed to the successful Normandy invasion; prelanding investigations delimited some deep-water "entrances" to coastal harbors and the location of suitable sites for 20 fighter airstrips near the front lines (King, 1951, p. 131–137).

Figure 14. A German underground-factory site of World War II, located in an unusual rock mass: loess deposits of the Danube River valley north of Linz, Austria. The soft-to-spongy loess effectively dissipated shock waves generated by the explosion of conventional bombs. One of the entrances to the abandoned underground system of large-scale interconnected openings is at the lower right; the entrance is now largely closed off. (Photograph by G. A. Kiersch, 1963.)

Some of the deep-water harbors along the coast are a consequence of its glacial history. Water in the English Channel was lowered 100 m, and the older beach deposits underwent erosion during the last glacial stages. Deep canyons were carved in the former off-shore, wave-cut rock platform; dunes, swamps, and fen peat formed on or behind the wave-cut platform. As the sea returned to the present level, the peats and swampy forests were submerged. Preinvasion study of maps of Normandy beaches indicated that some sectors were unsuitable for assault vehicles, an assessment confirmed on-site by commandos prior to the invasion. Professor Shotton, geologist of 21st Army Group, was largely responsible for delimiting the deep-water entrance to "Mulberry" harbor—an erosional gap or canyon in the old wave-cut platform.

The initial plan to land on the beaches of the Cotentin Peninsula in the Cherbourg area was dropped; this hilly area with its small fields, banks, and hedges was not suitable for a large number of airstrips (Rose and Pareyn, this volume). Instead the flat-lying Normandy terrain of Jurassic limestone provided large open areas covered by loess-like loam, an ideal geologic setting for the rapid construction of airstrips. The partially weathered limestone strata at depth afforded rapid surface drainage, and in the summer months the loess-like loam at the surface dried quickly to a firm soil (King, 1951, p. 131–137).

Following the Battle of Falaise in Normandy in mid-1944, the Allied armies moved so quickly across France that geology only again became a major factor when the frontlines were in Holland and at the German frontier in the spring of 1945 (King, 1951).

USSR forces

From the beginning of World War II, the Russian army employed a large body of geological specialists; German intelligence reported thousands were employed in military geology activities. In Russia as in France, geology was part of an officer's education. Hydrotechnical divisions and small units of geological specialists operated under the Army General Staff were responsible for all matters pertaining to water supply, geo-data maps, construction materials, and transport routes (Fox, 1949, p. 3).

Japanese forces

The Japanese military openly exchanged geological data with other major countries until 1937 when the practice stopped. The Japanese emphasized the use of botanical data as a guide to ground conditions (Smith, 1964, p. 319).

POST-WORLD WAR II

Ingenious presentations of geologically related data on terrain maps both by the German and Allied forces aroused interest among military specialists worldwide when analyzed after the war (Smith and Black, 1946; Wilson, 1948). Within the unique and very successful German Wehrmacht organization, the

Forschungesstaffel special unit had concentrated on terrain-evaluation maps, paying special attention to trafficability for tanks. These maps incorporated concepts developed by a team of physical geographers, plant ecologists, geologists, soil scientists, foresters, meteorologists, and others. The geological data base utilized aerial photographs and aerial reconnaissance combined with "ground-truth" sources to produce sophisticated terrain-evaluation maps.

The advent of World War II (1939–1945) brought about the proliferation of applied geology on a scale hitherto unimagined. Of these broadened activities, the application of geology to military operations, as developed in the European and Pacific regions, was the most important advancement in engineering geology during the 1940s and 1950s (Kiersch, 1955, p. 29–30). Furthermore, many important textbooks and publications that advanced the principles of engineering geology practice and military geology were released during and shortly after the war, including those by Blyth (1943), Bendel (1944), Trefethen (1949), Paige (1950), Trask (1950), Eckel (1951), Keil (1954), Kiersch (1955), and Schultz and Cleaves (1955).

The success of applied geology in serving World War II activities, both military and civil, led to establishing two new, related branches of the U.S. Geological Survey: the Engineering Geology Branch in late 1944 under E. B. Eckel, which began functioning in 1946, and the Military Geology Branch (MGB) in 1946 under C. B. Hunt and, later, E. S. Larsen and Frank Whitmore. The MGB was a formalization of the earlier Military Geology Unit (MGU of 1945). Still a third new branch, the Foreign Geology Branch, was established in 1946 under William Johnston and included many of the geologists who had served in Europe obtaining resource and related information for MGU. Functions of the MGB were closely coordinated with the Central Intelligence Agency, the U.S. Army Corps of Engineers, and the Army Map Service. In addition, the Alaskan Highway experience with permafrost and related difficulties led to establishing the Snow, Ice, and Permafrost Research Organization within the U.S. Corps of Engineers for enhancement of construction of all kinds in Arctic regions (Péwé, 1991, p. 277–298; Weeks and Brown, 1991, p. 333–350).

Similarly the U.S. Navy, through the Office of Naval Research and the Bureau of Ships, financed research in oceanography, submarine geology, and sedimentation, all topics for planning the support of amphibious operations and the control of beach erosion. Bates et al. (1982, p. 47–78) include an informative chapter on "Geophysics at War" with emphasis on oceanography.

Typical of the postwar stage of the MGU's activities in the Pacific region was a study of the Fukui earthquake on June 28, 1948, in Okinawa. The event provided an unusual opportunity for geologists and engineers to assess the importance of the seismic response of complex geological sites involving varied rock units and tectonic structures; the study provided new insights and data for design of earthquake-resistant structures (Collins and Foster, 1949).

The Korean War of 1950–1953 was largely served by members of the U.S. Geological Survey Military Geology Branch (MGB), successor to the Military Geology Unit, on duty in regions of Japan, Okinawa, and the Pacific Islands and is reviewed in the two papers by Cameron and one paper (Chapter 8) by Terman in this volume. Tunneling beneath fortifications as an attack and infiltration technique was again used by the North Korean forces.

Underground facilities

Offensive underground warfare was successful during the Civil War with tunnels beneath Confederate forces at Petersburg (Fig. 5) and in World War I to counterattack strongly fortified German positions (Figs. 6 and 7). Yet interest in geologic barriers did not become widespread among the armies of the world until World War II. Since then, the design and construction of underground chambers or deeply buried military facilities have involved a small segment of the military community (Kiersch, 1951, and O'Sullivan, 1961). More recently with the proliferation of precision-guided missiles, geologic barriers have taken on a new meaning relative to the vulnerability of an underground facility to conventional weapons. Moreover, the future role of strategic-geologic intelligence in planning future warfare is increasingly critical, as reviewed by Eastler et al. (this volume).

After 1945, realization of the destructive force of the atomic bomb and later the hydrogen bomb created concern that defense against their effects was nearly impossible. In response, the U.S. Army Corps of Engineers and government-sponsored research groups made tremendous strides in the design of protective construction to resist large-scale blasts, relying on knowledge of the geologic environs and physical properties of rock masses. Developments in destructive weapons dictated attention to underground locations for some military command centers and storage facilities. Any relocation of vital manufacturing plants must be highly selective and based on military importance, geology of alternate sites, safety, economic factors, and the operational needs of the facility.

Since the 1940s, several hundred post-war underground rock installations have been built in Sweden, Norway, France, Italy, and other European countries. Currently, more than 20 different kinds of underground installations have been or are being built in hard-rock formations, as shown on Figure 15 and described by Morfeldt (1984).

Broad geologic principles and their application to the location, construction, and operation of subterranean installations designed to resist modest-scale subsurface explosions of the early 1950s were reviewed by Kiersch (1949, 1951). The Underground Explosion Test Program of the U.S. Army Corps of Engineers (1947 to 1949) and the research of Engineering Research Associates, Minneapolis (ERA, 1952–1953), established the guidelines and principal geologic factors that impact the design of a large cavity. The Rand Corporation's underground construction symposium in 1959 reviewed the state-of-the-art design and construction of protective chambers (O'Sullivan, 1961).

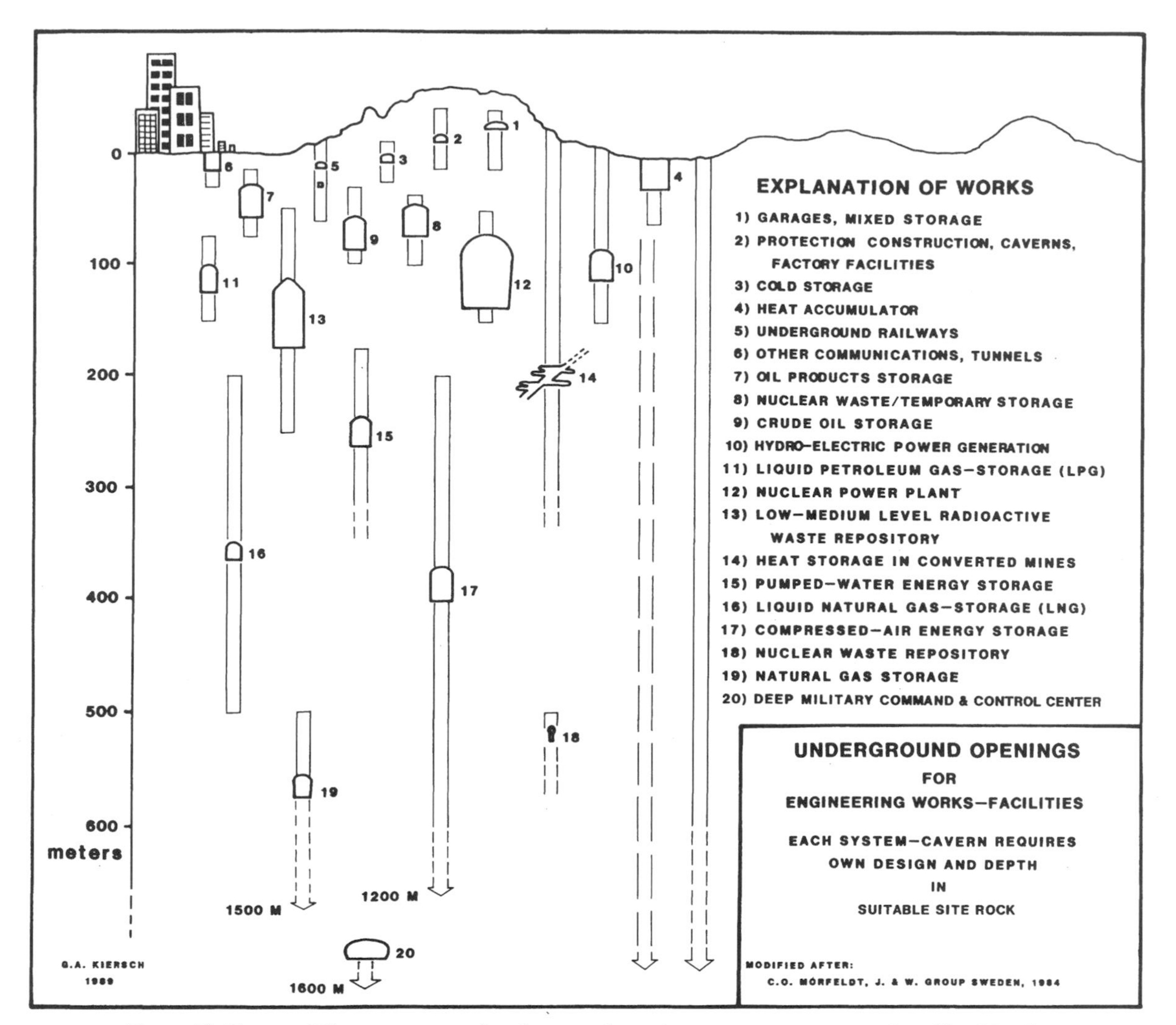

Figure 15. Twenty different systems of underground openings or caverns commonly utilized by the 1980s. Each type of engineered works requires its own depth location in a suitable rock mass. Crude-oil storage now extends to 1,500 m depth (James T. Neal, personal commun., 1996); deep military command and control centers may be at depths of 1,600 m or more. (After Morfeldt, 1984.)

A natural homogeneous medium capable of dissipating energy equally in all directions from a subsurface blast probably does not exist. Reaction of a rock mass to blast-induced stress is complicated by the rock's heterogeneity and inherent structural weaknesses that transmit stress unequally. Furthermore, an intact, hard mass of rock (limestone, igneous, or metamorphic rock) can retain a high residual stress so that the physical characteristics of an otherwise ideal rock mass may react differently to a conventional subsurface blast. At depth, the elastic limit of the rock is overcome by natural stresses. Cavities excavated below a critical depth will be situated in the zone of instability; stresses are active and openings require supports. Depth to the zone of instability varies with the rock type (Fig. 16); the "shear" strength of an average homogeneous sandstone may be exceeded at approximately 530 m, whereas that of an average homogeneous plutonic rock can be exceeded around 1,980 m. A substantial residual or active stress can render a site uneconomical.

The ideal cover for an underground installation may be a combination of competent high-velocity rock separated by low-velocity, weak rock. Formation boundaries in the overlying rock mass partially dissipate the explosive energy (Kiersch, 1951). Principles of underground design and construction, including the ability of a rock mass to resist explosive pressures, were reviewed by Duvall (in O'Sullivan, 1961, p. 313–335), and the U.S. Army Corps of Engineers (1961), and represent the state of the art in 1960. The current state of research on new designs to resist large-scale military blasts is described by Eastler et al. (this volume).

Two strategic U.S. military command-and-control centers were built underground in the 1950s and 1960s: the Omaha, Nebraska, Strategic Air Command Control Center, and the NORAD (North American Air Defense) Center in Cheyenne

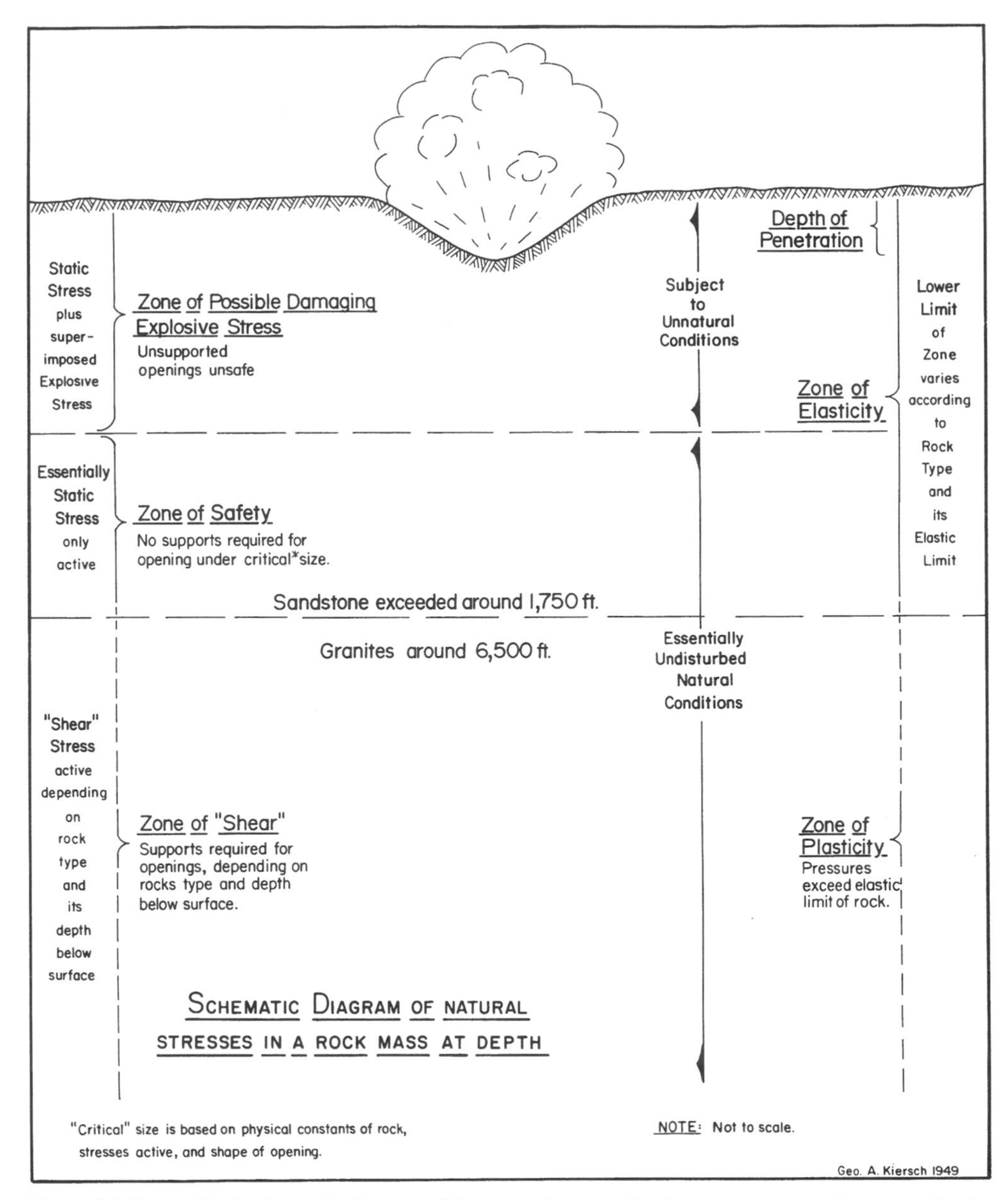

Figure 16. Generalized schematic diagram of the natural stresses in a homogeneous rock mass at depth, as related to the zone of stability vs. instability. (From Kiersch, 1951, fig. 1.)

Mountain near Colorado Springs, Colorado. Such critical projects required that particular attention be given to the geology because of the inherent site conditions and the large size of the underground chambers (Lane, 1971). In recent years, the military geology efforts by U.S. Geological Survey personnel have focused on the "geologic assessment at the sites of underground installations" (Leith, this volume).

Since the Korean War, much attention has been focused on the underground and tunneling by North Korean forces. They are known to have constructed several clandestine infiltration tunnels beneath the Demilitarized Zone (DMZ), such as Tunnel 2 and Tunnel 3 of the 1970s, driven into the granodiorite and gneiss bedrock (Cameron, Chapter 9, this volume). In 1989, Tunnel 4 was discovered in granodiorite by the use of cross-hole geophysics, site characterization by geologic mapping, and fracture analysis. The existence of additional tunnels is suspected (Uldrich, 1994, p. 2); some specialists estimate that the total number of tunnels approaches 50 (Colonel Allen Hatheway, U.S. Army [retired], personal commun., 1996).

Nuclear detections

During the 1950s, geoscientists of the Air Force Terrestrial Science Laboratory successfully distinguished between an artificial underground nuclear test signal and a natural seismic event (Haskell, 1957; Eckel, 1968). This major advancement for military intelligence subsequently led to establishing the VELA UNIFORM research project in October 1959 for the purpose of detecting underground nuclear detonations elsewhere in the world. Two companion activities were the VELA SIERRA program to detect high-altitude nuclear detonations and the VELA HOTEL program to detect high-altitude nuclear detonations by means of satellite-borne instruments (Bates, 1961). Other projects of the Air Force Terrestrial Science Laboratory are reviewed by the former director, Colonel Louis DeGoes (DeGoes and Neal, this volume), and by D. B. Krinsley (this volume).

CONCLUSION

Virtually all areas of engineering geology practice have been or may be utilized when confronted by the specialized problems of modern military activities. Geologic input has ranged from basic analysis of terrain for trafficability of troops and vehicles to analysis of supplies of ground water and construction materials, to the analysis of rock stresses and the geomechanics of underground protective construction. Education and training for service as a military geologist must be broadly based and founded on field experience and should include the development of awareness of the substantial geologic input to earlier campaigns of history.

ACKNOWLEDGMENTS

Assistance and input of colleagues have contributed to the breadth of this review. They include Colonel Louis DeGoes and Lieutenant Colonel James T. Neal, formerly of U.S. Air Force Terrestrial Science Laboratory; Dr. James F. Devine, assistant director, Dr. Clifford Nelson, historian, and Henry Zoller and Robert Bier, librarians, all of the U.S. Geological Survey, Reston; Robert Fickies, New York Geological Survey; Professor Arthur Bloom, Cornell (coral reefs); Professor Troy Péwé, Arizona State University (permafrost); Professor Richard Goodman, University of California, Berkeley, and Terzaghi Fellow, Norwegian Geotechnical Institute; Professor Lars O. Grande, Technical University (NTH), Trondheim (submarine pens); Professor Allen W. Hatheway, University of Missouri–Rolla (selected reference data); and John Uldrich (background on Petersburg tunneling). Kim Duffek of Kanoa Illustrations, Tucson, prepared some of the line drawings as did Lois Kain, Tucson. Jane Hoffmann of Road Runner Press in Tucson and Linda Clough of Kansas State University reproduced the text.

REFERENCES CITED

Aberdeen, E., 1945, Coral-reef sediments [abs.], *in* The activities of the Military Geology Unit of U.S. Geological Survey 1942–1945, Society of Economic Geology, v. 40, p. 586.

Barré, O., Commandant, 1897–1902, Cours de geographie: Croquis geographiques: Fontainebleau, Ecole d'Application de l'Artillerie et Genie, 2 volumes.

Baskerville, C. A., 1982, The foundation geology of New York City, *in* Legget, R. F., ed., Geology under cities: Geological Society of America, Reviews in Engineering Geology, v. V, p. 95–117.

Bates, C. C., 1961, The goals of project VELA: Geotimes, v. 6, p. 13–16.

Bates, C. C., Gaskell, T. F., and Rice, R. B., 1982, Geophysics in the affairs of man: New York, Pergamon Press, 492 p.

Bendel, L., 1944, Ingenieur-geologie: Wien, Springer-Verlag, Part I, 1948; Part II, 1949, 2nd ed.; Part I, 832 p; Part II, 832 p.

Berkey, C. P., 1933, New York City and vicinity, *in* 16th International Geological Congress Guidebook 9: Washington, D.C., U.S. Government Printing Office, p. 1–23, 36–39, 77–92, and 111–123.

Betz, F., editor, 1975, Environmental geology: New York, Van Nostrand Reinhold Benchmark Series, p. 95.

Betz, F., 1984, Military geology, *in* Finkl, C. W., Jr., ed., The encyclopedia of applied geology: New York, Van Nostrand Reinhold, p. 238–241.

Blyth, F. G. H., 1943, A geology for engineers: London, Edward Arnold and Co., 302 p.

Brooks, A. H., 1920, The use of geology on the western front: U.S. Geological Survey Professional Paper 128-D, p. 85–124.

Brown, A., 1961, Geology and the Gettysburg campaign: Pennsylvania Geological Survey Education Series, no. 5, 15 p.; Geotimes, 1961, v. 5, p. 8–12, 40–41.

Brown, A., 1963a, A geologist-general in the Civil War: Geotimes, v. 7, no. 7, p. 8–11.

Brown, A., 1963b, Geology in the Tullahoma campaign of 1863: Geotimes, v. 8, no. 1, p. 20–22, 53.

Brown, A., 1964, The Chickamauga campaign of 1863: Geotimes, v. 9, no. 3, p. 17–21.

Casagrande, L., 1952, Electro-osmotic stabilization of soils: Boston Society of Civil Engineering Journal, v. 39, p. 51–70 and 1949, Geotechnique, v. 1, p. 1959–1977.

Collins, J. J., and Foster, H. L., 1949, The Fukui earthquake, Hokuriku region, Japan, 28 June 1948; Volume I, Geology; Volume II, Engineering: Far East Command, U.S. Army General Headquarters, Geological Surveys Branch, Office of Engineers, v. I, 81 p., v. II, 205 p.

Cross, W., 1919, Geology in the World War: Geological Society of America Bulletin, v. 30, p. 165–188.

Dryden, L., 1945, Construction on coral reefs [abs.], *in* The activities of the Military Geology Unit of the U.S. Geological Survey 1942–1945; Society of Economic Geologists, v. 40, p. 589.

Eckel, E. B., 1945, Report on engineering geology in Germany (Underground factories and dams): Headquarters, U.S. Forces, European Theater, Infantry Section, Intelligence Division, Office Chief of Engineers, CIOS sub-section, APO 887, 48 p.

Eckel, E. B., 1951, Research needs in engineering geology; Presidential address: Denver, Proceedings of the Colorado Science Society, 11 p.

Eckel, E. B., editor, 1968, Nevada test site; Studies of geology and hydrology: Geological Society of America Memoir 110, 284 p.

Edmunds, C. S., 1945, Beach intelligence service, *in* The activities of the Military Geology Unit of the U.S. Geological Survey 1942–45: Society of Economic Geologists, v. 40, p. 590.

Erdmann, C. E., 1943, Application of geology to the principles of war: Geological Society of America Bulletin, v. 54, p. 1169–1194.

Fox, E. F., 1949, The use of military geologists in the Corps of Engineers: Study for Chief of Engineers, U.S. Army Engineers.

Haskell, N. R., 1957, An estimate of the maximum range of detectability of seismic signals: Bedford, Massachusetts, Terrestrial Science Laboratory Air Force Surveys and *in* Geophysics, no. 87, 42 p.

Hatheway, A. W., 1993, Biography of Charles Butler Hunt, geologist: Bulletin of the Association of Engineering Geologists, v. 30, p. 139–155.

Hunt, C. B., 1950, Military geology, *in* Paige, S., ed., Application of geology to engineering practice: Geological Society of America, Berkey Volume, p. 295–327.

Hunt, C. B., 1982, History of the military geology unit during World War II: Geologic Division Retirees Newsletter, U.S. Geological Survey, no. 8 (Nov.), p. 12–15.

Keil, K., 1954, Ingenieurgeologie und geotechnite: Halle, Seale, Wilhelm Knapp Verlag, 1132 p.

Kiersch, G. A., 1949, Underground space for American industry: Mining Engineering, v. 1, p. 20–25.

Kiersch, G. A., 1951, Engineering geology principles of subterranean installations: Economic Geology, v. 46, p. 208–222.

Kiersch, G. A., 1955, Engineering geology; History, scope and utilization: Colorado School of Mines Quarterly, v. 50, 123 p.

Kiersch, G. A., editor, 1991, The heritage of engineering geology; The first hundred years: Boulder, Colorado, Geological Society of America, Centennial Special Volume 3, 605 p. Sections on military geology, p. 33–34, 54–58.

Kieslinger, A., 1958, Restspannung und entspannung im gestein: Geologie und Bauwesen, v. 24, p. 97–112.

Kieslinger, A., 1960, Residual stress and relaxation in rocks: 21st International Geological Congress, Copenhagen, part 18, p. 270–276.

King, W. B. R., 1919, Geological work on the Western Front: Geographical Journal, v. 54, p. 201–221.

King, W. B. R., 1951, Influence of geology on military operations in northwest Europe: London, Advancement in Science, v. 8, p. 131–137.

Kranz, W., 1913, Militargeologie: Berlin, Kreigstech Zeitschrift, Officiere aller Waffen, v. 16, p. 464–471.

Lane, K. S., editor, 1971, Proceedings of Underground rock chambers Symposium, Phoenix, Arizona, January 13–14: American Society of Civil Engineers, 600 p.

M. F. W., 1899, Geological expedition of Dr. G. F. Becker to the Philippines: Science, v. 9, p. 722–723.

Marga, A., 1885, Commandant du Genie, Geographic militaire, pt. 1: Generalites, La France, 2 vols., and atlas, 4th ed., Paris 1885, pt. 2, Principaux é'tats de l' Europe, 3 vols., and Paris, 1 vol.

McPherson, J. M., ed., 1989, Battle chronicles of the Civil War: New York, Macmillian Publishing Co., p. 548–549.

Mead, W. J., 1941, Engineering geology, *in* Berkey, C. P., ed., Geology, 1888–1938; 50th Anniversary volume: Geological Society of America, p. 573–578.

Morfeldt, C. O., 1984, The influence of engineering geological data on the design of underground structures, and on the selection of construction methods, *in* General report, Theme III, International Symposium on Engineering Geology and Underground Construction: Lisbon, International Association of Engineering Geology, p. III-1, to III-24.

Muller, S. W., 1943, Permafrost or permanently frozen ground and related engineering problems: U.S. Army, Office, Chief of Engineers, Strategic Engineering Special Report 62, 231 p.

O'Sullivan, J. J., ed., 1961, Protective construction in a nuclear age: New York, Macmillan Co., 2 vols., 836 p.

Paige, S., editor, 1950, Applications of geology to engineering practice: Geological Society of America, Berkey Volume, 327 p.

Péwé, T. L., 1991, Permafrost, *in* Kiersch, G. A., ed., The heritage of engineering geology; The first hundred years: Boulder, Colorado, Geological Society of America, Centennial Special Volume 3, p. 277–298.

Portlock, J. R., 1859, A rudimentary treatise on geology: 4th edition, London.

Powell, W. H., 1989, The battle of the Petersburg Crater, *in* McPherson, J. M., ed., Battle chronicles of the Civil War, 1864: New York, Macmillan Publishing Co., p. 545–560.

Putnam, W. C., 1943, Map interpretation with military applications: New York, McGraw-Hill Book Co., 67 p.

Russell, R. D., 1950, Applications of sedimentation to naval problems, *in* Trask, P. D., ed., Applied sedimentation: New York, John Wiley & Sons, p. 656–665.

Schultz, J. R., and Cleaves, A. B., 1955, Geology in engineering: New York, John Wiley & Sons, 559 p.

Smith, J. S. C., 1964, Military application of geology: Transactions of Kansas Academy of Science, v. 67, p. 311–336.

Smith, P. S., 1918, The geologist in war times: U.S. Geological Survey's war time work: Economic Geology, v. 13, p. 392–399.

Smith, T. R., and Black, L. D., 1946, German geography: War work and present status; Geographic Review, Geographic Society, New York, v. 36, p. 398–408.

Trask, P. D., editor, 1950, Applied sedimentation: New York, John Wiley & Sons, 665 p.

Trefethen, J. M., 1949, Geology for engineers: New York, Van Nostrand Co., Inc., 629 p.

Uldrich, J., 1994, Position paper on U.S. Military Forces—South Korea: Pittsburgh, Praetorian Gate Press, 3 p.

U.S. Army Corps of Engineers, 1961, Design of underground installations in rock: U.S. Army Corps of Engineers EM 1110-345-431, 68 p.

Von Bulow, K., Kranz, W., and Sonne, E., 1938, Wehrgeologie: Leipzig, Quelle and Meyer, 178 p., 164 figs. (E.R.O. Translation T-23 by Kurt E. Lowe, Intelligence Branch, Office of the Chief of Engineers, 1943).

Wasmund, E., 1937, Wehrgeologie in ihrer Bedeutung fur die Landesverteidigung: Berlin, Mittler and Sohn, 103 p.

Weeks, W. F., and Brown, R. L., 1991, Snow and ice, *in* Kiersch, G. A., editor, The heritage of engineering geology: The first hundred years: Boulder, Colorado, Geological Society of America, Centennial Special Volume 3, p. 333–350.

Wilson, L. S., 1948, Geographic training for the post-war world: A proposal: Geographic Review, v. 38, p. 575–589.

Whitmore, F. C., 1950, Sedimentary materials in military geology, *in* Trask, P. D., editor, Applied sedimentation: New York, John Wiley & Sons, p. 635–665.

Whitmore, F. C., 1954, Military geology: The military engineer: v. XLVI, no. 331, p. 212.

Manuscript Accepted by the Society October 29, 1997

Printed in U.S.A.

Geological Society of America
Reviews in Engineering Geology, Volume XIII
1998

British military geologists through war and peace in the 19th and 20th centuries

Edward P. F. Rose
Department of Geology, Royal Holloway, University of London, Egham, Surrey TW20 0EX, United Kingdom
Michael S. Rosenbaum
Department of Geology, Imperial College of Science, Technology, and Medicine, Prince Consort Road, London SW7 2BP, United Kingdom

ABSTRACT

The first geologists employed in government service in Britain had military appointments: J. MacCulloch from 1809 to 1826 in England and Scotland, and J. W. Pringle followed by J. E. Portlock from 1826 to 1843 in Ireland. The founder of the British Geological Survey in 1835, and his successor as director-general in 1855, both had military origins. Several early influential members of the world's oldest geological society, founded in London in 1807, had military connections. From 1819 to about 1896 geology contributed to military education in Britain at the East India Company's military college, the Royal Military Academy, the Royal Military College, the Staff College, or the School of Military Engineering.

However, professional geologists were not strictly used as such in the British army until the 1914–1918 world war, and then they were primarily used in response to problems of static battlefield conditions on the western front in Europe. W. B. R. King guided development of potable ground-water supplies; T. W. E. David guided siting of mine tunnels and dugouts, and other geologists served with the Tunnelling Companies of the Engineer Corps.

Geologists were used more widely in the more mobile conflicts of the 1939–1945 world war: notably W. B. R. King in France and the United Kingdom, F. W. Shotton in North Africa and northwest Europe, and J. V. Stephens in Italy. These and others were all to some extent concerned with water supply, but increasingly geologists became involved in terrain assessment for military purposes (e.g., airfield sites, ground trafficability, quarrying of aggregates, and effects of aerial bombing).

In both wars there were but few British military geologists; most were granted Emergency Commissions in the Royal Engineers for their war service. Only since 1949 has the corps maintained continuity of geological expertise through a small team of reserve army officers. This team now provides support for regular forces in both peace and war.

INTRODUCTION

Four phases can be recognized in the development of military applications of geology in the British army:

(1). Early in the 19th century, there was a closer relationship between the army and some of the founding fathers of British geology than is commonly realized (Rose, 1996a). In consequence, by the middle of the century the relevance of geology to a military profession was recognized. Geology was taught as a distinct part of the curriculum at the East India Company's mili-

Rose, E. P. F., and Rosenbaum, M. S., 1998, British military geologists through war and peace in the 19th and 20th centuries, *in* Underwood, J. R., Jr., and Guth, P. L., eds., Military Geology in War and Peace: Boulder, Colorado, Geological Society of America Reviews in Engineering Geology, v. XIII.

tary college, Addiscombe, from 1819 to 1835, and again from 1845 to 1861; at the Royal Military Academy, Woolwich, from 1848 to 1868; at the Royal Military College, Sandhurst, from 1858 to 1870; at the Staff College, Camberley, from 1862 to at least 1882; and at the School of Military Engineering, Chatham, in the mid-1890s (Rose, 1997a).

(2). In the First World War, from 1915 onward, a few officers newly accepted into the expanding British army at the start of hostilities in Europe were given specific geologist appointments connected with water supply and with tunnelling. They were chosen on the basis of their earlier civilian geological experience and proven ability. Following the end of the war in 1918, all returned to civilian life.

(3). From the start of the Second World War in 1939, a few geologists were recruited or accepted from appropriate civilian occupations to fill specific military appointments connected with water supply and terrain assessment in Europe and North Africa. Most were granted Emergency Commissions in the Corps of Royal Engineers. All returned to civilian life soon after the close of hostilities in 1945.

(4). From 1949, the British army has been equipped not only with a military geological textbook (Anonymous, 1949; 1976) but also with a small team of geologists in the reserve army. These have provided geotechnical support to regular forces in both peace and armed conflict and maintained continuity of geological expertise within the Corps of Royal Engineers to the present day.

NINETEENTH CENTURY FOUNDATIONS

The first geologic map of any country to be based on an official survey and to be published by a government was a geologic map of Scotland by John MacCulloch (MacCulloch, 1836). Its author received military funding for most of his working life (Flinn, 1981): as a surgeon's mate in the Royal Artillery and also assistant chemist and assayist to the Board of Ordnance from 1795, and as chemist and assayist to the ordnance and also a lecturer at the Royal Military Academy, Woolwich, from 1804 to 1826. Moreover, his map developed as the by-product of two specific geological projects assigned by the Board of Ordnance—a military body. First, from 1809 to 1815 MacCulloch undertook fieldwork to find a limestone in Britain suitable for use as a millstone in gunpowder manufacture because the Napoleonic War had stopped importation of "Namur stone," a Carboniferous limestone from Belgium—use of an inferior stone from Ireland was inferred to have caused an explosion from sparks generated by its quartz content. Later, from 1814 to 1826, his task was to make a "minute geological and mineralogical" survey in support of the Trigonometrical Survey of Great Britain then in progress. Unacceptable differences had been found between trigonometrical and astronomical fixes of stations used in the primary triangulation, and a geological survey was necessary to help test the hypothesis that nearby hills or mountains, or inequalities of densities of rocks beneath the surface, might cause deflection of the plumb line used in calculating astronomical fixes. His efforts to convince the Board of Ordnance of the military necessity of a geologic map as such, however, proved unsuccessful. The office of ordnance chemist was abolished, and MacCulloch made redundant, on January 13, 1826, with the recommendation that he should "transfer his undivided attention and services to the geological survey under the Treasury" (cited in Flinn, 1981, p. 95).

One of the first government geological survey departments in any country was set up under military auspices in Ireland in 1826. As the Trigonometrical Survey conducted under the Board of Ordnance was extended from Great Britain into Ireland during 1824 under command of Major T. F. Colby, Royal Engineers, a geological branch was quickly established to complement the topographic survey (Bailey, 1952). This was directed between 1826 and 1828 by Captain J. W. Pringle, Royal Engineers (Davies, 1983; Rose, 1997b). A career soldier, Pringle had been wounded at the Battle of Waterloo on June 18, 1815—the only engineer officer listed as wounded in the *London Gazette*. From 1832 the geological branch was reactivated by Captain J. E. Portlock, Royal Engineers (Vetch, 1909). Portlock had served in Canada from 1814 to 1822, initially against the United States—in the siege of Fort Erie and then in Chippewa where he constructed fortified lines and the bridgehead. In 1825 he accompanied Colby to Ireland to assist with, and ultimately to direct, the triangulation survey. From 1834 he was tasked with compiling a descriptive memoir relating to geology, natural history, and productive economy. Preparation of the memoir as a whole was suspended in 1840 for financial reasons, but the geological achievements were published (Portlock, 1843) in the year that Portlock's work in Ireland came to an end. He returned to the ordinary duties of an engineer officer, finally retiring from the army in November 1857 in the rank of major-general.

However, the geological role of Royal Engineers in Ireland was not yet complete. In 1845, when H. T. De la Beche was formally appointed as director-general of the Geological Survey of Great Britain and Ireland, Captain Henry James, Royal Engineers, was appointed local director for Ireland (Bailey, 1952). James resigned the following year to superintend construction of the Admiralty Dockyard at Portsmouth, and thereafter he followed a career in the Ordnance Survey, serving as director from 1854 to 1875 and finally in the rank of lieutenant-general (Anonymous, 1877). With the appointment of his successor, Portlock's former chief assistant Thomas Oldham, direction of geological surveys in Ireland became thenceforth a civilian responsibility (Davies, 1983).

The British Geological Survey was founded in 1835, as a military initiative, by the master-general and Board of Ordnance (Bailey, 1952, p. 26). For its first 10 years it existed as the Ordnance Geological Survey, funded and controlled by the military. Moreover, although it became an entirely civilian organization in 1845, for 26 more years it was led by two successive directors-general who had received a military rather than a university education. H. T. De la Beche (the founder and, from 1845, director-general), as the son of an army officer, had appropriately entered the Junior Department of the Military College, then at

Great Marlow in Buckinghamshire, as a cadet in 1810. But there were problems of insubordination among the cadets, and documents at the Public Record Office note a speedy end to his military career: "Henry De la Beche, 8 October 1811, removed by order of the Commander-in-Chief" (McCartney, 1977, p. 4). Following his death in 1855, he was succeeded by R. I. Murchison. Murchison had also been trained at Great Marlow before being commissioned in 1807 as an ensign in the 36th (the Herefordshire) Regiment of Foot and before active service in the Peninsular War, notably at the Battle of Vimiera and the poetically celebrated retreat to Corunna (Geikie, 1875). He resigned from the army as a captain at the end of the Napoleonic wars after having transferred into the cavalry, the Inniskilling Dragoons, in an unsuccessful attempt to participate in the Waterloo campaign. However, both De la Beche and Murchison seem to have benefited from the Marlow curriculum, which, at least from 1808 (Smyth, 1961), consisted of geometry, trigonometry, surveying, theory of fortifications, military drawing, and modern history.

The Geological Society of London, founded in 1807 and so the world's oldest geological society, had a reserve army cavalryman as its first president, a veteran infantry officer as its first full-time employee, and Royal Engineer officers among the members of its early, formative councils. G. B. Greenough, the founding president 1807–1813 and president again 1818–1820 and 1833–1835, had considerable geological influence (Boulger, 1908). Additionally, for 16 years (1803–1819) he served in the reserve army, first in the ranks and then (from June 30, 1808) as a lieutenant in the Light Horse Volunteers of London and Westminster—a corps separate from the militia, yeomanry, and volunteer infantry whose officers were, most unusually, elected and promoted by ballot at a general meeting (Collyer and Pocock, 1843). He resigned his commission in 1819 as a public protest against the "Peterloo massacre" (Rose and Rosenbaum, 1993a; Rose, 1996b). William Lonsdale, the Society's librarian and curator or assistant secretary from 1829 until retirement in 1842, and recipient of its prestigious Wollaston medal in 1846 (Woodward, 1907), had a regular army background (Bailey, 1952). Commissioned as an ensign in the 4th (the King's Own) Regiment of Foot on February 1, 1810, *The Army List* records that Lonsdale served with his regiment in the Peninsular War where he was present at the Battle of Salamanca. He also served at Waterloo as a lieutenant—the only officer in his regiment not wounded according to Boase (1965) (but not Dalton, 1904). Retiring from the army on March 25, 1817, he remained on half pay throughout his geological career until his death in 1871. Royal Engineer officers who served on the early councils of the Society included T. F. Colby (1815–1818, 1819–1820, 1822–1825) until his departure to direct the Trigonometrical Survey in Ireland; J. W. Pringle (1831–1832, 1833–1834) following his return from Ireland; and J. E. Portlock (1849–1850, 1852–1862), who became president of the Society for 1857 and 1858.

Lectures on geology were given to potential British officers of the Honorable East India Company's army at its military seminary at Addiscombe, Surrey, from 1819 to 1835 by J. MacCulloch, and from 1845 to 1861 by D. T. Ansted. From 1848 to 1868, geological and mineralogical lectures were also given, by J. Tennant, to gentlemen cadets preparing for commissions in the Royal Artillery and the Royal Engineers at the Royal Military Academy, Woolwich (Rose, 1997a).

In consequence, it is not surprising to find that by the middle of the 19th century the relevance of geology to the military profession was increasingly recognized. Portlock (1849, p. 14) argued that "the Soldier . . . may find in Geology a most valuable guide in tracing his lines both of attack and defence," and his 116-page article on "geognosy and geology" in the authoritative *Aide-Mémoire to the Military Sciences* (Lewis et al., 1850, p. 77–182, plates 1–14) emphasized the point, although the article was slimmed to a mere 100 pages in the otherwise enlarged second edition (Lewis et al., 1860, p. 91–190, plates 1–14). An article entitled "importance of a knowledge of geology to military men" (Hutton, 1862) could not have been more specific.

Hutton was an experienced infantry officer: *The Army List* records purchase of his commission as an ensign in the 23rd (Royal Welch Fusiliers) Regiment of Foot on May 18, 1855, his promotion to lieutenant March 27, 1857, and his service in the Indian Campaign of 1857–1858, including the relief of Lucknow by Lord Clyde, defeat of the Gwalior Contingent at Cawnpore, capture of Lucknow, and operations north of the Goomtee. He entered the Staff College, Camberley, in 1860, prior to promotion to captain on March 2, 1862, and it is likely that his geological interests were encouraged there by T. Rupert Jones.

Jones had been appointed "lecturer on geology" at the Royal Military College, Sandhurst, in 1858. In 1862 he was appointed professor of geology at Sandhurst and subsequently the adjacent Staff College, Camberley—consequently resigning the post of assistant secretary to the Geological Society of London that he had held for the previous 11 years. He taught the relations of geology to topography, to questions of sanitation, and to water supply (Woodward, 1907) at Sandhurst until 1870, and at Camberley until 1882. When Jones retired, the chair of geology was abolished, and the teaching of geology as a distinct discipline was discontinued (T. A. Heathcote/P. E. Bendall, personal commun., 1994). Thus for 24 years geology formed part of the curriculum that provided either basic cadet training for potential infantry and cavalry officers in the British regular army or staff training for experienced young officers of all arms. The tradition then lapsed. Some geology lectures were given by professional servicemen at Camberley (Jones, 1898) and by a visiting professor at the School of Military Engineering, Chatham, (Edmonds, in King, 1919), but these practices were soon discontinued (Rose, 1997a).

THE 1914–1918 WORLD WAR

Effectively, use of geology within the British army began 35 years after routine officer training in the subject had ceased. Aubrey Strahan, the director of the Geological Survey of Great Britain, is credited by Howarth (1993, p. 640) with supervising "the preparation of a set of vital military geological maps of

Belgium and other war zones" at the outbreak of war in 1914. In response to a request from the deputy-director of works at the British Expeditionary Force General Headquarters, it was through Strahan's influence that the first professional geologist specifically to be recruited as such, W. B. R. King, was appointed to serve in the western European theater of the 1914–1918 world war (King, 1919; Brooks, 1920; Anonymous, 1922a; Rose and Rosenbaum, 1993a).

King had graduated from Cambridge with First Class Honors in geology in 1912 and joined the Geological Survey, quickly distinguishing himself as a field geologist in Wales. Then war broke out, and on September 21, 1914, he was commissioned in the Territorial Army, in the infantry, as a second lieutenant in the Seventh (Merioneth and Montgomery) Battalion of the Royal Welsh Fusiliers. But by April 1915 the concentration of troops and transport animals locked in near-static warfare across northern France made development of adequate supplies of potable water a military necessity. On Strahan's recommendation, King was nominated and quickly trained as a hydrogeologist for special appointment as a staff officer to the chief engineer, British Expeditionary Force. He was to serve in this post for the rest of the war, finally in the rank of captain.

As we have noted elsewhere (Rose and Rosenbaum, 1993a), on the basis of an early account (Anonymous, 1922a), over 470 borings for water were executed behind the western front by the British army during the war, and King supervised and interpreted many of these. The results were subsequently published for unrestricted use (King, 1921a, b). Most of the early work consisted of compiling water supply maps for those areas of northern France and Belgium occupied by German forces. Later work involved the selection of sites for boreholes, although "the majority of the sites for bore-holes were chosen more from military than geological considerations" (Anonymous, 1922a, p. 13). One great practical achievement in well technology at this time was the design of wire screens and sand filters that enabled yields of 4.5 m^3 per hour to be drawn from wells in the "Thanet" (Landenian: late Paleocene) sands of the Ypres salient—the only known aquifer in that area, and one that the local inhabitants had not previously been able to exploit fully by boreholes because of the extremely fine nature of these weakly cemented sands. Improved techniques for recording and mapping hydrogeological data were also developed as a military expedient (King, 1951).

In May 1916 another notable geologist arrived with the Mining Battalion of the Australian Corps: Major T. W. Edgeworth David (Fig. 1), a white-haired grandfather 58 years old. David had emigrated from Wales to Australia in 1882 to take up an appointment with the New South Wales Geological Survey. Subsequently he became professor of geography and geology in the University of Sydney and by 1916 was a geologist of very considerable distinction. He reached wartime France with the battalion of miners and geologists that he had himself helped to raise (Harvey, 1934; Carter and Browne, 1936; David, 1937; Mawson, 1949; Branagan, 1987; 1990). Initially he served as geological adviser to the controller of mines of the First, Second, and Third Armies, but later in 1916, following injury through an accident while inspecting a well shaft, he was posted to the Inspector of Mines Office at British Expeditionary Force General Headquarters, an office distinct from but close to the chief engineer's office that contained Bill King. In 1918 David was promoted to the rank of lieutenant-colonel.

Figure 1. Major, later Lieutenant-Colonel, T. W. E. David (born 1858, died 1934), the senior British military geologist of World War I. (From Branagan [1990] and Rose and Rosenbaum [1993a], courtesy of P. J. Davies and the Geologists' Association.)

As we have summarized elsewhere (Rose and Rosenbaum, 1993a), David's work related primarily to mining and dugouts. Mining, one of the oldest applications of engineering to the art of war, was employed on a vast and unprecedented scale on the western front. Beginning in 1915 (Anonymous, 1922b; Grieve and Newman, 1936), mine warfare reached its peak in 1916. Nine tunnelling companies were raised in 1915 and expanded to a total of 25 from 1916 to 1918. In 1916, 48 of 128 km of the front held by the British Expeditionary Force were protected by underground galleries, in several instances at more than one level. In some sectors, notably south of La Bassée, it was possible to walk along a continuous underground gallery in front of the British trenches for several kilometers (Pritchard, 1952). During 1916, nearly 1,500 mines were fired on both sides on the British-German front. The majority were not a part of any major or local surface attack but a part of the process of more-or-less continuous underground warfare. By the middle of 1916, the British had a total force of approximately 25,000 men actively engaged in mining, and David was at this significant time the geological adviser to the controller of mines. Through test boring and collation of geological data, David provided detailed information on the extent and depth of the Eocene "blue clay" (the most suitable mining stratum) in Flanders, the seasonal variation of water level in the Cretaceous Chalk, and other planning data. His crowning achievement in this aspect of his work was his contribution to preparation of the attack on the Messines-Wytschaete Ridge, originally planned for the summer of 1916 but postponed to June 7, 1917, to conform with general strategy. Initiation of the attack by simultaneous discharge of 19 mines in total containing nearly 450,000 kg of high explosive on a front of 16 km ranks as the greatest and most successful operation ever carried out in mine warfare. It is the sole instance in warfare of the explosion of heavy mines playing an outstanding part in the prosecution of an assault on a strongly defended position over a wide front. With the triumph of Messines, mining activity gradually declined, although tunnellers (and consequently geologists) played an important if less spectacular part in the preparations for the last great British offensive of 1917—Passchendaele.

But mining was effective only against a strongly held front line, and by the close of 1917 it became a recognized principle to hold the front by fire power rather than man power. Troops were deployed in depth and protected from bombardment by dugouts, subways, and other subsurface excavations. As offensive mining declined, so the tunnelling companies were increasingly diverted to the construction of dugouts. To guide siting, David prepared specialist geotechnical maps that were "probably the first environmental/engineering geology maps ever published" (Branagan, 1987, p. 43). Based on geological and hydrogeological information, these illustrated the relative suitability or otherwise of the ground for dugout construction. Two primary colors were used (Anonymous, 1922a, plate IV)—blue and red with intermediate shades. The redder the formation as indicated on the map, the drier it was for dugout purposes; the bluer, the wetter.

David was aided from September 1916 by Lieutenant C. Loftus Hills, formerly assistant government geologist of Tasmania, who supervised test boring for dugouts. Test borings along the frontline trenches from Nieuport on the Belgian coast to the River Somme alone totaled 10,000 m, largely under David's supervision. From October 1918, he was also assisted by Lieutenants G. A. Cook and C. S. Honman, who were lent to General Headquarters from the Australian Tunnelling Companies to aid in the preparation of geologic maps required for the army's advance into new territory.

Although the tunnelling companies were largely led by officers who were former miners and mining engineers, there were other geologists serving with them who were no doubt able to put their geotechnical expertise to good use. MacLeod (1988) cites Captain S. Hunter, a 53-year-old mining geologist from Victoria, formerly of the Geological Survey of Victoria, among the first 15 officers of the Australian Mining Battalion. Moreover, J. A. Douglas, later to become distinguished as a professor of geology at the University of Oxford, was one of the wartime officers quickly transferred from other duties to serve in the Royal Engineer tunnelling companies because of relevant pre-war expertise. A captain in the Gordon Highlanders, he was seconded to 172 Tunnelling Company, one of the first five formed in 1915 (Anonymous, 1981). Barrie (1962, p. 162–8) has recorded in some detail his role and courage in action at St. Eloi, where, in command of 600 men, he achieved conspicuous success although "He lacked the qualifications of most tunnelling company officers. He was not a mining or civil engineer—as the majority were; he was a geologist."

But greatest credit is deservedly given to Bill King and Edgeworth David for the recognition accorded to the value of military geology in the British army at this time. David's achievements in particular are documented by MacLeod (1988). To improve the allies' scant knowledge of the geology of Belgium and of their unfavorable position in the salient, he produced a completely new foundation of geological reference. Using drilling bores, he identified soils and underlying strata. Calling on advice from Belgian geologists, some of the first scientific water tables in Europe were compiled, enabling many British trenches and dugouts to be dug or relaid before winter rain and rising water levels made them uninhabitable. In July 1916, David began a series of lectures on the geological strata of the British front to the first Army School of Mines; by early 1917, he was using colored maps and vertical sections to show where quicksands prevailed and where trenches and tunnels might safely be dug. Within a year, David had persuaded General Headquarters to seek reports from its geological staff before undertaking any new operation or siting artillery concentrations; and by 1918, printed geologic maps with coordinates, replacing earlier drawings, were distributed through the controller of mines and the chief engineers of the four armies to all divisional staffs. But the war then ended, and in 1919 David returned to Australia, King to England, and all geologists to civilian occupations.

THE 1939–1945 WORLD WAR

Following declaration of war in 1939, W. B. R. King (Fig. 2) was called up from the Army Officers Emergency Reserve and

Figure 2. Major, later Lieutenant-Colonel, W. B. R. King (born 1889, died 1963), the senior British military geologist of World War II. (From Rose and Hughes [1993b], courtesy of C. A. M. King and the Institution of Royal Engineers.)

appointed to a Regular Army Emergency Commission in the Royal Engineers (Rose and Hughes, 1993a; Rose and Rosenbaum, 1993b). During the winter and spring of 1939–1940 he worked again as a military geologist on a variety of problems in northern France such as the siting of airfields, the provision of stone and gravel as construction materials, and water supply. In France he served as a local major on the staff of the engineer-in-chief, British Expeditionary Force, until he repatriated in 1940 with the Dunkirk evacuation. On return to the United Kingdom, he was attached to Northern Command for a year and then, from 1941 to 1943, to General Headquarters Home Forces. This was later to become 21 Army Group for the invasion of Normandy, and King's major role in World War II was to serve as geological adviser to the chief engineer (Rose and Pareyn, this volume) in 1943 as a lieutenant-colonel.

From 1940 to 1941 King was assisted in the United Kingdom by F. W. Shotton. At King's instigation, Shotton had joined the Army Officers Emergency Reserve in 1938, but it was not until 1940 that he was appointed to a Regular Army Emergency Commission in the Royal Engineers. For a while the two worked together undertaking ground investigations for anti-aircraft gun sites in England. However, in 1941 Shotton embarked for the Mediterranean to take responsibility under the director of works, Cairo, for all geological activities in North Africa and the Middle East. These dealt mainly with provision of ground-water supplies and technical direction of Royal Engineers well-drilling teams. Shotton ensured that ground-water resource evaluation—one of the main factors on which movement of the armies depended—was conducted by a geological analysis of the regional maps followed by test drilling (Shotton, 1945, 1946). Indeed, the success of the 8th Army's advance from El Alamein in October 1942 owed much to Shotton's careful hydrogeological studies. His elucidation of the structure of the Fuka Basin in western Egypt, the first of the water points to be developed when the advance started, is a well-documented example (Shotton, 1944; 1946). On the basis of geological advice it was possible to guarantee (1) pinpointing new borehole sites for water exactly along the axis of a syncline; (2) that these boreholes would be 27 m deep and would require 13–14 m of casing for the top part, which would be in clay; (3) promising a yield of 23 m^3 per hour from a 0.25 m diameter hole; and (4) that the water levels were suitable for air-lift installation, which could therefore be assembled in advance. In Egypt and Libya, ground water was generally derived from boreholes in carbonate sediments of Tertiary age, although it was also abstracted from boreholes in Quaternary sands and alluvium and from use of collection galleries in coastal sand dunes. Geophysical methods of field survey such as electrical resistivity traverses were used where appropriate, conducted by 42nd Geological Section of the South African Engineer Corps.

In 1943 Captain Shotton was recalled to London to join, and later succeed, Bill King at 21 Army Group (Rose and Pareyn, this volume). He was replaced in the Mediterranean area by J. V. Stephens, a veteran of the First World War and a British Geological Survey geologist (Rose and Hughes, 1993a; Rose and Rosenbaum, 1993b). Two weeks after appointment to a Regular Army Emergency Commission, on the General List, Stephens embarked for service in North Africa as an acting captain and staff officer (geologist). From September 9 he was permanently attached to 15 Army Group of the Combined Mediterranean Forces, a post that involved him in the Italian campaign of 1943 and 1944, including the Anzio landings and, incidentally, observation of the great eruption of Vesuvius. "Steve" ended the war as a major, his wartime activities largely unrecorded, although roles in potable water supply and in quarrying for road metal can be inferred.

King, Shotton, and Stephens had all been recommended for appointment by the director of the Geological Survey of Great Britain, E. B. Bailey. In 1943 Bailey landed in Gibraltar en route to give hydrogeological advice to the besieged garrison of Malta. The Rock of Gibraltar had been developed as a naval base and increasingly fortified since being ceded to Britain in 1713 at the end of the War of Spanish Succession. Provision of adequate potable water supplies posed a continuing problem. Moreover, a complex of tunnels and underground chambers had been intermittently excavated within the Rock since the late 18th century, and this was rapidly extended during the 1939–1945 war to a total length of nearly 50 km within the 5-km-long peninsula. But delays had been experienced in attempting to drive tunnels through zones of weak rock in the otherwise strong Jurassic dolomitic limestone that formed the bulk of the narrow peninsula. To guide future tunnelling and well drilling, Bailey recommended a full geological survey to be undertaken by a geologist by chance already stationed as a driver on the Rock. Sapper A. L. Greig, Royal Engineers, mapped the Rock at 1:5,280 scale, and thus became the only British noncommissioned military geologist ever to be appointed as such (Rose and Rosenbaum, 1990).

A few geologists were granted Regular Army Emergency Commissions in the Royal Engineers to command well-drilling teams. Those that have left a documented record include W. A. Macfadyen, W. T. Pickard, and A. K. Pringle. Macfadyen, a veteran infantry officer of the First World War who became first a petroleum geologist and then a hydrogeologist, served from 1941 in England, North Africa and the Mediterranean region—from 1943 as a major (Rose and Rosenbaum, 1993b). Pickard was commissioned in 1943 and posted to East Africa because the value of geology in determining water supply had been proved there by South African Engineer geologists in the Ethiopian and Eritrean campaigns (Pickard, 1946). Pringle was to serve in England, Normandy, and Belgium in 1944 (Rose and Pareyn, this volume).

Other geologists were put to use in aerial photographic interpretation. F. W. Anderson was a Territorial Army infantry officer seconded from the Geological Survey to the Ministry of Home Security to "carry out geological work in our Research and Experiments Department" (correspondence cited by Rose and Rosenbaum, 1993b, p. 102). By 1945 he was "in charge of the R & E contingent of the British Bombing Research Mission . . . engaged on field survey work in France" finally in the rank of local lieutenant-colonel. He participated in active service in Germany as well as Sicily, Italy, and France studying the effect of aerial bombing. N. L. Falcon was also a geologist employed in aerial photographic interpretation, ultimately as a lieutenant-colonel in the Intelligence Corps, but his function was not specifically geological.

By 1944 some geologists are known to have been appointed specifically to intelligence activities. Thus the Inter-Services Topographical Department in its preparations for the Normandy D-Day included at least three Royal Engineer geologists with Regular Army Emergency Commissions: J. L. Farrington (finally a major), D. R. A. Ponsford (also a major), and T. C. Phemister (a subaltern) (Rose and Pareyn, this volume). Other geologists were commissioned in the Royal Air Force and the Navy. Moreover, Simon (1957, p. 1567) records that four geologists were used in the Industry Section of the British Intelligence Centre located at R. A. F. Medmenham, Buckinghamshire, which later became the Allied Central Intelligence Unit. At least one geologist was a member of the Combined Strategic Target Committee whose function was to select bombing targets. Simon's conclusion on the basis of service with the American forces was that "geology is especially useful in the field of intelligence. Experience suggests that the British appreciate this more than we did."

At the end of the war, the Geological Museum in London (now part of the Natural History Museum) arranged a temporary exhibit to illustrate some of the contributions of British geologists to the war effort (Butler, 1947). This publicized the role of the British Geological Survey in home defence, through projects associated with ground-water supply, mining of resource materials (iron, non-ferrous metals such as lead and zinc, and coal), and underground construction to provide bomb-proof accommodation for factories, control rooms, and storehouses (Fig. 3). It also revealed the use of military geologists in operations overseas, by examples of geologic maps used during the war (in Italy, Albania, Austria, Indo-China, Thailand, and Borneo); 1:5000 scale "going" maps prepared in advance for the Normandy beachhead; cross-country movement maps for parts of Italy, France, Holland, and Germany; "airfield suitability" maps for northwestern France and North Africa; and ground-water supply maps for Italy, France, and Germany. Experience had shown that there were roles for geologists both in operational planning and field geology (Fig. 4) (Rose and Hughes, 1993a).

THE COLD WAR

Like their predecessors in 1918, the military geologists of the Second World War returned to their civilian occupations at the close of hostilities. A geologist appointment was occupied in Gibraltar from 1946 to 1948 by Lieutenant (later Captain) G. B. Alexander, Royal Engineers (Rose and Rosenbaum, 1990; 1992). Geologist expertise in Germany was maintained until 1952 by A. H. V. Smith at the Army Operational Research Group (Rose and Hughes, 1993b). However, with these exceptions, geologists were not used as such within the Regular Army.

But the wartime value of geology and of geologists had been noted. W. B. R. King, from 1943 Woodwardian Professor of Geology in the University of Cambridge, was appointed geological adviser to the War Office—the government department later to become the Ministry of Defence. Under his editorship and senior authorship, the first British manual on militarily applied geology was soon published (Anonymous, 1949). At the same time, under his influence, a Royal Engineers Pool of Geologists was set up within the Territorial Army. Seven geologists, all of them of proven or potential distinction, and all but one with previous wartime experience in the armed forces, were recruited from the staff of British universities, the Geological Survey, and British Petroleum (Rose and Hughes, 1993b). In 1953 the pool was transferred from the Territorial Army to the recently formed Army Emergency Reserve, where it continued to function for another 14 years. As the older members left, so they were replaced by younger members, but still recruitment was from geologists well established in their profession who had previous military experience, either through war service or peacetime conscription (two-year National Service). In 1967 the Geologists' Pool was merged with the Works Pool of Officers to form the Engineer Specialist Pool of the newly reorganized Territorial and Army Volunteer Reserve. This continued as such for 21 years, but since National Service had ended in 1959, recruit-

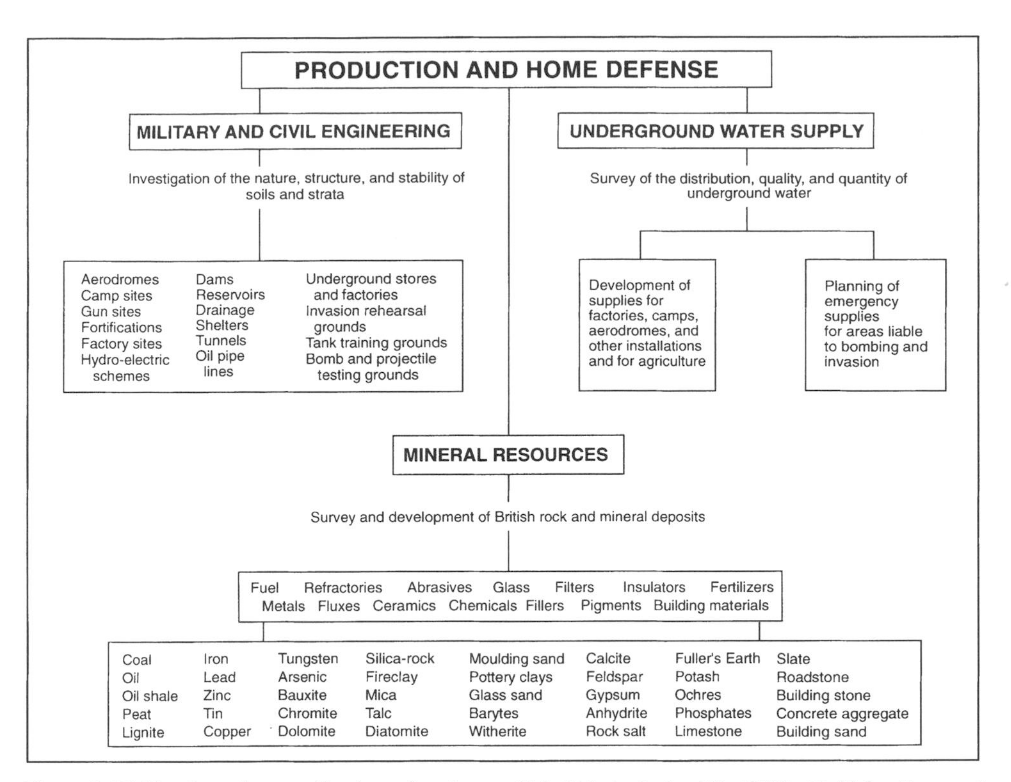

Figure 3. Fields of wartime application of geology within Britain during World War II. (After Rose and Hughes [1993a], redrawn from Butler [1947], courtesy of *The Museums Journal*.)

ment to maintain an establishment of about six geologists was increasingly from other units of the Territorial Army. From 1969, few recruits had Regular Army experience. Of the officers who achieved senior rank through long service, five were drawn from university staff, one from the Geological Survey, one from British Petroleum, and one from the British Museum (Natural History). Nine other geologist officers served within this period, recruited from a wider background, but most were unable to meet their optimum "training" commitment long term. Accordingly they achieved only short service and junior rank. In 1988 the Engineer Specialist Pool was renamed the Royal Engineers Specialist Advisory Team of the current Territorial Army (Rose, 1988; Rosenbaum, this volume).

Overall, the commitment of geologists in the reserve army was to average at least 15 days "training" each year, as well as to meet increasingly stringent requirements of physical fitness and military proficiency deemed necessary for survival on the potential battlefield. In practice, especially in the later years, military geologists often contributed more than 30 days service per year.

Initially, much of the training was undertaken in Germany. A particular study was made of "going" conditions on loessic soils in the area around Osnabrücke and Hameln. Studies of "going appreciation without access to site," based on geological information and aerial photographs, were continued and extended in the Pool when it transferred into the Army Emergency Reserve. However, the AER was reorganized in 1961, and following this there was an increase in visits abroad by individual geologists, e.g., to Malaya, East Africa, Canada, and Thailand, while still maintaining considerable effort into development of terrain evaluation systems (Rose and Hughes, 1993b). There were consultancy visits to develop potable water supplies for troops on exercise (e.g., in Libya) or during times of armed conflict (e.g., in Aden), and a geologist served on attachment with an Australian engineer squadron and later a Gurkha battalion during the conflict in Indonesia.

Following establishment of the Engineer Specialist Pool in 1967, military geologists were soon active in the preparation of a new textbook (Anonymous, 1976). They were also used on a range of tasks worldwide (Rose, 1978; Rose and Hughes, 1993c; Rose and Rosenbaum, 1993b). Geographic emphases changed with the perceived threat and the resources available.

In the Far East, the projects were related to ground-water development, slope stability, and quarrying for aggregates. Ground-water tasks took geologists to Thailand in 1969 and 1970, to Nepal in 1973 and 1976, and to Hong Kong in 1970, 1973, and 1977. Slope stability and quarrying in Hong Kong formed the basis of reports in 1977 and 1978, and in 1981 military geologists conducted a feasibility study that led to establishment of a geological survey section within the Hong Kong Government Public Works Department.

Within the Mediterranean region, military geological tasks in Libya came to an end in 1969 and those in Malta in 1974, but projects in Cyprus and Gibraltar have continued to the present day. In Libya, the need to locate potable ground-water supplies for British

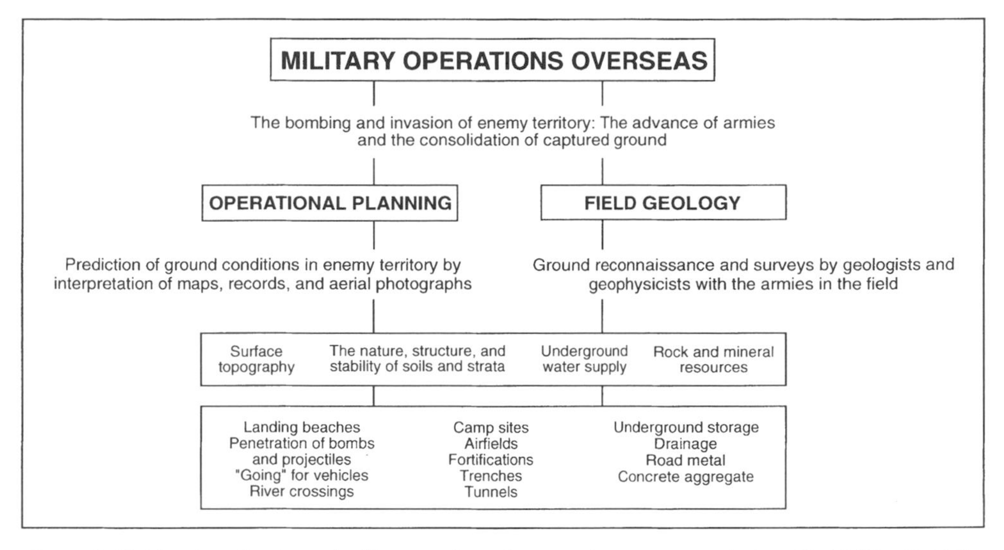

Figure 4. Fields of application of military geology in campaigns overseas during World War II. (After Rose and Hughes [1993a], redrawn from Butler [1947], courtesy of *The Museums Journal.*)

troops training in the area had generated a series of military geologist visits and reports through the 1960s, but with the change in government and consequent withdrawal of British troops, the need and opportunities for further work ceased. Malta, too, witnessed a withdrawal of British troops, but Royal Engineers lingered longer than most to provide expertise to the Maltese Government Public Works Department, and a military geologist contributed to several engineering geology studies. Cyprus has retained a British military presence to the present day, and provided scope for a diverse range of geotechnical tasks such as ground-water supply, quarrying for road metal, and cliff stability. Geologists have even served on brief attachment with United Nations forces on the island. Gibraltar has also retained a British military presence, and although conspicuously smaller than Cyprus (6 rather than 9,251 km^2), it has nevertheless provided a significant range and number of military geologist tasks, principally relating to ground-water development, road construction, slope stability, quarrying, movement of beach sand, and above all, tunnel stability. These studies have, as a by-product, generated the first detailed geologic map to be published for the peninsula (Rosenbaum and Rose, 1991) and numerous recent publications.

There have also been studies in the United Kingdom and in other parts of the world, notably South Arabia, Oman, and Belize, relating to quarrying and well drilling. However, from 1977 much effort went into providing support for 1st (British) Corps in Germany. Military geologists provided the technical guidance necessary for preparation of Terrain Engineering Characteristics maps, liaised with Bundeswehr geologists, and trained regularly for a perceived operational role. Yet when "war" came in 1982, it was not in Germany but the South Atlantic. The Falklands conflict brought a requirement for military geologist advice initially on terrain appreciation and potable water supply, and later on quarrying (Rose and Hughes, 1993c; Rose and Rosenbaum, 1993b). Details of these and more recent activities by British military geologists form the subject of a separate communication (Rosenbaum, this volume).

CONCLUSION

The 19th century enthusiasm to incorporate geology in military education, most broadly at Sandhurst and Camberley, proved impermanent. Perception of geology's direct military relevance waned: "The useless subject of Geology at length disappeared [from the curriculum at Camberley] in 1882, with the retirement of Professor Jones, whose reduction had been recommended in 1868, but the terms of whose engagement had enabled him to linger on for fourteen years" (Godwin-Austen, 1927, p. 204). "In 1882 the retirement of the Professor of Geology, Rupert T. Jones [sic], at last enabled the study of that subject to be confined to its practical use for military purposes" (Bond, 1972, p. 138). Even that residue was short-lived, for Edmonds (in King, 1919, p. 217) notes that long before the First World War "the study of geology in the Army appears to have gradually dropped out," as the remaining lectures on the subject ceased both at Camberley and the School of Military Engineering, Chatham.

Perception of geology's practical use for British military purposes changed significantly through the work of David and King on the western front in Europe during the 1914–1918 War. The *History of the Corps of Royal Engineers* records that "Great importance should be attached to the appointment . . . to the Staff of the Engineer-in-Chief of two very qualified and exceptionally efficient and hard-working geologists" (Pritchard, 1952, p. 501). In consequence, at the end of the war proposals were made for a future wartime military geologist organization in the British army, and for peacetime "training of a small staff for this special work" so that "whatever and whenever the next war may be, let us not be unprepared, either in the training or organization of a geological staff" (Strahan, in King, 1919, p. 220).

The staff, it seems, was never trained or organized, but the expertise lived on with King to be available from the start of the 1939–1945 war. Largely through the initiative of King and the director of the British Geological Survey, the British army was able to make effective use of military geologists in the United Kingdom, North Africa, and Italy. Minor but significant use was made elsewhere, notably in Gibraltar and East Africa. Very significant use was made both in the planning and operational aspects of the liberation of western Europe (Rose and Pareyn, this volume), and the chief engineer of 21 Army Group subsequently recorded with pride that "we had, fortunately, long appreciated the importance of geology in modern war" (Inglis, 1946, p. 94).

In consequence, from 1949 a pool of 6–8 geologist officers was maintained within the British army within the reserve forces of the Royal Engineers. Initially within the Territorial Army, the Geologists' Pool was transferred to the Army Emergency Reserve in 1953. In 1967, the Geologists' and Works Pools were merged to form the Engineer Specialist Pool of the newly created Territorial and Army Volunteer Reserve. In 1988 this pool became the Royal Engineers Specialist Advisory Team of the present Territorial Army. For nearly 50 years British reserve army geologists have served widely and cost-effectively in times both of peace and of armed conflict (Rose, 1978; Rose and Hughes, 1993b, c; Rose and Rosenbaum, 1993b; Rosenbaum, this volume). They continue to do so.

ACKNOWLEDGMENTS

Parts of this paper, as indicated above, are derived from earlier papers by the two authors where data and sources are documented more fully. Additionally, E.P.F.R. gratefully acknowledges help to locate new data from the chief librarian and archivist of the British Geological Survey (G. McKenna); the librarians and the honorary archivist (J. Thackray) of the Geological Society, London; librarians at the Royal Engineers Corps Library, Chatham; the curator of the Royal Military Academy Sandhurst Collection (T. A. Heathcote); the librarian of the Staff College, Camberley (P. E. Bendall); and from correspondence with W. S. McKerrow concerning the war service of Oxford geologists. Daniel B. Krinsley and J. Ponder Henley kindly refereed the original draft of the manuscript.

REFERENCES CITED

Anonymous, 1877, Obituary: The late Lieut.-General Sir Henry James: Royal Engineers Journal, v. 7, p. 54.

Anonymous, 1922a, The work of the Royal Engineers in the European war, 1914–19: Geological work on the Western Front: Chatham, Institution of Royal Engineers, 71 p., 8 pls., 19 figs.

Anonymous, 1922b, The work of the Royal Engineers in the European War, 1914–19: Military mining: Chatham, Institution of Royal Engineers, 148 p., 93 pls/photos.

Anonymous, 1949, Military engineering Vol. XV: Application of geology: War Office Code 8287: London, His Majesty's Stationery Office, 216 p., maps A–G.

Anonymous, 1976, Military engineering Vol. XV: Applied geology for engineers: Army Code No. 71044: London, Her Majesty's Stationery Office, 378 p., 2 maps.

Anonymous, 1981, Who was who. Vol. VII 1971–1980: London, Black, 889 p.

Bailey, E. B., 1952, Geological Survey of Great Britain: London, Murby, 278 p.

Barrie, A., 1962, War underground: London, Muller, 272 p.

Boase, F., 1965, Modern English biography. Vol. II–I to Q: London, Cass & Co., 1775 p.

Bond, B. J., 1972, The Victorian army and the Staff College, 1854–1914: London, Eyre Methuen, 350 p.

Boulger, G. S., 1908, Greenough, George Bellas (1778–1855), *in* Stephen, L., and Lee, S., eds., Dictionary of national biography. Vol. VIII Glover–Harriot: London, Smith, Elder & Co., p. 524–525.

Branagan, D. F., 1987, The Australian Mining Corps in World War I: Australian Institute of Mining and Metallurgy Bulletin and Proceedings, v. 292, p. 40–4.

Branagan, D. F., editor, 1990, Sir Edgeworth David memorial oration: Parkville, Australasian Institute of Mining and Metallurgy, 74 p.

Brooks, A. H., 1920, The use of geology on the Western Front: U.S. Geological Survey Professional Paper 128-D, p. 85–124.

Butler, A. J., 1947, War-time geology: Special exhibits in the Geological Museum, South Kensington: Museums Journal, v. 46, p. 233–238.

Carter, H. J., and Browne, W. R., 1936, Tannatt William Edgeworth David, 1858–1934: Linnean Society of New South Wales Proceedings, v. 61, p. 342–347.

Collyer, J. N., and Pocock, J. I., 1843, An historical record of the Light Horse Volunteers of London and Westminster: London, Wright, 279 p.

Dalton, C., 1904, The Waterloo Roll Call, with biographical notes and anecdotes, 2nd ed.: London, Eyre and Spottiswoode, 296 p.

David, M. E., 1937, Professor David: The life of Sir Edgeworth David: London, Arnold, 320 p.

Davies, G. L. Herries, 1983, Sheets of many colours: The mapping of Ireland's rocks: Dublin, Royal Dublin Society, 242 p.

Flinn, D., 1981, John MacCulloch, M. D., F. R. S., and his geological map of Scotland: His years in the Ordnance, 1795–1826: Royal Society of London Notes and Records, v. 36, p. 83–101.

Geikie, A., 1875, Life of Sir Roderick I. Murchison: London, Murray, v. 1, 387 p., v. 2, 375 p.

Godwin-Austen, A. R., 1927, The Staff and the Staff College: London, Constable & Co., 323 p.

Grieve, G., and Newman, E., 1936, Tunnellers, the story of tunnelling companies Royal Engineers, during the World War: London, Jenkins, 334 p.

Harvey, R. N., 1934, Lieut.-Colonel Sir Tannatt William Edgeworth David: Royal Engineers Journal, v. 48, p. 616–619.

Howarth, R. J., 1993, Strahan, Sir Aubrey (1852–1928), *in* Nicholls, C. S., ed., Dictionary of national biography: Missing persons: Oxford, Oxford University Press, p. 640–641.

Hutton, F. W., 1862, Importance of a knowledge of geology to military men: Royal United Service Institution Journal, v. 6, p. 342–360.

Inglis, J. D., 1946, The work of the Royal Engineers in north-west Europe, 1944–45: Royal Engineers Journal, v. 60, p. 92–112. (Reprinted from Royal United Service Institution Journal, v. 91 [for 1946], p. 176–195).

Jones, T. R., 1898, Charles Cooper-King: Geological Society of London Quarterly Journal, v. 54, p. lxxiv–lxxvii.

King, W. B. R., 1919, Geological work on the Western Front: Geographical Journal, v. 54, p. 201–215; discussion p. 215–221.

King, W. B. R., 1921a, Résultats des sondages exécutés par les armées britanniques dans la Nord de la France: Annales de la Société géologique du Nord, Lille, v. 45, p. 9–26.

King, W. B. R., 1921b, The surface of the marls of the Middle Chalk in the Somme valley and the neighbouring districts of northern France, and the effect on the hydrology: Geological Society of London Quarterly Journal, v. 77, p. 135–143.

King, W. B. R., 1951, The recording of hydrogeological data: Yorkshire Geological Society Proceedings, v. 28, p. 112–116.

Lewis, G. G., Jones, H. D., Nelson, R. J., Larcom, T. A., De Moleyns, E. C., and Williams, J., eds., 1846–1852, Aide-mémoire to the military sciences: London, Weale, v. 1 (1846) 545 p., v. 2 (1850) 546 p. + 14 pls, v. 3 (1852) 794 p.

Lewis, G. G., Jones, H. D., Larcom, T. A., Williams, J., and Binney, C. R., eds., 1853–1862, Aide-mémoire to the military sciences, 2nd edition : London, Weale, v. 1 (1853) 555 p., v. 2 (1860) 589 p., v. 3 (1862) 800 p.

MacCulloch, J., 1836, Geological map of Scotland: London, Arrowsmith, by order of the Lords of the Treasury: scale 4 inches: 1 mile, 4 sheets, hand colored.

MacLeod, R., 1988, Phantom soldiers: Australian tunnellers on the western front, 1916–18: Australian War Memorial Journal, no. 13, p. 31–43.

Mawson, D., 1949, David, Sir (Tannatt William) Edgeworth (1858–1934), *in* Legg, L. G. W., ed., Dictionary of national biography 1931–1940: London, Oxford University Press, p. 212–213.

McCartney, P. J., 1977, Henry De la Beche: Observations on an observer: Cardiff, Friends of the National Museum of Wales, 77 p.

Pickard, W. T., 1946, Geological work in the East Africa Command: East African Engineer, July 1946, p. 17–19.

Portlock, J. E., 1843, Report on the geology of the county of Londonderry, and parts of Tyrone and Fermanagh: Dublin, Milliken, 784 p., 38 pls.

Portlock, J. E., 1849, A rudimentary treatise on geology: For the use of beginners: London, Weale, 182 p.

Pritchard, H. L., editor, 1952, History of the Corps of Royal Engineers. Vol. V. The Home Front, France, Flanders and Italy in the First World War: Chatham, Institution of Royal Engineers, 728 p., 5 maps.

Rose, E. P. F., 1978, Engineering geology and the Royal Engineers: Royal Engineers Journal, v. 92, p. 38–44.

Rose, E. P. F., 1988, The Royal Engineers Specialist Advisory Team (V): Royal Engineers Journal, v. 102, p. 291–292.

Rose, E. P. F., 1996a, Geologists and the army in nineteenth century Britain: a scientific and educational symbiosis?: Geologists' Association Proceedings, v. 107, p. 129–141.

Rose, E. P. F., 1996b, The military service of G. B. Greenough, founder president of the Geological Society, *in* Sparks, R. S. J., French, W. J., and Howarth, R. J., organizers, Applied Geoscience 15–18 April 1996 Abstracts: London, Geological Society, p. 26.

Rose, E. P. F., 1997a, Geological training for British army officers: A long-lost cause?: Royal Engineers Journal, v. 111, p. 23–29.

Rose, E. P. F., 1997b, 'John W. Pringle (c. 1793–1861) and Ordnance Survey geological mapping in Ireland' by Wyse Jackson (1997): Biographical comment: Geologists' Association Proceedings, v. 108, p.157.

Rose, E. P. F., and Hughes, N. F., 1993a, Sapper Geology: Part 1. Lessons learnt from world war: Royal Engineers Journal, v. 107, p. 27–33.

Rose, E. P. F., and Hughes, N. F., 1993b, Sapper Geology: Part 2. Geologist pools in the reserve army: Royal Engineers Journal, v. 107, p. 173–181.

Rose, E. P. F., and Hughes, N. F., 1993c, Sapper Geology: Part 3. Engineer Specialist Pool geologists: Royal Engineers Journal, v. 107, p. 306–316.

Rose, E. P. F., and Rosenbaum, M. S., 1990, Royal Engineer geologists and the geology of Gibraltar: Gibraltar, Gibraltar Museum, 55 p. (Reprinted from the Royal Engineers Journal, v. 103 [for 1989], p. 142–151, 248–259; v. 104 [for 1990], p. 61–76, 128–144.)

Rose, E. P. F., and Rosenbaum, M. S., 1992, Geology of Gibraltar: School of Military Survey Miscellaneous Map 45 (published 1991) and its historical background: Royal Engineers Journal, v. 106, p. 168–173.

Rose, E. P. F., and Rosenbaum, M. S., 1993a, British military geologists: The formative years to the end of the First World War: Geologists' Association Proceedings, v. 104, p. 41–49.

Rose, E. P. F., and Rosenbaum, M. S., 1993b, British military geologists: Through the Second World War to the end of the Cold War: Geologists' Association Proceedings, v. 104, p. 95–108.

Rosenbaum, M. S., and Rose, E. P. F., 1991, Geology of Gibraltar: School of Military Survey Miscellaneous Map 45, scales 1:10,000 and 1:20,000, 1 sheet (2 sides).

Shotton, F. W., 1944, The Fuka basin: Royal Engineers Journal, v. 58, p. 107–109.

Shotton, F. W., 1945, Water in the desert: Inaugural Lecture, University of Sheffield, 19 p.

Shotton, F. W., 1946, Water supply in the Middle East campaigns: Water and Water Engineering, v. 49, p. 218–226, 257–263, 427–436, 477–486, 529–540.

Simon, L. J., 1957, Additional notes of the use of geology in the European theater of operations during World War II: Geological Society of America Bulletin, v. 68, p. 1567.

Smyth, J., 1961, Sandhurst: London, Weidenfield & Nicholson, 311 p.

Vetch, R. H., 1909, Portlock, Joseph Ellison (1794–1864), *in* Lee, S., ed., Dictionary of national biography. Vol. XVI Pocock-Robins: London, Smith, Elder & Co., p. 197–198.

Woodward, H. B., 1907, The history of the Geological Society of London: London, Geological Society, 336 p.

Manuscript Accepted by the Society October 29, 1997

Printed in U.S.A.

Geological Society of America
Reviews in Engineering Geology, Volume XIII
1998

American geologists at war: World War I

Walter E. Pittman
History and Social Science Department, University of West Alabama, Livingston, Alabama 35470

ABSTRACT

During World War I the combatants committed the total resources of their nations in this first great total war. This came to include geological expertise. The original use of geologists on the battlefield was to locate potable water supplies; later employments were an outgrowth of the stalemate on the battlefield. Mine warfare quickly developed as the belligerents tried to tunnel under the formidable trench systems. Geologists in uniform provided assistance for these efforts and came to be valued for their professional advice.

More uses were quickly found for geologists. Trafficability studies of terrain, predictions of stream and river heights, sources of construction materials, and location of water supplies were important missions. Later, as both sides learned to communicate through ground-loop telephony, ground-conductivity studies became important.

By the time the United States entered the war in 1917, mine warfare had been neutralized by countermining, and no further active mine operations were undertaken. The U.S. Army sent 10 geologists (three more were en route on November 11, 1918), a mining regiment, and a water supply regiment of engineers to support the American Expeditionary Force. Most geologic work was in terrain studies and in mapping, water supply, and soil trafficability studies. In the United States, other geologists worked to discover sources of scarce raw materials. American geologists generally were disappointed, however, at the contributions they were able to make to the war effort, whether in France or America.

GEOLOGY IN THE EUROPEAN WAR 1914–1917

When war broke out in August 1914, none of the belligerents was ready for the totality of the war that ensued. No country was prepared for the total commitment required. Particularly, no one foresaw the important effects that science and technology would have on the battlefields of World War I. The potential military uses of geology were probably less foreseen than those of the other sciences.

Very little attempt had been made to integrate geology with warfare prior to 1914. Several military academies, for example, West Point, taught an introductory geology course as part of a general-education program. Some military engineering academies offered a little more, usually topography or geomorphology or physiographic geology as it was then named. In Britain, Professor T. Rupert Jones taught geology to cadets at the Royal Military College from 1858 to 1870, and to higher level officers at the Staff College until 1882 (Rose and Rosenbaum, this volume). He was succeeded at the Staff College by the more practical Lieutenant Colonel Chester Cooper King, who taught there from 1886 to 1898. However, the general pattern among all the armies was to ignore the sciences, including geology, and the contributions they might make to modern warfare (Brooks, 1921a).

The war did not bring any immediate change to this attitude, at least for geology. On the Allied side, the first use of a geologist occurred when the British, fighting in the chalk lowlands of northern France and Belgium, called upon the British Geological Survey to supply an expert to locate water supplies for the army in a region largely lacking in surface-water resources. It was, said British Major General W. A. Liddell, "practically a desert,"

Pittman, W. E., 1998, American geologists at war: World War I, *in* Underwood, J. R., Jr., and Guth, P. L., eds., Military Geology in War and Peace: Boulder, Colorado, Geological Society of America Reviews in Engineering Geology, v. XIII.

and it was impossible to take 300,000 men and 100,000 horses across it without secure water supplies (King, 1919, discussion). Consequently, in April 1915, a hydrogeologist was appointed to the engineer staff at British Expeditionary Force (BEF) General Headquarters: Lieutenant (later Captain) W. B. R. King, formerly a field geologist with the British Geological Survey (Rose and Hughes, 1993; and Rose and Rosenbaum, 1993, this volume). In May 1916, other geologists arrived with an Australian mining battalion, notably Major (later Lieutenant Colonel) T. W. Edgeworth David, in civilian life a professor of geography and geology at the University of Sydney and a pioneer explorer of Antarctica, and Lieutenant Loftus Hills, formerly assistant government geologist of Tasmania. On September 25, 1916, David was seriously injured by a fall while inspecting a well shaft, and when he recovered he too was assigned as a staff officer to BEF General Headquarters. By chance rather than design, this facilitated close cooperation with King in a nearby office. Loftus Hills had command of a "special light boring plant" used for site investigations from September 25, 1916, until almost the end of the war. In total, some 10,000 m of test bores, for dugouts and so on, were made under his direction, largely in front-line positions (King, 1919).

David actually was in a separate command from King, who served under the engineer-in-chief of the BEF. David served under the Inspector of Mines because, on the battlefields of Europe, military mining, after centuries of absence, suddenly became important again. This developed from the tactical stalemate that began in late 1914 and persisted throughout the war. The power of defensive weapon systems, incorporating machine guns, barbed wire, entrenchments, and quick-firing cannons and rifles (deadly at long ranges), made any movement across the battlefield suicidal. Both sides dug a maze of entrenchments that grew steadily deeper and more complex and eventually stretched, almost unbroken, from Switzerland to the North Sea. The lines would remain nearly unchanged through four years of war, despite the slaughter of millions of men wasted in futile offensives. In many places the lines were only a few hundred feet apart. It must have occurred to many fertile minds on both sides that, while only death awaited anyone moving on the shell-scarred surface, they might be able to get at their enemies by tunneling underneath the ground. The Germans apparently began the process, but the British responded promptly and began mining on a large scale (Brooks, 1920a).

Thus was reborn, on the largest scale in history, a military technology that had been largely dormant for years. Most of the mines were driven 30–60 m, but some of the French mines were as long as 180 m (Brooks, 1920b). The mines were driven under enemy trench lines and chambers were hollowed out and filled with explosives that, when detonated, would collapse a section of trench line. Into the gap poured waiting assault troops. The mining had to be done with the greatest speed, secrecy, and silence if surprise were to be obtained. Only hand tools could be used, and waste had to be secretly removed and hidden from aerial observers. To the normal dangers of mining, even in peacetime a hazardous profession, were added the perils of war. The best defense against mining was countermining—the digging of another tunnel under the one being driven toward a trench line and the emplacing and detonating of explosives under it, resulting in the destruction of the mine and sometimes the miners. Sometimes British or German miners dug into one another's tunnels, deliberately or accidentally, and vicious little underground battles resulted. Poison gases were used underground and sometimes seeped into mines from the surface through adits and mine shafts; shells sometimes penetrated into mine galleries. Even mine entrances could be deadly, for the tunnels usually originated in or close to the front lines (Trounce, 1918).

As mining warfare spread in 1915, the British recruited English, South African, and Australian miners to support the BEF. Some 40,000 miners, including the French, eventually were at work (DeBeque, 1919). Geologists like David were added to provide their expertise to the miners. British General Liddell, who came up with the idea of the mining companies, had to be persuaded to add the geologists. He was a professional engineer and later admitted that he had not known that geology had anything to do with mining. The geologists also found employment in locating entrenchments and dugout sites that would be safe from shellfire but not subject to surface or underground flooding (King, 1919).

Mine warfare gradually disappeared by 1918. Both sides discovered that the defense could react too quickly after the explosion of a mine for a breakthrough to be successful. Countermining also eroded effectiveness. Even the explosion of the massive Messines Ridge mine of June 7, 1917, failed to achieve its tactical goals. The British had implanted 19 mines totaling 435,375 kg. When they were detonated, "gigantic roses" with "crimson petals" shaped like the underground explosive chambers rose slowly into the air (King, 1919). London and Paris were shaken by the explosion. Huge craters, some as large as 38 m deep and 57 m across, were created. The Germans, however, prevented any real British gains, and BEF casualties were heavy in the tortured landscape. From this high point, the use of mine warfare declined on both sides, although all the belligerents maintained their capabilities. The process was simply too time consuming and costly for the realized gains (DeBeque, 1919).

Britain's ally, France, never formally incorporated geology into its military structure. Apparently French engineers dug their mines, sometimes in crystalline mountain rocks, and bored their water tubes according to their time-honored peacetime method of trusting luck and their excellent published geological surveys. The Italians, likewise, never integrated geology into their military but relied upon civilian advisers, particularly in the development of a huge 335 m mine driven under Austrian lines in the hard rock Alps and detonated with great loss to the Austrians. Nothing is known of the military mining efforts of the Russians or the Serbs (Brooks, 1921b).

One French development, the geophone, proved successful in war and spurred civilian research after the war ended. Both sides discovered quickly that they could often hear one another's digging efforts through the earth, and special listening devices were devised by all the combatants. French physicists created the geophone, in which sound waves transiting the solid earth were translated into air pressure vibrations using a suspended heavy mass acting as a diaphragm; these vibrations then were picked up by a sort of stethoscope or microphone. The French soon developed the technique of using multiple geophones to localize sound sources by triangulation. The use of geophones helped to neutralize offensive mining, at least by the Germans (Ball, 1919). After the war, serious efforts were made by the U.S. Bureau of Mines (USBM) to develop an improved geophone for mine rescue work (Leighton, 1922).

Similarly, the search for secure communications led to the development of electromagnetic telegraph systems using a ground-return path and frequencies of 300–1,200 Hz, which were relatively high at the time. The ranges were limited to a few hundred to a few thousand feet and were easily intercepted but were a great improvement over anything else available. The most famous of these was the French T.P.S. (*telegraphie par sol*), which was copied by the U.S. Army. It was quickly learned that the conductivity of different kinds of rock varied greatly, and geologists were soon providing maps to their armies (British, French, German, U.S.) showing relative conductivity (Brooks, 1921b).

Like the other belligerents, Germany entered World War I without an organized geological service, despite the efforts of a few inspired reformers. One of these, Captain Walter Kranz, was a military engineer in the Imperial German Army and a trained geologist. In a thoughtful article (Kranz, 1913), he called for the creation of a geological corps of officers for the general staff who would be augmented by civilian geologists in time of war. Kranz pointed out that the other disciplines of modern science were rapidly being incorporated into warfare, whereas geologists were assigned to the general staff. His article stimulated some discussion among geologists but no reaction from the general staff before the onset of hostilities.

Apparently, however, long before 1914 the general staff was receiving high-quality professional advice from the eminent Albrecht Penck, one of the founders of geomorphology (Brooks, 1921b). Part of the credit for Germany's greatest victory in World War I, that of the Mazurian Lakes (Tannenburg) in August 1914, has been given to the unknown Germans, perhaps including Penck, who carried out the terrain analysis of this wilderness region of eastern Germany when war plans were being developed before the war. They had determined, from the types of aquatic plants growing on the surface of the lakes, which of the lakes had hard bottoms and were passable, and which were boggy. The Russian Second Army was driven into a morass of swamps and lakes and annihilated by the Germans who encircled it, moving on firmer ground (Cross, 1919; Stone, 1975). The victory can be attributed to the intelligent use of interior lines of communications by the German army, but perhaps teutonic geologists also played a role.

With the advent of hostilities the German geologists soon found themselves fulfilling wartime roles. Leopold von Werbeke, director of the Geological Survey of Alsace-Lorraine, visited the Lorraine front in the spring of 1915 to advise the army on ground-water sources. In the summer of 1915, Hans Phillipp, formerly a professor of geology at Greifswald, then serving as an army private, surveyed the water resources of the Moselle basin. By November 1915, four geologists were working under army auspices in Lorraine, and by February 1916, 20 German geologists were headquartered at Metz (Brooks, 1921b).

During this same summer of 1915, Captain Kranz established a geologic office at the University of Lille in northern France, and Dr. August Leppla did the same at Brussels in Belgium. Over time they were augmented with a mix of civilian geologists and geologists in uniform. These offices evolved to a formal military organization late in 1917.

Geology was a staff function, part of the department of military surveys under the administrative control of a general staff officer (*Chef des Kriegsvermessungwesens*). Operational control was vested at army or army group level in a survey staff officer (*Stabsoffizier der Vermessungwesens*), who usually was a major and who was assigned to the army, or army group, staff. The geologists reported to this engineering staff officer through a geologist in uniform (*Gruppeleiter*), usually a captain. They were distributed in small functional groups (*Geologen Stellen*), depending upon the tactical and strategic needs at the moment. Each section comprised geologic professionals, at that time in uniform, and was commanded by a proficient geologist who might be a commissioned officer, a noncommissioned officer, or even a private. In fact, a distinctive rank of private, the *Kriegsgeologan*, was created to encompass these specialists. Nevertheless, the relatively low rank accorded geologists created problems for them in the hierarchical military system. It also said much about the low prestige of geology in the eyes of the German army. To support the field sections, three geologic intelligence sections (*Geologische Arbeitstelle*) were established at Lille, Brussels, and Metz and staffed by older reservists who did office, library, and laboratory work (Brooks, 1921b).

The field sections were distributed along the battle lines in their areas of responsibility; the number of sections is unknown. After the U.S. Army took the St. Mihiel salient, five geologic-section offices and 20 men distributed about 20 km apart were found in that area. Some 60 men were assigned to the Lille area under Captain Karl Regelman. Alfred H. Brooks estimated that more than 100 geologists in the German army were assigned to military operations on the western front and at least 34, mostly civilians, were engaged in economic geology. The geologic section offices captured by the Allies proved to be so comfortable and well equipped that they evoked envy; their offices were connected by telephone. Their work paralleled that of their enemy specialists across the line. The numbers and functions of those German geologists assigned to eastern front are unknown (Brooks, 1921b).

The first need for professional geological advice arose from the necessity for a safe and secure water supply. Next came the requirement to support engineering work, fortifications, roads, and so on. Finally, geologists were used in an attempt to predict battlefield conditions from the soil, terrain, hydrology, geology, and climatic factors that would be encountered. The German geologists provided their army with a series of geologic maps at a scale of 1:25,000 that depicted the geology of areas of military interest in militarily relevant and understandable terms. Although overly detailed by American standards, the German maps were of excellent quality. Late in the war, the Germans began to issue "military geologic" maps on a 1:50,000 scale that closely resembled the American and French trafficability maps.

The Austro-Hungarian army followed the lead of its senior ally. Geologists were attached to the general staff, and apparently a geologic corps of specialists was organized, although little is known of its work. However, it is known that the general staff had enlisted prominent geologists of the Empire (Tietze, Stüche, Waagen, and Vetters) to help survey the most probable war sites east of the Carpathians and along the upper Adriatic even before 1914. The main purpose was to locate secure water sources (Cross, 1919).

AMERICA ENTERS THE WAR

America's entry into the war in 1917 came after the earlier belligerents had already incorporated geologists into their military systems. The U.S. Army, having little time to plan, wisely copied the programs of their senior allies, principally the British. Shortly after the declaration of war in April, American staff officers led by their commander, General John J. Pershing, hurried to Europe to survey the situation and consult with their new allies. Among these Americans were two engineers, General S. A. Cheney and Colonel Ernest Graves, who closely studied the British experience. They were trying to determine the special engineering requirements of the nascent American Expeditionary Force (AEF) so that specialized military units could be created, staffed, trained, equipped, and rushed to France to meet those requirements. They found two areas where specialized skills were needed; mining and water supply. They also recognized the need for geological support for these operations and arranged for geologists to be incorporated into the AEF engineering staff. Lieutenant Colonel (then Major) Alfred H. Brooks, a peacetime U.S. Geological Survey geologist, was given the task of organizing the geologic support for the AEF (Fig. 1). Brooks was assigned to the office of the Chief Engineer at Tours rather than to AEF operational headquarters at Chaumont, a clear indication that the geologists were not regarded as militarily important (Chief Engineer, U.S. Army, 1919). Initially, Brooks reported to the division of frontline engineering, indicating the importance incorrectly attached to mining by the American authorities. Later the geological staff was placed under the division of engineering intelligence, indicating that the topographic mapping and trafficability studies had become predominant (Brooks, 1921b). Eventually, 13 geologists would be assigned, as geologists, to the AEF. Three did not arrive before the end of the war, and four others arrived only in its final stages (Chief Engineer, U.S. Army, 1919).

Brooks found his geologists in many places. Major Morris La Croix was an instructor in the Army Engineer School, Lieutenant R. S. Knappen was a topographic officer in a railroad regiment, Lieutenant H. F. Crooks was an engineer in a highway-construction regiment, Lieutenant Wallace Lee was in the Field Artillery, and Lieutenant Kirk Bryan was a private working as a draftsman (Fig. 2; Brooks, 1921b). Most of these worked out of General Headquarters (GHQ) of the AEF at Tours, but geologists were later assigned to the field armies. La Croix and Lee were assigned to the staff of the U.S. 1st Army and Knappen to the 2nd Army. More personnel, five for each field army, were authorized, but the war ended before the assignment of personnel was accomplished (Chief Engineer, U.S. Army, 1919; Brooks, 1920a; Brooks, 1921b).

The mining requirement envisioned by AEF planners never developed. By the time America entered the war, the belligerents had effectively neutralized one another's combat mining efforts. To meet this incorrectly anticipated need, the U.S. Army created a specialized mining regiment, the 27th Engineers. This regiment was authorized six companies, each with five officers and 250 men and a 43-man headquarters company. Civilian miners were recruited through trade journal advertisements and letters to mine operators. Because of the war, U.S. mines were at full production and recruiting was slow. The regiment was finally filled out with draftees (Engineers, U.S. 27th, 1918a). Many of these proved

Figure 1. Lieutenant Colonel Alfred H. Brooks. (Photograph courtesy of the U.S. Geological Survey.)

Figure 2. Two U.S. Geological Survey scientists who served in France in WWI. Left, Lieutenant R. S. Knappen, 2nd U.S. Army, AEF. Right, Lieutenant Kirk Bryan, General Headquarters AEF. (Photographs courtesy of the U.S. Geological Survey.)

unsuitable for strenuous mining work, and in one company more than half could not even speak English. Western hard-rock miners were preferred over eastern coal miners as "more intelligent"; most came from the Joplin, Missouri, region (Engineers, U.S. 27th, 1918b). Among the officers were civil, mining, mechanical, and electrical engineers and a miscellany of others, but no geologists (Engineers, U.S. 27th, n.d.). Despite the changed situation, the miners gave good service to their country in a variety of roles. They engaged in road building, quarrying, stevedoring, and one company won a citation for bridge building under intense shellfire in the critical spring days of 1918 (MacGlashan, n.d.).

The 26th Engineers was a specialized water-supply regiment consisting of 1,434 men and 51 officers organized as seven water-supply companies and one headquarters company. No geologists were assigned (Engineers, U.S. 26th, 1919). The men were recruited primarily from oil-well drillers. One regiment proved insufficient for the work load, a second was being formed when the war ended, and a third was planned (Pratt, 1919). Brooks' staff worked closely and successfully with drillers throughout the war in locating potable water sources. One success came early when the U.S. geologists were able to predict correctly an artesian water source that provided adequate water (more than four million gallons per day) for the major American base at Bourdeaux where water was scarce, even in peacetime. In all, 18 general reports and 14 specialized ones on water supply covering some 14,939 km^2 were produced by Brooks' staff (Chief Engineer, U.S. Army, 1919).

As American involvement in combat increased, topographic mapping became the primary function of the geological staff who worked closely with the topographic engineers. More than five million maps were printed and issued to AEF forces from July to November 1918. Some were based on aerial photography, which was the first time this technique was used, at least by Americans. After the war, the maps based on aerial surveys were compared to the actual ground and were found to be very accurate (Chief Engineer, U.S. Army, 1919). For the Meuse-Argonne offensive, the geologists prepared a map and brief report, derived from French geological publications, showing the nature of the subsoil and the expected trafficability of the battle area. In support of this, the largest American operation of the war, some 255 maps and 25 overlays were produced by a staff working 12–18 hours a day (Chief Engineer, U.S. 1st Army, 1918).

Although the contribution of American geologists to military operations was valuable, it was a disappointment to the geologists who sought a larger role. After the war, Brooks outlined the areas of military operations where geologic information was useful: fortifications, maneuvering (trafficability), water resources, transportation, construction, mineral resources (in the area of operations), earth telegraphy, camps, cantonments and bases, and areas of possible inundation. He called for the peacetime collection of data in potential areas of operation, a reserve staff of geologists and engineers, and a vast expansion of geologic education at all levels of school, both civilian and military. Reflecting on the limited battlefield use of geologists, Brooks asserted his belief that it would be in the area of mineral resources that geologists could make their greatest contribution. However, geologists at home in the United States were facing the same sort of frustrations that Brooks encountered in Europe (Brooks, 1921a).

As the bloody stalemate in Europe had continued from 1914 to 1917, the chances of American involvement had grown steadily. Belatedly, some preparations to mobilize the domestic economy were made as hostilities grew inexorably nearer. In the spring of 1916, at the request of President Woodrow Wilson, the National Academy of Sciences organized the National Research Council (NRC) to mobilize America's scientific skills for war. J. M. Clarke initially chaired the subcommittee for geology, but when the NRC was reorganized and put on a full wartime footing in January 1918, John C. Merriam became the chairman of the Division of Geology and Geography. Actually, Merriam served for a time as head of the NRC and Whitman Cross as chairman of the division (Cross, 1919).

The geologists quickly found, to their patriotic dismay, that little awareness existed in government of the contributions that geology and geography could make to the war effort, either military or civilian. It was only by repeated personal persistence that individuals were able to find a valuable part to play in the great drama unfolding around them, and most of those who were successful used their talents in areas other than geology (De Wolf, 1919).

The failure to be accorded an important role in this first great total war was a tremendous shock to the professional pride of American geologists. During and after the war, the literature was filled with soul-searching questions as to whether geology had any real meaning to modern life, either for its educational value or its utility. If it did, why was it so difficult for geology and geologists to become part of the war effort? The answer, geologists felt, was a lack of education, specifically education in geology. Few American precollege students had studied geology (0.5%), and the number had fallen in recent years. It was rare even at the college level. Of the 512 colleges selected for the Army Training Corps (ATC), presumably the best schools, 43% had no geology courses. In 173 colleges, geology was taught part-time by a specialist in another area, and in 58 schools, only one geologist was available to offer courses. At America's better colleges, only about 12% could offer geology courses beyond the introductory level (Berkey, 1918).

Despite the general lack of appreciation of the value of their field, patriotic geologists strove to find ways to make meaningful contributions. R. A. F. Penrose wrote a pamphlet, "What a geologist can do in the War," under the auspices of the NRC. He claimed that geologists had unique knowledge of terrain, structure, drainage, water supply, topography, and mapping. This was, as Bateman (1917) pointed out, exactly the type of knowledge required by artillery officers who were then needed in large numbers.

The suggestions were followed up promptly. To supply the large number of young officers that were expected to be needed in what then appeared to be a long war, the U.S. Army undertook training its own by establishing the Army Training Corps (ATC). Some 512 of the better quality colleges and universities were utilized to provide a shortened basic college program combined with military training. The academic program consisted primarily of science, mathematics, and engineering subjects of direct relevance to prospective army officers. Through the NRC, geology won a foothold in the ATC curriculum, but only after intensive lobbying efforts. Special courses were devised and textbooks created by the best minds in the field on the understanding that the courses would be incorporated into the required ATC program (Berkey, 1918). Even so, the weakness of the geologists' position can be seen in the fact that the course was sold as primarily being a course in topography (Gregory, 1918). Eventually, the geological course became an elective for which only 10% of the student officers opted. Some academic geologists and geographers did teach the required course in surveying and map making, which was similar to the civilian course given prospective railroad engineers (Berkey, 1918).

To try to gain a foothold for geology in the ATC program, a special NRC subcommittee was formed in June 1918 under C. P. Berkey. Their recommendation, that the geology course be made a mandatory part of the ATC curriculum, was forwarded to the Secretary of War where it languished until the war ended. Two books, written under Geological Society of America auspices and published with donated money, were distributed to ATC libraries. These were W. M. Davis' *Handbook of Northern France* and Douglas W. Johnson's *Topography and strategy in the War*. The war ended without a required geology course in the army's training program.

American geologists also found less employment for their patriotic energies than they expected in the search for raw materials for the war effort. There were few scarcities; the minerals needed for war were the same ones used by the domestic economy, and the supply proved generally adequate for wartime needs (Pogue, 1917). The big exception was petroleum. The U.S. Geological Survey, already gutted for manpower by the 216 of its staff in uniform, abandoned most nonessential work to concentrate on locating and developing new energy resources. The survey succeeded in locating oil in Oklahoma and natural gas in West Virginia for the navy; some overseas exploration was done. U.S. Geological Survey scientists investigated nitrate deposits in Guatemala (Smith, 1918) and chromite and manganese deposits in Cuba. The U.S. Geological Survey also cooperated with the U.S. Bureau of Mines (USBM) to locate domestic sources of other minerals that were expected to be in short supply, for example, tin and graphite, and also to relieve the strain on critically short shipping by curtailing mineral imports (Bastin, 1918).

Probably the most important U.S. Geological Survey function was the collection of data on minerals needed by the War Industries Board (WIB) in its management of the wartime economy. The U.S. Geological Survey worked in cooperation with the USBM and the Bureau of Commerce. It was quickly found that the prewar production reports on an annual or semiannual basis were inadequate and that monthly or weekly reports were needed; new data also were required. "Stocks on hand" was not a meaningful concept in peacetime but became urgent during the war. Data also were collected on the costs of production to enable the WIB to fix commodity prices and to determine which mines to close or keep open. In addition, information was collected to aid control of mineral exports. The U.S. Geological Survey also advised the War Finance Committee, which controlled all wartime major industrial investments, on mine investments (Bastin, 1918). All of these activities might have become critically important in a lengthy war, but the war ended so soon that American industry never fully came into war production, which became something of a scandal itself.

Professional geologists made other contributions more directly related to the war. The U.S. Geological Survey historically had acted, in an informal way, as consulting geologists to the army and navy; this role expanded during World War I, particularly in regard to the siting of training cantonments and military bases (Berkey, 1918). This function was formalized under the NRC with F. W. De Wolf heading a special subcommittee. One major result was a 12-volume study, *Report on Materials and Facilities for Rapid Road and Fortification Construction*, that had been planned before the entrance of the United States into World War I. This report might have become critically important had the war spread to North America (Cross, 1919).

CONCLUSION

Overall, the experience of American geologists in the war effort was unsatisfactory to them. Whether in France or at home, they found their specialty unappreciated by the larger world. According to the president of the Geological Society of America, this forced geological professionals to take a hard look at the practical value of their profession and to take a much deeper interest in education. The world learned the value of science from the war. Now, he said, strong efforts must be made toward "restoring geology as a science to its proper place" in schools, government, and research (Cross, 1919).

ACKNOWLEDGMENTS

Edward P. F. Rose of the University of London made numerous invaluable suggestions that greatly enhanced the quality of this paper and for which the author is grateful.

REFERENCES CITED

Ball, H. S., 1919, The work of the miner on the western front, 1915–1918: Transactions of the Institution of Mining and Metallurgy, v. 28, p. 199–204.

Bastin, E. S., 1918, War time mineral activities in Washington: Economic Geology, v. 13, p. 524–537.

Bateman, A. M., 1917, The geologist in wartime—the training of artillery officers: Economic Geology, v. 12, p. 628–631.

Berkey, C. P., 1918, Engineering geology in and after the war: Geological Society of America Bulletin, v. 30, p. 81–82.

Brooks, A. H., 1920a, Military mining: Occasional Papers, n. 62, The Engineer School, U.S. Army p. 5–43.

Brooks, A. H., 1920b, Military mining in France: Engineering and Mining Journal, v. 109, p. 606–610.

Brooks, A. H., 1921a, Application of geology in war behind lines and at front: Engineering and Mining Journal, v. 109, p. 764.

Brooks, A. H., 1921b, The use of geology on the Western Front: U.S. Geological Survey Professional Paper 128, p. 85–125.

Chief Engineer, U.S. 1st Army, 1918, Report of the Chief Engineer, 1st Army, AEF, On the engineer operations in the St. Mihiel and Meuse-Argonne offensives—1918: Occasional Papers, n. 69, The Engineer School, U.S. Army, April, 1929, p. 140.

Chief Engineer, U.S. Army, 1919, Historical report of the Chief Engineer, American Expeditionary Forces 1917–1919: Washington, Government Printing Office, p. 125–131.

Cross, W., 1919, Geology in the World War and after: Geological Society of America Bulletin, v. 30, p. 165–188.

DeBeque, G. R., 1919, Mine warfare: Appendix No. 22 to the Historical Report of the Chief Engineer of the AEF 1917–1919, Washington, Government Printing Office.

De Wolf, F. W., 1919, The outlook for geology and geography: School Science and Mathematics, v. XIX, p. 391–397.

Engineers, U.S. 26th, 1919, Historical–technical report from September 10, 1917 to 1st January 1919: National Archives, p. 1–32.

Engineers, U.S. 27th, 1918a, Historical–technical report to November 1, 1918: National Archives, p. 1–12.

Engineers, U.S. 27th, 1918b, Historical–technical report, December 1918: National Archives, p. 1–17.

Engineers, U.S. 27th, n.d., Individual biographies (of officers): National Archives, p. 1–9.

Gregory, H. E., editor, 1918, Military geology and topography: New Haven, Yale University Press, p. iii–xiii.

King, W. B. R., 1919, Geological work on the Western Front: The Geographical Journal, v. 54, p. 201–221.

Kranz, W., 1913, Militargeologie: Kriegstechnische Zeitschrift, v. 16, p. 464–471.

Leighton, A., 1922, Application of the geophone to mining operations: U.S. Bureau of Mines, Technical Paper 277, p. 1–33.

MacGlashan, A., n.d., Memo: To the officers and men of Co. B, 27th Engineers, National Archives, 1 p.

Pogue, J. E., 1917, Military geology: Science, v. 46, n. 1175, p. 8–9.

Pratt, A. H., 1919, Water supply for a field army: Professional Memoirs, Corps of Engineers, U.S. Army, v. XI, p. 730–741.

Rose, E. P. F., and Hughes, N. F., 1993, Sapper geology: Lessons learnt from world war; Royal Engineers Journal, v. 107, p. 27–33.

Rose, E. P. F., and Rosenbaum, M., 1993, British military geologists: The formative years to the end of the First World War: Proceedings of the Geologists' Association, v. 104, p. 41–49.

Smith, P. S. 1918, The Geologist in war times: The United States Geological Survey's war work: Economic Geology, v. 13, p. 392–399.

Stone, N., 1975, The Eastern Front, 1914–1917: New York, Charles Scribner's Sons, p. 44–69.

Trounce, H. D. 1918, Fighting the Boche underground: New York, Charles Scribner's Sons, p. 1–229.

Manuscript Accepted by the Society October 29, 1997

Printed in U.S.A.

Geological Society of America
Reviews in Engineering Geology, Volume XIII
1998

Military Geology Unit of the U.S. Geological Survey during World War II

Maurice J. Terman
Scientist Emeritus, International Programs, Office of the Chief Geologist, U.S. Geological Survey, 917 National Center, Reston, Virginia, 20192

ABSTRACT

On June 24, 1942, the temporary Military Geology Unit of the U.S. Geological Survey was formalized after the U.S. Army Corps of Engineers requested them to prepare terrain intelligence studies to meet wartime priorities. The entire Military Geology Unit wartime roster was 114 professionals, including 88 geologists, 11 soil scientists, and 15 other specialists; 14 were women. Assisting staff (illustrators, typists, photographers, and others) totaled 43. The unit produced 313 studies, including 140 major terrain folios, 42 other major special reports, and 131 minor studies. These reports contain about 5,000 maps, 4,000 photographs and figures, 2,500 large tables, and 140 terrain diagrams. Most products were designed in the beginning for general strategic planning in Washington and later for detailed strategic planning overseas; they utilized graphics and nontechnical, telegraphic-style tabular texts.

The Military Geology Unit's principal effort was the preparation of the terrain folios titled Strategic Engineering Studies. They varied somewhat in content and format, but the key components usually were introduction, terrain appreciation, rivers, road and airfield construction, construction materials, and water resources. The folios, produced at an average rate of about one per week and at an average cost of $2,500, were compiled from scientific journals, books, maps, and photographs available in the Washington area by a team of 3 to 8 scientists; 8 to 12 teams might be working concurrently. MGU personnel took great pride in never having missed a delivery deadline.

In 1944, 5 Military Geology Unit consultants were sent to Europe and 5-man teams were assigned to the Southwest Pacific Area and to the Central Pacific Area. Each team produced large-scale terrain reports, mostly from aerial photographs, and consulted with engineer units and tactical officers in the field. By the end of the war, a consolidated 20-man team worked in Manila at the headquarters of the Armed Forces Pacific.

INTRODUCTION

For more than 50 years, the U.S. Geological Survey has had an active program in military geology. This administrative overview is of the earliest formal phase of that program, i.e., the Military Geology Unit (MGU), from its initiation in 1942 through the remainder of World War II to 1945. I joined the unit after the war so I have relied heavily on the remaining unpublished administrative files, on prior publications of Charles B. Hunt, particularly his excellent summary entitled *Military Geology* (Hunt, 1950), and on *The Military Geology Unit* (U.S. Geological Survey and Corps of Engineers, U.S. Army, 1945). This overview focuses on the varied nature and scope of the MGU activities through the war years rather than on the details of the MGU reports.

ORIGIN

In February 1942, the Board of Economic Warfare asked geologists of the U.S. Geological Survey to prepare quick sum-

Terman, M. J., 1998, Military Geology Unit of the U.S. Geological Survey during World War II, *in* Underwood, J. R., Jr., and Guth, P. L., eds., Military Geology in War and Peace: Boulder, Colorado, Geological Society of America Reviews in Engineering Geology, v. XIII.

mary reports on the strategic minerals of African countries. As described by one of the participating scientists (Lincoln Page, personal communication, 1994), on the day of the request, eight scientists assembled all of the library data that they could find on one country—Sierra Leone, and worked throughout the night to finish the report. Several scientists searched through the library documents marking applicable source material, others abstracted the pertinent data, and others organized the information by commodity titles and stitched together a draft narrative accompanied by appropriate graphics. Finally, a sleeping secretary was awakened to transform the edited manuscript into a finished product. The success of their effort led, later that spring, to the preparation by the U.S. Geological Survey of an additional 30 country reports.

At the same time, several scientists, notably Wilmot H. Bradley and Charles B. Hunt, began to experiment with some pilot projects on the application of geoscience to military needs. From data in the U.S. Geological Survey library, Hunt compiled sample research reports for three quadrangles in Morocco to illustrate the type of information that geoscientists could provide to the U.S. Army Corps of Engineers. The Office of the Chief of Engineers (OCE) found the products interesting but then requested studies in areas that could be field checked, such as in southern California for checking by OCE, and in Madagascar for checking by the British who were operating there. After the studies were completed, OCE was impressed by the California report, and the British were enthusiastic about the Madagascar report, stating that they wished that they had had it prior to that operation. In October 1943, they established a unit similar to the MGU, the Inter-Service Topographic Department (ISTD).

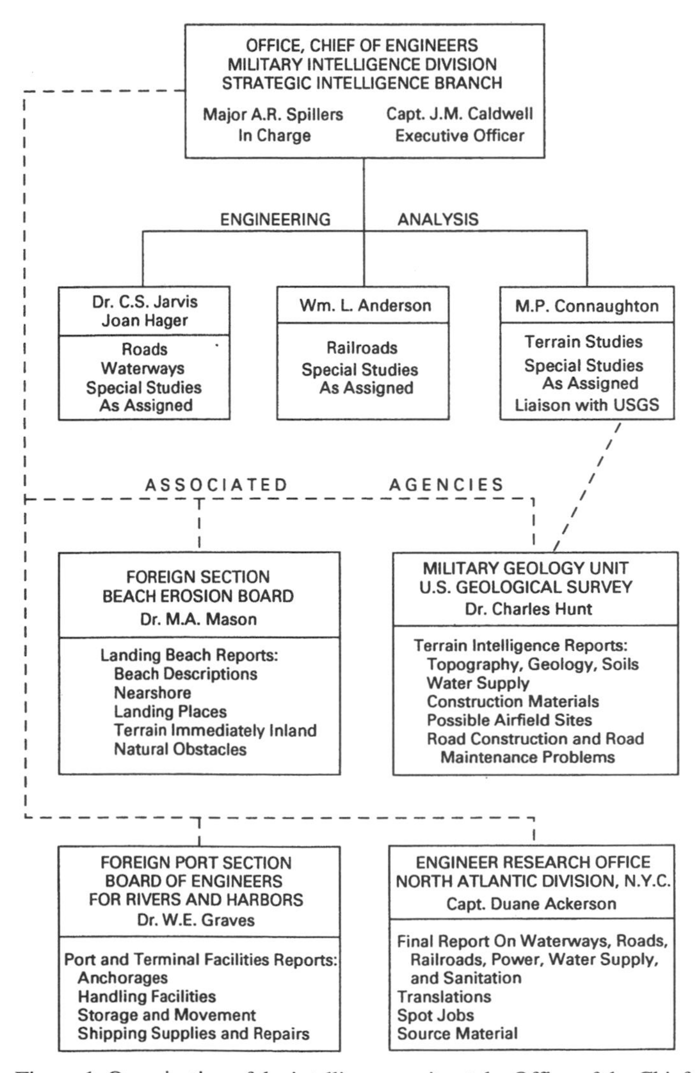

Figure 1. Organization of the intelligence units at the Office of the Chief of Engineers, U.S. Army Corps of Engineers, during World War II.

MGU ORGANIZATION, PERSONNEL, AND PRODUCTS

On June 24, 1942, the U.S. Geological Survey formalized the MGU, and subsequently a cooperative agreement was signed between the U.S. Geological Survey and the U.S. Army Corps of Engineers for the MGU "to prepare geological reports of strategic areas designated by the Corps" for which the Corps of Engineers would transfer $30,000 to the U.S. Geological Survey during FY1943 (fiscal year 1943). That letter initiated a very workable relationship that lasted for three decades.

The MGU's contact at the OCE was through the Strategic Intelligence Branch (SIB) of the Military Intelligence Division (Fig. 1). Other agencies associated with SIB provided expertise on ports and on beaches, so the MGU's responsibilities were normally from the beaches inland.

The initial MGU roster included l0 geologists, a typist, and an illustrator; this group grew steadily to plateau at about 100 people in 1945 (Fig. 2). The entire wartime roster of 157 personnel was composed of 114 professionals (Table 1), including

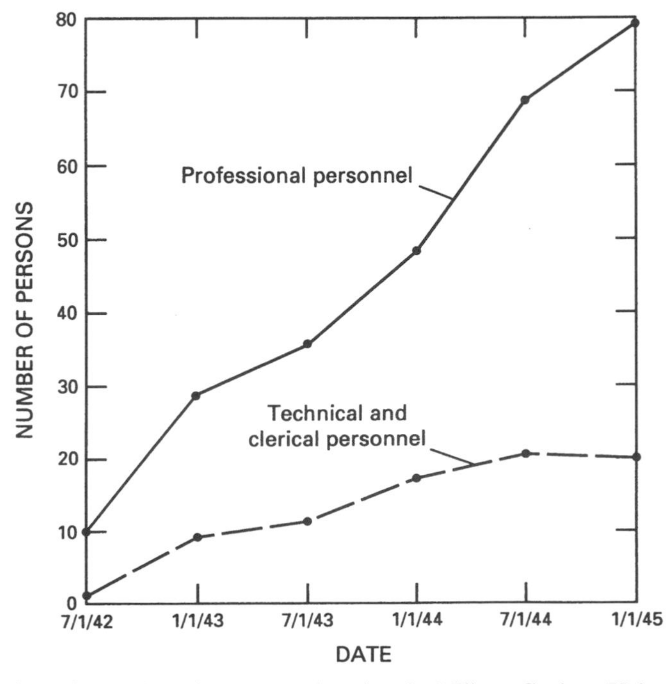

Figure 2. Number of persons assigned to the Military Geology Unit at six-month intervals from July 1, 1942, to January 1, 1945 (only times for which complete records are available).

TABLE 1. PROFESSIONAL SCIENTISTS IN THE MILITARY GEOLOGY UNIT DURING WORLD WAR II (2/42–10/45)*

GEOLOGISTS

Name	Former Affiliation
Aberdeen, Esther J.	U.S. Geological Survey
Ahrens, Thomas P.	U.S. Bureau of Agricultural Economics
Albritton, Claude C., Jr.	Southern Methodist University
Allen, Alice S.	U.S. Geological Survey
Balsley, James S., Jr.	U.S. Geological Survey
Behre, Charles H.	Columbia University
Betz, Frederick, Jr.	U.S. Bureau of Mines
Bleyberg, Arnold H.	The Texas Company
Bradley, Wilmot H.	U.S. Geological Survey
Brown, Glenn F.	U.S. Geological Survey
Bryson, Robert P.	U.S. Geological Survey
Buie, Major Bennett F.	U.S. Army
Butler, Arthur P.	U.S. Geological Survey
Capps, Stephen R.	U.S. Geological Survey
Cassel, Dorothy A.	None
Collins, John	U.S. Geological Survey
Dane, Carle H.	U.S. Geological Survey
de Laguna, Wallace	Queens College
Denny, Charles S.	Wesleyan University
Dryden, Lincoln	Byrn Mawr College
Eckel, Edwin B.	U.S. Geological Survey
Elias, Maxim M.	U.S. Engineer Corps
Erdmann, Charles E.	U.S. Geological Survey
Ferguson, Henry G.	U.S. Geological Survey
Fitzsimmons, J. Paul	U.S. Geological Survey
Foxhall, Harold B.	U.S. Geological Survey
Fryxell, Fritiof M.	Augustana College
Gardner, Julia A.	U.S. Geological Survey
Garrels, Robert M.	Northwestern University
Gilluly, James A.	Univeristy of California at Los Angeles
Goldman, M. I.	U.S. Geological Survey
Hack, John T.	Hofstra College
Hill, Mary E.	Smith College (student)
Hilpert, Lowell S.	U.S. Geological Survey
Howland, Arthur L.	Northwestern University
Hsu, Ginn-Tze	Nanking Central University
Huff, Lyman C.	U.S. Geological Survey
Hunt, Charles B.	U.S. Geological Survey
James, Harold L.	U.S. Geological Survey
Jeffords, R. M.	U.S. Geological Survey
Johnson, Charles G.	Illinois Geological Survey
Juan, Vei-Chow	University of Chicago (student)
Kemmer, George H.	War Department
King, Philip B.	U.S. Geological Survey
Krauskopf, Konrad B.	Stanford University
Larsen, Esper, 3rd	Saranac Laboratory
Lemmon, Dwight M.	U.S. Geological Survey
Lewis, G. Edward	Yale University
Lohman, Kenneth E.	U.S. Geological Survey
McKee, Edith	Northwestern University (student)
McKelvey, Vincent E.	U.S. Geological Survey
Morrill, Charlotte	Byrn Mawr College (student)
Morrison, Harriet W.	None
Morrison, Roger B.	U.S. Geological Survey
Moss, John H.	Harvard University (student)
Muller, Siemon W.	Stanford University
Myers, W. Bradley	U.S. Geological Survey
Neuschel, Sherman K.	U.S. Geological Survey
Nicol, Allan	California Department of Highways
Noble, Levi S.	U.S. Geological Survey
Nye, S. Spencer	Private business
Putnam, William C.	University of California at Los Angeles
Ray, Louis L., Jr.	Michigan State College
Reed, Robert W.	The Texas Company
Richards, Ralph	U.S. Geological Survey
Richmond, Gerald M.	U.S. Engineer Corps
Roberts, George	U.S. Navy
Roberts, Ralph J.	U.S. Geological Survey
Rodgers, John	U.S. Geological Survey
Sampson, Edward	Princeton University
Sayre, A. Nelson	U.S. Geological Survey
Schmidt, Ruth A. M.	Columbia University
Shenon, Philip J.	Consulting geologist
Smith, H. T. U.	University of Kansas
Smith, J. Fred	Texas A. and M. College
Spencer, Lt. Frank N.	U.S. Army
Stadnichenko, Taisa	U.S. Geological Survey
Stauffer, Hans K.	Bataafsche Petrol. Mij.
Stewart, Ralph E.	U.S. Geological Survey
Swenson, Frank A.	U.S. Geological Survey
Theis, Vernon C.	U.S. Geological Survey
Varnes, Helen D.	Northwestern University (student)
Waring, Gerald A.	U.S. Geological Survey
White, Walter S.	U.S. Geological Survey
Whitmore, Frank C., Jr.	Rhode Island State College
Wiese, John H.	U.S. Geological Survey
Wyckoff, Dorothy	Bryn Mawr College
Yen, T. C.	U.S. National Museum

SOIL SCIENTISTS

Name	Former Affiliation
Baldwin, Mark	U.S. Agricultural Research Administration
Bodman, Geoffrey B.	University of California
Cady, John G.	University of Idaho
Fruhauf, Bedrich	U.S. Navy
Li, Lien-Chieh	China Geological Survey
Mason, CWO David D.	U.S. Army
Miller, John T.	U.S. Agricultural Research Administration
Pendleton, Robert L.	War Department, General Staff
Sokoloff, Vladimir P.	University of California at Los Angeles
Striker, Marion M.	U.S. Agricultural Research Administration
Thorp, James	U.S. Agricultural Research Administration

TABLE 1. PROFESSIONAL SCIENTISTS IN THE MILITARY GEOLOGY UNIT DURING WORLD WAR II (2/42–10/45)* (continued)

OTHER PROFESSIONALS Name	Speciality	OTHER PROFESSIONALS Name	Speciality
Andresen, Ruth	Bibliographer	Jussen, Virginia	Editor
Andron, Bette	Editor	Lewis, Regina	Librarian
Blach, Fredrich	Sanitary engineer	MacArthur, Mary E. Y.	Editor
Creagh, Agnes	Bibliographer	Pangborn, Mark	Librarian
Foxworthy, Fred W.	Forester	Reed, Ruth Y.	Terrain specialist
Gruber, Z. I.	Librarian	Siegrist, Marie	Research analyst
Guyton, William F.	Hydraulic engineer	Stephenson, Phillip M.	Highway engineer
Heather, Leroy W.	Topographic engineer		

*After U.S. Geological Survey and Corps of Engineers, U.S. Army, 1945.

88 geologists, 11 soil scientists, 6 bibliographers, 5 engineers, 3 editors, and 1 forester; 14 of these were women. The assisting staff totaled 43, including illustrators, typists, photographers, and reviewers; at any one time, this assisting staff made up about 20–25% of the MGU.

The list of the 88 geologists that served in the MGU (Table 1) contains a number of scientists who later became prominent U.S. Geological Survey professionals, including a director, an assistant director, several chief geologists and assistant chief geologists, and others who became well known in the geologic community. The MGU recruitment at the U.S. Geological Survey and at many universities was successful in attracting professionals with broad-ranging competence and demonstrated ability to work with others. Since the MGU hired no translators, facility in at least two languages was essential; many of the professional staff had some competence in three or four languages, some in six or seven, and the MGU was especially fortunate in recruiting three Chinese scientists.

Through most of the war, the table of organization for the MGU was as diagrammed in Figure 3. The flow of typical project efforts was generally as follows: (1) The OCE established assignments, priorities, and deadlines, some as early as the next day. (2) The research supervisor organized and scheduled project teams. (3) The bibliographers searched out basic source materials, usually geologic texts and maps that were borrowed from the U.S. Geological Survey library and the libraries of Congress, War Department, Department of Agriculture, and even from private companies. Where these sources were meager, photographs acquired great significance, particularly those at such places as the National Geographic Society. (4) Project teams of 3 to 8 scientists synthesized and interpreted the available source materials and compiled the studies; 8 to 12 teams sometimes worked concurrently. Graphics were the key products. Maps and photographs were essential, and terrain diagrams rapidly won great favor. In most studies, each map was rated for reliability on a scale from A to E, reflecting the compiler's evaluation of the quality of the basic data. Map legends were explained in simple, nontechnical, telegraphic-style explanatory tables. This technique was subsequently copied by other U.S. agencies. (5) The reports were carefully edited by geologists before being drafted and typed. (6) Completed reports then were reviewed by nongeologists. (7) Finally, completed studies in photo-negative form were forwarded by the MGU chief to the Army Map Service for printing.

Frank C. Whitmore, who first served in MGU as an editor, stated (personal communication, 1996) that

> This editor/reviewer system was unique. The reviewers' task was to read the report from the point of view of the nonscientific reader. The reviewers' office habitually worked at night. At the end of the day when the editors had assembled the final copy, it would be passed to the reviewers who would work on it all night. In the morning, the chief editor would be confronted with the errors that had been unearthed. Crow would be eaten, corrections made, and the report would go back to the photographers.

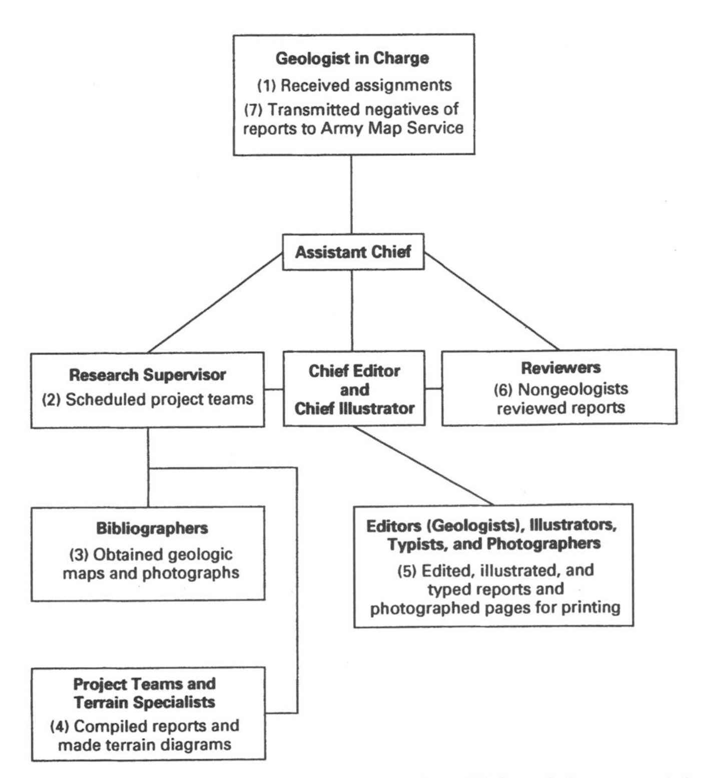

Figure 3. Organization of the Military Geology Unit and the sequential steps in the production of a report (see steps 1 to 7 in text).

One embarrassing error that slipped through stated: "The valleys are wide and inhabited by Bedouins with date palms in their bottoms."

The principal focus throughout the war was the preparation of Strategic Engineering Studies (SES) that covered specific geographic regions or even entire countries. These folios, mostly 19 in × 24 in (48 cm × 61 cm), were produced at an average of almost one per week and cost about $2,500. In 1942, the folios had about 12 black-and-white maps and 10 tables; three copies were prepared for general strategic planners in Washington. By 1945, some folios contained more than 100 maps, many of them multicolored, and 50 tables; on occasion 500 or more copies were printed for detailed strategic planning overseas. During busy periods, MGU members put in as many as 75 hours a week or as many as 36 consecutive hours. The MGU was proud that in three years it never missed a delivery deadline.

Frank C. Whitmore also stated (personal communication, 1996) that "It is hard to describe the *esprit de corps* of MGU. There were constant vigorous arguments about subject matter because we were making important interpretations on insufficient data, but I recall no personal animosity despite the pressure and the close quarters in which we worked—we were too busy." These rush jobs led to the slogan: "In Military Geology, an intelligent quick decision is better than delay in search of the ideal." Of course, the project teams shortened this to "Don't think—act!" (Sommers, 1945).

The MGU produced 313 strategic reports (Hunt, 1950), including 140 major terrain folios for the U.S. Army Corps of Engineers; 42 other major studies, mostly for the U.S. Army Air Corps (24 Special Intelligence Reports, 9 Commodity Reports, 9 Bomb Target Reports); and 131 minor, usually quick reports for other organizations (e.g., the first 31 of that total were for the Board of Economic Warfare). Together these reports contained about 5,000 maps, 4,000 photographs and figures, 140 terrain diagrams, and 2,500 tables. The formats of the SES folios varied somewhat, but the key components usually were those listed in Table 2.

TABLE 2. KEY COMPONENTS OF THE STRATEGIC ENGINEERING STUDIES OF THE MILITARY GEOLOGY UNIT*

Introduction and Summary	**Road and Airfield Construction**
Sources	Site selection
Methodology	Foundation problems
Reliability	
Terrain diagram	**Construction Materials**
	Soil types
Terrain Appreciation	Rock types
Topography	
Ground conditions	**Water Resources**
Trafficability	Surface water
Cover and concealment	Ground water
	Quality of water
Rivers	
Valley descriptions	
Regimen	
Barrier problems	

*Basic-data maps of soils or geology and resources were generally considered "too scientific" for easy comprehension by the military users of these reports.

SEQUENTIAL PHASES OF THE WORK

During FY1943, the budget was adjusted from the original agreement of $30,000 to $115,000 and the work expanded to include 53 SES folios and 23 other reports, focused mainly within the Mediterranean region. An excellent example of the folios designed for detailed planning overseas was the SES series for Sicily. The request for them was received on May 9, 1943. The MGU spent 10 days compiling 36 tables and 56 maps for 1:100,000-scale quadrangles, including terrain (16 sheets), water resources (16 sheets), possible construction problems for roads (3 sheets) and airfields (9 sheets), terrain diagrams (2), other maps (10), and photographs (39). The Army Map Service spent another 10 days printing and delivering 600 copies to North Africa by the June 1 deadline. After the Sicilian campaign, General Davidson, the chief engineer for the operation, stated that the maps "proved to be accurate and complete, were at times indispensable, and in many cases possessed more information than the natives had themselves" (Sommers, 1945).

The next fiscal year, from July 1943 to June 1944, the area of interest of the MGU shifted to the equatorial Pacific, the budget increased to $223,000, and the work included the preparation of 31 SES folios and about 50 other reports. By FY1945, most of the MGU's effort was directed to the northern Pacific, the budget again increased, this time to $400,000, and the Washington-based MGU staff produced another 56 SES folios and about 30 other reports. Meanwhile, in early 1945 as the Allies were breaking out of the Normandy beachhead, the European command made a final request for a quick compilation of 65 1:100,000-scale trafficability maps for Germany; 18 were completed in five days and then 12 were finished each week for the next month. Also during FY1945, five consultants were assigned to Europe: two to the ISTD in England and three to the European Theater of Operations.

The most significant change in MGU operations was the organization of field teams for the Pacific region: one five-man team joined the Southwest Pacific Area (SWPA) Headquarters as it moved from Australia through New Guinea to the Philippines, and another five-man team was assigned to the Central Pacific Area (CPA) Headquarters in Honolulu. That year, each of these teams prepared about 100 special, large-scale reports, mostly from analysis of aerial photographs, and the scientists consulted directly with engineer units and tactical officers. A. Nelson Sayre, a member of the SWPA team, subsequently was awarded the Medal of Freedom for his contributions during the spring and summer of 1945.

In the first few months of FY1946 (the calendar-year months of July to September 1945), the Pacific field teams were consolidated at the headquarters of the Armed Forces Pacific (AFPAC) in Manila and were expanded rapidly; at the end of the war in September, the engineer intelligence unit included 20 MGU scientific consultants.

Figure 4. The logo for the Military Geology Unit during World War II. (When the logo was drawn, the presumption was that it would be named a Section, the standard operational U.S. Geological Survey element, whereas in actual fact it did not become a Section until after World War II.)

EVALUATION OF THE MGU

The MGU proved to be beneficial, and one of the keys to its effectiveness was that it was reactive to its users. The MGU also was as competitive as any other intelligence agency in dealing with costs, production, and quality. It had an excellent cost/benefit ratio; less than $750,000 had paid for all the products requested during the war. The reliability of the MGU products was largely dependent on the amount of the available basic geologic data. In Europe, published sources were voluminous, whereas in the Pacific, the meager published data had to be supplemented by the interpretation from aerial photography. Such scientific detective work and the secrecy surrounding production schedules and the location of designated study areas were essential ingredients of the MGU research, and these were conveyed by the MGU logo (Fig. 4), drawn in the early stages of the war when the organizational name had not been formalized.

Not surprisingly, military personnel generally were not technically qualified to determine the earth-science factors bearing on military operations. Therefore, the MGU took over the sole responsibility for determining the type and scope of the study reports and maintained complete freedom of initiative. The military provided the priorities, the absolutely necessary criteria, and the deadlines (speed was always paramount), and the OCE and the MGU worked closely together to develop the best presentation techniques. The folio model, and particularly the terrain diagram, proved most effective.

From these experiences, it is well understood that terrain intelligence always will be needed in wartime, but the U.S. Geological Survey also learned the important lesson that terrain data can be equally vital in peacetime activities. Thus, a major heritage of World War II was the creation of the U.S. Geological Survey Engineering Geology Section late in 1945 and its staffing by experienced MGU scientists; the U.S. Geological Survey engineering geology effort still thrives.

ACKNOWLEDGMENTS

James R. Underwood, Jr., and Peter L. Guth are commended for their planning and implementation of the GSA Symposium on Military Geology in War and Peace and for their continuing patience and encouragement in collecting and editing the manuscripts. Special thanks are due for the review comments provided by Lawrence D. Bonham and particularly by Frank C. Whitmore, Jr.

REFERENCES CITED

Hunt, C. B., 1950, Military geology, *in* Paige, S., ed., Application of geology to engineering practice: Geological Society of America, Berkey Volume, p. 295–327.

Sommers, M., 1945, The Army's pet prophets: Saturday Evening Post, March 24, 4 p.

U.S. Geological Survey, and Corps of Engineers, U.S. Army, 1945, The Military Geology Unit, Geological Society of America, Pittsburgh Meeting (pamphlet), 22 p.

Manuscript accepted by the Society October 29, 1997

Printed in U.S.A.

Geological Society of America
Reviews in Engineering Geology, Volume XIII
1998

British applications of military geology for 'Operation Overlord' and the battle in Normandy, France, 1944

Edward P. F. Rose
Department of Geology, Royal Holloway, University of London, Egham, Surrey TW20 0EX, United Kingdom
Claude Pareyn
Professeur émérite de Géologie, Université de Caen, 31 Rue de Jersey, 14000 Caen, France

ABSTRACT

British geologists participated for more than a year in the planning of "Operation Overlord," the Allied invasion of northwest France. Following D-Day on June 6, 1944, they contributed to the subsequent 11-month operational phase in western Europe, including the initial 3-month battle for Normandy.

Beachhead maps were prepared prior to the invasion at 1:5,000 scale from published topographic and geologic maps, aerial photographs, and secret ground reconnaissance. They indicated the character of the beaches and cliffs, distribution of different surface sediments, and other factors likely to affect cross-beach mobility. Airfield suitability maps were made to indicate the distribution within enemy territory of candidate areas for the rapid construction of airfields. After the invasion, between June 7 and August 13, 1944, 20 airstrips, mostly 1,100–1,500 m in length, were completed in the British occupied area of Normandy. Geological information was used to guide the systematic development of road metal. Initially, weak Jurassic limestones were quarried, as at Creully; later, stronger Paleozoic quartzites were worked, as at Mouen, southwest of Caen. Stone produced by the Royal Engineers in Normandy quickly rose to a peak monthly total of more than 140,000 tonnes in August 1944. Water supply intelligence and the control of well siting and drilling were geologist's responsibilities. In 1st Corps area, about 50 water points were established, with 12 operational at any one time. Water in Normandy was obtained largely from rivers and existing wells, supplemented by 33 new boreholes.

Geologists were also used to assess the effects of aerial bombing; soil conditions affecting cross-country vehicular movement; ground conditions for river crossings; and the nature of the sea floor beneath the English Channel. Normandy thus provides a case history of British military geology "par excellence."

INTRODUCTION

Historical aspects of Operation Overlord, the allied amphibious assault on the coast of northern France on June 6, 1944 (D-Day), and the subsequent three-month battle for control of Normandy, have been extensively documented (e.g., by Montgomery, 1946; Ellis et al., 1962; Hastings, 1984; Collier, 1992). Yet few of these accounts mention the role of geology and of military geologists in influencing the course of events. Kaye (1957), Snyder (1957), and Simon (1957) have provided a brief American perspective. Shotton (1947) has provided a far briefer British one, amplified only recently by Rosenbaum (1990), Rose and Hughes (1993), Rose and Rosenbaum (1993b), Pareyn (1994), and Rose and Pareyn (1994; 1995; 1996a, b).

Rose, E. P. F., and Pareyn, C., 1998, British applications of military geology for 'Operation Overlord' and the battle in Normandy, France, 1944, *in* Underwood, J. R., Jr., and Guth, P. L., eds., Military Geology in War and Peace: Boulder, Colorado, Geological Society of America Reviews in Engineering Geology, v. XIII.

It is the role of British geologists, especially in the Anglo-Canadian sector of operations, that is more comprehensively reviewed here. Unlike all other major operations in World War II in which American troops took part or planned to take part, the engineers were not furnished with folios prepared by the Military Geology Unit based upon the U.S. Geological Survey. For "the Normandy landings . . . the work was done by British geologists" (Anonymous, 1945c, p. 355). Yet in contrast to the armies both of America and of Germany, which were supported by relatively large military geological organizations (Hunt, 1950; Rose, 1980; Rose et al., 1996), Britain employed only a few military geologists. Their success was thus influenced to a greater extent by individual qualities. Brief biographical details provided here indicate these in terms of academic excellence and professional experience.

Units of measure used below are expressed both as Imperial units (for consistency with historical sources) and as metric units (in accordance with current international practice).

PLANNING

According to Shotton (1947), effective planning began approximately a year prior to the invasion. Major-General [Sir] J. D. Inglis, chief engineer of 21 Army Group, the allied invasion force, recorded that (Inglis, 1946, p. 177–178):

> At that time, the plan was to land on the beaches of the Cotentin Peninsula with the immediate object of capturing Cherbourg. Aeroplane photographs, however, showed that the peninsula was hilly in most parts and full of small fields separated by banks and hedges. In such country, it would obviously be very difficult to guarantee the early completion of a large number of airfields. In fact, it would be extremely difficult even to select with any certainty potential sites for airfields. It was this factor which largely turned our eyes towards the beaches between Caen and Carentan. We had, fortunately, long appreciated the importance of geology in modern war, and at that time had the services of Professor King, who had been with us since the beginning of the war . . . Amongst other extremely valuable advice Professor King pointed out that between Caen and Bayeux there was a patch of country which was not only gently undulating but also possessed a top soil which was particularly suitable for airfields because of its excellent drainage qualities. This was, in fact, one of the main factors which led to the selection of the beaches eventually used.

W. B. R. King, professor of geology in the University of London at the start of the war, and in the University of Cambridge from October 1943, effectively served as geological adviser to 21 Army Group and the British army's engineer-in-chief (Pakenham-Walsh, 1958, p. 323). With distinguished military geological experience dating from the First World War, he was granted a Regular Army Emergency Commission in the Royal Engineers in 1939, quickly promoted to temporary major, and in 1943 promoted to lieutenant-colonel (Rose and Hughes, 1993; Rose and Rosenbaum, 1993a, b). In 1943 he was joined and soon succeeded on the 21 Army Group planning staff by Captain (later Major) F. W. Shotton, returning from active service in North Africa (Rose and Rosenbaum, this volume). (After the war, Shotton became a professor of geology first at the University of Sheffield and soon afterward at the University of Birmingham.)

Shotton joined a team led by the distinguished physicist and crystallographer J. D. Bernal, a professor of physics at Birkbeck College in the University of London who served during the war as a scientific adviser to the chief of combined operations. The team's activities included detailed study of the literature pertaining to the geology and history of the Normandy coast; laboratory study of beach processes; examination of aerial photographs; and study of comparable regions in England (Bernal, 1955; Arditi, 1994).

Their work was complemented by that of another unit. "As early as 1940, Churchill had urged the creation of an Inter-Services Topographical Unit [Department]. Headed by Professor Freddie Wills, a retiring don, and Colonel Sam Bassett of the Royal Marines, the unit's first remit, in the quiet haven of the Ashmolean Museum, Oxford, was the topography of all coastal zones—but by 1943 the war had narrowed this down to one stretch of coastline" (Collier, 1992, p. 104). By then ISTD contained several geologists, at least three commissioned as Royal Engineer officers (Rose and Hughes, 1993; Rose and Rosenbaum, 1993b). J. L. Farrington, a graduate of the University of British Columbia, had pre-war experience in the geological interpretation of aerial photographs (Farrington, 1936). T. C. Phemister, then professor of geology and mineralogy at the University of Aberdeen but formerly an associate professor at British Columbia, was also used in air photographic interpretation. Additionally, he "made contact with colleagues in France, particularly at the University of Rennes" (Tait, 1984, p. 177). Geologists at Rennes were supportive of the Resistance, and the university awarded Phemister an honorary doctorate in 1947, soon after the end of the war. The third "sapper" was D. R. A. Ponsford, an experienced exploration geologist who was to serve postwar as a district geologist with the British Geological Survey.

GEOLOGY OF NORMANDY

In 1944 the Geographical Section, general staff, of the British army copied and printed in full color, for general allied distribution, all the 1:80,000 scale geologic maps of the Carte Géologique de la France (Kaye, 1957). Sheets published between 1895 and 1926 cover Normandy. (More recent maps, with descriptive memoirs, are now available at 1:50,000 scale.) Publications by Bigot (1930) and others describe the local geology; Doré et al. (1987) provide a postwar regional guide.

To the west, the Cotentin peninsula with Cherbourg at its northern extremity forms part of the Amorican massif. Folded Paleozoic and Precambrian (Brioverian) rocks are continuous with those of Brittany (Fig. 1). In spite of the general flatness and low altitude (generally <100 m), the surface undulates as a consequence of stream erosion (the main rivers in deep valleys) and differential weathering (resistant strata forming distinct ridges). The area is wooded and characterized by small fields separated by banks and hedges—the bocage.

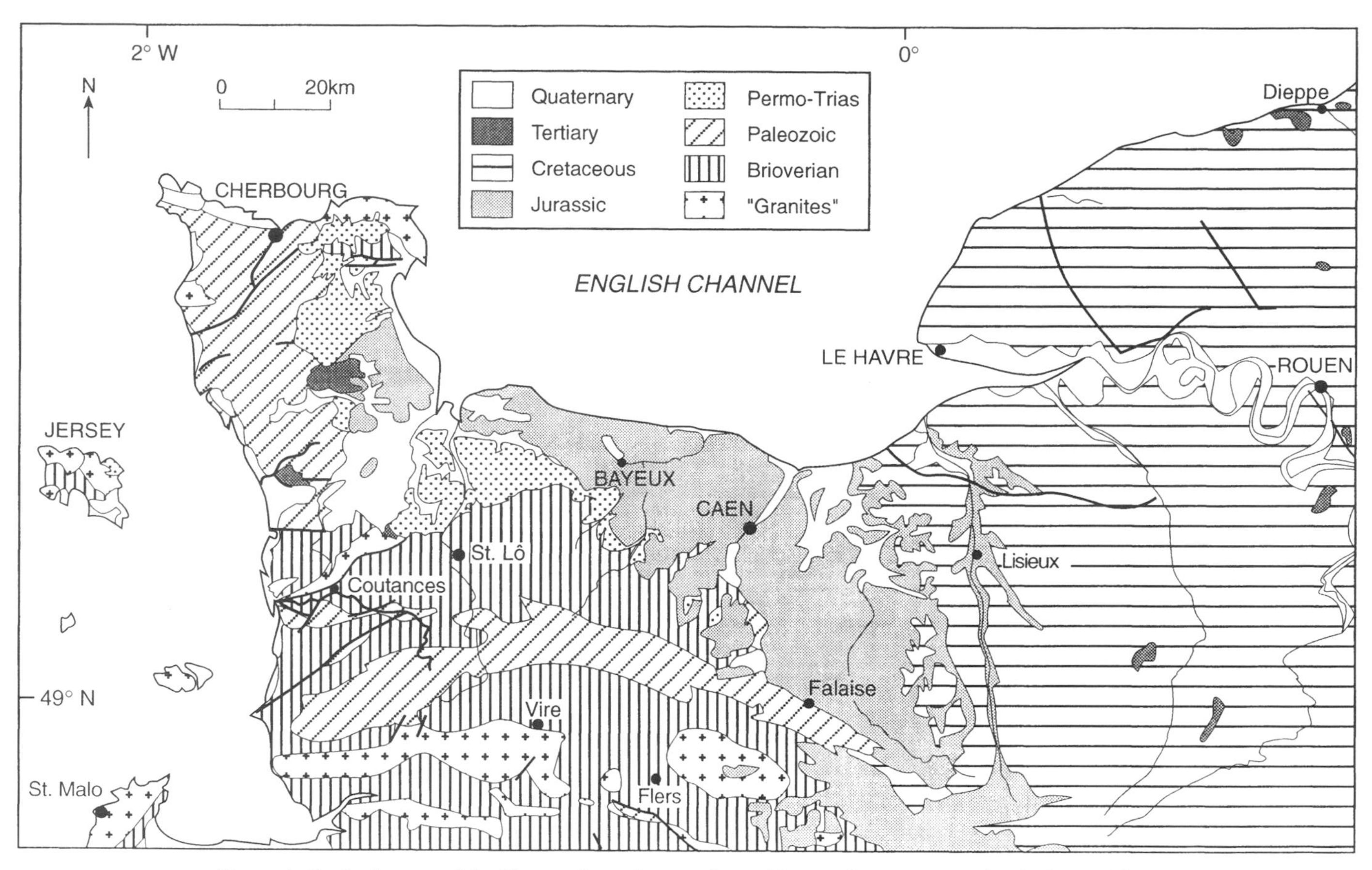

Figure 1. Geologic map of the Normandy region, northwest France. Quaternary omitted where <10 m thick. (After Doré et al., 1987.) (Figs. 1, 3, 5, 8, 9, and 11 are from Rose and Pareyn [1995], courtesy of Blackwell Publications.)

To the east, the plateau of Calvados forms part of the western margin of the Paris basin. Mesozoic rocks, ranging from the Trias to the Cretaceous and including a nearly continuous Jurassic succession, unconformably overlie the Precambrian-Paleozoic basement and dip gently northeastward at about 12 m per kilometer—the nearly horizontal strata are disrupted by minor folds and faults. Most of the plateau between Bayeux and Caen is underlain by Middle Jurassic (Bathonian-Bajocian) shallow-water limestones with subordinate marls (Fig. 2)—a sequence some 100 m thick—with Upper Jurassic clays and limestones cropping out farther to the east and north.

Various Quaternary deposits lie unconformably upon the Mesozoic. The "Limons des plateaux" form a widespread cover of loess, 2.5 m thick on average but up to 5 m thick locally, deposited during the dry phases of the Weichselian glaciation (Rioult et al., 1989). It was this well-sorted, fine-grained (median 35 microns), exceptionally homogeneous deposit that provided the qualities required for rapid construction of military airfields (King, 1951a): excellent drainage in wet weather; firm, even surface in dry weather with lower shrinkage and cracking than a "true clay"; and development of a widely consistent arable soil cultivated in large open fields without hedges or ditches. (The anticipated summer-season dry weather conditions were significant because when wet, weathered loess "puddles-up" to a fluid mud almost impassable to concentrated or heavy traffic.)

BEACHES

Once the Calvados coast (Fig. 3) had been selected for invasion, the character of the beaches and coastal cliffs and of the soils and topography immediately inland where the first phase of the battle would be fought were of particular concern (Butler, 1947). It was important that vehicles followed firm routes over the actual landing beaches (Fig. 4). Accordingly, the beaches were analyzed in detail not only with regard to configuration and slope but also to the distribution of the patchy peat, clay, sand, and shingle that formed the surface. Maps on the scale of 1:5,000 were prepared from existing records, aerial photography, and secret ground reconnaissance.

Bernal (1955) has described detailed literature searches in this respect. Publications by Edmond Hue (e.g., Hue, 1916; 1938) were especially valuable, for they documented the position of erratic blocks of rock and patches of subsurface peat, some of which were only intermittently exposed by storms and exceptional tidal conditions.

Aerial photographs provided information on natural as well

as artificial obstacles. Shotton, as noted by Rose and Rosenbaum (1993b), recounted after the war how patches of dark peat could be recognized after storm movement of sand, and how the load-bearing properties of some beach areas could be estimated from the depth of wheel marks left by German carts transporting defense stores. Some photographs were taken obliquely by aircraft flying along the beach at altitudes as low as 50 feet (16 m).

To confirm details, the beaches were covertly sampled. Midget submarine X-craft operated by COPP (Combined Operations Beach Reconnaissance and Assault Pilotage Parties) carried commando-trained volunteers close to the beaches. Their task was to swim ashore and auger soft sediment and collect samples of "stone" (Scott-Bowden, 1994). According to Evans (1945, p. 305) "a party armed with spades were landed one night towards the close of this long period of preparation. The necessary samples were obtained without them being detected, and these were rushed back to [the British Geological Survey at] South Kensington where they were analyzed and pronounced suitable for landing our mechanised armies on the day scheduled as D-Day."

Testing and training prior to D-Day were conducted on the Brancaster beaches of Norfolk, eastern England, where similar geological conditions had been identified. Beach processes were studied in "numerous laboratories," particularly by R. A. Bagnold (Bernal, 1955, p. 537), who achieved distinction both as a soldier and as an expert on the movement of sand (Rose and Rosenbaum, 1993b; Rose and Pareyn, 1996a). General beach morphology—an offshore wave-cut platform of solid rock cut intermittently by deep navigable passages, bordered landward by patches (often extensive) of plastic clay and peat overlain by a veneer of modern beach sand of variable thickness, and finally a storm beach of pebbles behind which modern fen-like deposits were forming—was explicable in terms of the Pleistocene geological history of the region (King, 1951a), especially changes in relative sea level. The rock platform, the low-lying coastal areas of Juno and Gold beaches, the cliffs of Arromanches, and the contrasts with Omaha and Utah beaches have all recently been illustrated and described by Michel (1991), whose book relates variation in beach and cliff morphology to the local geology, but such a convenient publication was not available at the time.

AIRFIELDS

The total time individual fighter aircraft could spend over Normandy, if based in the United Kingdom, was severely limited by the need to return for refueling (Wilson and Nowers, 1994; Mitchell, 1994). Air superiority over the battlefield was to be provided by Spitfires and Hurricanes for protection against enemy

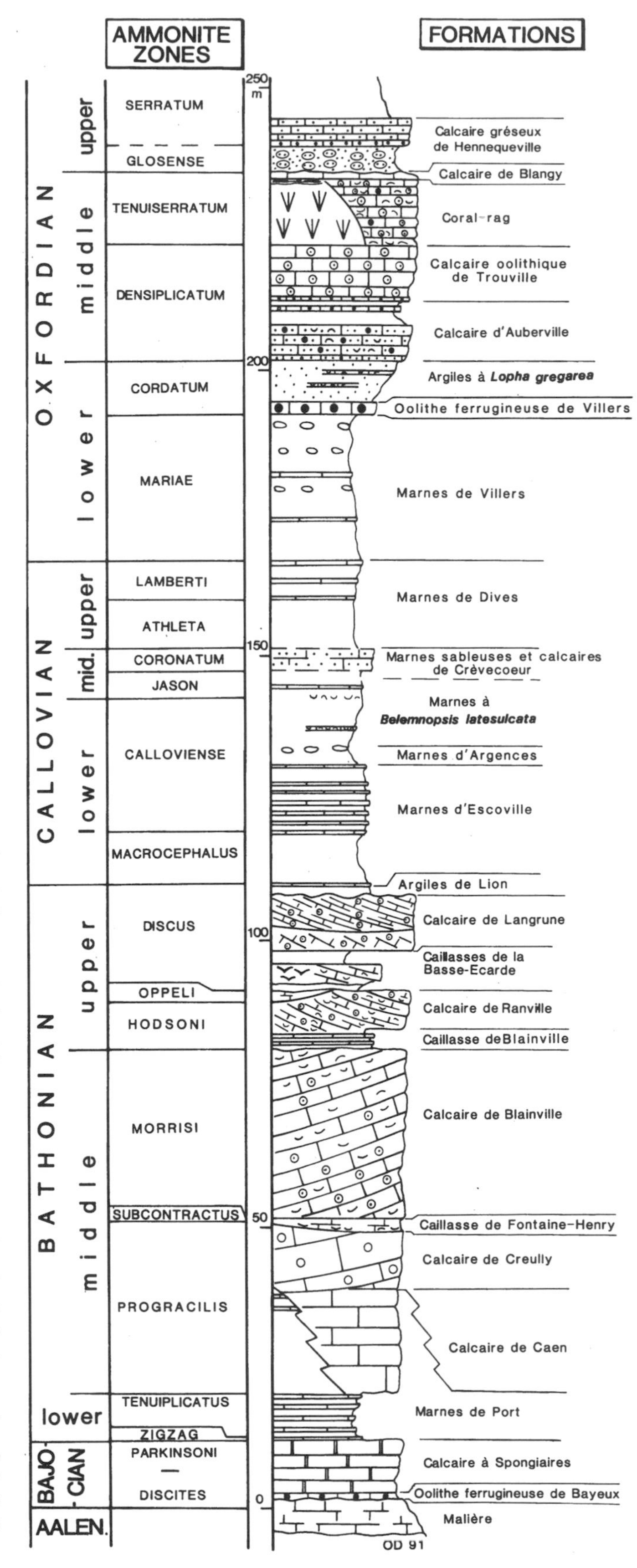

Figure 2. Generalized Middle-Upper Jurassic succession in Normandy, about 250 m thick in total, indicating relative weather resistance. Limestones (calcaires) of Bathonian age underlie most of the plateau between Bayeux and Caen, the Anglo-Canadian invasion sector (Fig. 3). (From Rioult et al., 1991.)

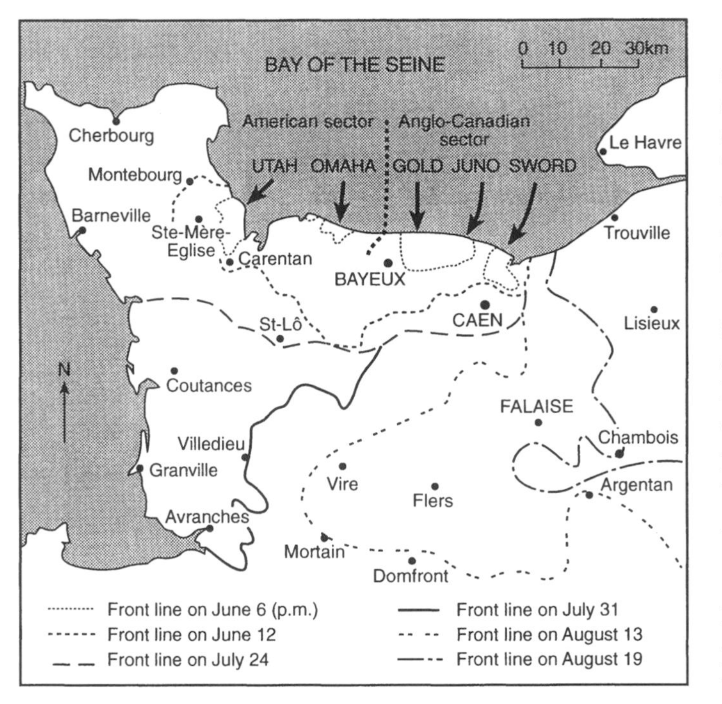

Figure 3. Map indicating the American (Utah, Omaha) and Anglo-Canadian (Gold, Juno, Sword) invasion beaches and the Allied advance across northwestern France through Normandy, 1944. (After Desquesnes, 1993.)

aircraft; rocket-firing Typhoons for offensive action against enemy tanks and transport in the immediate battle area; and Mosquito and Mustang aircraft for operational reconnaissance. One hundred miles (160 km) from base was the estimated effective limit of operational use. It was therefore essential that temporary landing strips be constructed within Normandy at the earliest possible stage of battle:

(1) Emergency landing strips: flat ground, roughly graded, with a minimum length of 1,800 ft (550 m).

(2) Refueling and rearming strips: a minimum length of 3,600 ft (1,100 m), and two marshaling areas.

(3) Advanced landing grounds: a minimum length of 3,600 ft (1,100 m) for fighters and 5,000 ft (1,520 m) for fighter bombers, with dispersal facilities for 54 aircraft.

To build them, one airfield construction wing Royal Air Force and five Airfield Construction Groups Royal Engineers were created and trained, the latter by converting existing Royal Engineer and Pioneer Corps units. Raising and training started as early as the summer of 1942. Each of the Royal Engineer groups comprised a headquarters plus two so-called Road Construction Companies, both of these having a pioneer company attached and under operational command. The plant available included crawler tractors, motor graders, scrapers, rollers, dump trucks, and transporters. Since members of these units had little experience in the actual construction of airfields, practical exercises were held near Fairford in southern England, where conditions of topography and geology were similar to those expected during the early stages of conflict in Normandy.

The initial plan called for one emergency landing strip to be available by the end of D-Day, the first refueling and rearming strip by D+3, and five advanced landing grounds by D+8. In the event, the first emergency landing strip was operational on June 7, five more airfields were progressively finished by June 17 (D+11), and a total of 11 had been taken over for Royal Air Force operation by the end of the month. The total reached 20 airfields by August 13 (Table 1 and Fig. 5).

Sites for a number of strips had been preselected by the planning staff, using geologist expertise. Construction groups were authorized to change locations "if those selected proved unsuitable. In the event the intelligence proved very accurate and no important changes had to be made" (Pakenham-Walsh, 1958, p. 374). This is perhaps remarkable in that the characteristics of Normandy soil were very imperfectly known at the time. Detailed maps of the geomorphology and the superficial deposits (Journaux, 1971; 1973) were not available until long after the war.

Advanced parties from the Airfield Construction Groups landed on D-Day, with the main bodies of each unit plus their plant, equipment, and stores following during the next two to three days. A vivid personal account of the construction of B3, the advanced landing ground at Sainte-Croix, has been given by Mitchell (1994).

To form a firm runway surface able to carry loads, topsoil was removed by scraper and subsoil compacted at optimum moisture content using sheep's foot and wobble wheel rollers. As predicted, the loessic soils drained rapidly in wet weather. However, the combination of fine weather with the agitation of fine-grained loessic soils produced dense clouds of dust during early operational use. Some airfields were sprayed with water to minimize the dust hazard to visibility and engine wear. Where possible, later airfields were constructed with minimum disturbance to any grass surface, accepting a considerable roughness in dispersal areas, but the only real solution was to operate from surfaced runways. Square Mesh Track (SMT) was in general use as a ground surfac-

Figure 4. Men and vehicles landing across a beach in Normandy, June 7, 1944. (Copyright, The Imperial War Museum, London: photograph CL57.)

TABLE 1. BRITISH AIRFIELDS IN NORMANDY 1944*

Code Number	Locality Name	Length (m)	Completion Date
B1	Asnelles	550	7 June
B2	Bazenville	1,520	13 June
B3	Sainte-Croix	1,100	11 June
B4	Bény-sur-Mer	1,100	15 June
B5	Camilly	1,520	17 June
B6	Coulombs	1,520	16 June
B7	Martragny	1,100	26 June
B8	Sommervieu	1,100	21 June
B9	Lantheuil	1,100	21 June
B10	Plumetot	1,100	24 June
B11	Longues	1,100	26 June
B12	Ellon	1,520	18 July
B14	Amblie	1,100	3 July
B15	Ryes	1,100	6 July
B16	Villons-les-Buissons	1,100	31 July
B17	Carpiquet	1,600	8 August
B18	Cristot	1,100	6 August
B19	Lingèvres	1,520	8 August
B21	Sainte-Honorine-de-Ducy	1,520	13 August

*Additionally, an unnumbered 1,000 by 100 yard (915 by 92 m) airstrip was completed at Saint Aubin d'Arquenay by July 19 for the removal of gliders used by 6 Airborn Division on D-Day. B13 was not built by U.K. forces; B20 was built by First Canadian Army.
See Figure 5 for diagram of airstrip orientation and geographical position.
A second (untracked) strip was constructed parallel and adjacent to the first (tracked) strip at B3, B5, B6, B7, and B10, to preserve the tracked strip for use in wet weather, provide a wider strip for quick take-off when urgent scrambling was necessary, and make it more difficult for bombing to put the airfield out of commission.
From Rose and Pareyn, 1996a; basic data from Panet, 1945; Anonymous, 1944, 1945a–d.

ing material (Fig. 6). At Sainte-Croix it was laid over hessian (burlap) soaked with fuel oil. Additionally, Prefabricated Bitumenized Surfacing (PBS) had been developed by the Canadians (from a British origin) as an aid to rapid surfacing (Fig. 7). It was laid from long, continuous rolls, and the strips were then sealed together to maintain the airfield's bearing capacity. Several bridgehead airfields surfaced with this material had been planned and should have been nearing completion by June 11, but could not even be started since their sites were still under enemy fire. By the end of August, 23 airfields were in use, 10 of which had runways made with SMT, 2 with PBS, 2 with repaired concrete, and 9 without any special runway material (Buchanan, 1953).

In his memoirs, General Karl Koller, chief of the German Air Staff (cited in Mitchell, 1994, p. 45) noted that "there are many reasons why Germany lost the War . . . what was decisive in itself was the loss of air supremacy." From D-Day to May 5, 1945, this was facilitated by construction or repair of 125 airfields, in total length equivalent to 2,020 mi (3,250 km) of 20-ft (6-m) wide roadway—effectively 1,260 mi (2,028 km) of new graded but unsurfaced road, 360 mi (579 km) of surfaced road, 260 mi (418 km) of repairs to cratered road, and 140 mi (225 km) to cratered earth road. Approximately 17,000 troops of all ranks were employed almost continuously on airfield construction in 21 Army Group, 6,000 in army units, and the remainder in Royal Air Force works units (Anonymous, 1945a).

The deputy chief engineer of Airfield Construction recorded (Panet, 1945, p. 1) among the lessons learned:

> The value of detailed topographical and geological study in selecting airfield sites in the beach-head area was fully proved . . . . Geological study enabled calculated risks to be taken in forecasting the rate of import of airfield surfacing stores. Based on geological information provision was made for surfacing only 50% of airfields in the British sector as against 75% in the U.S. sector, where more clay was expected. This enabled the stores demand in the British sector to be reduced by over 400 tons a day for the first 30 days of invasion, when every ton was of great importance. Even with this 50% reduction, airfield stores made up nearly 25% of the total planned tonnage of Engineer stores to D+30 . . . . The most valuable forms of specialised Int[elligence] were the 1/100,000 geological overprints . . . prepared by the Geologist.

QUARRIES

In the 1939–45 war armies achieved a high state of mechanization. In many instances the network of roads behind the forward troops could not cope with the traffic, and roads had to be strengthened or widened, or new roads had to be built. On D-Day, 8,851 vehicles landed in Normandy. By D+50 there were 152,499 military vehicles in an area little more than 20 mi broad and 10 mi deep (32 by 16 km) (Anonymous, 1945b). The weight of some of the vehicles and the passage of those equipped with caterpillar tracks proved destructive to road surfaces, and road maintenance became a serious problem. The armies, too, needed hard standings for their stores depots, vehicle and gun parks, and runways for their supporting aircraft. All this meant a requirement for aggregates—in large quantities (Williams, 1950).

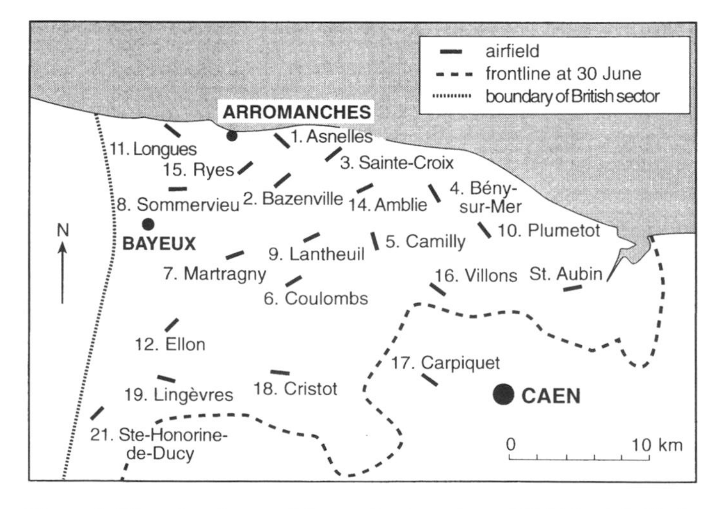

Figure 5. Map showing position and orientation of British airfields in Normandy, 1944. For lengths and completion dates, see Table 1. (After Anonymous, 1945a; Rioult et al., 1994.)

Figure 6. A British landing strip in Normandy being constructed with Square Mesh Track, on August 10, 1944. (Copyright, The Imperial War Museum, London: photograph CL710.)

Figure 7. Landing strip in Normandy being constructed from Prefabricated Bitumenized Surfacing. Note the flat surface of the Calvados plateau, ideal for temporary airfield construction, and the dust being generated from dry loessic soil by a Typhoon taxiing across an "unpaved" area. (Copyright, The Imperial War Museum, London: photograph CL468.)

Stone in Normandy and the subsequent campaign was provided by the Quarry Group Royal Engineers, formed in March 1944. Commanded by a lieutenant-colonel, this comprised a Headquarter Group and five Quarrying Companies, initially with a strength of just over 900 troops of all ranks. The constituent companies had been formed at various dates during 1940–1943, largely by recruitment from the quarrying industry in the United Kingdom. Each company comprised four officers plus 173 men of other ranks (including 52 quarrymen, 50 drivers, and various tradesmen), supplemented by approximately 100 unskilled men (from the Pioneer Corps, prisoners of war, or local civilians). The plant and equipment scale approximated to that listed in Table 2.

The first Quarrying Company (No. 853) began landing on the beaches at Courseulles on June 18 with its equipment and within a few days was quarrying near Creully (Fig. 8). The other companies soon followed (No. 856 on June 29), and by August 6 the whole group was in Normandy, less the major part of one company torpedoed crossing the Channel. In June, 12,905 tons (>13,000 tonnes) of rock were produced, rising to 107,270 tons (3,460 per day) in July, with up to 2,000 tons per day from Creully. In August, output in France peaked at 145,498 tons (147,826 tonnes) (Fig. 9), 4,694 tons per day.

Initially, production was started in existing quarries at Creully, Esquay, Douvres, Reviers, Carpiquet South, Blay, and Tilly-sur-Seulles. Later a new quarry was opened at Ver-sur-Mer. Apart from Esquay, a Quaternary sand deposit, all these quarries were within Jurassic limestones. They produced rock that was weak and generally unsuitable for road repair (Fig. 10). However, within the restrictions imposed by the front line and sector boundaries (Figs. 3, 8), there were no alternatives. At Creully and comparable sites the rock quarried was a Middle Jurassic shallow-water limestone with well-developed cross bedding. Cementation was variable, and horizons previously worked for

TABLE 2. MAIN ITEMS OF PLANT AND EQUIPMENT FOR QUARRYING COMPANY, ROYAL ENGINEERS*

Item	Number
Excavators, $^5/_8$ or $^3/_4$ cubic yard	
Base machines	3
Face-shovel equipments	3
Dragline equipment	1
Angle dozer (7-foot blade)	1
Compressors, mobile	
200 cubic feet capacity	7
100 cubic feet capacity	1
Rock drills	
Drifters, 4-inch	3
Jackhammers, 50–60 lb.	3
Jackhammers, 25–35 lb.	8
Drill sharpener and furnace	1
Drill steels	
1$^1/_4$-inch, round	3 tons
1-inch, hexagonal	6 tons
$^7/_8$-inch, hexagonal	6 tons
Crushing and grinding units, mobile	
10 tons/hour	5
25 tons/hour ("Iowa")	1
Vehicles	
Dumpers, 3-cubic yard (Muirhill)	12
Tippers, 3-ton	20
Trucks, 15-cwt	2
Van, utility	1

*From Williams, 1950, with Imperial units of measure as in the original.

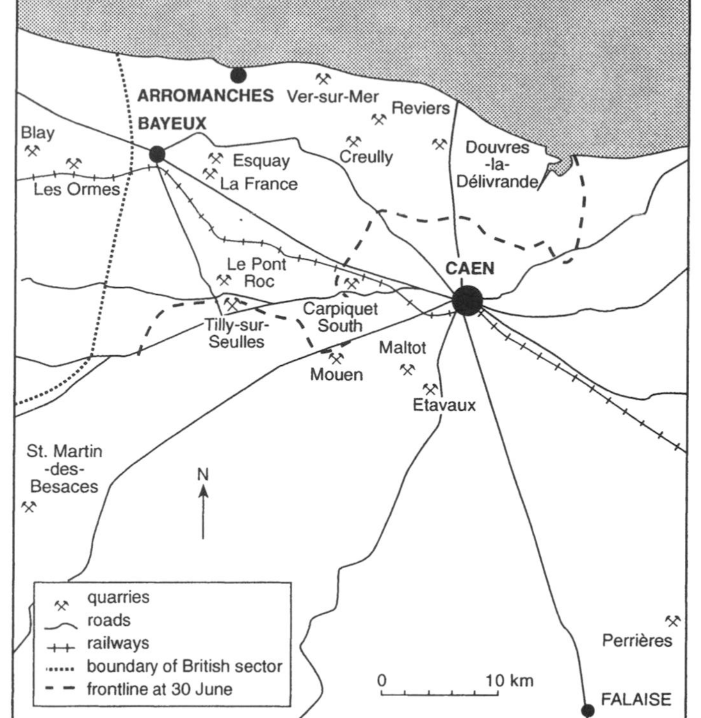

Figure 8. Map showing positions of main roads and military quarries in Normandy, 1944; cf. Figures 1 and 2 for geologic setting. (After Williams, 1950.)

building stone or rough walling lacked continuity. These strata dipped gently to the northeast (Fig. 1), exposing Lower Jurassic strata toward the southwest margin of the Jurassic outcrop. Quarries in these Liassic limestones were developed at Blay, Les Ormes, and Tilly-sur-Seulles—working well-cemented limestone beds 0.10–0.35 m thick alternating with thinner marls.

Good roadstone became available only with the capture of Mouen, southwest of Caen, toward the end of July. This provided access to strong, steeply dipping Cambrian quartzites flanking a northwest prolongation of the Precambrian (Brioverian) metasedimentary outcrop.

Weak rock quarries were set aside as those yielding stronger rocks became accessible with the progress of battle. Creully, Douvres, Reviers, Ver-sur-Mer, Blay, and Tilly were successively abandoned and the quartzite or well-cemented Paleozoic sandstone quarries at Perrières, Etavaux, Maltot, and St. Martin-des-Besaces brought to production. Choice of these particular quarries from the many that became available with the expansion of the beachhead was dictated mainly by considerations of proximity to areas most needing stone and the rapidity with which production could be started. The Perrières quarry was the best (in a strong Ordovician sandstone, but Grès Armoricain rather than "Grès de May" as claimed by Williams, 1950). It was well served by road and rail according to Williams (1950), although unlike the nearby Grès Armoricain quarry at Vignats, there never has been a direct rail link. Production began at the end of August and continued until May 1945, supplying British and American forces with nearly 80,000 tons (81,000 tonnes) of stone, nearly all crushed, from this source. However, as the Allied armies advanced, so French quarries were replaced as sources of aggregate by those in Belgium and finally Germany (Fig. 9). In total, 49 different quarry sites were operated during the 11-month campaign in France, Belgium, Holland, and Germany. In this period the group supplied a total of 1,800,000 yd^3 (1,376,000 m^3) of crushed rock, with an average output of about 5 yd^3 (3.8 m^3) per man per day.

"Of obvious importance in planning stone production and the allocation of Quarry Group's plant and man-power to the best advantage was knowledge of the stone resources of the areas occupied and likely to be occupied by 21 Army Group. Field reconnaissance yielded much useful information of a detailed nature, but for an overall picture geological maps proved invaluable" (Williams, 1950, p. 223).

WATER SUPPLY

In the planning stage of Operation Overlord, geological activities included the preparation of water intelligence maps, and water supply intelligence continued and developed after D-Day, together with operational control of well drilling (Shotton, 1947). Maps on the scale of "1/50,000 or thereabouts" (King, 1951b,

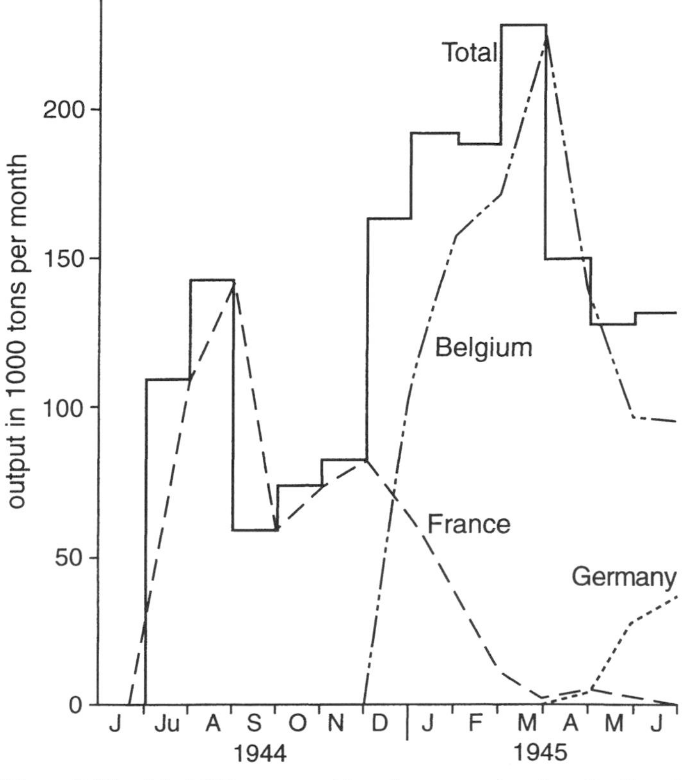

Figure 9. Monthly 1,000-ton quantities of stone produced or distributed by the Quarry Group, Royal Engineers, from June 1944 to June 1945, in total (solid line) and progressively from sources in France, Belgium and finally Germany. (After Anonymous, 1945a.)

Figure 10. Military quarrying of Middle Jurassic limestone near Carpiquet, August 1944. Calcaire de Caen (Bathonian); cf. Figure 2 for stratigraphic context. (Copyright, The Imperial War Museum, London: photograph CL811.)

p. 115) were prepared for all the bridgehead areas of Normandy before D-Day, showing the main aquifers divided into three zones: (1) where small springs might be expected but where boring was unlikely to produce large supplies; (2) the main outcrop area with, if possible, water-table contours; and (3) subsurface contours on the top or base of the aquifer to indicate depth of boring. A distinction was made between ground water expected to be of good quality and that expected to be saline. Explanatory notes were provided, including geological terms. As acknowledged by King, these maps were a development of the Ground Water Inventory Maps of the Ground Water Provinces of the United States.

By D+1 the existing water supply network in the British-held area had virtually ceased to operate because of failure of the electricity supply to the pumps from the power station in Caen, still in German hands. However, west of the River Orne (Fig. 11) water was available from small rivers and existing deep wells. A network of water purification and storage points was speedily established. These comprised mobile pumping sets, mobile filtration and chlorination plants, and sectional steel storage tanks. Output varied widely—10,000 gal (~45,500 l) per day was considered the minimum realistic figure—but by noon on D+1 one water point with a capacity of 20,000 gal (91,000 l) per day was open, with steadily increasing output in subsequent days (Fig. 12). Eight water points had been established in the area covered by 1 Corps by D+4, and these were supplemented by local wells where possible. During the first week, each division required ~50,000 gal (227,000 l) per day from Royal Engineer sources (Wilson and Nowers, 1994). Ultimately about 50 water points were established, 12 of which were operated at any one time—adequate traffic circuits were the most important operational factor.

Existing wells were supplemented by new wells. No. 8 Boring Section Royal Engineers was deployed within Line of Communication Troops during the build-up phase following the initial

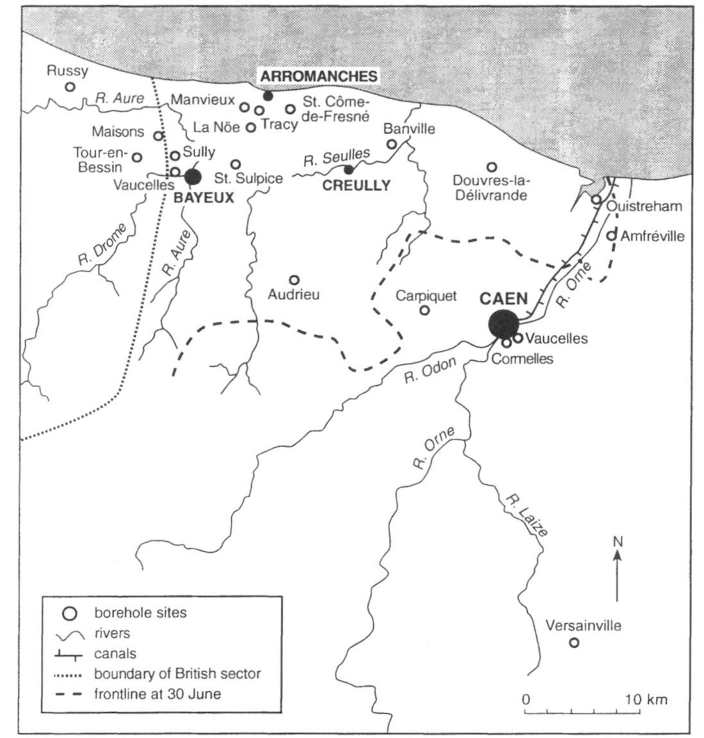

Figure 11. Map showing positions of main rivers and military borehole sites, Normandy 1944. See Table 3 for borehole depths and yields: cf. Figures 1 and 2 for geologic setting. (Borehole data from Shotton, 1945; Bigot, 1947.)

Figure 12. Royal Engineers water point on the River Seulles at Creully, which supplied on average 80,000 gal (364,000 l) per day during the battle. (Copyright, The Imperial War Museum, London: photograph B6988.)

assault (i.e., from D+5 onward). This unit was commanded by a geologist, Lieutenant A. K. Pringle. A first class honors graduate of Glasgow University, Pringle had spent the early years of his career exploring for oil and minerals in the Middle East, Australia, and New Guinea (Munro, 1985), but had returned in 1940 to volunteer for war service. Commissioned initially in the Royal

Electrical and Mechanical Engineers, he served as a radio officer (radar) until transferred to the Royal Engineers and to well drilling duties in England, France, and Belgium (E. G. Bell, personal commun., 1994). He returned to civilian life in December 1944, first to the Anglo-Iranian Oil Company, then to Aberdeen University, and finally to the chair of Applied Geology in Strathclyde University, Scotland.

Borehole summary data were recorded by Shotton (1945) for the Inter-Service Topographic Department, and after the war given in a paper by Bigot (1947). Shotton numbered the boreholes 1–35; nos. 8 and 33 are missing—presumably the two Belgian boreholes mentioned in his covering letter.

Six shallow holes (his nos. 1–6) were put down in alluvium near the River Drome (Fig. 11), in ground unfavorable for deep boreholes. Five of these were intended for hospital supply along a pipeline system that otherwise would have required filtration plants at the river. The gravel of the alluvium effected a natural filtration, reducing the number of filtration pumps and plants required. Yield possibilities were limited, and the holes were not used when the demand for water had grown to about 1 million gal (4.5 million l) per day.

Two boreholes (Shotton's nos. 27, 28) penetrated the Cretaceous, to the east of the area shown in Figure 11. Prospects were calculated to be problematic in the Upper Chalk. To make certain of obtaining a sufficient yield, it was determined to drill 500 ft (152 m) or more into the Lower Chalk (Cenomanian). The first hole was abandoned as crooked at 300 ft (91 m), the second abandoned at 377 ft (115 m) due to operational moves. Both wells were virtually dry.

Most boreholes, however, were put down in Jurassic strata (Table 3; compare Fig. 11 with Fig. 1). One (that at Le Haut d'Audrieu) Shotton recorded as being sunk by Second Canadian Drilling Company, which did not have geological advice. By implication the others were sunk by 8 Boring Section, under his control. All penetrated Middle Jurassic oolitic and/or bioclastic limestones and subordinate clays (the sequence generally Bathonian and Bajocian in age), most boreholes extending into the underlying Liassic marly limestones. One (Versainville Château) reportedly terminated in Precambrian (Brioverian) St. Lô Phyllites beneath the Lias.

The importance of water, and of the Royal Engineers, in contributing to the Allied victory in Normandy has now been emphasized by Rioult et al. (1994).

TABLE 3. ROYAL ENGINEER BOREHOLES FOR GROUND WATER THROUGH JURASSIC STRATA IN NORMANDY, 1944*

Locality	Depth (m)	Yield (m^3h)
1. Sully	34.75	18.00
2. Amfréville	100.60	3.00
3. Tracy-sur-Mer (La Noë)	97.50	0.45
4. St.-Côme-de-Fresné (Buhot No. 1)	105.15	11.80
5. Maisons	26.00	13.50
6. Tracy-sur-Mer (La Rosière)	38.00	18.00
7. St.-Côme-de-Fresné (Buhot No. 2)	24.40	13.60
8. Banville	45.70	14.50
9. Douvres-la-Délivrande	39.60	22.70
10. Tour-en-Bessin (Grivilly)	20.70	nil.
11. Manvieux	45.70	nil.†
12. Ouistreham	33.00	31.90
13. Tour-en-Bessin (Le Coudray)	25.00	9.10
14. Audrieu (Le Haut d'Audrieu)	58.00	3.40
15. Caen (Vaucelles)	73.25	28.70
16. Vaucelles	7.00	13.60
17. Cormelles	62.50	14.00
18. Audrieu (Le Bas d'Audrieu)	30.50	9.00
19. Versainville (Château)	70.10	4.50
20. Russy (Russy No. 1)	28.35	3.20
21. Russy (Russy No. 2)	18.30	2.70
22. Russy (Russy No. 3)	18.30	2.70
23. Carpiquet	81.40	4.10
24. St. Sulpice	15.25	1.34
25. Audrieu	45.70	18.20

*From Rose and Pareyn, 1996b; basic data from Shotton, 1945; Bigot, 1947. According to Shotton, most yields were calculated from pumping tests carried out during 24 hours.
See Figure 11 for geographic position and lower part of Figure 2 for generalized Middle-Lower Jurassic sequence penetrated by most boreholes.
†Abandoned before completion due to operational necessities.

CONCLUSION

Shotton (1947) has recorded how the work of geologists associated with the invasion of northwest Europe fell naturally into two periods: the planning stage, approximately a year prior to the invasion, and the operational stage of the following 11 months.

During the planning period, the two most important lines of work were the study of invasion beaches and the soil conditions of possible airfield sites. Other geological activities were the study of the cliffs of the invasion area, the provision of information on the foundations of enemy defenses for their effective bombing, the preparation of water intelligence maps, information on sources of road metal, sand, and gravel, the submarine geology of ports, and the detailed study of certain rivers with a view to assault crossings. W. B. R. King had earlier been asked for his opinion on the nature of the submarine floor of the English Channel, across which pipelines were laid to supply fuel to allied forces later in the campaign (Sutton, 1978). Moreover, Simon (1957) has noted that four geologists were used in the industry section of the British Intelligence Centre located at R. A. F. Medmenham, Buckinghamshire, which later became the Allied Central Intelligence Unit. Also, at least one geologist was a member of the Combined Strategic Target Committee, whose function was to select bombing targets.

During the operational period, Shotton landed in France about three and a half weeks after D-Day (written commun., 1989). There he was to provide further water supply intelligence and to control well drilling, provide information on the soil and drainage of possible airfield sites ahead of the advancing armies, and the systematic development of road metal resources (often visiting sites with the Quarry Group's commanding officer, Lieutenant-Colonel A. R. O. Williams, although without direct control over quarrying as such). Additionally, the infantry officer Major (later Lieutenant-Colonel) F. W. Anderson, a geologist seconded from the British Geological Survey, served with and eventually commanded the Research and Experiments Department of the British Bombing Research Mission, including service in Normandy to study the effects of aerial bombing (Rose and Rosenbaum, 1993b). From criteria such as the profile of bomb craters, the crater rim, and the texture of the aureole of material that fell beyond the rim, it was possible to infer soil conditions: Cohesive soils and bedrock showed U-shaped craters, steep rims, and lumpy aureoles; noncohesive sands and gravels showed V-shaped craters, moderate rims, and smooth-textured aureoles; and compressible soils such as peat showed narrow craters, swollen rims, and small aureoles (Stone, 1970). In the absence of other information, bomb crater morphology proved a useful guide to soil characteristics, and so to ground trafficability and to foundation strengths.

Following the breakout from the beachhead on July 25 (D+49), allied armies advanced 400 mi (640 km) in 42 days. This brought new tasks for Shotton (and other geologists, such as D. R. A. Ponsford, who later accompanied him), notably in development of cross-country movement maps and prediction of river bed and approach characteristics for the final attack across the Rhine. Eight months from the breakout, the war ended. British military geologists had contributed in a diversity of significant ways to the final victory.

ACKNOWLEDGMENTS

Grateful thanks are due to G. McKenna, chief librarian and archivist of the British Geological Survey, for drawing our attention to Shotton's summary of Normandy boreholes; to Professor G. Downie (University of Aberdeen), E. G. Bell (University of Strathclyde), and W. Cawthorne (Geological Society of London) for helping to trace wartime career details of Professors T. C. Phemister and A. K. Pringle; to Brigadier E. R. Holmes for correspondence about the Inter-Services Topographical Department; to staff at the Royal Engineers Corps Library, Chatham, the Imperial War Museum, London, and the Public Record Office, Kew, for access to archive material; to L. Blything (Royal Holloway) for preparation of the figures; and to Allen F. Agnew and an anonymous referee for commenting on the manuscript as first submitted.

REFERENCES CITED

Anonymous, 1944, 1 Corps engineer report. Normandy 6th June 1944 to 31st July 1944: Royal Engineers Library, accession no. 2776, 18 sections (unpaginated), drawings A–M, unpublished.

Anonymous, 1945a, The administrative history of the operations of 21 Army Group on the continent of Europe 6 June 1944–8 May 1945: Germany, BAOR, 151 p., appendices A–U.

Anonymous, 1945b, Operation 'Overlord'—An account of the RE works contribution: Chief Engineer's Branch, HQ Second Army; Royal Engineers Library, accession no. 12609, 63 p., 30 appendices, unpublished.

Anonymous, 1945c, The operations of 2nd Army, 21 Army Group, in France, Belgium and Holland 1944–45: Royal Engineers Library, accession no. 8507HQ, 311 p., unpublished.

Anonymous, 1945d, The history of airfield construction [in northwest Europe, 1944–1945]: Royal Engineers Library, accession no. 11466, 16 p., unpublished.

Anonymous, 1945e, US Geologists in the War: Discovery, v. 6, p. 355–356.

Arditi, S., 1994, La préparation scientifique du débarquement de Juin 1944 : La Recherche, v. 25, p. 691–693.

Bernal, J. D., 1955, The scientific preparation for the Normandy landings of June 1944: Actes du 74 Congrès de l'Association française pour l'avancement des Sciences, 15–22 Juillet 1955. Publications de l'Université de Caen, p. 536–542.

Bigot, A., 1930, Sketch of the geology of lower Normandy: Geologists' Association Proceedings, v. 41, p. 363–395.

Bigot, A., 1947, Forages pour recherches d'eau dans le Calvados; VII—Forages de l'Armée Anglaise en 1944: Bulletin de la Société Linnéenne de Normandie, sér. 9, v. 5, p. 130–133.

Buchanan, A. G. B., 1953, The Second World War 1939–1945. Army works services and engineer stores: London, The War Office, 231 p.

Butler, A. J., 1947, War-time geology: Special exhibits in the Geological Museum, South Kensington: Museums Journal, v. 46, p. 233–238.

Collier, R., 1992, D-Day June 6, 1944. The Normandy landings: London, Cassell, 224 p.

Desquesnes, R., 1993, Normandy 1944: The invasion, the battle, everyday life: Mémorial de Caen, Editions Ouest-France, 237 p.

Doré, F., Juignet, P., Larsonneur, C., Pareyn, C., and Rioult, M., 1987, Normandie, Maine; Guides Géologiques Régionaux, 2nd ed: Paris, Masson, 216 p.

Ellis, L. F., with Allen, G. R. G., Warhurst, A. E., and Robb, J., 1962, Victory in the west. Vol. 1. The battle of Normandy: London, Her Majesty's Stationery Office, 595 p.

Evans, W. D., 1945, War-time geology: Discovery, v. 6, p. 300–305.

Farrington, J. L., 1936, A geological application of aerial survey: Geological Society of South Africa Transactions, v. 38, p. 57–71, 22 pls.

Hastings, M., 1984, Overlord: D-Day and the battle for Normandy: London, Guild Publishing, 368 p.

Hue, E., 1916, Les blocs erratiques des environs de Luc-sur-mer (Calvados): Le Mans, Le Mannoyer, 38 p.

Hue, E., 1938, Contribution à l'étude du Quaternaire. Plage surélevée de Luc-sur-Mer: Bulletin de la Société préhistorique française, v. 10, p. 403–408.

Hunt, C. B., 1950, Military geology, *in* Paige, S., ed., Application of geology to engineering practice: Boulder, Colorado, Geological Society of America, Berkey Volume, p. 295–327.

Inglis, J. D., 1946, The work of the Royal Engineers in north-west Europe, 1944–45: Royal United Service Institution Journal, v. 91, p. 176–195. (Reprinted in the Royal Engineers Journal, v. 60, for 1946, p. 92–112.)

Journaux, A., 1971, Caen. Cartes des Formations Superficielles et Cartes Géomorphologiques de Basse-Normandie: Caen, CNRS, scale 1:50,000, 2 sheets.

Journaux, A., 1973, Bayeux-Courseulles. Cartes des Formations Superficielles et Cartes Géomorphologiques de Basse-Normandie: Caen, CNRS, scale 1:50,000, 2 sheets.

Kaye, C. A., 1957, Military geology in the United States sector of the European theater of operations during World War II: Geological Society of America Bulletin, v. 68, p. 47–54.

King, W. B. R., 1951a, The influence of geology on military operations in northwest Europe: Advancement of Science, v. 30, p. 131–137.

King, W. B. R., 1951b, The recording of hydrogeological data: Yorkshire Geological Society Proceedings, v. 28, p. 112–116.

Michel, F., 1991, Les côtes de France. Paysages et géologie: Orléans, Editions du BRGM, 160 p.

Mitchell, T., 1994, Construction of B3 ALG airfield, Normandy, June 1944: Royal Engineers Journal, v. 108, p. 35–45.

Montgomery, F. M. The Viscount, 1946, 21 Army Group: Normandy to the Baltic: Germany, BAOR, 279 p.

Munro, S. K., 1985, Alexander Kerr Pringle: Edinburgh Geological Society Proceedings, v. 15, p. 5–6.

Pakenham-Walsh, R. P., 1958, History of the Royal Engineers Vol. IX. 1938–1948: Chatham, Institution of Royal Engineers, 644 p.

Panet, H. de L., 1945, Lessons on the campaign in Europe 1944–45. Airfield construction: Royal Engineers Library, accession no. 4562WD, 29 p., unpublished.

Pareyn, C., 1994, Des plages de la retraite de Dunkerque aux plages du débarquement de 1944; Le rôle des officiers géologues du Royal Engineers Corps de l'armée britannique: Techniques, Sciences, Méthodes. Revue de l'Association Général des Hygiénistes et Techniciens municipaux, no. 7–8, p. 431–435.

Rioult, M., Coutard, J. P., de la Querière, P., Helluin, M., Larsonneur, C., Pellerin, J., and Provost, M., 1989, Notice explicative de la feuille Caen à 1/50,000: Orléans, Editions du BRGM, 104 p.

Rioult, M., Dugué, O., Jan du Chêne, R., Ponsot, C., Fily, G., Moron, J.-M., and Vail, P. R., 1991, Outcrop sequence stratigraphy of the Anglo-Paris basin, Middle to Upper Jurassic (Normandy, Maine, Dorset): Bulletin du Centre de Recherches Exploration-Production Elf Aquitaine, v. 15, p. 101–194.

Rioult, M., Bauduin, P., Heintz, A., Poisson, E., and Quéré, G., 1994, L'eau à la source de la victoire 1944: Caen, Agence de L'Eau Seine-Normandie, 112 p.

Rose, E. P. F., 1980, German military geologists in the Second World War: Royal Engineers Journal, v. 94, p. 14–16.

Rose, E. P. F., and Hughes, N. F., 1993, Sapper Geology: Part 1. Lessons learnt from world war: Royal Engineers Journal, v. 107, p. 27–33.

Rose, E. P. F., and Pareyn, C., 1994, British applications of military geology for "Operation Overlord" and the battle in Normandy, France 1944: Geological Society of America Abstracts with Programs, v. 26, no. 7, p. A275.

Rose, E. P. F. and Pareyn, C., 1995, Geology and the liberation of Normandy, France, 1944: Geology Today, v. 11, p. 58–63.

Rose, E. P. F. and Pareyn, C., 1996a, Roles of sapper geologists in the liberation of Normandy, 1944. Part 1. Operational planning, beaches and airfields: Royal Engineers Journal, v. 110, p. 36–42.

Rose, E. P. F. and Pareyn, C., 1996b, Roles of sapper geologists in the liberation of Normandy, 1944. Part 2. Quarries, water supply, bombing and cross-country movement: Royal Engineers Journal, v. 110, p. 138–144.

Rose, E. P. F., and Rosenbaum, M. S., 1993a, British military geologists: The formative years to the end of the First World War: Geologists' Association Proceedings, v. 104, p. 41–49.

Rose, E. P. F., and Rosenbaum, M. S., 1993b, British military geologists: Through the Second World War to the end of the Cold War: Geologists' Association Proceedings, v. 104, p. 95–108.

Rose, E. P. F., Häusler, H., and Willig, D., 1996, A comparison of British and German military applications of geology in world war, *in* Sparks, R. S. J., French, W. J., and Howarth, R. J., organizers, Applied Geoscience 15–18 April 1996 Abstracts: London, Geological Society, p. 29.

Rosenbaum, M. S., 1990, Geologists at war: The D-Day operations and subsequent advance: Geologists' Association Proceedings, v. 101, p. 163–165.

Scott-Bowden, L., 1994, COPP to Normandy 1943/44—a personal account of part of the story: Royal Engineers Journal, v. 108, p. 10–18.

Shotton, F. W., 1945, Summary of Normandy boreholes for water. 5 p. plus covering letter dated 21 May 1945, ref. 21 AGp 8779 WK (British Geological Survey archive document, unpublished).

Shotton, F. W., 1947, Geological work in the invasion of North-West Europe: Geological Society of London Quarterly Journal, v. 102, p. v.

Simon, L. J., 1957, Additional notes of the use of geology in the European theater of operations during World War II: Geological Society of America Bulletin, v. 68, p. 1567.

Snyder, C. T., 1957, Use of geology in planning the Normandy invasion: Geological Society of America Bulletin, v. 68, p. 1565.

Stone, I. R., 1970, Military geology in the two world wars: British Army Review, v. 34, p. 63–67.

Sutton, J. S., 1978, Geologists in the Second World War: New Scientist, v. 78, p. 831–833. (Reprinted in the Royal Engineers Journal, v. 94 [for 1980], p. 9–13).

Tait, E. A., 1984, Thomas C. Phemister: Aberdeen University Review, 1983–84, p. 177–178.

Williams, A. R. O., 1950, The Royal Engineer quarrying companies in the northwest European campaign: Quarry Managers' Journal, v. 34, p. 214–225, 2 maps.

Wilson, C., and Nowers, J., 1994, D-Day June 1944 and the Royal Engineers: Chatham, The Royal Engineers Museum, 14 p.

MANUSCRIPT ACCEPTED BY THE SOCIETY OCTOBER 29, 1997

Printed in U.S.A.

Geological Society of America
Reviews in Engineering Geology, Volume XIII
1998

Engineer intelligence and the Pacific geologic mapping program

Gilbert Corwin†
Military Geology Branch, U.S. Geological Survey

ABSTRACT

Lack of terrain data contributed significantly to the high costs of lives and operations during the Pacific campaign of World War II. After the war the U.S. Army Corps of Engineers contracted with the Military Geology Branch of the U.S. Geological Survey to gather detailed terrain information about the occupied islands under direct U.S. jurisdiction in the event they or comparable oceanic islands became sites of future military operations. The U.S. Geological Survey established a headquarters in Tokyo and initiated field studies of Okinawa during 1946. Subsequent detailed studies were launched at the Palau Islands (1947), Yap Islands (1947), Saipan (1948), Tinian (1949), Guam (1951), Pagan, Marianas Islands (1954), Truk (1954), Ishigaki and Miyako (1955), and the Marshall Islands (reconnaissance, 1951). Initial plans for detailed studies of all mandated islands were abandoned for lack of time, but members of the field parties briefly visited nearly all. Field teams included geologists, hydrologists, soils scientists, a plant ecologist, and a climatologist. The Tokyo office gathered and translated existing Japanese literature about the islands; more than 600 articles were translated. A by-product was the establishment of a joint U.S.–Japanese project to compile and publish a series of 1:250,000 geologic maps of formerly held Japanese territories, including Korea, Manchuria, northeast China, southern Sakhalin Island, and the Kuriles.

Results of the field studies were published in a series of military geology folios composed of both basic and interpretive chapters. U.S. Geological Survey professional papers presented many of the scientific results.

INTRODUCTION

The collection of basic engineering data on the world's terrain and interpretation of these data for future military use, both strategic and tactical, are among the most important functions of Engineer Intelligence during times of peace. Research to improve methods of collecting and evaluating the data is an important aspect of these functions. Successful military operations during times of emergency depend largely upon the amount and quality of data that have been accumulated and upon the adaptability of methods that have been devised for making the data available and useful.

†Deceased

The worldwide mapping program under supervision of the Intelligence Division, Office of the Chief of Engineers, U.S. Army, provides an outstanding example of a program to collect and present basic terrain data. The occultation program, use of the tellurometer, experimentation with other new surveying techniques, and cooperative mapping programs were attempts to improve the quality of maps. Terrain analysis and area intelligence programs, carried out in collaboration with other agencies of the U.S. government, complemented the mapping program and were necessary adjuncts for effective use of maps in planning future operations.

The scope and objectives of engineer terrain analysis and area intelligence programs are varied. The area of interest for a program may be large or small, strategically important or rela-

Corwin, G., 1998, Engineer intelligence and the Pacific geologic mapping program, *in* Underwood, J. R., Jr., and Guth, P. L., eds., Military Geology in War and Peace: Boulder, Colorado, Geological Society of America Reviews in Engineering Geology, v. XIII.

tively unimportant, densely populated or uninhabited, arctic or tropic, forest or desert, mountainous or flat, accessible or inaccessible. The subjects of interest may be restricted to a single topic, such as sources of water for a small installation on a remote island, or they may cover many topics, such as those in National Intelligence Survey reports produced for the Central Intelligence Agency. With advancing technology and an accumulation of basic knowledge, the scope and objectives are constantly changing and continuously broadening.

Like mapping, terrain analyses at various scales are desirable and in some cases essential for maximum military use. For strategic and high-level planning, one or a few small-scale maps and a concise report on a large area are generally sufficient. Studies at intermediate scales are well suited to operational planning at theater level and to meet many specific problems, such as location of suitable airfield sites, layout of road networks, and conditions for cross-country movement in continental areas of relatively uniform conditions. Terrain studies prepared by the Area Analysis Division at the U.S. Army Map Service, Far Fast, and by similar organizations in the United States and European Theater are examples. Large-scale terrain studies are needed in areas of particular strategic importance, of intensive land utilization, and of active and potential military development, such as, for example, urban areas, specific airfield sites, missile-launching sites, landing beaches, and small strategically located islands that may serve as naval, air force, or intelligence bases. Further, intensive large-scale studies may aid in the preparation of analyses used in smaller-scale studies of areas for which information is limited.

The availability of information on a given area may limit the scale on which studies can be made. During World War II, the practicability of producing such studies on little-known areas, especially in the Western Pacific, was proven. Large-scale efforts, that is, studies at 1:75,000 scale and larger, are necessarily confined to areas that are accessible or for which information is abundant. Because of the effort involved, such studies have been confined to places of particular strategic or military importance and to areas having conditions that are representative of inaccessible regions. The detailed reports on Okinawa prepared by the Military Geology Branch illustrate a study of a strategic area; investigation of permafrost in Alaska is an example of a study on conditions that are common in large inaccessible areas.

Because subjects are diverse and requirements are constantly changing, research specialists of many types are needed for engineer terrain analyses. Geographers, climatologists, oceanographers, geologists, botanists, soils scientists, hydrologists, hydrographers, and construction and soils engineers are representative of those who can contribute; highly trained specialists and authorities within these and other fields are often required to solve specific problems. Assistants and technicians of many types must be available to support the specialists in the collection and compilation of the basic data and in presentation of the results.

WESTERN PACIFIC ISLANDS

Background

The small islands of the Western Pacific Ocean (Fig. 1), which extend over an area that nearly equals all of North America in size, were ideal for the testing and development of large-scale engineer terrain intelligence and intelligence techniques. Prior to World War II, the islands had been under the control of various governments, and at the beginning of the war, most had been under Japanese control for 25 years or more. They were enemy territory that had been closed to visitors for more than 10 years, and information concerning the islands was scattered and incomplete. When the islands became "stepping-stones" to the Battle of Japan, joint intelligence forces of the U.S. army and navy undertook intensive studies of them. Early campaigns supported by a minimum of intelligence were commonly very costly in both manpower and equipment. The battle for Tarawa in the Gilbert Islands, where lack of information on reefs and tides resulted in great losses, is a grim example. Later campaigns benefited not only from experience gained in the earlier campaigns but also from much more effective and complete information collected by the intelligence forces. Because much of the intelligence data came from obscure sources, time was needed to assemble and interpret it. Terrain intelligence teams operating in Washington, D.C., Hawaii, and the southwest Pacific and teams with the fleet forces were major contributors. Although the intelligence was vastly improved over that produced earlier, there still were errors and some misinterpretations.

Strategic Engineering Studies (SESs) prepared by the U.S. Geological Survey for the Chief of Engineers and reproduced by the U.S. Army Map Service were significant contributions to the intelligence effort. The studies, based on fragmentary information from many sources, were used as guides in planning and accomplishing campaigns and later in developing bases. They also formed the basis for terrain analyses within the Theater of Operations. These studies contained descriptions and evaluations of the terrain, water supplies, sites for road and airfield construction, engineering properties of soils, and sources of construction materials. Aerial photography and basic scientific documents on geography, climate, and especially geology supplied most data for these reports. One of the most successful studies of an island that actually was invaded was based almost entirely on a single Japanese geologic report (Hanzawa, 1935).

In the present era of long-range bombers, atomic weapons, guided missiles, and great troop mobility, it is unlikely that the small western Pacific Islands will ever again have the importance of stepping-stones that they had during World War II. However, in a world that is mostly covered with water, islands will continue to serve as strategic bases of operations. The varied terrain conditions of the Pacific islands are representative of other islands in tropical and subtropical areas throughout the world.

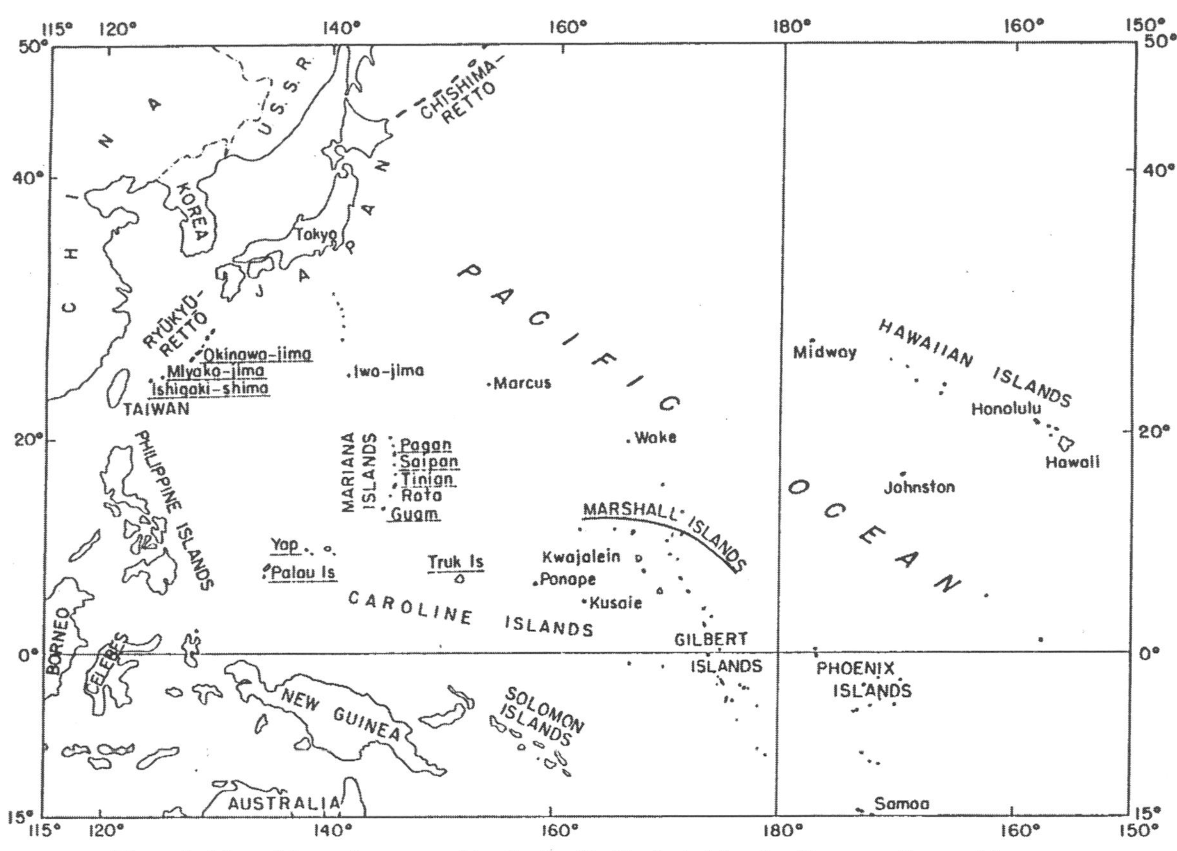

Figure 1. Map of the region covered by the Pacific Geologic Mapping Program. Geographic names were current as of 1958 when the program was drawing to a close.

Pacific Geologic Mapping Program

After World War II, the Pacific Geologic Mapping Program, a cooperative project of the Office of the Chief of Engineers and the U.S. Geological Survey, was established to make detailed studies of the terrain and resources of Western Pacific islands. Originally, objectives of this program were to evaluate studies made during the war and to collect basic information required for the development and use of military bases on the islands in the event that the islands again became strategic bases of operation. Following initiation of the program in March 1946, the original scope and objectives were modified and expanded to include research on problems peculiar to islands and to the tropics, and experimentation on methods of preparing and presenting large-scale terrain analyses.

Under the Pacific Geologic Mapping Program, 11 island groups were studied in detail (Table 1). Special-purpose investigations requested by island commander and civil administrations were made on other island groups. Many of the remaining islands under U.S. administration were visited and briefly reconnoitered.

The selection of islands for detailed study was based on their importance as strategic military bases (Okinawa, Guam, Saipan), possible use as future bases or supporting bases (Tinian, Miyako,

TABLE 1. FIELD STUDIES OF THE PACIFIC GEOLOGIC MAPPING PROGRAM

Area	Field Work	Field Personnel Specialities*
Okinawa	April 46–March 49	A, B
Palau	February 47–November 48	A, B
Yap	July 47–July 48 (Geology) November 55–June 56 (Soils)	A, B, (D), G
Saipan	September 48–August 49	A, B
Tinian	September 49–July 51	A, B, (D)
Guam	August 51–December 54	A, B, C, (D), E, F
Northern Marshalls	November 51–April 52 July 52–August 52	A, C
Pagan	July 54–September 54	A
Truk	November 54–July 55	A, B, (D), E
Ishigaki	June 55–October 56	A, B, C
Miyako	November 55–September 56	A, B, C

*A = Geology; B = Soils science; C = Botany; (D) = Climatology (independent or supporting study incorporated in report); E = Hydrology; F = Marine geology; G = Mining.

Ishigaki), the presence of mineral resources (Okinawa, Palau, Yap), and the conditions as representative of island types (Okinawa, Guam, Palau, Yap, Truk, Pagan, northern Marshall Islands). Okinawa and Guam are large islands with varied conditions; Pagan, all but two of the Palau Islands, the Truk Islands, and the Marshall Islands are comparatively small, and each island or group of islands is characterized by a given combination of representative conditions. Some of the islands are mountainous (northern Okinawa, Ishigaki, Pagan, some islands of the Palau group, the central islands of Truk); others are low and relatively flat (Marshall Islands, Tinian, the reef islands of Truk, Miyako, southern Okinawa). One has recently active volcanoes (Pagan); several are old deeply dissected volcanoes (Truk, the northern Palau Islands, southern Guam); others form portions of modern coral reefs (perimeter islands of Truk, Marshall Islands); still others are remnants of old deeply dissected mountain masses (northern Okinawa, Ishigaki). All islands have offshore coral reefs characteristic of most tropical islands; reef conditions, however, differ greatly between the island groups.

The first detailed field study was begun on Okinawa during April 1946. It was staffed by officer and civilian geologists assigned to the Office of the Engineer, Far East Command, by geologists of the U.S. Geological Survey, and by soils scientists of the Department of Agriculture. Subsequent field parties were made up entirely of personnel of the U.S. Geological Survey. Dates of field work and types of specialists who constituted the field parties are summarized in Table 1. The last field studies, on Yap, Ishigaki, and Miyako, were completed during 1956.

Results of the field studies and subsequent laboratory and office analyses were published in a series of preliminary reports on various special subjects and in comprehensive military geology reports. Basic information and samples collected in the course of the studies furnished material for articles published in 18 scientific publications and eight U.S. Geological Survey Professional Papers. A list of the military geology reports is given in Appendix A.

Military geology report

The scope of the comprehensive military geology report is broad and covers all, or nearly all, subjects normally treated in Engineer area intelligence files and reports. Emphasis and manner of presentation of the various subjects differ for each report, partly because of differing conditions on the various islands and partly because of specialties, experience, and qualification of personnel assigned to each project. Changing scope and objectives of the program have also influenced content of the various reports.

Comprehensive military geology reports of that era were divided into three major parts: basic aspects, tactical aspects, and engineering aspects. Basic aspects include comprehensive descriptions of the terrain, coasts, beaches, offshore conditions, climate, vegetation, soils, geology, man-made features, and special geophysical phenomena. Special emphasis was placed on these subjects because they formed the basis of interpretations in the other parts and contained confirmed data that could be used in subsequent detailed studies or reinterpretations regardless of the requirements and criteria to be evaluated.

The emphasis on basic aspects was particularly critical in view of rapidly advancing technology; interpretations and evaluations based on current equipment and requirements may be practically unusable tomorrow or within a few years. Examples of such changes since World War II are plentiful. Prior to and during World War II, nearly all volcanic rock was considered suitable for use as concrete aggregate because the rock is hard and crushes uniformly; it was sufficient for the engineer to know that "trap rock" was available. Today such information is not enough. The engineer must know if the volcanic rock contains opal or unstable glass that will react with cement to cause disintegration of the concrete, and the engineer also must know if it is vesicular—a factor that is important in determining the structural weight and strength of the concrete. Petrographic descriptions of the rocks, based on a microscopic examination, include these and other basic data on rock composition and nature. The methods are rapid and inexpensive, and the results are easily tabulated as basic data. Required terrain and foundation conditions for heliports, missile bases, and many other kinds of military installations would have been difficult to nearly impossible to evaluate 10 years ago.

Military requirements for engineer intelligence data are extremely varied and difficult to predict. It would be impractical, in fact nearly impossible, to anticipate, evaluate, and interpret the basic aspects of the terrain in terms of all potential uses for information. For example, on Guam the Materials Testing Laboratory of the Navy Public Works Office was assigned the task of determining the grasses and shrubs most suitable for use by the Division of Buildings and Grounds to improve the appearance of installations. On Saipan, land classification maps were required to settle claims of inhabitants whose property was being used by the naval civil administration. In the southern Ryukyus, a resettlement program was initiated to alleviate overpopulation on Okinawa; many resettlement areas in the Yaeyama group proved unsatisfactory because of improper agricultural conditions. Preliminary copies of basic soils and vegetation maps prepared for the Military Geology Reports were furnished commanders to help them solve these particular problems, each a military problem requiring large-scale engineer intelligence data and analyses.

Part II of the reports, tactical aspects, was an interpretive analysis of the terrain in terms of approaches to the island (both amphibious and airborne), cross-country movement, and, in some cases, cover, concealment, observation, and fields of fire. Generally, emphasis in this part was on existing conditions that could be expected to affect military operations. This emphasis on conditions was based on the philosophy that knowing conditions, the military planner will adapt or develop equipment and weapons that will operate most effectively under those conditions. The analyses of conditions was supplemented with interpretations in terms of current equipment and tactics, including landings from assault boats, assault aircraft, and helicopters; movement by foot and mechanized troops; and use of standard artillery, mortars, and bombs.

From the practical engineer's standpoint, part III, engineering aspects, was the meat of the Military Geology Report. In later reports, this part was divided into chapters on sources and uses of construction materials, suitability of the terrain for construction, and water resources. The first of these subjects was a straightforward presentation of rock and soils materials available for construction. It was based mostly on interpretation of the basic soils and geology sections of the reports and on engineer test results of representative soil and rock samples. The second subject evaluated the terrain and foundation conditions for construction of roads and airfields, heavy structures, underground installations, and miscellaneous types of construction. Emphasis was again on conditions, with supplemental evaluation based on present economic criteria. There is no such word as "unsuited" in the vocabulary of the construction engineer. Knowing the conditions, he can construct almost any reasonable type of installation almost anywhere. Three-thousand-meter runways with cleared approaches could be built on every large Pacific island studied. On most, however, the conditions make such construction impractical because of costs and effort involved. It was the planner, who must coordinate with the comptroller, for whom the supplemental analysis was made. Determination of construction conditions involved evaluation of nearly all basic data collected and contained in basic sections of the reports.

On small islands, the final topic of engineering aspects was vitally important. Water resources were commonly limited and required special techniques for development and conservation. Most streams were short or, on limestone and some volcanic islands, nonexistent. During dry seasons or droughts, the flow of the short streams diminished greatly or ceased; only a few comparatively long streams on the larger islands were dependable sources of surface water throughout the year.

The concept of a thick lens of ground water beneath islands become widely known and accepted during the Pacific Geologic Mapping Program. Although a valuable concept in studies of island hydrology, it may be complicated by many conditions, and on many islands it cannot be applied at all. This was true for nearly all islands studied under the Pacific Geologic Mapping Program. On Pagan, a volcanic island that is almost ideally suited to development of a lens because of its size and the high rates of rainfall and infiltration, thermal convection associated with a recently active volcano causes mixing of salt and fresh water and has greatly modified or destroyed much of the lens. On Angaur in the Palau Islands, free mixing of salt and fresh water in caverns connecting lakes near the center of the island with the ocean has prevented development of an extensive lens. On most larger islands, impermeable rocks above, at, and below sea level make the lens useless as a practical water source.

Of the special requests for information submitted to the U.S. Geological Survey Military Geology Branch, problems of water resources and supply on remote islands were among the most numerous. Many of these came from the air force, which had constructed air bases requiring much water on larger islands, and a network of radar stations with fewer but essential requirements on numerous small, high islands. Principles determined in the detailed studies of the selected islands were useful in solving problems and recommending plans for development of water resources on these islands.

Field studies and preparation of reports

To collect data and prepare the basic part of the various military geology reports, the Pacific Geologic Mapping Program relied most on trained geologists and soils scientists. On several projects, studies of the geologists and soils scientists were supplemented with studies by a vegetation expert, hydrologists, and a climatologist of the U.S. Weather Bureau. All were briefed in military requirements; about half had previous military intelligence training. Because of difficulties in obtaining enough qualified personnel, most parties consisted of one to six scientists at a given time. Native helpers and camp hands provided most subprofessional help in the field. Geologists served as project leaders and coordinators in the preparation of the reports.

Logistic support for the field was provided by the army, navy, air force, and civil administrations on the various islands. An excellent example of interservice cooperation in support of a project was illustrated by experiences of the field party that spent 10 weeks on the island of Pagan in the Northern Marianas Islands. The 29th Engineer Battalion (Base Topo) provided camping equipment including tents, cots, stoves, and mess gear. The air force on Guam furnished radios, jeeps, gasoline, oil, commissary, and PX items. The navy civil administration made arrangements for the stay of the party on the island and provided transportation of the party and equipment to the island. Communications with Guam were provided through radio contact with planes of the 54th Weather Reconnaissance Squadron. Later, samples collected by the party were tested by the Materials Testing Laboratory, Public Works Office, U.S. Navy on Guam.

In the field the rocks, soils, and vegetation were mapped on aerial photographs. Notes on the terrain, rocks, soils, vegetation, fauna, and cultural features were taken concurrently and referenced to the photographs. At nights and on rainy days, the field information was transferred from the field photographs and notes to a preliminary field map to check completeness of coverage and relationships of the various units. Transfer at this stage was generally done by eye. Representative samples of rocks, soils, vegetation, and water were collected during the field mapping and prepared at night for later study and analyses. Special large samples for engineer testing were collected separately and shipped to the nearest army or navy testing laboratory. A jeep-mounted soil auger was used on three islands to collect subsurface information.

During routine field mapping, the scientists worked alone or in pairs. They generally were accompanied by a native guide who assisted them in collecting samples, cut trails, and helped carry equipment. Efforts of the individual workers were coordinated, reviewed, and checked by the party chief who had general responsibility for the mapping.

Following the field work, the parties returned to headquarters in Tokyo or Washington to prepare final reports. Basic maps were replotted using photogrammetric instruments and checked for accuracy and completeness; interpretive maps were prepared from the basic maps and field notes; field samples were studied and sent to others for analyses; and finally all field summaries and results of laboratory data were assembled in an integrated text, in tables, and in marginal data for the maps. Climatologic information, some of which was collected by the field parties, and results of library research on history, geophysical phenomena, and previous studies of the islands were incorporated in the reports during the report writing stage.

The finished reports were submitted to critics who checked them for general completeness, consistency, and accuracy of interpretation. Following revision by authors, based on the critique, the reports were turned over to the Publications Unit, which prepared them for final Engineer review and printing. Details of the report were checked and cross-checked for consistency; maps and text were edited for repetition, grammar, clarity, and accuracy; tables were rearranged to fit the format of the report; and the maps were drafted for reproduction. Professional personnel were required for these tasks as an understanding of the subjects was essential to maintain accuracy of basic scientific detail and of interpretations. Because of emphasis on accuracy and detail, each stage in the preparation of a final Military Geology Report required time.

CONCLUSIONS

The Pacific Geologic Mapping Program was an experiment in large-scale terrain analysis and Engineer-style area intelligence. Those who were closely associated with this program felt it was a successful experiment. Much basic data of immediate and potential value to the military on these and similar islands were obtained. Techniques of investigation and methods of presentation that should be of great value in future large-scale studies have been developed. Scientific data collected in the course of the program provided, and will continue to provide, contributions to the knowledge of the earth on which we live.

ACKNOWLEDGMENTS

This paper represents an outgrowth of a report prepared for the Intelligence Division, Far East Command, U.S. army, in 1958. Joshua I. Tracey, Jr., and Frank C. Whitmore, Jr., contributed reviews that greatly helped recast the paper for the current audience. Special thanks are extended to Joshua I. Tracey, Jr., for compiling Appendix 2.

APPENDIX 1. REMINISCENCES OF THE PACIFIC GEOLOGICAL SURVEYS PROGRAM

This appendix contains the text of Corwin's presentation at the 1994 Geological Society of America meeting, which he was unable to incorporate into this paper before his death.

Since accepting the invitation to tell you about our western Pacific Islands military studies, I've puzzled over how much I could tell you about this complex, multifaceted 15-year program in 15 minutes, a program that involved either directly or indirectly more than 150 geologists, hydrologists, soils scientists, botanists, climatologists, and other professionals, some of whom were or have become well known. The program produced a five-foot bookshelf of military geology and special reports published by the army, a couple of feet of U.S. Geological Survey professional reports and outside journal articles, several feet of translations of Japanese reports, a stack of inside reports, and, of special note, a series of geologic maps covering the Kurile Islands, the southern half of Sakhalin Island, all of Korea and Manchuria, and much of northeastern China at a scale of 1:250,000. For my part, in early 1946, I lucked into a very minor role in program preparations, became a member of the first field study groups on Okinawa, was transferred to the second field project where I had a major role, was the leader of a four-man team in studies of Pagan—a volcanic island in the northern Mariana Islands—and in the ninth year of the program became its fourth chief. My three predecessors as chiefs are no longer with us; I've lost track of my successor, R. C. Kepferle, who was left with the job of closing our Tokyo office in 1960. Retired for the past 12 years, I prefer to reminisce rather than provide you with a scholarly talk.

As World War II ended, a proposal for detailed terrain studies of the Western Pacific Islands under U.S. jurisdiction was submitted to and accepted by the Intelligence Division of the Army Corps of Engineers. The initial proposal envisioned evaluation of studies made during the war and experimental large-scale basic investigations and mapping of all islands under military jurisdiction with interpretation of the results in the event that they, or comparable islands elsewhere, became involved in future conflicts. This goal subsequently was reduced to studies of representative island types and modified to include research on problems peculiar to islands and to the tropics. These detailed studies included mapping the geology, soils and vegetation at scales of 1:75,000 and larger.

The U.S. Geological Survey assistant chief geologist at the time, Harry Ladd, was dispatched to evaluate operational and logistic needs for launching the studies. At Okinawa, he stayed in the tent of an officer with whom I worked, and that is when I met Harry and learned of the proposed program. On Harry's second visit, he was accompanied by John Rodgers, and I wangled the job of giving them a tour of the island, a tour that was made memorable in part by Ladd's ability to sleep in the uncomfortable back seat of the jeep on a bumpy road.

A few weeks later, in March 1946, the program was launched with Quentin Singewald in charge of the Tokyo headquarters and with John Rodgers and a small team of army and U.S. Geological Survey geologists headed for Okinawa to begin mapping the island's geology. It took me about six months to work my way out of the military and onto the Okinawa team.

The second field project, launched in early 1947 by a party of four, was in the Palau Islands where the army engineers had a field unit establishing topographic control. By August, the U.S. Geological Survey party had dwindled to one person, Sam Goldich, who was there to study the bauxite deposits and whose time commitment was nearly up. The other three who had started systematic mapping of the islands hadn't gotten very far. One, Charley Johnson, had headed for the Yap Islands to undertake an individual project promised in his employment contract; the other two had fulfilled their six-month commitments, a period that was far too brief to accommodate the time and delays involved in getting to and from the islands. The need for longer commitments was clear; finding people to accept them provided a challenge to the home front in Washington.

I was pulled off Okinawa and sent to Palau to help Sam Goldich finish his work and hold the fort until a new party arrived. Sam did finish and headed for home. Then the army survey unit pulled out, withdrawing support of natives who were serving as field helpers and our means of getting around the islands. Getting the U.S. Geological Survey to approve my hiring the natives at island wage levels between 35 cents and one dollar a day gave the Washington home front another challenge. The Tokyo office arranged to have the army unit leave some of its unneeded equipment and supplies to me. I ended up with three landing craft, only one of which was marginally operational, a weapons carrier that was operational and two jeeps that weren't, and a number of drums of diesel oil and gasoline. On getting settled at the officer's quarters of the navy civil administration, I took off in the weapons carrier to do some mapping, only to discover that I was being followed by one of our jeeps loaded with our six native workers. I had hired an outstanding mechanic to whom I paid the top permissible wage of one dollar a day and who later lost his life in an accident aboard our largest landing craft, the only major casualty of our Pacific program.

A few weeks later, a new field crew began to dribble in, and we were able to finish mapping the geology and soils of the island by the end of 1948. 1 don't believe any of the other island projects had problems to match those of Palau, and I could go on talking about the problems for another hour. Before leaving Palau, however, I must tell of one more incident that I feel has historic importance. Most of you knew or at least have heard of Preston Cloud, and some of you are probably aware that he wrote his doctoral dissertation at Yale on a series of Paleozoic limestones in Texas. Pres joined the Pacific Program to lead planned studies of Guam. On arrival there, he was asked to come to Palau to review our efforts, a week with me for the northern volcanic islands and a week with our party chief, Arnold Mason, for the southern high limestone islands. For our review of the islands, we used our large twin-engine landing craft. On reaching the north end of the volcanic islands, Pres requested a visit to Kyangel Atoll, which has a shallow lagoon and lies north of, and is separated from, the main Palau barrier reef by a deep channel. On passing through the boat channel on the west side of the atoll, Pres suggested that we throw out a rope, put on our face masks, and have the boat pull us across the lagoon. We had not gone far before we passed over a coral mound. Up came Pres's face from the water shouting "Gil! Gil! Did you see that! Just like we have in the Silurian!" Never before or since have I seen anyone so excited. Fifteen years later, Pres was instrumental in establishing the Survey's marine program; I firmly believe that it was the coral mound that sparked his interest in marine geology.

The Tokyo headquarters of our program was initially located in the Office of the Engineer, about two blocks from MacArthur's headquarters, but was later moved to an area within the north Tokyo suburbs occupied by the Army Map Service, Far East. The Army Map Service, Far East, had a printing plant and provided us with other services, including employment of a supporting Japanese staff. The headquarters was charged with providing liaison with our military sponsors, with planning and fulfilling logistic and administrative needs of the field parties, and with collecting Japanese information concerning the islands and preparing it for use by the field parties. Sherman K. Neuschel replaced Quentin Singewald as chief. When I took over the office I had specific instruction of our Washington boss, Frank Whitmore [chief, Military Geology Branch, 1946–1959], to reduce administrative staff.

In addition to the administrative personnel, the staff included report editors, a librarian, a translation unit of bilingual Japanese under the direction of an American editor, and a Japanese drafting unit with an American chief. From time to time the staff was called on to provide geologic consulting services to the army engineer. I especially remember a one-week trip to Iwo Jima with Helen Foster to investigate a 58-minute phreatic eruption. Our translation unit included some respected Japanese geologists, and by the end of the program had completed more than 600 translations, mostly concerning the islands that we were studying.

Helen Foster of the Tokyo office directed preparation of our first major military report: an annotated bibliography of more than 4,000 articles, mostly by Japanese authors, pertaining to the geology and soils of the island territories. While collecting information for the bibliography and for the field teams, Helen and others of our Tokyo staff met and became friends with many Japanese geologists. This led to a proposed joint Tokyo Geographical Society–U.S. Geological Survey project to compile and publish a geologic map series of areas under former jurisdiction of the Japanese. The army agreed to furnish unclassified base material and print the maps. Members of our translation unit provided liaison with the compilers, and, although of second priority, the drafting of color separations for the geologic overprints became the principal activity of our drafting unit. Two 100-sheet editions of the final maps were printed, one utilizing the unclassified bases for presentation to, and distribution by, the Japanese and the other with added base data for military and U.S. Geological Survey distribution.

From the outset and through the first 10 years of the island mapping effort, the planned scope and goals of the final island reports remained fuzzy. At the outset we were instructed to produce sound base maps accompanied by texts that the engineers could understand and that could later be transformed into U.S. Geological Survey Professional Papers. Later we were told to include military engineering and tactical interpretations of the maps and data. On requesting guidance, we found ourselves on our own. Wheels spun. Chapters of the final reports were written, rejected, and rewritten. Progress was dreadfully slow, and our army engineer sponsors were annoyed. Some preliminary reports concerning special investigations were published but not much else. The scope of the program was reduced. Finally in 1954 I was asked to lead a four-man team to study Pagan in the northern Mariana Islands.

The island of Pagan has an active volcano, the geology is relatively straightforward, and there are no soils to muddy the picture. I was fortunate in having two geologists with strong military geology backgrounds assigned to the team. On our return from the island to Washington, they proposed a report format and wrote interpretive sections that passed muster quickly. My own section on the basic geology suffered in the hands of the editor, but we had a completed report in record time. The approach was applied to our unfinished reports and to those in progress, and the final reports began to roll.

APPENDIX 2. MILITARY GEOLOGY REPORTS OF THE PACIFIC ISLANDS GEOLOGIC MAPPING PROGRAM

1. Military Geology of Okinawa-jima, Ryūkyū-rettō

Vol. 1 Introduction and Engineering Aspects, A. H. Nicol, D. E. Flint, and R. A. Saplis: Tokyo, Intelligence Division, Office of the Engineer, Headquarters U.S. Army Japan with personnel of the U.S. Geological Survey, 1957, 35 p., 19 pls., 1 fig.

Vol. 2 Water Resources, C. C. Cameron, D. E. Flint, and R. A. Saplis: Tokyo, Intelligence Division, Office of the Engineer, Headquarters U.S. Army Japan with personnel of the U.S. Geological Survey, 1958, 82 p., 2 pls., 4 maps, 5 figs.

Vol. 3 Soils, C. Stensland: Tokyo, Intelligence Division, Office of the Engineer, Headquarters U.S. Army Japan with personnel of the U.S. Geological Survey, 1957, 56 p., Soils map (11 sheets), 2 figs.

Vol. 5 Geology, D. E. Flint, R. A. Saplis, and G. Corwin: Tokyo, Intelligence Division, Office of the Engineer, Headquarters, U.S. Army Pacific with personnel of the U.S. Geological Survey, 1959, 88 p., 2 pls., 11 figs.

2. Military Geology of Palau Islands, Caroline Islands, A. C. Mason, G. Corwin, C. L. Rogers, and P. O. Elmquist, geologists; A. J. Vessel and R. J. McCracken, soil scientists: Tokyo, Intelligence Division, Office of the Engineer, Headquarters U.S. Army Forces Far East and Eighth U.S. Army (Rear), with personnel of the U.S. Geological Survey, 1956, 285 p., 32 pls. (including 13 map pls.), 12 figs.

3. Military Geology of Saipan, Mariana Islands

Vol. 1 Introduction and Engineering Aspects, P. E. Cloud, A. H. Nicol, R. G. Schmidt, H. W. Burke, geologists; R. J. McCracken and R. E.

Zarza, soil scientists: Tokyo, Intelligence Division, Office of the Engineer, Headquarters U.S. Army Forces Far East and Eighth U.S. Army with personnel of the U.S. Geological Survey, 1955, 67 p., 19 pls. (including 5 map pls.).

Vol. 2 Water Resources, D. A., Davis; includes Soils Map of Saipan, Mariana Islands, R. J. McCracken and R. E. Zarza, and Geologic Map of Saipan, Mariana Islands, P. E. Cloud, Jr., R. G. Schmidt, and H. W. Burke: Tokyo, Intelligence Division, Office of the Engineer, Headquarters U.S. Army Pacific with personnel of the U.S. Geological Survey, 1958, 96 p., 3 pls., 14 figs.

4. Military Geology of Tinian, Mariana Islands, D. B. Doan, H. W. Burke, H. G. May, C. H. Stensland; Part I, Description of terrain and environment, Part II, Engineering aspects: Tokyo, Intelligence Division, Office of the Engineer, Headquarters U.S. Army Pacific with personnel of the U.S. Geological Survey, 1960, 148 p., 5 maps, 10 figs., 16 photographs.

5. Military Geology of Yap Islands, Caroline Islands, C. G. Johnson, R. J., Alvis, and R. L. Hetzler; Part I, General description of terrain and environment, Part II, Engineering aspects of the terrain: Tokyo, Intelligence Division, Office of the Engineer, Headquarters U.S. Army Pacific with personnel of the U.S. Geological Survey, 1960, 164 p., 5 maps, 10 figs., 64 photographs.

6. Military Geology of Guam, Mariana Islands, J. I. Tracey, Jr., C. H. Stensland, D. B. Doan, H. G. May, S. O. Schlanger, and J. T. Stark; Part I, Description of terrain and environment, Part II, Engineering aspects of geology and soil: Tokyo, Intelligence Division, Office of the Engineer, Headquarters, U.S. Army Pacific with personnel of the U.S. Geological Survey, 1959, 292 p., 19 figs., 49 pls. (including maps).

Water Resources Supplement, P. E. Ward and J. W. Brookhart: Tokyo, Intelligence and Mapping Division, Office of the Engineer, Headquarters U.S. Army Pacific with personnel of the U.S. Geological Survey, 1962, 182 p., 5 figs., 1 map.

7. Military Geography of the Northern Marshalls, F. R. Fosberg, F. S. MacNeil, and T. Arnow: Tokyo, Intelligence Division, Office of the Engineer, Headquarters U.S. Army Forces Far East and Eighth U.S. Army with personnel of the U.S. Geological Survey, 1956, 320 p., 56 figs., 33 pls.

8. Military Geology of Truk Islands, J. T. Stark, J. E. Paseur, R. L. Hay, H. G. May, and E. D. Patterson, with section on climate by D. I. Blumenstock, and section on water resources by M. H. Carson: Tokyo, Intelligence Division, Office of the Engineer, Headquarters U.S. Army Pacific with personnel of the U.S. Geological Survey, 1958, 205 p., 16 figs., 66 pls., 50 maps.

Water Resources Supplement, S. Valenciano and K. J. Takasaki: Tokyo, Intelligence Division, Office of the Engineer, Headquarters U.S. Army Pacific with personnel of the U.S. Geological Survey, 1959, 81 p., 5 figs., 8 maps.

9. Military Geology of Pagan, Mariana Islands, G. Corwin, L. D., Bonham, M. J. Terman, and G. W. Viele: Tokyo, Intelligence Division, Office of the Engineer, Headquarters U.S. Army Japan, 1957, 259 p., 18 figs., 41 pls., 8 maps.

10. Military Geology of Ishigaki-shima, Ryūkyū-rettō, H. L. Foster, C. H. Stensland, H. G. May, F. R. Rosberg, and R. J. Alvis, Part I, General description of the terrain, Part II, Engineering aspects of the terrain: Tokyo, Intelligence Division, Office of the Engineer, Headquarters U.S. Army Pacific with personnel of the U.S. Geological Survey, 1960, 323 p., 6 figs., 7 map pls., 77 photographic pls.

11. Military Geology of the Miyako Archipelago, Ryūkyū-rettō, D. B. Doan, J. E. Paseur, and F. R. Rosberg, Part I, Basic aspects, Part II, Engineering aspects: Tokyo, Intelligence Division, Office of the Engineer, Headquarters U.S. Army Pacific with personnel of the U.S. Geological Survey, 1960, 214 p., 8 figs., 6 maps, 85 photographs.

REFERENCES CITED

Hanzawa, S., 1935, Topography and geology of the Riuku Islands: Tohoku Imperial University Science Reports, 2d series (Geology), v. 17, p. 1–61, pls. 1–20.

MANUSCRIPT ACCEPTED BY THE SOCIETY OCTOBER 29, 1997

Printed in U.S.A.

Geological Society of America
Reviews in Engineering Geology, Volume XIII
1998

Military Geology Branch of the U.S. Geological Survey from 1945 to 1972

Maurice J. Terman
Scientist Emeritus, International Programs, Office of the Chief Geologist, U.S. Geological Survey, 917 National Center, Reston, Virginia, 20192

ABSTRACT

After World War II, the Military Geology Unit of the U.S. Geological Survey was transformed into the Military Geology Branch, which recruited about 150 younger scientists over time to continue the compilation of terrain intelligence on a global scale. Source materials were the scientific journals, books, maps, and photographs available in the Washington area, and most topics were presented on maps accompanied by succinct tables and short narrative texts. The following principal administrative units and research programs evolved: (1) Strategic Studies Section (1945–1972; funded by the U.S. Army Corps of Engineers and later by the Defense Intelligence Agency) was the major production unit. Its principal responsibility was to contribute to the comprehensive small-scale National Intelligence Surveys Program. Another significant section effort included the Pacific Engineer Intelligence Program, which involved the compilation in Washington of seven series of 1:250,000-scale thematic maps for most of Southeast Asia. (2) Pacific Field Program (1945–1962; Corps of Engineers) was a research and mapping program in areas formerly occupied by Japan. (3) Alaska Terrain and Permafrost Section (1947–1965; Corps of Engineers and U.S. Geological Survey) conducted field studies on surficial geology and permafrost phenomena in Alaska and other Arctic areas and compiled reports with maps and engineering interpretations, most at a 1:250,000 scale of Alaska. (4) European Field Program, principally undertaken by the U.S.G.S. Team (Europe) (1953–1964; U.S. Army Europe), published a series of 24 1:250,000-scale maps of military engineering geology for western Germany and 131 1:100,000-scale maps of cross-country movement for all of Germany, and the team served as a consultant. (5) Austere Landing-Site Program (1956–1970; U.S. Air Force) compiled large-scale studies on arid lands, both inside and outside the United States and on Arctic ice-free land. (6) Special Intelligence Element (established 1959; Corps of Engineers and Defense Intelligence Agency) continues to serve as geoscientific consultants to the special intelligence community. (7) Nuclear-Test Detection Program (principally 1962–1972; Advanced Research Projects Agency) compiled studies assisting in the interpretation of global seismic signals, particularly the five-volume 1:5,000,000-scale Atlas of Asia and Eastern Europe.

Terman, M. J., 1998, Military Geology Branch of the U.S. Geological Survey from 1945 to 1972, *in* Underwood, J. R., Jr., and Guth, P. L., eds., Military Geology in War and Peace: Boulder, Colorado, Geological Society of America Reviews in Engineering Geology, v. XIII.

INTRODUCTION

After World War II, as most of the 100 or so compilers of military terrain studies quickly returned to their prewar duties, the Military Geology Unit (MGU) became for the first time a formal part of the U.S. Geological Survey organizational structure, and its name changed to the Military Geology Section. On January 1, 1949 (Survey Order No. 171), it became the Military Geology Branch (MGB); the specific dates for these name changes are not critical to this history, and the name Branch will be used throughout this paper. Some constituent elements will be systematically called sections.

During the next quarter century, a large number of tasks were undertaken for a variety of sponsors, mostly within the Department of Defense (DOD). The following very brief overview of the MGB will highlight the principal administrative units and their research programs—over time some 150 younger and less experienced scientists were recruited. Table 1 identifies each major unit of the MGB and lists some of their key products as cited in a comprehensive 135-page U.S. Geological Survey Open-File Report compiled in 1981 by Selma Bonham (1997). Table 2 cites the manpower and funding for each major unit of the MGB during FY1964 (fiscal year 1964) (July 1, 1963, to June 30, 1964), if applicable, as identified from an unpublished 26-page memorandum dated January 13, 1964, to the chief, MGB, by an ad hoc review committee chaired by William L. Newman. Personnel and fiscal data for other years are not readily available, and thus this FY1964 information serves as a single representative cross section of the MGB activities, albeit obviously misleading in that some earlier functions had ceased and the staffing and funding probably reached their peak during that year. One major change in support occurred in 1963 when the principal long-term MGB/DOD contact shifted from the U.S. Army Corps of Engineers to the Defense Intelligence Agency (DIA).

STRATEGIC STUDIES SECTION

The Strategic Studies Section (SSS) continued the compilation of terrain intelligence studies as in WWII but considered all nations and not just those where military operations were probable. During the next quarter century, the organization of the section (Table 3) was always geared to the compilation and production of maps, tables, and short texts that synthesized geoscience data from the scientific journals, books, maps, and photographs available in the Washington area, and then provided a military evaluation of that basic data. The 10 Joint Army-Navy Intelligence Studies served as a bridge between the wartime folios and the massive new program known as the National Intelligence Surveys (NIS). These mostly small-scale studies of individual countries or geographic regions were the principal activity of the MGB for years; in FY1964, the SSS spent 56 man-years on NIS and completed 17 studies. Altogether, the SSS probably compiled more than 30 NIS studies for the new intelligence agencies, and the most common contribution was the compilation of the maps and texts for a study

TABLE 1. MILITARY GEOLOGY BRANCH REPORTS AND MAPS*

Years	Report or Map Types	Countries or Regions	Reports or Maps
STRATEGIC STUDIES SECTION			
1945–1972	Special reports and miscellaneous papers	>40	291
1946–1948	Joint Army-Navy Intelligence Studies	7	10
1948–1966	National Intelligence Surveys	>30	>30
1949–1965	Basic map compilations (mostly 1:1,000,000 scale)		
	Geologic maps	34	502
	Geomorphologic maps	10	17
	Hydrologic maps	11	116
	Lithologic maps	13	18
	Soil maps (1:15,000,000 scale)	World	11
	Vegetation maps	12	83
1954–1962	Engineer Intelligence Studies	11	10
1957–1964	Engineer Intelligence Guides	1	10
1957–1965	Lunar surface materials studies	1	>4
1959–1965	Pacific Engineer Intelligence Program (1:250,000)	7	423
1960–1965	Tactical Commander's Terrain Analyses	n.d.	n.d.
PACIFIC FIELD PROGRAM			
1945	Armed Forces Pacific Research and Reports Branch	3	45
1946–1952	Supreme Commander for the Allied Powers Natural Resources Section	5	230
1946–1962	Military Geology Branch Office in Tokyo		
	Research and Analysis Geology, Manchuria and Korea	3	166
	Korea Terrain Analyses (1:250,000)	1	38
	Pacific Geological Surveys reports	11	11
	Special reports and miscellaneous papers	14	247
ALASKA TERRAIN AND PERMAFROST SECTION			
1947–1965	Special reports and miscellaneous papers	5	62
1947–1960	Engineer Intelligence Studies	5	16
1954–1964	Geologic maps (mostly 1:250,000 scale)		
	Alaska	1	134
	Greenland	1	59
	Norway	1	12
1960–1964	Basic Terrain Studies (1,250,000 scale)	1	22
1960–1965	Terrain studies of exercise areas (large scale)	1	6
1961–1964	Terrain Study of Alaska (1,2,500,000 scale)	1	12
EUROPEAN FIELD PROGRAM			
1948–1949	Campbell Project reports	13	15
1953–1964	USGS Team (Europe) in Heidelberg		
	Cross-Country Movement (1:100,000 scale)	1	131
	Military Engineering Geology (1:250,000 scale)	1	24
	Special reports	5	35
AUSTERE LANDING-SITE PROGRAM			
1957–1963	Ice-free arctic studies	4	25
1961–1970	Arid land studies	10	31
SPECIAL INTELLIGENCE ELEMENT		n.d.	n.d.
NUCLEAR-TEST DETECTION PROGRAM			
1948–1949	Seismic array and explosion site studies	7	15
1962–1972	Advanced Research Projects Agency reports	6	90

*Abstracted from Bonham, 1997; n.d. = no data.

TABLE 2. MILITARY GEOLOGY BRANCH FY1964 FUNDING*

A. FUND ALLOCATION (listed in decreasing order)

Fund Provider	Allocation	Overhead	Project Funds
Defense Intelligence Agency	$1,390,000	$319,800	$1,070,200
Advanced Research Projects Agency	250,000	62,500	187,500
Air Force Cambridge Research Lab	131,000	24,600	106,400
U.S. Geological Survey	20,000	n.a.	20,000
U.S. Army Europe	16,100	3,700	12,400
Carryover funds	....	....	8,729
Totals	$1,807,100	$410,600†	$1,405,229

B. FUND DISTRIBUTION (listed in decreasing order)

Unit	Man-Years	Project Funds	Fund Provider
Strategic Studies Section	86	$841,400	Defense Intelligence Agency
Nuclear-Test Detection Program	13	187,500	Advanced Research Projects Agency
Alaska Terrain and Permafrost Section	11	171,800	Defense Intelligence Agency + U.S.G.S.
Austere Landing-Site Program	7	115,129	Air Force Cambridge Research Lab
Special Intelligence Element	5	55,000	Defense Intelligence Agency
Pacific Field Program	2	22,000	Defense Intelligence Agency
European Field Program	1	12,400	U.S. Army Europe
Totals	125§	$1,405,229**	

*After Ad Hoc Review Committee, 1964; n.a. = not available.
†Includes $312,400 for Division, $60,100 for Bureau, and $38,100 for Publications.
§Military Geology Branch office: Nine salaries paid by Geologic Division from overhead.
**$1,247,629 for salaries and $157,600 for other expenses (supplies, equipment, travel, etc.).

of topography (Section 24) for each area. Table 4 is a standard table of contents of such a study; the SSS was not responsible for the analysis of culture features.

The World Soil Geography Unit of the Soil Survey, Soil Conservation Service, Department of Agriculture, working under a cooperative reimbursable agreement with the U.S. Geological Survey, produced the analysis of engineering soils and state of ground (soil conditions as affected by weather). This unit initiated the practice of compiling 1:1,000,000-scale uniform soil maps of the world from which they derived their NIS or other small-scale studies. Eventually compilers of other unpublished disciplinary data followed suit and produced maps with standard legends for geology, geomorphology, hydrology, lithology, and vegetation.

Personnel from the SSS also completed a number of other assignments. Some experienced MGB scientists cooperated with the U.S. Department of Army staff to prepare in 1952 a technical manual that provides an excellent and important compendium on geology and its military applications (U.S. Department of the Army, 1952). From 1946 to about 1958, the MGB professionals also intermittently conducted field mapping at large scales on military reservations in the United States, and subsequently prepared military geology reports for use in training and in project activities undertaken by the post engineers. The first such study was comprised of five maps on Fort Knox and vicinity (U.S. Geological Survey, 1946). Another such effort on the U.S. Air Force Academy in Colorado was published as a U.S. Geological Survey Professional Paper (Varnes and Scott, 1967). Beginning in 1954, the MGB compiled 17 Engineer Intelligence Studies, mostly medium-scale syntheses some of which involved field work in the United States, for example, the 79-page report with 40 maps on Fort Benning and vicinity (U.S. Geological Survey, Military Geology Branch, 1958); these studies were requested and prepared for the more specialized users at the U.S. Army Corps of Engineers.

In the early 1960s, the Pacific Engineer Intelligence Program in the SSS compiled thematic overlays for the new 1:250,000-scale topographic sheets of Southeast Asia; the four themes were water resources, construction materials, suitability for military operations (cross-country movement and airborne operations), and suitability for road and airfield construction; the countries covered included Burma, Cambodia, Indonesia, Laos, Malaysia, Thailand, and Vietnam. In FY1964, the SSS spent five man-years on this effort. Also in the early 1960s, the SSS prepared the Tactical Commander's Terrain Analyses, most at a scale of 1:50,000 for a variety of areas, and in FY1964, three and one-half man-years were devoted to this task. As a change of pace, the section was involved in a number of interesting short-term assignments, such as an intermittent effort starting in 1957 to map and interpret the lunar surface materials (one man-year during FY1964), including a special study of the surface of the Moon (Hackman and Mason, 1961).

The SSS had a budget of $841,400, underwriting 86 man-

TABLE 3. ORGANIZATION AND FUNDING OF THE STRATEGIC STUDIES SECTION*

Unit		Man-Years	Funding
Administration		3	$26,000
Source Materials:			
Administration	2		....
Photographs	2		....
Documents	3		....
Total		7	55,400
Geology Research:			
Supervisors	5		....
Geoscientists	31		....
Total		36	375,000
Staff Botanists		2	26,000
World Soils Group		15	195,000
Report Processing:			
Editors/reviewers	5		....
Draftsmen	14		....
Typists	4		....
Total		23	164,000
Grand total		86	$841,400

*Ad Hoc Review Committee Report, 1964.

years of effort in FY1964 (Table 3). These funds covered 3 man-years of administration, 7 man-years of searching out source materials, 23 man-years of report processing (editors/reviewers, draftsmen, and typists), and 53 man-years of scientific research, 15 of which were by the World Soil Geography Unit and 2 by staff botanists. The source-materials and report-processing units of the SSS also assisted many of the other Washington-based MGB programs.

PACIFIC FIELD PROGRAM

At the end of the war, 20 MGU scientists were stationed at the Armed Forces Pacific Headquarters in Manila. In late 1945, these scientists compiled some 20 short papers on miscellaneous northwestern Pacific locations (Table 1). Also, a three-man team was sent from Manila to Korea with the Eighth Army; they prepared a series of 24 short reports principally on mineral resources. But by the end of 1945, most of the MGU scientists moved to Tokyo to serve with the Headquarters of the Supreme Commander for the Allied Powers; these scientists were assigned to the Mining and Geology Division of the Natural Resources Section. Frank C. Whitmore, one of those scientists, stated (personal communication, 1996): "With this move, the duties of the MGU scientists changed from terrain intelligence to the inventory and management of natural resources and thus played a critical role in the Allied policy of restoring the Japanese economy." The MGB in Washington also recruited a variety of disciplinary specialists from other government agencies such that a total of 40 MGB specialists were at the Natural Resources Section between 1946 and 1952; they produced 155 reports, which included many resource studies and one comprehensive volume of the natural resources of Japan (Supreme Commander for the Allied Powers, Natural Resource Section, 1949), the first such synthesis in English. These reports and 75 other miscellaneous studies covered not only Japanese minerals and fuels, but also soils, agriculture, forestry, metallurgy, hydrology, and many living resources (fish, birds, and so on).

In 1946, four MGB geologists were assigned by Washington to the XXIV Corps in Korea. Frank C. Whitmore, a member of this team from Tokyo, stated (personal communication, 1996):

> Under the 56-year Japanese occupation, the Geological Survey of Korea had been staffed by Japanese whose policy did not allow professional education for Koreans. To ensure continuity while Koreans were being trained, one of the MGB geologists, David Gallagher, was appointed director of the Korean Survey. Members of the Korean team also prepared reports on the roads, railroads, and ports of Korea, based upon field reconnaissance and examination of data in Korean offices. The report on the port of Inchon was of particular interest later during the Korean War.

This team and subsequent MGB assignees prepared studies focused mainly on resources and compiled 38 1:250,000-scale terrain-analysis sheets and large-scale maps of key areas.

The MGB was very proud to have four of its scientists (Thomas A. Hendricks, Charles G. Johnson, John Rodgers, and Frank C. Whitmore, Jr.) awarded the Medal of Freedom for their significant contributions during their service in the Pacific region.

In March 1946, the MGB created a new branch office in Tokyo to undertake the Pacific Geological Mapping Program, a research and mapping effort in areas formerly held by Japan (see Corwin, this volume). This new office worked with the Tokyo Geographical Society and Japanese geologists returning home

TABLE 4. TYPICAL TABLE OF CONTENTS FOR THE SECTION ON TOPOGRAPHY IN NATIONAL INTELLIGENCE STUDIES

A. General
- 1. Summary
- 2. Glossary

B. Descriptive Analysis
- 1. Landforms, relief, and drainage pattern
- 2. Drainage characteristics
- 3. Water resources
  - a. Surface water
  - b. Ground water
- 4. Soils
- 5. Rock types
- 6. Vegetation
- 7. State of ground
- 8. Culture features
- 9. Special physical phenomena

C. Military Evaluation
- 1. Cross-country movement
- 2. Constructional aspects
  - a. Suitability for airfields
  - b. Suitability for roads
  - c. Suitability for underground installations

D. Comments on Principal Sources

from overseas or wartime assignments to compile 1:250,000-scale geologic maps; the project, entitled Research and Analysis, Geology of Manchuria and Korea, eventually compiled 149 multi-colored maps of Sakhalin, Kurile Islands, Korea, and Manchuria, and 17 monchrome sheets of South China. Some of these maps subsequently were published by the Tokyo Geographical Society. Other Japanese scientists became involved in producing more than 600 translations of useful articles from the Japanese geo-science literature, and the MGB staff assembled comprehensive annotated bibliographies.

At the same time, American scientists were sent to many Pacific sites, usually on short visits, and prepared 247 special reports and miscellaneous papers on the natural settings and resources. One hydrologist produced brief summaries on the water supply of 29 Japanese and 3 Korean cities. But the particular focus of the Pacific Geological Mapping Program was on 11 island groups in the Pacific representing the complete range of settings from coral-reef atolls to rugged active volcanoes. For each island group, a multidisciplinary team prepared a large-scale terrain analysis similar in content to the small-scale strategic studies prepared in Washington; in fact, two scientists from the SSS were included in the four-man team that mapped one of the first island groups completed (Corwin et al., 1957) and helped establish the format for most of the subsequent Pacific military geology reports. In the subsequent decades, these monographs on the islands have proven useful to the inhabitants. Although the mapping program ended in 1962, in FY1964 (Table 2) the DOD still was supporting effort on continued Pacific-island consultancies.

ALASKA TERRAIN AND PERMAFROST SECTION

The Alaska Terrain and Permafrost Section carried out field studies resulting in the compilation of geologic maps and terrain interpretations of Alaska and other Arctic areas. Frank C. Whitmore, branch chief in 1948, noted (personal communication, 1996):

> For many years, the U.S. Geological Survey functions in Alaska had been conducted by the Alaskan Branch, but in 1948 that branch was abolished, and its functions were distributed among other units in the U.S. Geological Survey. The study of permafrost was assigned to the MGB where it was combined with the preparation of reports on Arctic terrain designed for military use. This program was unique in the MGB because it was jointly funded by the U.S. Geological Survey and the U.S. Army Corps of Engineers.

Most of the 205 geologic maps and 38 studies (Table 1) produced by this section were at a scale of 1:250,000, although the six annual wintertime exercise-area studies were at a larger scale, and the set of 12 maps of the Terrain Study of Alaska was at a scale of 1:2,500,000 (U.S. Geological Survey, Military Geology Branch 1961–1964). Some 72 other special reports and miscellaneous papers were prepared between 1947 and 1972. The supporting Department of Defense annual budget reached a high of $185,000, underwriting 18 man-years of effort; in FY 1964 (Table 2) that had decreased to $151,800 covering 11 man-years, but was bolstered by $20,000 directly from the U.S. Geological Survey budget.

EUROPEAN FIELD PROGRAM

In 1948–1949, the U.S. Army European Command completed the Campbell Project, which constituted terrain studies of Germany, the Baltic countries, and nine nations of Eastern Europe. These studies (Table 1), compiled by European authors working under MGB supervisors based in Salzburg, Austria, were comprehensive and covered a wide range of military geology topics; the four volumes on Germany totaled 1,344 pages, and six other studies exceeded 300 pages each; the maps generally were at the scale of 1:1,000,000.

From 1954 to 1964, generally four to six American scientists of the MGB and the Soil Conservation Service, together with several German scientists and some very competent enlisted personnel, constituted the USGS Team (Europe) based at the Engineer Intelligence Center of the Heidelberg headquarters of the U.S. Army Europe (USAREUR). Their principal focus was the publication of 131 maps entitled Cross-Country Movement at a scale of 1:100,000 for all of Germany, and of 24 maps entitled Military Engineering Geology at a scale of 1:250,000 for western Germany. Each of these maps was a complete document unto itself. The maps for cross-country movement (USAREUR Engineer Intelligence Center, 1962a) showed trafficability classes based on soil and slope, overprinted with obstacle effects of very steep slopes, forest, and surface drainage; the text on the back of the map summarized climate and described terrain regions. The maps for military engineering geology (USAREUR Engineer Intelligence Center, 1962b) showed units of natural materials with critical surface drainage and culture features at a scale of 1:250,000. Summary maps at a scale of 1:1,000,000 showing landforms, vegetation, and overall construction aspects were also on the front of the sheet; complete descriptions and engineering evaluations were tabulated on the back of the sheet for both natural materials and military geographic regions.

The basic continuing program for preparing these maps was punctuated by the preparation of 34 quick Special Reports at the request of U.S. Army Europe and the North Atlantic Treaty Organization. By FY1964 (Table 2), the last year of operation in Europe, the team had been reduced to only one consultant at the Heidelberg headquarters and was funded at $12,400.

AUSTERE LANDING-SITE PROGRAM

The Austere Landing-Site Program, under the direction of the Air Force Cambridge Research Center, prepared site studies (Table 1) based on ice-free land research in the Arctic and on arid lands, both inside and outside the United States. Candidate landing sites were selected by analysis of aerial photography as being at least 1,500 m long with clear approaches. Subsequent onsite studies examined micro-relief, drainage, soils, engineering evaluations, and, in the Arctic, frost features and depth to permafrost. Those sites with acceptable parameters were subjected, where possible, to C124 and C130 test landings (see Krinsley, this volume). Desert sites were field checked in Afghanistan, Australia, Chile, India, Iran, Pakistan, Turkey, and the United States (see Neal, Chapters 11 and 18, this volume; Krinsley, 1970). These austere-site assignments were valued by personnel in the MGB because they afforded interesting opportunities for overseas field work. In FY1964 (Table 2), the air force provided $115,129 for seven man-years of site studies.

SPECIAL INTELLIGENCE ELEMENT

Beginning in 1959, the Defense Intelligence Agency requested the U.S. Geological Survey to seek out volunteers for two-year assignments to work at the agency in a cloistered environment and compile diverse and extremely important, highly classified studies; these scientists also served as geoscience consultants to the special intelligence community and to the highest government levels. In FY1964 (Table 2), as in many other years, DIA provided $55,000 for five man-years of effort. After 1972, this was the only unit to continue its operations; in the 1990s, it still functions at about the same level but with greater funding from a variety of sources (see Leith and Matzko, this volume).

NUCLEAR-TEST DETECTION PROGRAM

In 1948–1949, the Corps of Engineers funded some initial studies of explosion sites and sites suitable for seismic arrays in the United States and other countries to help detection of nuclear explosions (Table 1). But beginning in 1962, MGB scientists compiled a variety of studies under the direction of the Advanced Research Projects Agency, principally to assist in the more sophisticated interpretation of global seismic signals. During the next decade, 18 special reports were prepared and 11 geoscience maps at the scale of 1:10,000,000 were compiled of the U.S.S.R.; these maps generally presented a single critical feature, such as salt deposits, caves, thick sediments, epicenters, permafrost, fossil fuels and mines, and oil and gas deposits. In 1963–1966, an intensive effort produced a five-volume Atlas of Asia and Eastern Europe at a scale of 1:5,000,000 (U.S. Geological Survey, 1966–1969); these volumes covered terrain regions, tectonics, seismicity, features affecting testing (mining sites, salt deposits, and so on), and crust and mantle conditions. In FY1964 (Table 2), the agency provided $187,500 for 13 man-years of effort focused on this atlas. Subsequently in 1974, a Tectonic Map of China and Mongolia composited from this atlas was published by the Geological Society of America (Terman, 1974). Also in the late 1960s, the Advanced Research Projects Agency funded MGB research on the environments at U.S. and U.S.S.R. nuclear-explosion sites, particularly dealing with explosions for peaceful purposes (Terman, 1973).

MGB PRODUCTS AND FUNDING

The summary of the MGB reports and maps (Table 1) prepared between 1945 and 1972 is remarkable in its variety and scope. It is estimated that the total number of products, both recorded (Bonham, 1997) and unrecorded, exceeds 3,000, most of which are maps with succinct scientific explanations and/or engineering interpretations. The MGB contributions to the National Intelligence Surveys covered most of Eurasia; the more detailed studies were focused at various times through the years on Japan, Korea, Pacific islands, Alaska, Germany, U.S.S.R., and Southeast Asia. The record of accomplishments, although diminished after 1972, continues today.

The approximate $1,400,000 of FY1964 was the maximum annual budget that reached the MGB. A summary of the FY1964 funding (Table 2) interrelates the source of funds with the manpower and the budgets for the separate MGB programs and units described above. The principal funder in FY1964 was the Defense Intelligence Agency (77%), and the principal user of the funds for salaries and operational expenses was the Strategic Studies Section (58%). By 1966, most funding had ceased and many of the MGB functions were being taken over by other intelligence agencies. The MGB personnel were gradually reassigned until all but the Special Intelligence Element were discontinued by 1972.

PROBLEMS AND RECOMMENDATIONS

Some of the problems in peacetime MGB operations were obvious. The classification of the products meant that the scientists functioned within a security system that isolated them from the remainder of the U.S. Geological Survey, and thus contacts and relationships were limited. Most assignments did not require any indepth professional specialization; the majority of the successful individuals became adept at a variety of basic or interpretive studies, but did not need, and were not encouraged, to become real experts in any particular aspect of the military applications. The products commonly constituted comprehensive analyses of terrain but rarely required any geoscience research to answer the questions posed by the military sponsors.

Almost all new employees did their apprenticeship within the Strategic Studies Section and some spent much of their MGB career doing library research. Field work was the exception and generally was limited to assignments for the smaller programs in the Pacific, Europe, Alaska, or on austere sites. Frank C. Whitmore, the long-time branch chief, noted (personal communication, 1996): "The impetus for the field programs in the Pacific, Alaska, and the Arctic was the lack of knowledge of those environments that had become apparent during terrain intelligence work in World War II. This fieldwork was planned with a twofold objective: the publication of unclassified applied geology reports and maps by the Corps of Engineers and the publication of the scientific results of the work by the U.S. Geological Survey in scientific journals." Unfortunately, the great bulk of the MGB products was not publishable except within the classified community and then most often without attribution; it was hard to compete within the U.S. Geological Survey for promotions or even for recognition without a publication record. For such reasons, the average individual scientist eagerly sought reassignment to another U.S. Geological Survey branch after a minimum tour of duty with the MGB.

The following recommendations are offered if such a unit is to be reconstructed again in the future to meet the needs of DOD. First, choose an organizational name, such as "Strategic Geology Unit," that has a more positive connotation. Then carefully establish parallel DOD and U.S. Geological Survey project activities such that there are contemporaneous military and civilian components with any specific participant able to contribute to both and thereby escape from security isolation and nonpublishable anonymity; rarely does a military request for information not have a parallel need in the nonmilitary world. Scientists should be encouraged to seek out appropriate training that would permit immediate specialization in some aspect of terrain-data collection, integration, and interpretation; such new expertise promotes more effective professional growth both in the individual and in the unit.

Today any basic areal data collected should be digitized and become part of global standardized data bases within a comprehensive geographic information system (GIS); this objective would build on the first steps taken by the World Soil Geography Unit and the MGB during the 1950s. The current sophisticated software capabilities also should permit evolution of very rapid terrain-analysis programs to provide answers to the most complex application problems, whether they be military or civilian. Such global data bases and the resultant GIS could be the foundation for future environmentally sound, sustainable-development scenarios for many countries and regions and might be a worthwhile task for the U.S. Geological Survey to propose to undertake even on its own initiative.

ACKNOWLEDGMENTS

This paper relied mostly on generally unavailable material, much of which I found in my own file collected over the 23 years that I served with the MGB. The U.S. Geological Survey historian, Cliff Nelson, also graciously shared his materials on military geology. But the two most important documents were cited in the introduction, and the U.S. Geological Survey Open-File Report citing the MGB reports and maps (Bonham, 1997) is an absolute must for anyone that wants to understand the full breadth and scope of the MGB contributions over the years. The paper has benefited from knowledgeable readings by Daniel B. Krinsley and Joshua I. Tracey, and most significantly from the very careful review by Frank C. Whitmore, Jr., who served for 13 years (1946–1959) as the branch chief of the MGB and has a unique knowledge of its activities and its personnel. He made a number of suggested improvements, the most substantive of which have been quoted directly from his comments. Anyone interested in a greater understanding of the content and nature of past and possible future MGB reports should seek out his relevant publications (Whitmore, 1954, 1960).

REFERENCES CITED

Bonham, S., compiler, 1997, Reports and maps of the Military Geology Unit, 1942–1975: U.S. Geological Survey Open-File Report 97-175, 135 p. (Complied by S. Bonham in 1981; edited by W. Leith in 1997.)

Corwin, G., Bonham L. D., Terman, M. J., and Viele, G. W., 1957, Military geology of Pagan, Mariana Islands: U.S. Army Japan, Office of the Engineer, Pacific Geological Surveys Reports, 259 p., 8 maps, 3 cross sections, scale 1:25,000.

Hackman, R. J., and Mason, A. C., 1961, Engineer special study of the surface of the Moon: U.S. Geological Survey Miscellaneous Geologic Investigations Map I-351, 4 sheets, scale 1:3,800,000.

Krinsley, D. B., 1970, A geomorphological and paleoclimatological study of the playas of Iran: Bedford, Massachusetts, Air Force Cambridge Research Laboratories, Research Reports, 2 parts, 815 p., maps.

Supreme Commander for the Allied Powers, Natural Resources Section, 1949, Japanese natural resources, a comprehensive survey: Tokyo, Hosokawa Printing Co., 559 p., supplemental volume has 9 folded maps.

Terman, M. J., 1973, Nuclear-explosion petroleum-stimulation projects, United States and USSR: American Association of Petroleum Geologists Bulletin, v. 57, no. 6, p. 990–1026.

Terman, M. J., principal compiler, 1974, Tectonic map of China and Mongolia: Geological Society of America Map and Chart Series MCH-004, 2 sheets (map, cross sections), scale 1:5,000,000.

USAREUR [U.S. Army Europe] Engineer Intelligence Center, 1962a, Cross-country movement, Berlin: USAREUR Engineer Intelligence Center Series M6425S CCM, Sheet 5046, map and text, scale 1:100,000.

USAREUR [U.S. Army Europe] Engineer Intelligence Center, 1962b, Military engineering geology, Western Europe 1:250,000, Halle: USAREUR Engineer Intelligence Center Series M5011, Sheet NM 32-3, map and text, scale 1:250,000.

U.S. Department of the Army, 1952, Geology and its military applications: U.S. Department of the Army Technical Manual TM 5-545, 356 p.

U.S. Geological Survey, 1946, [Terrain study of the United States] Fort Knox and vicinity: U.S. Geological Survey, 5 sheets, scale 1:50,000. (Construction: materials and suitability of terrain, geology, soils, tank trafficability, terrain diagram.)

U.S. Geological Survey, 1966–1969, Atlas of Asia and eastern Europe to support detection of underground nuclear testing: [Washington, D.C.] Advanced Research Projects Agency, 5 volumes with 7 maps each, scale 1:5,000,000. (Terrain and tectonic regions, tectonics, seismicity, features affecting underground nuclear testing, and crust and mantle conditions.)

U.S. Geological Survey, Military Geology Branch, 1958, Terrain study of Fort Benning and vicinity: U.S. Army Corps of Engineers, Engineer Intelligence Study (EIS) 211, 79 p., 6 maps at 1:100,000 scale, 34 maps at 1:25,000 scale.

U.S. Geological Survey, Military Geology Branch, compiler, 1961–1964, Terrain study of Alaska: U.S. Army Corps of Engineers, Engineer Intelligence Study (EIS) 301, scale 1:2,500,000, 12 maps. (I, Climate; II, Physiographic regions; III, Soils; IV, Rock types; V, Vegetation; VI, Water resources, VII, State of ground; VIII, Permafrost; IX, Cross-country movement; X, Airfield and road construction; XI, Airborne operations; XII, Terrain Summary.)

Varnes, D. J., and Scott, G. R., 1967, General and engineering geology of the United States Air Force Academy site in Colorado: U.S. Geological Survey Professional Paper 551, 93 p., map, scale 1:12,000.

Whitmore, F. C., Jr., 1954, Military geology: Military Engineer, v. 46, no. 311, p. 212–215.

Whitmore, F. C., Jr., 1960, Terrain intelligence and current military concepts: American Journal of Science, v. 258A, Bradley Volume, p. 375–387.

Manuscript Accepted by the Society October 29, 1997

Printed in U.S.A.

Geological Society of America
Reviews in Engineering Geology, Volume XIII
1998

Dearly bought ridges, steep access valleys, and staging grounds: The military geology of the eastern DMZ, central Korean Peninsula

C. P. Cameron
Department of Geology, Box 5044, University of Southern Mississippi, Hattiesburg, Mississippi 39406-5044

ABSTRACT

Steep and broken terrains of the T'aebaeksanmaek dominate east-central Korea and the Demilitarized Zone (DMZ). These terrains are controlled by geologic elements, prediction of which can be used advantageously by the military commander. Narrow valleys and ridge slopes are controlled variously by trends of fault zones, major joint sets, and metamorphic foliations. The trends of these elements follow predictable patterns and have influenced the evolution of harsh, rugged terrain that complicates mechanized infantry mobility and logistic resupply. Slope-stability problems arise from weathering on both natural slopes and oversteepened road cuts. Of equal significance in terrain analysis of this region are the distributions of intrusive granite bodies and their history of weathering and erosion during the Quaternary. The granite masses commonly weather deeper and are more easily eroded than the quartzose schist and gneiss they intrude. Resistant metamorphic rocks generally form ridgelines and hill crests, whereas granite plutons characteristically form wide, bowl-shaped valleys and depressions. By an unfortunate coincidence of geology and geopolitics, many of these granite-floored features are located just north of the 38th parallel; they provided important staging grounds for the North Korean (NK) invasion of South Korea in June 1950.

Severely contested high points of these terrains were won in a series of costly Allied ground operations during the summer and fall of 1951. The significance of these operations, conducted while negotiations at Panmunjon bogged down, escaped some authors of Korean War history who disagreed about whether the enormous human sacrifice needed to secure the ridges was worthwhile. Terrain analysis of this sector clearly shows that with these victories Allied forces secured the strategic high ground controlling north-south access routes and assured that the final line of demarcation would not follow the original border along the 38th parallel. The Allies seized the staging grounds from which the NK II Corps divisions had launched their invasion of the central and eastern sectors. In "elbowing" the line northward, the Allies also secured the mountainous salient that bounds the 1950 "invasion corridor" on its eastern flank. Arguably, their sacrifice bought long-term peace on the Korean Peninsula. Worthwhile indeed!

Cameron, C. P., 1998, Dearly bought ridges, steep access valleys, and staging grounds: The military geology of the eastern DMZ, central Korean Peninsula, *in* Underwood, J. R., Jr., and Guth, P. L., eds., Military Geology in War and Peace: Boulder, Colorado, Geological Society of America Reviews in Engineering Geology, v. XIII.

The terrain is to be assessed in terms of distance, difficulty or ease of travel, dimension, and safety.

—Sun Tzu, *The Art of War*, circa 300 B.C.

INTRODUCTION

Military commanders know that terrain and its variations control the battlefield; that geology controls terrain is, unfortunately, not as well appreciated. Even less understood is the predictable nature of the controlling geologic factors in some terrains, factors that can be used advantageously in tactical and strategic planning. The Korean Peninsula and the history of the savage war fought there during the period 1950–1953 offer many excellent examples of this, but especially striking are those presented by the battlefields of east-central Korea where the terrain is dominated by a north-northwesterly trending mountain range, the T'aebaeksanmaek. From Kumhwa and Hwachon eastward, the elevations of the T'aebaeksanmaek rise sharply. Especially steep slopes commonly occur on the west and south sides of twisted ridgelines. The deeply incised streams and rivers of the region form as significant a barrier to easy access as do the mountain ridges, particularly where the drainages are dammed and impound large reservoirs (Fig. 1). In the second year of the Korean War a series of battles were fought for control of this ter-

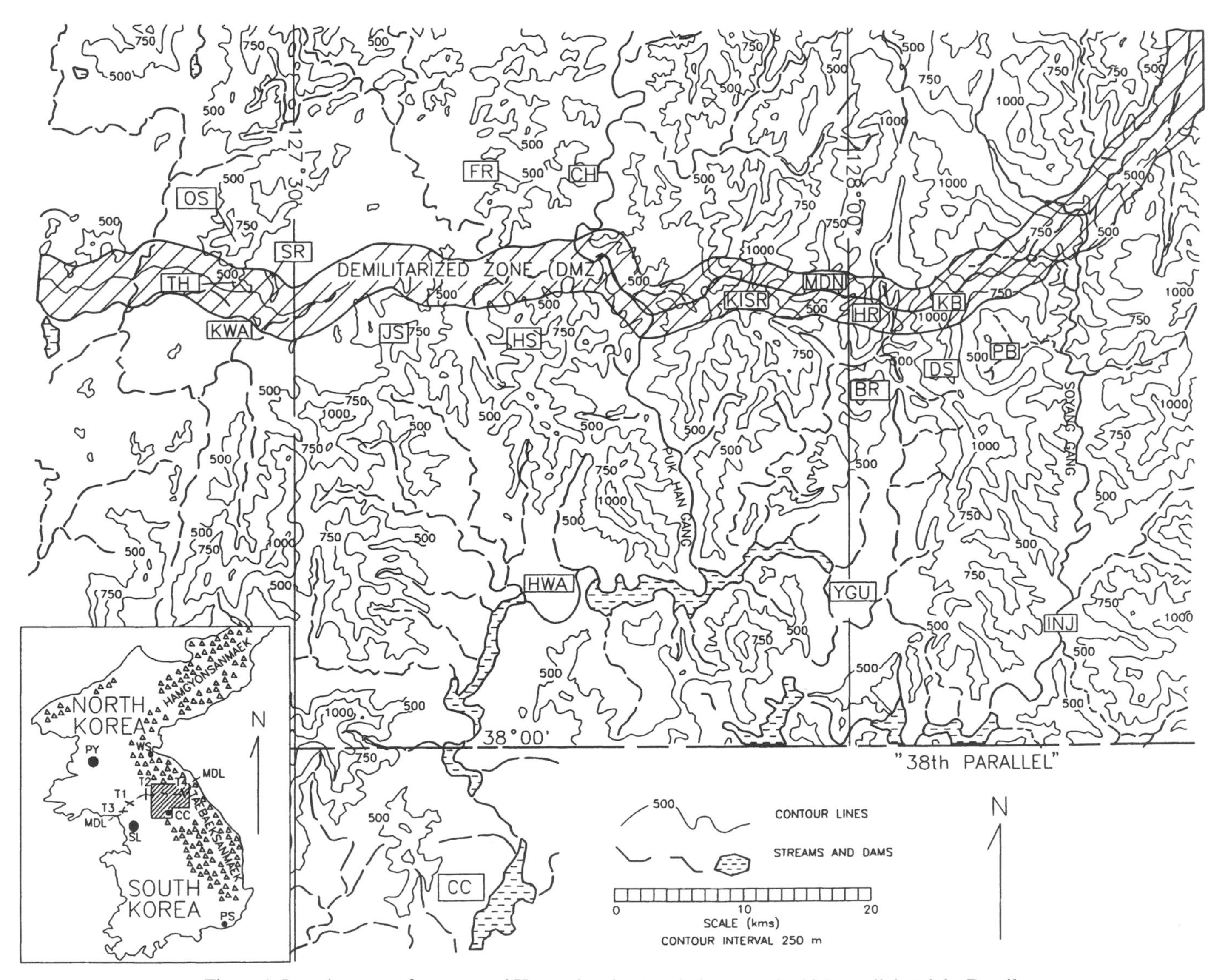

Figure 1. Location map of east-central Korea showing terrain between the 38th parallel and the Demilitarized Zone (DMZ). Abbreviations: MDL—Military Demarcation Line; T1, T2, T3, and T4—Clandestine Tunnels 1, 2, 3, and 4, respectively; SL—Seoul; PY—Pyongyang; WS—Wonson; PS—Pusan; CC—Chunchon; KWA—Kumhwa; HWA—Hwachon; YGU—Yang gu; INJ—Inje; OS—Osongsan; TH—Triangle Hill; SR—Sniper Ridge; FR—Finger Ridge; CH—Capitol Hill; JS—Jeogguensan (Hill 1073); HS—Heuinbausan (Hill 1179); KISR—Kim'il Sung Ridge; MDN—Mundûng-ni; HR—Heartbreak Ridge; KB—Kach'il Bong; BR—Bloody Ridge; DS—Daeusan; PB—Punchbowl.

rain. Staged in bowl-shaped depressions, reached by perilous traverses through steep valleys, and fought on the towering ridgelines, these battles became part of the military heritage both of the United States and of the Republic of Korea: Punchbowl, Bloody Ridge, Heartbreak Ridge, Hill 1220, Capitol Hill and Finger Ridge, and Triangle Hill and Sniper Ridge. The names resonate yet today, as descriptive as they are poignant to those who have walked the terrain.

Many of these hills and ridgelines became the northern border of the Republic of Korea. Although one of the issues during two years of frustrating talks, the position of the sinuating 248 km Military Demarcation Line (MDL) and its flanking 4-km-wide buffer, the Demilitarized Zone (DMZ), was not so much negotiated at Panmunjon as won and held at enormous sacrifice by Allied forces. This border has stood as a defensible entity for more than 43 years. As a measure of its remarkable growth and maturity, Republic of Korea Army divisions, backed by units of U.S. Forces Korea (USFK) who play a supporting role, are now responsible for the defense of this ground.

The objectives of this paper are to (1) document the geologic elements that control terrain in the DMZ sector of east-central Korea; (2) emphasize how these terrain features dominated military action in this area; and (3) note the importance of the victorious, albeit costly, battles for the ridge lines that shifted these terrain conditions to the long-term advantage of Allied forces.

Geographic names and usage

My own notes, maps, and experience in the field played a large part in the compilation of this paper. I relied extensively on detailed topographic maps of the central peninsula that use Korean terms exclusively to describe common terrain features. Use of the suffixes "_sanmaek" (mountain range), "_san" (mountain), "bong" (prominent, usually rounded, hilltop), and "gang" (river or stream) is as much a matter of convenience for me as it is for those who wish to locate these features on the published map sheets. In deference to the nation that was, for a considerable time, my gracious host, I have also adopted the terms "West Sea of Korea" and "East Sea of Korea," while at the same time maintaining an absolutely neutral stance by parenthetically including the traditional names, "Yellow Sea" and "Sea of Japan," respectively.

REGIONAL GEOLOGY

The regional geology of the central Korean Peninsula is dominated by an ancient polymetamorphic shield composed of the Sino-Korean and Yangtse microplates, uplifted and intruded by granitoid rocks during the Jurassic Daebo and the Cretaceous Bulguksa orogenies, respectively. Literature accounts of the historical geology of this region document a 2.8-billion-year record of repeated cycles of sedimentation, orogeny, metamorphism, and igneous intrusion. Thorough and well-referenced treatises that detail the geological evolution of the Korean Peninsula include Lee (1987) and Paek et al. (1993).

Mapping

The geology of the east-central Korean DMZ sector shown in Figure 2 is a digitized version of the 1981 geologic map of the peninsula produced by the Korean Institute of Energy and Resources (KIER) (Hyun et al., 1981). It has been altered in part to reflect new information from a 1993 geologic map published by the North Korean Geological Research Institute (GRI). Other additions and changes were made in the interests of space and simplicity and according to my own mapping and field work done for the U.S. Eighth Army Tunnel Neutralization Team (EUSA-TNT).

For example, the "Punchbowl Granite" has been added because it does not appear on any Korean or Japanese maps of the region. A major fault in the Puk Han Gang area has been shifted to reflect field evidence. Finally, GRI mapping indicates that the terrains to the northeast of the Kumwha sector of the DMZ, assigned by KIER to the Ordovician Limestone Series, are all part of the upper and middle Proterozoic Sangweon System (GRI, 1993).

Very little detailed geologic mapping has been done near the DMZ since the work of the Japanese in the 1920s and 1930s. Until recently, nonmilitary access was restricted for security reasons and, except for military infrastructure supported by small garrison towns, the area was quite underdeveloped. The geology of the Punchbowl was discussed by Cameron (1993, this volume). Other accounts of the detailed geology of portions of this sector are contained in several reports and maps, mostly unpublished, completed for the EUSA-TNT by me and several of my colleagues, for example, Cameron and Ballard (1992).

Lithology and structure

The "basement" of this part of central Korea is an ancient metamorphic terrane called either the Kyônggi Basement Metamorphic Complex or the Kyônggi Gneiss Complex, the age of which is Archean (Kim, 1970, 1973; Sang and Hee, 1984; Kim and Na, 1987). This basement complex comprises layered, very hard, metamorphic rocks, many of which had a sedimentary origin, and nonfoliated but commonly lineated metamorphic rocks of apparent igneous origin. Metamorphic grade commonly is upper greenschist and amphibolite facies developed both under high-temperature and high-pressure conditions and low-temperature and high-pressure conditions.

Specific lithologies include banded quartz-feldspar-biotite-garnet gneiss, migmatite, biotite-garnet gneiss, subordinate lenses of biotite schist, quartzite, and crystalline carbonates (marbles) and calc-silicates. Coarse and very coarse, generally nonfoliated quartzo-feldspathic (gray-blue plagioclase) biotite-garnet "orthogneiss" also is present in the eastern Punchbowl sector. The metamorphic terranes are interlaced by amphibolite masses that are dark, hard-lineated rocks that have both concordant and discordant contacts with the host Proterozoic gneiss. Most amphibolite probably is the product of volcanism and

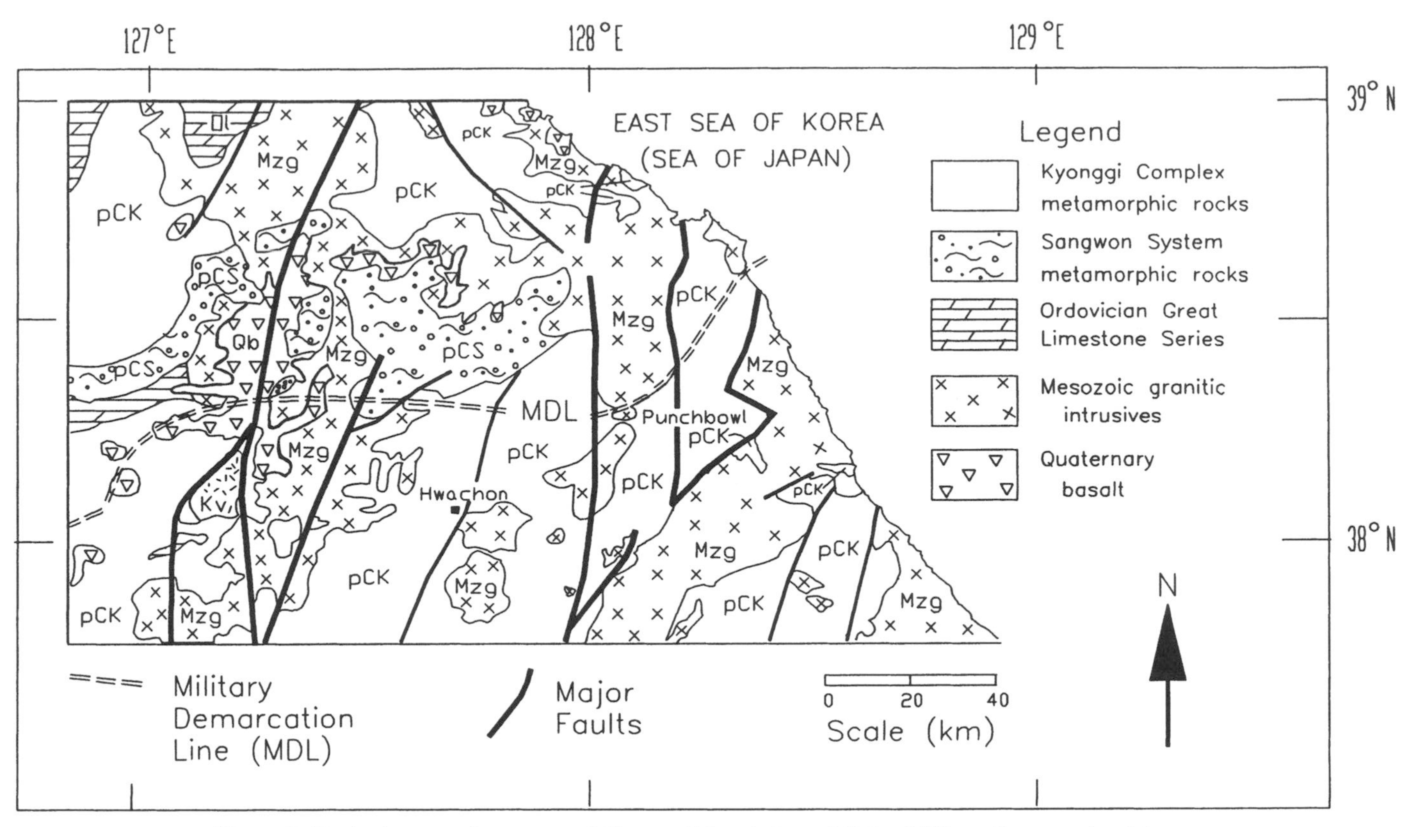

Figure 2. Geologic map of east-central Korea. Abbreviations: MDL—Military Demarcation Line; pCK—Precambrian Kyônggi Gneiss Complex; pCS—Precambrian Sangwôn System; Mzg—Mesozoic Granite; Ol—Ordovician Great Limestone Series; Qb—Quaternary basalt.

mafic (basaltic) intrusion during an early Precambrian orogenic event or series of events.

The Kyônggi Gneiss Complex was affected by at least one, and quite possibly two, later orogenic movements about 1400 and 900 Ma. Subsequently, the complex was invaded extensively by granitic stocks and batholiths during the Triassic Songrim disturbance and the Jurassic Daebo orogeny, 230–140 Ma. This regional intrusive episode was responsible for the creation of calc-silicate schists in the Sadang-u Series (Late Proterozoic Sangweon System) in and adjacent to the Kumwha area. Later still, the explosive volcanism that signaled the culmination of major uplifts of the Sino-Korean and Yangtse plate or plates was followed by intrusion of late-kinematic granite plutons during the Late Cretaceous Bulguksa disturbance, 90–65 Ma.

Faulting in these rock masses spans most of their history. Late crustal movements, apparently during the Bulguksa disturbance, led to reactivation of an ancient tectonic fault zone and the formation of a north-northeast–south-southwest trending rift, the "Chugaryeong Rift Valley," between Wonsan and Seoul. Basalt erupted along this feature during the Late Cretaceous. During the Pleistocene, 11 extensive basalt flows spread along a series of grabens in a reactivated portion of the rift northeast of Seoul. These basalt flows extended mostly to the southwest from their eruptive centers, but some also flowed southeastward down the Pukhan Gang into the T'aebaeksanmaek; basalt floors the river and is sparsely exposed along its banks near the DMZ.

Physiography

The physiographic form of Korea as a peninsula is a relatively late phenomenon. During the Late Cretaceous, terranes now part of Japan separated from Korea and moved eastward (Paek et al., 1993, p. 504–518). Full development of the East Sea of Korea (Sea of Japan) probably occurred during the Oligocene and Miocene epochs. During the late Miocene and early Pliocene, uplifts of as much as 200 m along the T'aebaeksanmaek on the east coast established current drainage patterns in the central portion of the peninsula. Renewed severe dissection of elevated regions followed and set the stage for development of the modern West Sea of Korea (Yellow Sea). Although the West Sea of Korea was the site of intracontinental basin development during the Paleogene and Miocene, it did not subside sufficiently to allow an oceanic connection until late Pliocene or early Pleistocene. The present outline of the Korean Peninsula and its classic western ria coast are a function of Pleistocene and Holocene sea-level change.

Pleistocene effects

Two key elements of Pleistocene history in this region played a crucial role in the development of terrain features, and hence military geology, in the Korean Peninsula. The first is that the peninsula was not glaciated. The only report of glacial deposits in the peninsula refers to a "glacial deposit and cirque" reported on Kwanmobong (2,541 m) in North Korea (Sasa, 1938). The lack of glaciation coupled with an apparently wet climate produced deep chemical weathering of intrusive Mesozoic granite plutons particularly where they are crossed by major fault zones. Saprolite accumulations tens of meters thick are not uncommon in faulted and sheared granitoid masses of the west-central peninsula. Well-developed grus and deep sandy soils also are developed over much of the granite in the eastern sector of the DMZ, notably, over the Punchbowl, Yang-gu, and Chunchon granite bodies, and over those parts of intrusive granitic masses situated at relatively low elevations in the terrain west of Hill 1073 (Joegguensan).

The second element is the eustatic sea-level changes during the Pleistocene that played a significant role in terrain development. Major low stands drastically increased hydraulic energy in stream systems already incised along structurally controlled avenues of erosion. Continued severe dissection of the peninsula and selective erosion of the deeply weathered granite plutons combined to produce a terrain of alternately hard and soft rocks, generally high and steep in the east, low and rolling in the west.

MILITARY GEOLOGY

Key elements that influence considerations in military geology and terrain analysis in this region include: (1) geologic controls on access routes and mobility of ground forces; (2) geologic controls on staging areas; (3) geologic controls on slope stability, particularly along access routes; (4) geologic and terrain controls on tunnel construction and stability; and (5) geologic and terrain controls on surface and subsurface hydrology. Additionally, both terrain aspects and rock-mass physical properties have important implications regarding strike capabilities of sea and air forces, and the effectiveness of the weapons that they deliver to targets on or beneath the ground surface.

The steep and broken terrains of the eastern DMZ sector are segmented by long, narrow valleys that historically have provided north-south access through this region. These access routes connect a series of broad bowl-shaped and, in places, elongated topographic basins that long have been the sites of substantial farming communities and centers of local government, commerce, and communication. In some places, for example, the Kumhwa and Chunchon areas, road networks coexisted with strategically important railroads. These terrain features, shown in Figure 1, control the disposition and mobility of military ground forces in this region.

Further influence on overland mobility is exerted by rock and soil-mass slope stability, which is a problem wherever road cuts are oversteepened relative to fracture and foliation plane orientation. The nature of the terrain and the occupation of strongly defended fixed positions by opposing forces also gave rise to an underground tunneling campaign by North Korea designed both to conceal and protect military assets from aerial bombardment and to provide avenues for clandestine infiltration of forward fire-support elements and special forces.

Access routes

The control of structural geologic elements on mountain routes amenable to mechanized access can be demonstrated by comparing lineations created by faults and joints with trends derived from the regional road network. The results for the eastern DMZ sector are shown in Figures 3, 4, and 5.

Landsat lineaments from Kang (1979, 1984) were digitized for the entire central Korean Peninsula, and the orientation data were processed by a Rockware Inc. computer program ("Rose") to produce rose frequency diagrams (Fig. 3). Linear trends of the regional access-road network also were digitized and processed (Fig. 4). Fracture data from outcrop studies in the northern Punchbowl, including measurements from depths 10–300 m underground in Tunnel 4 and its countertunnel, were processed similarly.

The compiled results show only minor statistical differences in the three data sets, particularly when the data are filtered to show the main orientation directions. The statistical similarity between data sets of the east-northeast orientations is particularly pronounced. Work at Tunnel 4 showed that the east-northeast orientation is dominant in granite; whereas all three directions are present in the metamorphic terranes. First- and second-order fractures exert the major control on road access on this portion of the DMZ and are very important with respect to providing east-west crossing routes to link major valleys with northerly and northwesterly trends.

Fractures also control drainage in the area, particularly the incised channels of the Soyang Gang and Pukhan Gang. Because these systems are dammed and, for at least part of the year impound large reservoirs, they form significant barriers to overland transportation and mobility. Many of the narrow bounding roads around these reservoirs are forced to hug the sides of steep valley walls where they are constantly threatened with closure by landslides, commonly during the July-September monsoon season.

Other than along the main corridors, overland travel in this terrain is slow and difficult because of steep topographic gradients and rapid change in slope morphology, especially in terrains formed of the Precambrian metamorphic rocks. There, slope orientation and gradient are controlled chiefly by planes of jointing and metamorphic foliation. Rapid change in terrain morphology is a function of intersecting joint and metamorphic foliation planes having moderate to steep inclinations with orientation trends in every quadrant of the compass. A 20-km-trip over unimproved secondary roads can require as long as 90 min, and for heavy transport (2.5 tons or greater), considerably longer than that. Bad weather, especially in the winter, can render many of the winding secondary roads along steep ridge flanks impassable for weeks at a time for anything heavier than jeeps and light trucks. This allows commanders to saturate the main corridors with defensive fighting and artillery positions, as was illustrated

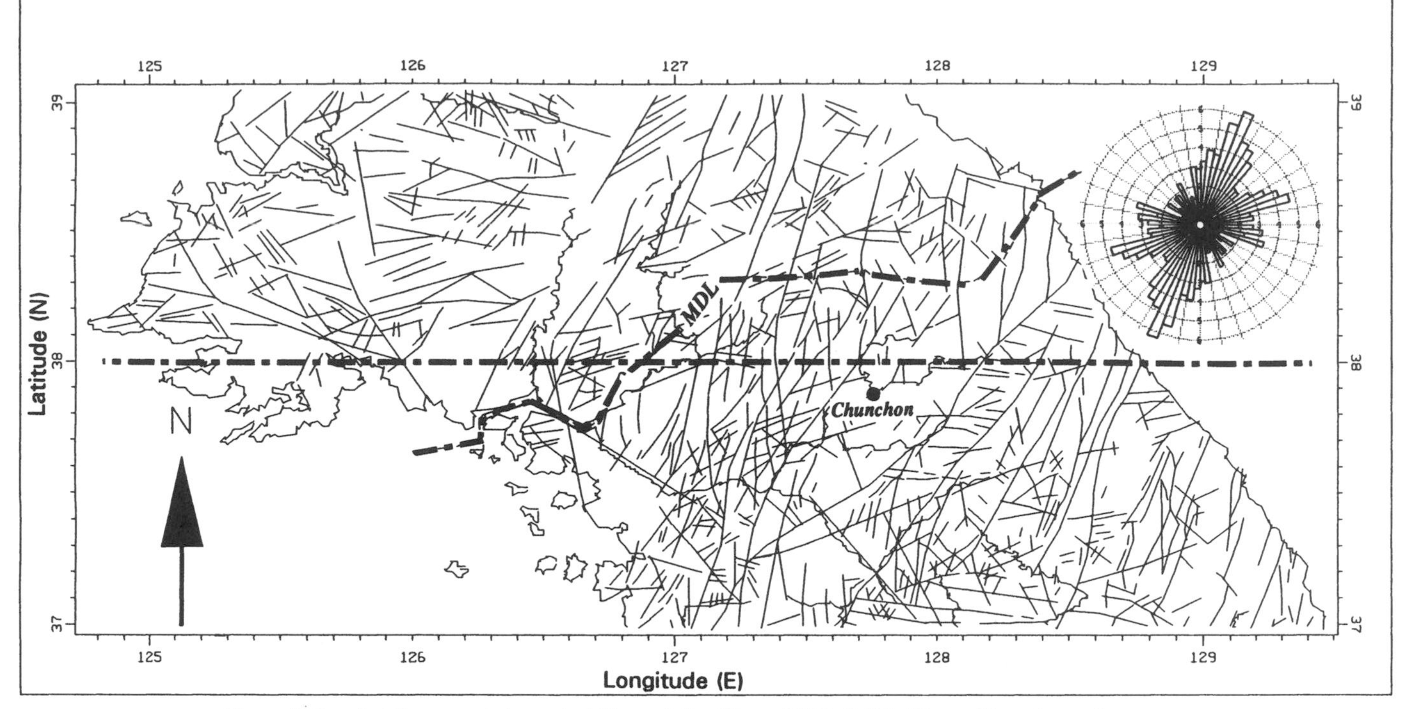

Figure 3. Landsat lineaments in central Korea (after Kang, 1979, 1984). The unfiltered rose frequency diagram in the upper right shows the major directional trends of all linears. Abbreviation: MDL—Military Demarcation Line.

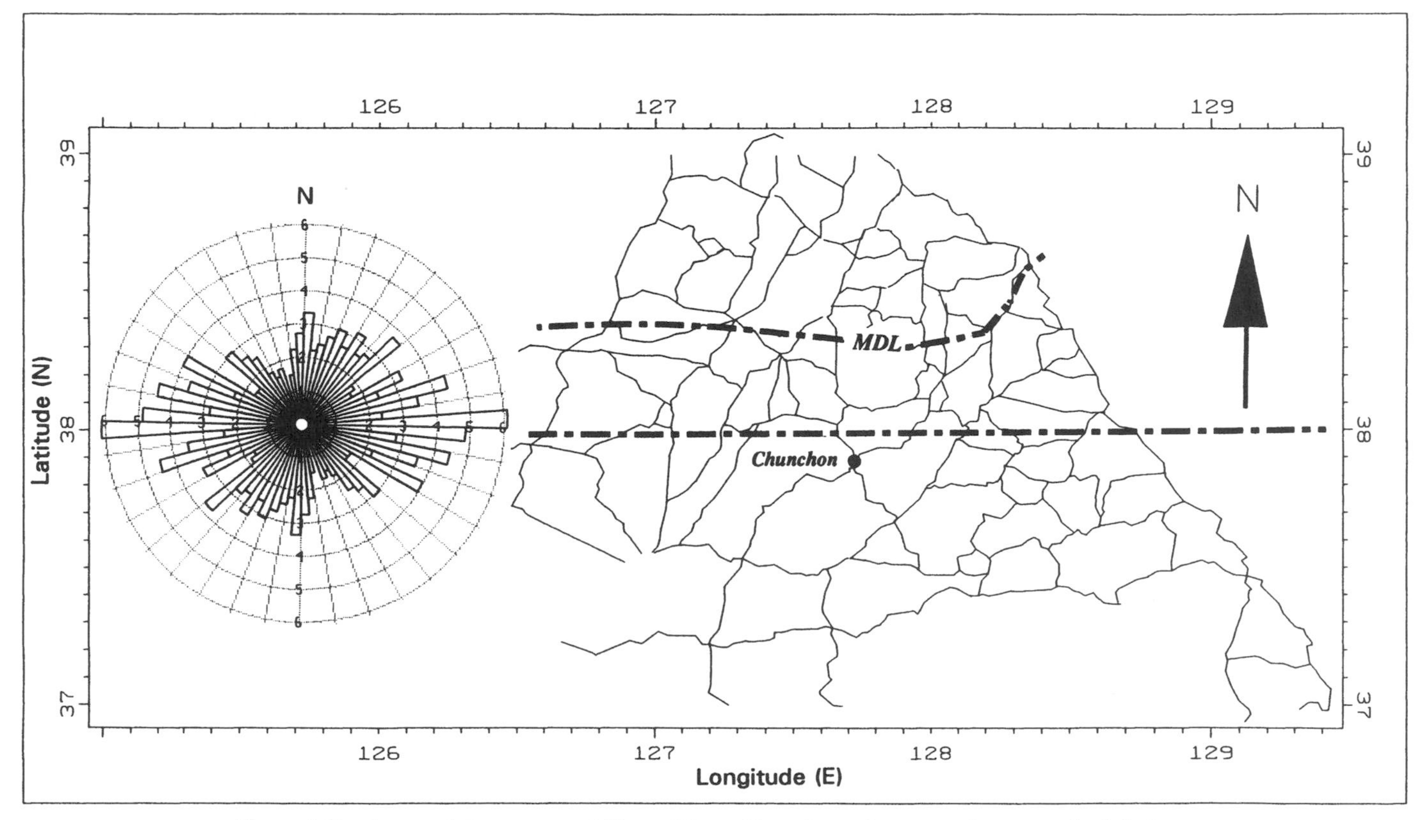

Figure 4. Road network in east-central Korea. The unfiltered rose frequency diagram at the left shows directional trends of all linear road segments in east-central Korea. Abbreviation: MDL—Military Demarcation Line.

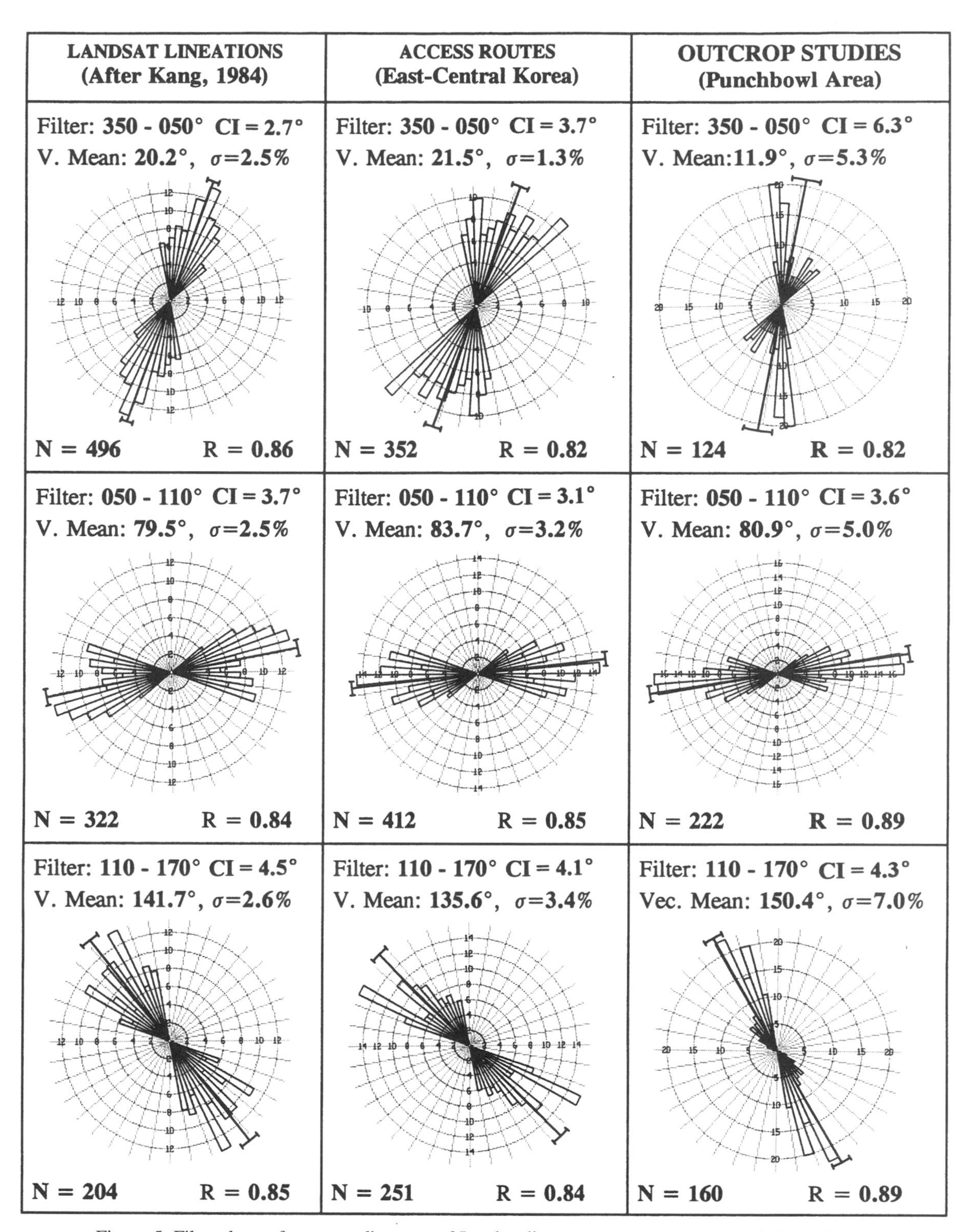

Figure 5. Filtered rose frequency diagrams of Landsat lineaments, access routes, and discontinuity measurements (surface and subterranean) at the Punchbowl. Included are filter parameters, vector mean directions (V. Mean), confidence interval around the vector mean (CI), standard deviation of the frequency percent in rose petals (σ), sample population (N), and a measure of the dispersion around the vector mean, (R).

during the opening days of the Korean War in the battle for the city of Chunchon.

Chunchon is situated just south of the 38th parallel in a wide, granite-floored symmetrical valley into which winds a series of roads from the north, south, east, and west. Capture of Chunchon by the afternoon of the first day was the main objective of the II Corps of the North Korean Peoples Army (NKPA or Inmun Gun). As in other sectors, the NKPA attack was to follow the major roads from staging areas south and southwest along their lines of departure (Fig. 6).

On the morning of June 25, 1950, the 4th and 6th Regiments of the NK 2nd Division, staged in an open-valley complex immediately northwest of Hwachon, launched their surprise assault on the central sector with a two-pronged attack on Chunchon. Fighting from well-prepared positions on the heights surrounding the city, the 7th Regiment of the Republic of Korea Army (ROKA) 6th Division inflicted severe casualties on the NK forces, principally with artillery barrage and counterbattery fire. The NK advance in the center was stopped in its tracks and held up for more than three days. The enemy lost the advantage of surprise, suffered a casualty rate of more than 40 percent, and lost about half of its heavy artillery. The ROKA defenders were forced to retreat on the fourth day when they were flanked by armor and infantry elements of the NK 7th Division that had crossed the 38th parallel, without significant opposition, from staging grounds at Inje and Yang gu (Appleman, 1961, p. 19–28).

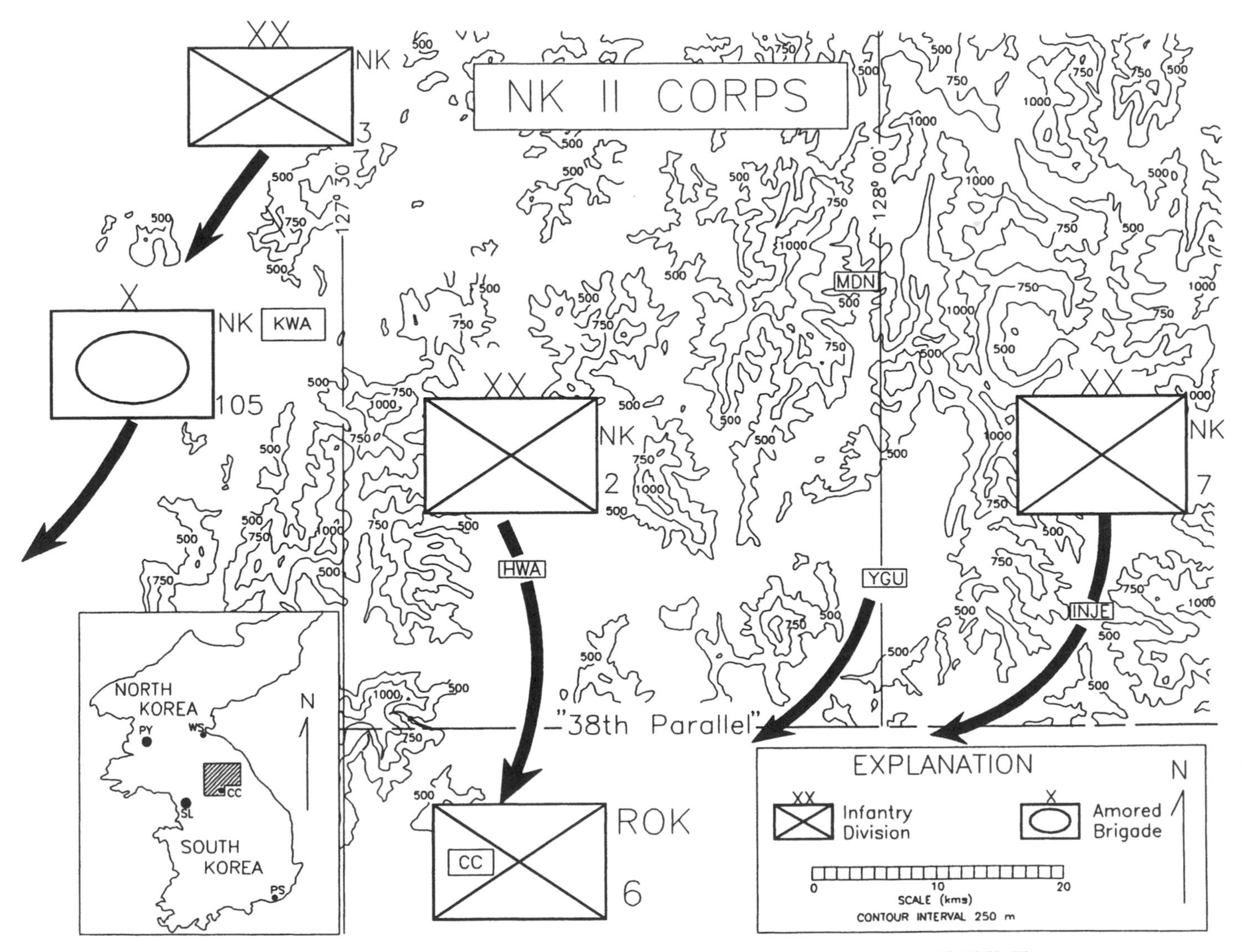

Figure 6. The North Korean attack across the 38th parallel in the eastern mountains on June 25, 1950. The attack was launched from topographically suitable staging grounds near local and regional population centers, for example, Hwachon and Inje. Abbreviations: SL—Seoul; PY—Pyongyang; WS—Wonson; PS—Pusan; CC—Chunchon; KWA—Kumhwa; HWA—Hwachon; YGU—Yang-gu; MDN—Mundûng-ni; INJE—Inje.

Staging grounds

Wide, bowl-shaped, topographic basins occur at the Punchbowl, at Yang gu and Chunchon, just south of Hwachon, and in a broad area to the immediate south and east of Kumhwa (Fig. 1). These depressions are a function of lithology, specifically the occurrence of Mesozoic intrusive granitic rocks. These rocks have a mineralogy and mode of emplacement that make them prone to rapid chemical weathering and erosion relative to the more resistant quartzose schist and gneiss that they intrude (Fig. 7).

The broad nature of most of these features coupled with gentle topographic gradients at their lower elevations and adequate supplies of water make them ideal for staging purposes, all other factors being equal, for example, proximity to enemy positions. This ground allows for the location of headquarters, communication and aviation facilities, training and rest areas, and hospitals. These areas also support substantial crops, for example, rice, that can provide a significant portion of the food requirements of permanently resident military forces.

Some granite masses are easily eroded and form topographic basins in the central portion of the peninsula, but others do not. This leads to the speculation that the finer-grained, more easily erodible masses are related to the Cretaceous Bulguksa event, whereas resistant granite that forms rocky, exfoliated ground is more likely to be associated with the Jurassic Daebo orogeny (Cameron, this volume). This contention is most likely an oversimplification of a difficult problem. In some instances, severe tectonism of these rock masses played a significant role in differential weathering and erosion. Selective, late-stage hydrothermal alterations, for example, extensive saussuritization of plagioclase, may also have predisposed portions of the intrusive masses to accelerated rates of weathering.

The use of texture and geotechnical properties to assign absolute ages to rocks is hardly recommended; petrography, isotope geochemistry, and tectonic analysis are necessary to resolve any relationship between the timing of granite emplacement and the character of granite weathering in the central Korean Peninsula. This is not an academic exercise because depth of weathering and soils development are important aspects of military geology and terrain analysis. Geologic studies, today as in the past, provide important information for terrain analysis by satellite images or aerial photographs. In many places, rock-mass physical properties must be projected into areas of denied access where such information aids in estimates of weapons effects, particularly with respect to ground penetration. Rock-mass physical properties are important parameters in models used to evaluate various effects of near-surface and underground detonations, including those involving the generation, coupling, and propagation of seismic waves.

Other terrain suitable for staging developed as a function of tectonism and subsequent erosion of metamorphic rocks in and around the intersections of major fault systems, for example, the staging terrains near Inje and immediately northwest of Hwachon.

Figure 7. The northwest Punchbowl Rim above its weathered and eroded granodiorite floor. The photograph was taken from a small hill of granite in the north-central Punchbowl floor. In the foreground are resistant granite corestones excavated from a large bulldozed cut. Total relief shown is approximately 750 m. Clandestine Tunnel 4 penetrated into the Republic of Korea by crossing underneath the rim in the right-center of the photograph (Cameron, this volume).

Slope stability

The Precambrian rocks of the Korean Peninsula were structurally deformed in at least three major orogenies and affected to a lesser extent in at least two others. Uplift during the Late Cretaceous and Tertiary periods resulted in deep dissection and the creation of a steep, high-relief terrain in much of eastern Korea (Fig. 8). Weathering and erosion, aided by nonglaciation and relatively high precipitation during the Pleis-

Figure 8. Steep ridges north of Yang-gu along the DMZ are formed of schist and gneiss. Severe dissection along the steep valleys as well as slope aspect are controlled by rock-mass discontinuities, for example, faults, major foliation planes, and joints. Particularly well illustrated in this photograph are the slope aspects in the background hills, including triangular facets, formed by regional joints and foliation. The cleared fenceline along the foreground ridgeline helps protect observation posts overlooking the Military Demarcation Line (MDL) in the DMZ.

tocene and Holocene, is pervasive along fractures and foliation planes where the Precambrian rock masses are exposed or present in the shallow subsurface. Deforestation and denudation during the Japanese occupation and the Korean War resulted in half a century of accelerated soil erosion and weathering along rock-mass discontinuities such as bedding, foliation, fault, and joint planes. The logistics and training requirements of large-scale military infrastructure had an impact on this region. For example, unstable talus fans on steep bedrock slopes commonly are the product of the construction of ridgeline roads and other military infrastructure. The T'aebaeksanmaek has recovered to the extent that young and surprisingly thick vegetation now covers slopes cut and blasted bare during wartime actions. This recovery notwithstanding, mass movement is very common and rock-slope stability is a problem for engineered infrastructure along oversteepened road cuts, railway cuts, and around dam abutments.

Rock slides or falls pose the most significant hazard to mechanized mobility in this area. Rock slides or falls commonly occur where moderately steep (30°–60°) joint and foliation planes crop out on steep (45°–70°) excavated cuts oriented at right angles to dip directions of discontinuities. Both planar failures and wedge failures, resulting from the intersection of discontinuities that combine both perpendicular and parallel trends relative to excavated cuts, are common.

The role of foliation planes deserves special mention. Much of the metamorphic basement rock in this region is of sedimentary origin. Original boundaries between beds of quartz-rich versus quartz-poor lithologies are preserved as major foliation planes. Differential stress during folding, metamorphism, and uplift caused foliation plane slip or shear, or both. At the same time, low shear-strength clay minerals, e.g., chlorite, crystallized and filled the dilated discontinuities. In other places, simple and complex shear zones developed along bounding surfaces marked by graphite-rich phyllonites. The foliation planes thus formed are particularly prone to failure when oriented at the critical attitudes described in the previous paragraph. These discontinuities tend to be smoother, that is, unstepped, and better lubricated by graphite and clay minerals such as chlorite and smectite than those of most joint sets.

Downhill toppling of jointed rock masses is not uncommon along steep ridgelines, particularly along those in the Punchbowl, Heartbreak Ridge, and the Hill 1220 areas. Where this phenomenon is pronounced, fresh scree fans cover portions of the slopes beneath the ridgeline and attest to the unstable nature of the rock mass and the ongoing nature of the mass wasting.

Comparatively fewer problems with slope stability and rock falls or slides occur in the terrains underlain by granite in this part of Korea. Chemical weathering of granite produces sandy soils and grus and a more subdued terrain with gentler topographic gradients, particularly south of the DMZ. The granite has relatively fewer planes of discontinuity than the metamorphic rock and these characteristically are widely spaced; joint apertures generally are very rough and interlocking. Even where weathering is moderate and the granite relatively fresh, the nature of the rock mass and its pronounced lithologic homogeneity is such that its "stand-up" character is excellent, even on near-vertical excavated cuts.

Triggering mechanisms for rock falls and slides in this region include ground vibration by any mechanism, for example, microseismic activity, thunder associated with cloudbursts, construction blasting, explosive ordinance demolition, or artillery ordinance impacts. All of these mechanisms are most effective following the prolonged rainfalls common in the summer months or following spring thawing of snow and ice.

The fall of 1989 provided a good example when the only low-elevation road that provided access to the Punchbowl was blocked by a significant rock slide (Fig. 9). Reconstruction of the slide event showed that an east-northeast–trending road cut was oversteepened to about 65°NW during road construction that involved widening and preparation for asphalt sealing. No measures such as benching or emplacement of rockbolts or anchors were employed to prevent failure of the hard, well-jointed, sheared, crystalline gneiss and migmatite rock mass. The oversteepened road cut was excavated in a northwest-facing slope controlled by a major joint set developed along moderately dipping (45°–55°) regional foliation. Following 7–10 days of rain at the close of the monsoon season, planar and wedge failure along these discontinuities brought down tons of rocks, many as huge blocks, and cut access along this important route for several weeks. Anecdotal evidence suggests that the slide may have been triggered by explosive-ordinance demolition in the adjacent Soyang Gang.

Underground war. If the east-central Korean Peninsula demonstrates one single aspect of military geology, it can be found in the adage: "You don't have to occupy some terrains in

Figure 9. This rockslide involved large slabs of metamorphic rock that blocked the low-elevation route to the Punchbowl north of Inje during the fall of 1989. Oversteepening of the road cut resulted in failure along planes controlled by regional joints that dip moderately from right to left in the photograph. This joint set parallels the metamorphic foliation in the area, a factor that appreciably degrades rock quality along some joint planes with respect to ground water inflow and the presence of low-strength aperture fillings, for example, chlorite.

order to control them." The dominating ridgelines of the eastern DMZ sector offer sweeping vistas to those fortunate enough to occupy them. Correspondingly, the same terrain can frustrate the ambitions of mechanized infantry who, lacking in adequate air cover and forced to follow the low and narrow defiles, can be decimated by artillery fire controlled by ridgeline observers.

Empirical evidence gathered over the 40 years following the Korean Armistice Agreement suggests that the NKPA battle plan for reconquest of South Korea calls for a multipronged light and mechanized infantry assault preceded and later supported by intensive artillery barrages. Without air superiority to direct artillery fire, this plan must presuppose that tactically advantageous ridges and hill tops can be occupied in the hours preceding such an attack and, at the earliest moment, that counterbattery fire can be neutralized. Part of the latter proposition involves cutting communications within, along, and between defending divisional boundaries. Because overland assault would at the very least cost the element of surprise, the North embarked on the one strategy that historically has highlighted the vulnerability of ground forces defending fixed fortified positions: tunneling and underground infiltration of forces.

Then. The military use of extensive underground facilities was introduced by the Chinese Communist Forces (CCF) after they entered the Korean War. Allied air superiority made defense of NKPA and CCF ground troops from air attack largely a matter of camouflage and extensive use of underground fortifications. "The tunnel became a great Chinese Institution" said a senior member of Marshall Peng's staff wryly (Hastings, 1987, p. 276). One must wonder what he would have thought of Ho Chi Minh's efforts under the battlefields of Vietnam during the succeeding two decades! In the eastern sector, the use of underground works by the CCF and NKPA is reasonably well documented, albeit much of the information is from unpublished military records and somewhat anecdotal with respect to technical considerations. Of particular concern to the allies were tunnels having well-concealed portals protected from artillery and air strikes, excavated in steep north-facing reverse slopes. The tunnels commonly crossed underneath mountains to gun positions opening on the south front. These installations as well as company-sized bunkers were deep, beyond the effective penetration depths of conventional weapons. The underground fortifications and their network of firing positions formed a line rarely seen in modern warfare. Excavated 10 times as deep as any dug on the western front in World War I, many of these fortifications were constructed in granite and metamorphic bedrock; it was reasoned by the communist forces that they might have to withstand nuclear attack (Fehrenbach, 1963, p. 500–506).

Old mines and prospect adits from the era of Japanese occupation played a role there and elsewhere along the furiously contested frontlines of mid-1951. In several accounts of the battles for Bloody and Heartbreak ridges, notable reference exists to the enemy's use of the Mandae fluorite mine at Mundûng-ni for concealment and logistic resupply (Fig. 6). Accounts of heroic efforts of U.S. 2nd Division Engineers to clear a route through a narrow defile blocked by boulders, mines, and booby traps on the west flank of Heartbreak Ridge prefaced the subsequent race north by the tanks of Baker Company (72nd Tank) to take and destroy the Mandae underground works. In the same valley, Baker Company destroyed (or neutralized) 350 bunkers and other underground installations in 10 sorties over five days (Fehrenbach, 1963, p. 517–526). Continued bombardment from the air and close support by Marine Corsairs sending napalm, rockets, and machine gun bullets into those bunkers had failed to neutralize them; the Inmun Gun had learned to dig. Another generation would continue the tradition two decades hence.

Now. It is a matter of historical record that the NKPA was ordered to build infiltration tunnels of considerable offensive potential in the early 1970s (Breen, 1993; Atkeson, 1995). Allied forces discovered and neutralized some of this effort, capturing clandestine Tunnels 1, 2, 3, and 4 (Fig. 1). These tunnels crossed the Military Demarcation Line (MDL) and penetrated well into the southern margin of the DMZ (Bettencourt, 1988; Cameron, this volume). The number of built tunnels yet to be discovered is uncertain; however, the situation with respect to the objectives of NKPA Clandestine Tunnel 4 is pertinent to this discussion.

NKPA Clandestine Tunnel 4 was constructed under the northwest rim of the Punchbowl, the floor of which is situated to the east and northeast of its highest points, Kach'il Bong (1,252 m) and Daeu-San (1,172 m), respectively. The details of the geology and exploration case history are discussed in Cameron (this volume). Based on his mapping of the Punchbowl and Tunnel 4, the author has concluded the following:

1. It is unlikely that the NKPA intended to break out large numbers of infantry onto the floor of the Punchbowl itself. The high ground formed by resistant gneiss rimming this feature makes its gently rolling, granite floor a potential "killing field."

2. The likely objective of Tunnel 4 was the Punchbowl western rim, especially its extension south of Deausan. A clandestine force occupying this ridgeline has long east-west line-of-sight control of two major north-south access routes that lead to the local population centers of Inje and Yang-gu, then west and south to the dams, power, and transportation infrastructure at the regionally important centers of Hwachon and Chunchon. Such a force could direct fire not only to support attacking infantry but also to prevent ROKA reinforcements from reaching strategic defensive positions.

3. Tunnel 4 was constructed at depths ranging to 300 m underneath the Punchbowl rim. At these target depths and in steep terrain, the defender is at a distinct disadvantage in attempting to locate a clandestine tunnel. Only good intelligence combined with advanced technology and dogged persistence over a long period of time can neutralize such threats.

4. Tunnel 4 was constructed in both Mesozoic granite and Precambrian metamorphic rocks, unlike Tunnels 2 and 3 that were constructed only in granite; this was the first real proof that the undertakings of the 1970s and later were not confined to a preferred lithology. Follow-up mapping of Tunnel 4 revealed that, as at Tunnel 2, NKPA engineers used a major north-

northwest–trending joint set to speed tunnel construction and to maximize tunnel stability (Cameron, this volume).

Surface and subsurface hydrology

The large reservoirs pooled by dams on the Pukhan Gang and Soyang Gang systems provide the major supply of water to the eastern and central sectors of the country. These reservoirs have historic strategic importance as major suppliers of water for irrigation, hydroelectricity, and commercial use. In addition to its other absurdities, the politically convenient but otherwise arbitrary partitioning of Korea along the 38th parallel split these systems between North and South Korea. The South lost considerable hydroelectric power, the North considerable storage. Not surprisingly, the dams and surrounding rail and road network at the regional center of Chunchon were a major objective of the NK 2nd Division at the opening of the Korean war.

Local surface supplies of water in the deeply weathered granite-floored open valleys and staging areas are adequate for permanently resident military forces and the production of local crops. Most local tributary streams are dammed; replenishment occurs as a function of spring meltwater and summer monsoon rains. Ground-water supplies, on the other hand, generally are inadequate in this region, except in local situations. Porosity is controlled mostly by such discontinuities as foliation and joints in the metamorphic rock. Although saturated zones occur at relatively shallow depths (0–10 m) in most boreholes, ground-water flow is very slow. Most boreholes having 13-cm diameters and depths in the 100–300 m range can be pumped down in 30–50 minutes. Porosity is similarly controlled by fractures in the intrusive granite. The water table commonly is positioned between highly weathered and moderately to slightly weathered rock masses. Some intergranular porosity occurs in the weathered granite, but permeability is low because of the high percentage of clays in these zones.

The best aquifers are those formed by the intersections of faults in quartzose metamorphic rocks where local fracture-intersection frequency is high and clay-mineral occurrence low. Even where such situations exist and can be tapped, ridgeline positions and fortifications generally must be supplied with water from lower elevations where catchment and recharge areas are enlarged. This supply is generally accomplished by pumping from boreholes situated in a good local aquifer or from a local cistern trapping surface runoff.

DISCUSSION

The focus of military geology of the east-central sector of Korea is first and foremost the use of difficult terrain. Ample recorded evidence shows that both General Matthew B. Ridgeway and his Eighth Army Commander, General James Van Fleet, had full appreciation of the dominating nature of the mountains that controlled their battlefront in the spring of 1951 (Schnabel, 1972, p. 397–402). The forward divisions of the Eighth Army were deployed along "Kansas" and "Topeka," defensive phase lines established in favorable terrains to check the CCF offensive of that spring. A year after the sudden onset of the Korean War, communist and U.N. armies faced each other across this front, both having traversed the length and breadth of the peninsula in campaigns fought at appalling cost to both sides.

The Inmun Gun and the CCF had lost the best of their veteran forces during their southern assaults and subsequent retreats, albeit they were reinforced by tough, younger troops of equal fanaticism and bravery. U.N. forces, on the other hand, had learned to fight and win in Korea but ultimately were directed by politicians apprehensive about the possibility of war with the Soviet Union in western Europe. Allied forces, responsible to an electorate weary of the high-attrition struggle necessary to win decisive victory and largely indifferent to the fate of a small land half a world away, were directed by those who were inclined to negotiate peace in Korea. Of such stuff are stalemates made, particularly when armies meet in steep, rugged terrains and narrow defiles.

In the eastern sector, Van Fleet decided to increase the defensive strength of the Kansas Line and its outpost screen to the west, "Wyoming," by occupying the dominating mountain ranges to their immediate north and deploying his units to utilize fully the water barriers that occurred within them (Fig. 10). Although restricted by General Ridgeway's directives from engaging in any major offensives, Van Fleet issued a series of Letters of Instruction (LOI) to his corps commanders ordering them to establish patrol bases ahead of his main lines of resistance. He then pressed to be allowed to "elbow" his lines forward at least 16 km into new, steep, ridgeline positions. This, he argued, would remove disadvantageous southerly "sags" along his front and improve the Allies defensive position further with respect to terms being discussed in Panmunjon (Hermes, 1966, p. 72). An approving Ridgeway readily agreed to this course of action. His decision shifted the nature of the war from one of fluid, back-and-forth movement to one characterized by the media and others as a deadly children's game of "king of the mountain." Given the vaguely derogatory nature of the latter commentary, considerable polemic is encountered among media personnel of the time and military historians as to whether the high cost in human life required to secure the forward ridgelines was indeed worthwhile (Fehrenbach, 1963, p. 500–526; Hastings, 1987, p. 330–344).

The battles necessary to secure the ridgelines from the Punchbowl to the Kumhwa during the long summer and fall of 1951 were as savage as any fought during the war. The key high points were contested bitterly everywhere along the line. Figure 10 shows the position of the frontlines on July 1. A clear indication of the intensity of the high-attrition struggle is seen in the rotation of major Allied and enemy units to areas of frontline engagements. Entire divisions rotated on and off the line during the summer and fall campaigns, as can be seen by comparison with Figure 11, which shows the dispositions of the frontline units at the end of October 1951. By the end of the summer the 1st Marine Division, having started the summer

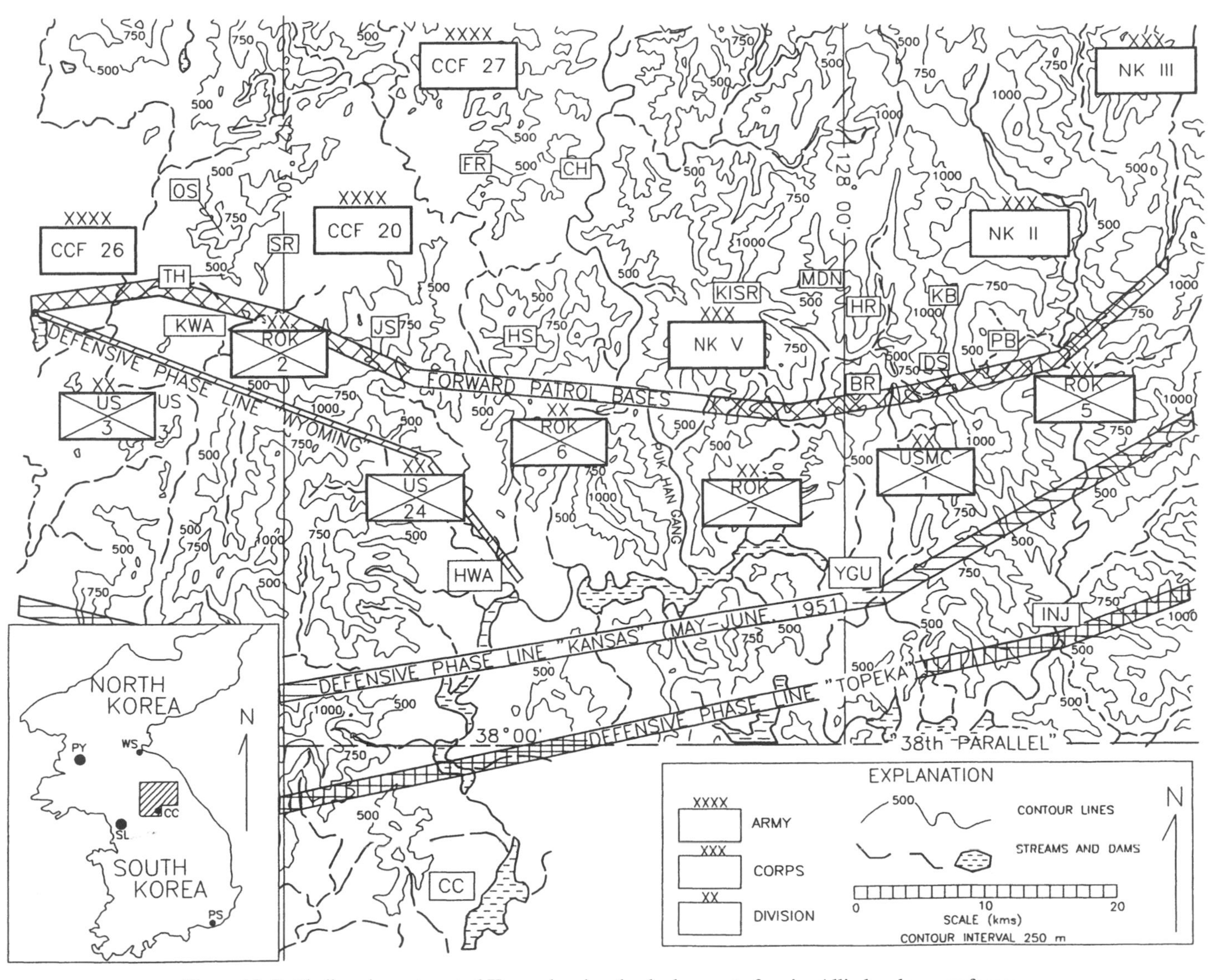

Figure 10. Battle lines in east-central Korea showing the deployment of major Allied and enemy forces prior to the summer and early fall operations of 1951. Abbreviations: SL—Seoul; PY—Pyongyang; WS—Wonson; PS—Pusan; CC—Chunchon; KWA—Kumhwa; HWA—Hwachon; YGU—Yang gu; MDN—Mundûng-ni; INJ—Inje; OS—Osongsan; TH—Triangle Hill; SR—Sniper Ridge; FR—Finger Ridge; CH—Capitol Hill; JS—Jeogguensan (Hill 1073); HS—Heuinbausan (Hill 1179); KISR—Kim'il Sung Ridge; HR—Heartbreak Ridge; KB—Kach'il Bong; BR—Bloody Ridge; DS—Daeusan; PB—Punchbowl; US—U.S. Army; USMC—U. S. Marine Corps; ROK—Republic of Korea Army; NK—North Korea Peoples Army; CCF—Chinese Communist Forces.

encamped just north of Yang-gu, occupied ground to the northeast of the Punchbowl whose northern rim they had successfully gained from the NK 1st Division. The U.S. 2nd Infantry Division (U.S. 2nd ID) moved into the line to support the ROK 5th Division in its battle for Bloody Ridge. Soldiers of the U.S. 2nd ID not only helped take Bloody Ridge but campaigned on through the torrid summer. Together with their ROK counterparts and European U.N. contingents, they bore the brunt of the fighting in the struggle for Heartbreak Ridge and Hill 1220 on the Kim'il Sung Ridge. Having fought the good fight they turned over their sector to the U.S. 7th Division on September 23. Hard success in October also was won farther west as the U.S. 24th Division, flanked by the ROKA 2nd and 6th Divisions, fought their way north to take the Capitol Hill and Finger Ridge complexes. Van Fleet had his 16 km of defensible ground; the sags were gone from his front.

The enemy resisted mightily, and its use of difficult terrain was superb. The Inmun Gun and CCF seemed to have had as great an appreciation for the value of these mountain strongholds as the Allied forces. Having had the advantage of being able to start a war from the open staging grounds and corridors just north of the 38th parallel, the communist forces perhaps had a

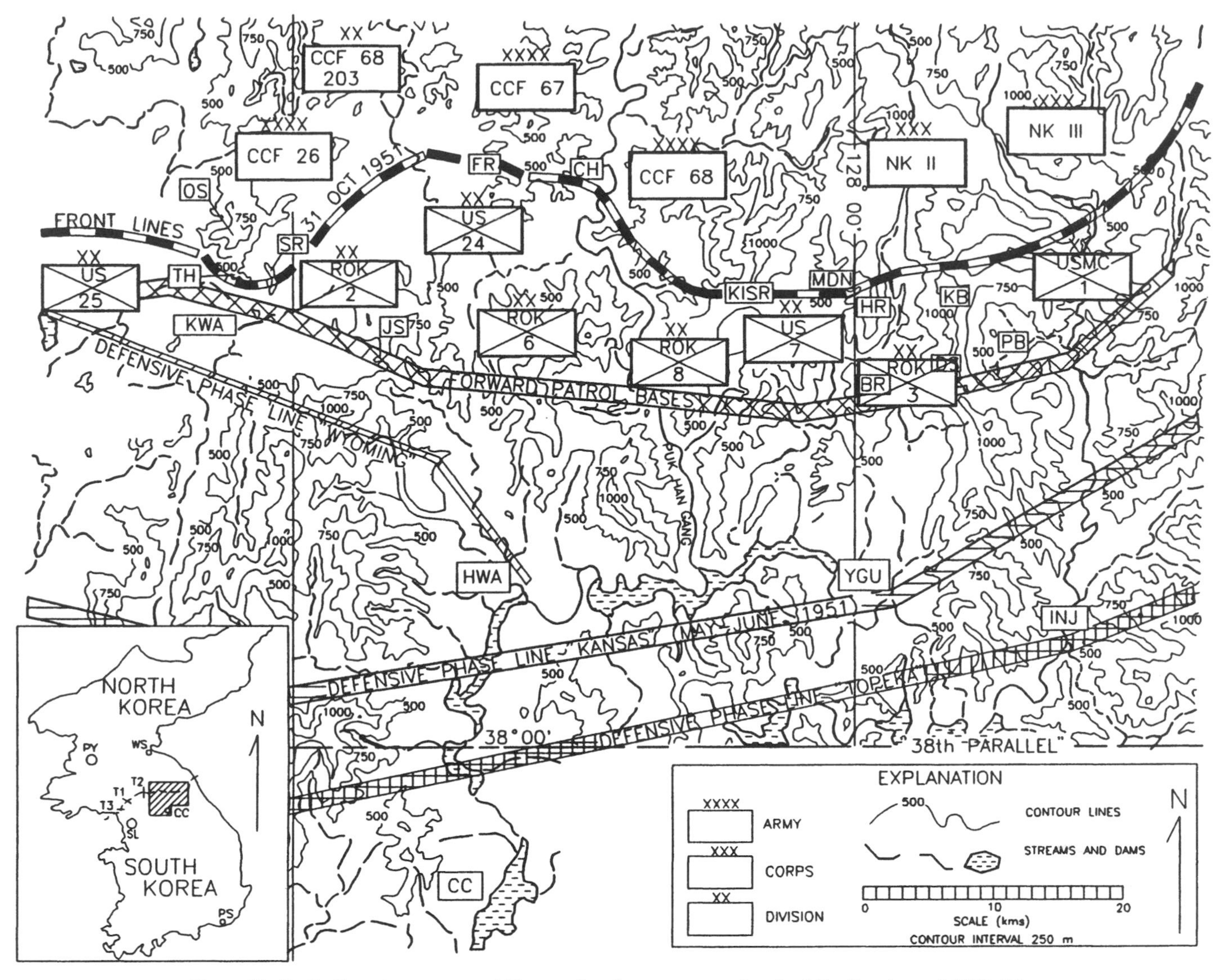

Figure 11. Battle lines in east-central Korea after the summer and early fall offensives of 1951. The frontline, straightened when the Allies withdrew from Finger Ridge and Capitol Hill in 1952, ultimately became the northern border of the Republic of Korea. Abbreviations same as Figure 10.

unique perspective of what the loss of these terrains would mean in the long run.

The Allies were successful in securing most of the major objectives of the spring through fall operations of 1951. However, the events of the succeeding year showed the limits to which the enemy could be pushed in yielding valuable terrain. For example, the ROKA forces that had fought so hard to take and hold the Finger Ridge and Capitol Hill terrains were forced ultimately to yield those features a year later. Experience showed that the Allied forces would have to mount exceedingly costly operations, at corps level, to secure the flanking mountainous terrain above the twisted valley of the Pukhan Gang and its tributary streams. In the end the terrain was judged too costly for the anticipated gain. A similar situation developed near Kumhwa in the fall of 1952.

The eastern Chorwon Valley in the Kumhwa area (Fig. 11) is dominated by a high massive hill called Osongsan (Hill 1062). The CCF had built strong fortifications on and within this feature. They also took advantage of Japanese-era underground mines, which were common in the area, for concealment and staging purposes. Attempting to improve the defensive position of the line established by the 1951 operations, Van Fleet recommended adoption of IX Corps plan "Showdown" (Hermes, 1966, p. 310–318). The main objective of Showdown was to take the Hill 598 (Triangle Hill)–Sniper Ridge terrain immediately to the south of Osongsan with elements of the U.S. 7th Division and the ROKA 2nd Division. The 135th Regiment, 45th Division, CCF 15th Army, fighting with excellent knowledge of the sharp ridgelines and using well-prepared positions and deep tunnels, refused to yield despite lengthy aerial and artillery bombardment and intense ground pressure from the two Allied divisions. After more than a month of bitter fighting, and having sustained more than 9,000 casualties, the Allies broke off the action. At this stage of the war, and in the absence of concerted corps-level offensives, these were "ridges too far."

Despite these setbacks, the Allies secured the strategic high ground controlling north-south access routes from the Soyang Gang to the Pukhan Gang and on to the eastern Chorwon Valley. They established a defensible front and assured that the final line of demarcation would not follow the original border at the 38th parallel. They also seized the staging grounds from which the NK II Corps divisions had launched their invasion of the central and eastern sectors, and in "elbowing" the line northward secured the mountainous salient that bounds the Chorwan "invasion corridor" on its eastern flank. Ultimately, what the Allies achieved in taking and holding the terrains north of the 38th parallel in east-central Korea was *a secure stalemate that holds to this day. Steep, broken, mountainous terrains will not yield better results often, and this should be appreciated by military commanders.* History shows that this zealously protected stalemate led to a long-term peace and a prosperous, developing economy in a new democracy—the Republic of South Korea. That such an enterprise as Van Fleet's was worthwhile can scarcely be denied.

CONCLUSIONS

Military operations in severe terrains such as the T'aebaek-sanmaek in eastern Korea present unique and challenging problems. Many would contend that such terrain should not be the focus of extensive military operations at all. This opinion is justifiable in many respects but is somewhat naive given the history of war and the propensity of nations to engage their forces irrespective of the suitability of terrain. The prudent military commander must learn to read difficult terrain to judge accurately what is achievable at reasonable cost and when to avoid "ridges too far." Because these terrains are unforgiving, objectives must be clearly defined and then pursued swiftly and decisively. Appreciation of the geological controls of terrain assists in maximizing their use in military operations in much the same way that such understanding provides major insights for planning in civilian land-use projects. The analysis of the eastern DMZ of Korea provides the following lessons.

(1) The steep, narrow valleys and ridge slopes of the eastern sector of the DMZ are controlled by the orientation of fault zones, major joint sets, and metamorphic foliation. The trends of these elements follow predictable patterns; however, they nonetheless limit mechanized transport and logistic resupply, as do slope stability problems arising from oversteepened roadway cuts and military infrastructure construction.

(2) Detailed analyses of fracture discontinuities applied to the prediction, prevention, or initiation of rock-slope landslides along the main access routes should be a continuing concern to the military engineer interested in mechanized troop mobility and support in this sector of the peninsula.

(3) Steep ridgelines are, in most places, underlain by schist and quartzose gneiss of the Kyônggi Metamorphic Basement Complex in this sector of the DMZ.

(4) Broad, bowl-shaped, topographic depressions in this sector mostly are underlain by intrusive Mesozoic granitoid rock. These features provide staging and headquarters areas and provide natural locations for troop concentration points, fixed and rotary-wing aircraft facilities, and large-vehicle and rail depots.

(5) The igneous and metamorphic rocks of this sector generally are suitable for tunnel construction, support requirements, and maintenance needs. The military use of tunnels and other underground installations will remain a threat as long as fixed, fortified, defensive lines along the Korean DMZ remain a reality.

(6) Costly allied victories during the spring, summer, and fall of 1951 won staging grounds from which the NKPA II Corps divisions earlier had launched their invasion of the central and eastern sectors of South Korea. In a series of savage engagements fought over severe terrain, the U.N. forces "elbowed" the Kansas Line northward and secured the difficult mountainous salient that bounds the Chorwan "invasion corridor" on its eastern flank. Their sacrifice bought long-term stalemate and peace on the Korean Peninsula.

When the position is such that neither side will gain by making the first move it is called temporizing ground.

—Sun Tzu, *The Art of War*, circa 300 B.C.

ACKNOWLEDGMENTS

Part of the data described herein was collected in the field during the author's tenure as Project Geologist for the U.S. Eighth Army Tunnel Neutralization Team (EUSA-TNT-J2), funding for which was provided by the U.S. Army Belvoir Research, Development, and Engineering Center (BRDEC) through a Broad Agency Grant Contract (DACA39-90-K-0029) issued by U.S. Army Engineer Waterways Experiment Station (WES). Sincere appreciation is extend to Ray Dennis (BRDEC) and to Bob Ballard (WES) for their long-term backing and encouragement during the course of this work. Field work in the DMZ sector would not have been possible without the enthusiastic support of the officers and enlisted soldiers of EUSA-TNT(J-2), the Republic of Korea Army Tunnel Detection Section (ROKA-TDS-G2), and the ROKA soldiers who man the frontline divisions.

REFERENCES CITED

Appleman, R. E., 1961, The United States Army in the Korean War: South to the Naktong, north to the Yalu (June–November, 1950): Washington D.C., U.S. Government Printing Office, 813 p.

Atkeson, E. B., 1995, The North Korean military threat in perspective: Arlington, Virginia, Association of the U.S. Army Institute of Land Warfare Paper No. 21, 21 p.

Bettencourt, V. M., Jr., 1988, A theoretically based evaluation of the Korean tunnel search, *in* Proceedings, Third Technical Symposium on Tunnel Detection, Golden, Colorado, January 1988: Fort Belvoir, Virginia, Combat Engineering Directorate, U.S. Army Belvoir Research Development and Engineering Center, p. 262–283.

Breen, M., 1993, Korea's tunnel vision: U.S. searches DMZ for enemy underfoot: Washington Times, April 8, 1993, p. A1.

Cameron, C. P., 1993, Exploration case history and engineering geology of "Tunnel-4," northern Punchbowl sector, Korean Demilitarized Zone, *in* Proceedings, Fourth Tunnel Detection Symposium on Subsurface Exploration Technology: Fort Belvoir, Virginia, Combat Engineering Directorate, U.S. Army Belvoir Research, Development, and Engineering Center, p. 595–615.

Cameron, C. P., and Ballard, R. F., 1992, Geology of the Chu'Dong area, Demilitarized Zone sector, Kumwha-Gun, (Kangwon-Do), Republic of Korea: Vicksburg, Mississippi, Technical Report GL-92-4, U.S. Army Engineer Waterways Experiment Station, 35 p.

Fehrenbach, T. R., 1963, This kind of war: A study in unpreparedness: New York, Macmillan Company, 689 p.

Geological Research Institute Academy of Sciences, 1993, Geological map of Korea: Pyongyang, Foreign Languages Books Publishing House, Democratic Peoples Republic of Korea, scale 1:1,000,000, 1 sheet.

Hastings, M., 1987, The Korean War: New York, Simon and Schuster, 389 p.

Hermes, W. G., 1966, The United States Army in the Korean War: Truce tent and fighting front: Washington D.C., U.S. Government Printing Office, 571 p.

Hyun, B. K., Um, S. H., and Chun, H. Y., compilers, 1981, Geologic map of Korea: Korea Institute of Energy and Resources (KIER), scale 1:1,000,000, 1 sheet.

Kang, P. C., 1979, Geological analysis of Landsat imagery of South Korea (I): The Journal of the Geological Society of Korea, v. 15, p. 109–126.

Kang, P. C., 1984, Lineament map of Korea from Landsat imagery, *in* A study on remote sensing application for the tectonic framework of the Korean Peninsula [Ph.D. thesis]: Seoul, University of Korea, scale 1:1,000,000, 1 sheet.

Kim, H. S., and Na, K. C., 1987, Part IV: Igneous activity, *in* Lee, D. S., ed., The geology of Korea: Seoul, The Kyohak-Sa Publishing Co., p. 289–344.

Kim, O. K., 1970, Geology and tectonics of the mid-central region of South Korea: Journal of the Korean Institute of Mining Geology, v. 2, p. 73–90.

Kim, O. K., 1973; The stratigraphy and geologic structure of the metamorphic complex in the northwestern area of the Kyônggi Massif: Journal of the Korean Institute of Mining Geology, v. 6, p. 201–218.

Lee, D.-S., ed., 1987, The geology of Korea: Seoul, The Kyohak-Sa Publishing Co., 514 p.

Paek, R. J., Kan, H. G., Jon, G. P., Kim, Y. M., and Kim, Y. H., editors, 1993, The geology of Korea: Pyongyang, Foreign Languages Books Publishing House, Democratic Peoples Republic of Korea, 619 p.

Sang, H. O., and Hee, Y. C., 1984, Geological evolution and tectonic classification of Korea: Seoul, Republic of Korea, Korean Institute of Energy and Resources Report 86-7, p. 35–90.

Sasa, Y. O., 1938, Glaciated topography in the Kambo Massif, Tyosen (Korea), *in* Proceedings of Hokkaido University: Hokkaido, Japan, v. 4, nos. 1–2(J).

Schnabel, J. F., 1972, The United States Army in the Korean War: Policy and direction: The first year: Washington D.C., U.S. Government Printing Office, 443 p.

MANUSCRIPT ACCEPTED BY THE SOCIETY OCTOBER 19, 1997

Printed in U.S.A.

Geological Society of America
Reviews in Engineering Geology, Volume XIII
1998

Clandestine Tunnel-4, northern Punchbowl, Korean Demilitarized Zone

C. P. Cameron
Department of Geology, Box 5044, University of Southern Mississippi, Hattiesburg, Mississippi 39406-5044

ABSTRACT

The Punchbowl is a large topographic basin in the rugged T'aebaek Mountains (T'aebaeksanmaek) of east-central Korea. Its steep northern rim is part of the Korean Demilitarized Zone (DMZ). This terrain was the target of various infiltration attempts in the postwar period and its potential to host clandestine tunnels was indicated by Allied intelligence studies of the early 1970s. In December 1989, at a depth of 145 m in intrusive granodiorite, a coherent cross-borehole electromagnetic anomaly with a signature indicative of an air-filled cavity was recorded by tunnel-search teams using continuous wave and pulsed ground-probing borehole-radar systems. Natural air-filled cavities at this depth were precluded by geological considerations and ground-water elevations. Target-evaluation drilling followed and, on December 24 at 0130 hours, a drill bit penetrated "Tunnel-4." A borehole television camera provided clear images of artifacts of human tunnel-construction activity. Two and a half months later, Korean engineers, using a Wirth 3-m tunnel-boring machine (TBM), gained access to the tunnel and confirmed its nature and purpose.

Mapping and surveying of the tunnel revealed that (1) the tunnel was constructed by drill and blast methods in Mesozoic intrusive granodiorite, similar to that hosting previously identified DMZ Tunnel-2 and Tunnel-3, and in Precambrian gneiss; (2) the tunnel slopes up to the south at an average grade of 2.3%, ranging from a low of 0.91% to a high of 3.5%; (3) the average cross-sectional dimensions of the tunnel, 1.6 m × 1.6 m, are somewhat smaller than other hardrock tunnels discovered elsewhere in the DMZ; and (4) standard rock-mass rating schemes do not accurately predict tunnel-construction difficulty for hardrock tunnels smaller than 2 m × 2 m in cross section.

INTRODUCTION

The Punchbowl is aptly named. It is a large topographic basin whose steep northern rim forms part of the southern boundary of the Korean Demilitarized Zone (DMZ) in the rugged T'aebaeksan of east-central Korea (Fig. 1). Almost circular in plan, the basin has a rim-to-rim diameter of about 7.5 km. The floor of the basin has elevations approximately 400–500 m above mean sea level (MSL). Summit elevations along the dramatic western and northern rim are in the range of 1,000–1,242 m, whereas those of the gentler eastern and southern rims attain elevations of 800–1,000 m. The Punchbowl gained worldwide attention and a place in military history during July and August of 1951 when high summits along its western rim, especially Taeu San and Kach'il Bong, were major objectives of a series of allied ground operations.

The battle for the Punchbowl was followed by Allied victories at similar steep ridges and peaks 9 km directly west at Bloody Ridge and Heartbreak Ridge, and at the Kim 'il Sung Ridge (Hill 1220) 9 km farther west. With these victories, Allied forces secured the strategic high ground controlling north-south access routes from the Soyang-Gang to the Pukhan-Gang, and they stabilized the northern border of the Republic of South Korea at its current DMZ position.

During the early 1970s, defector reports, direct observations

Cameron, C. P., 1998, Clandestine Tunnel-4, northern Punchbowl, Korean Demilitarized Zone, *in* Underwood, J. R., Jr., and Guth, P. L., eds., Military Geology in War and Peace: Boulder, Colorado, Geological Society of America Reviews in Engineering Geology, v. XIII.

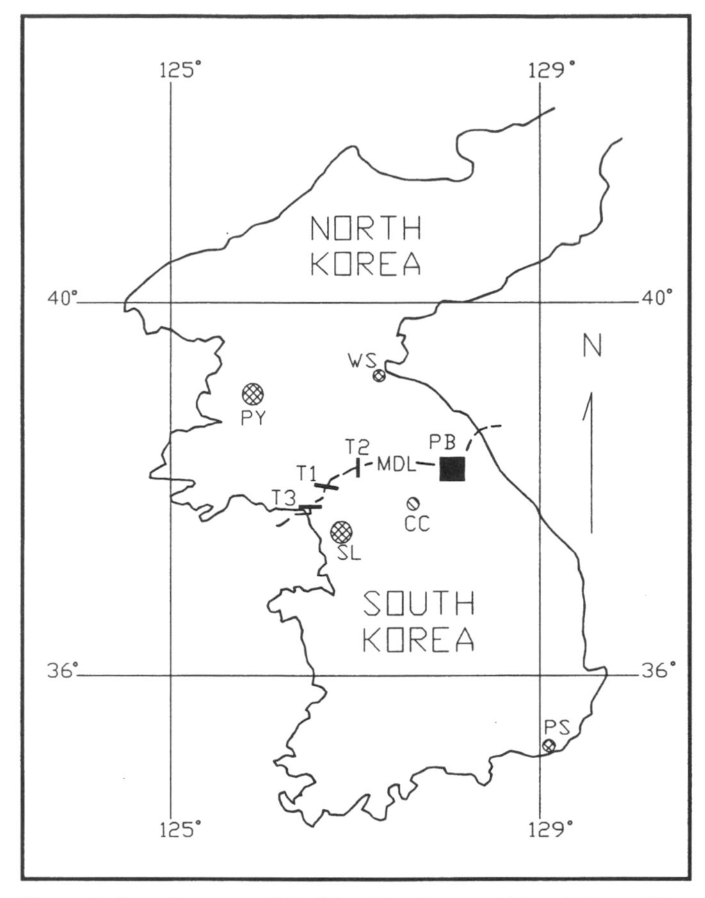

Figure 1. Location map of the Punchbowl area. Abbreviations: PB—Punchbowl; MDL—Military Demarcation Line; SL—Seoul; CC—Chunchon; PS—Pusan; PY—Pyongyang; WS—Wonsan; T1, T2, T3—Clandestine Tunnels 1, 2, and 3, respectively.

by Allied troops in the DMZ, and counterintelligence studies led to the discovery and capture of clandestine infiltration tunnels constructed by North Korean (NK) units in the central and western sectors of the DMZ (Tunnels 1, 2, and 3). Tunnel-1 was a shallow, 1-m-diameter passageway constructed, for the most part, in soils and weathered rock, and clearly designed for the covert infiltration of a relatively small number of NK personnel. Tunnels 2 and 3, both approximately 2 m in diameter, were constructed to depths of 75–85 m in hard, relatively fresh, unweathered granite. Postdiscovery analysis indicates that these tunnels were intended to allow the passage southward of significantly large numbers of NK troops, probably NK special forces, with specific missions as forward artillery observers, saboteurs, and assassins, just prior to the outbreak of major hostilities.

The steep terrains fronting the northern and northwestern Punchbowl rim were the target of repeated infiltration attempts during the early and middle postwar years. During the 1970s, suspicious movements of ground personnel and sounds indicative of underground construction by drill and blast methods emphasized the possibility that the area hosted NK clandestine tunnels. Subsequent pursuit by Allied military and civilian teams combined drilling, cross-hole geophysics, and site characterization including geologic mapping and fracture analysis.

GEOLOGY

Accounts of the historical geology of central Korea document a 2.8-billion-year rock record, in many places incomplete, of repeated cycles of sedimentation, metamorphism, orogeny, and igneous intrusion, (Lee, 1987; Paek et al., 1993). The Punchbowl area is underlain by Kyônggi Metamorphic Basement Complex, an early Precambrian gneiss and migmatite terrane that was uplifted during the intrusion of a Bulguksa-Event or a Daebo-Event granitic pluton of Late Cretaceous and Jurassic ages, respectively. The intrusion resulted in the creation or reactivation of a ring- and radial-fault pattern. The roof of the pluton was stripped of the intrusive mass by erosion or crypto-volcanic venting, or both. The granitic rock, being more prone to weathering and erosion by virtue of its composition, mode of emplacement, and crystallization history, preferentially eroded to create a bowl-like topographic basin rimmed by resistant Precambrian gneiss. The metamorphic rocks in the immediate environs of the Punchbowl form roof pendants to large intrusive granitoids at depth. Regional geological analysis suggests that the entire area, including terranes across the Military Demarcation Line (MDL) in North Korea, is underlain by granite below elevations of 300–400 m (MSL). Figures 2, 3, and 4 illustrate details of the geology of the northern Punchbowl.

Lithology

Kyônggi Gneiss Complex. Rocks of the Kyônggi Gneiss Complex rimming the Punchbowl are in contact with intrusive "Punchbowl Granite" at elevations above 900–1,000 m along the western, northwestern, and northern rims and are at variable elevations of 550–900 m along the northeastern and eastern rims. The metamorphic rocks comprise a suite of high-grade paragneiss and orthogneiss of amphibolite and upper amphibolite facies. The following classification for megascopic rock-mass description was developed by the author during field mapping of the northern Punchbowl during the fall and winter of 1989 (Cameron, 1990, 1993).

(1) Foliated porphyroblastic and banded paragneiss. This is typically a medium- and coarse-grained, hard and very hard, quartz-feldspar-biotite-garnet (hornblende) paragenesis. This is the most common metamorphic rock in the study area; volumetrically it constitutes 50–80% of the paragneiss section at any given location along the western and northern Punchbowl rims.

(2) Migmatite. Quartzose and quartzo-feldspathic gneiss grades to amphibolitic and mafic varieties, the resulting admixed assemblage forming a classic migmatite or mixed-rock terrane. The migmatite is commonly characterized by fine bands or laminations, giving the rock a lit-par-lit fabric.

(3) Granitic gneiss. Part of the Precambrian terrane, this rock mass is formed of generally nonfoliated, lineated, porphyroblas-

tic (2–6-cm porphyroblasts), plagioclase-biotite-garnet crystalline gneiss. The plagioclase is commonly gray and blue-gray with pronounced twinning. The rock may be the product of metamorphism of a Precambrian intrusive. Alternatively, differential anatexis of calcareous metasediments could produce such a mineral assemblage. These rocks form an extensive outcrop belt along the eastern side of the Punchbowl.

Punchbowl Granite. The term "Punchbowl Granite" is used herein to define the massive, jointed, medium–coarse-grained, biotite-hornblende granite and granodiorite pluton that form the lower slopes and floor of the Punchbowl. Larger intrusive granite masses that occupy terranes both to the north and south apparently connect, via petrogenetic continuity or intrusive-fault contact, with the Punchbowl Granite at depth. The Punchbowl Granite is separated from the other intrusive masses by ridges of gneiss that crop out at elevations ranging from 700–1,200 m. The Punchbowl Granite comprises two recognizable petrological phases, although in places their relationship is blurred and transitional in the field.

(1) A main intrusive phase composed of medium- and coarse-grained leucocratic, gray, biotite-hornblende granodiorite and subordinate granite. Where observed in the countertunnel, the granodiorite contains common and occasionally abundant mafic, rounded, fine-grained xenoliths commonly 10–40 cm in diameter.

(2) A marginal contact phase, also intrusive, composed of fine- to medium-grained hornblende granodiorite and trachyte and

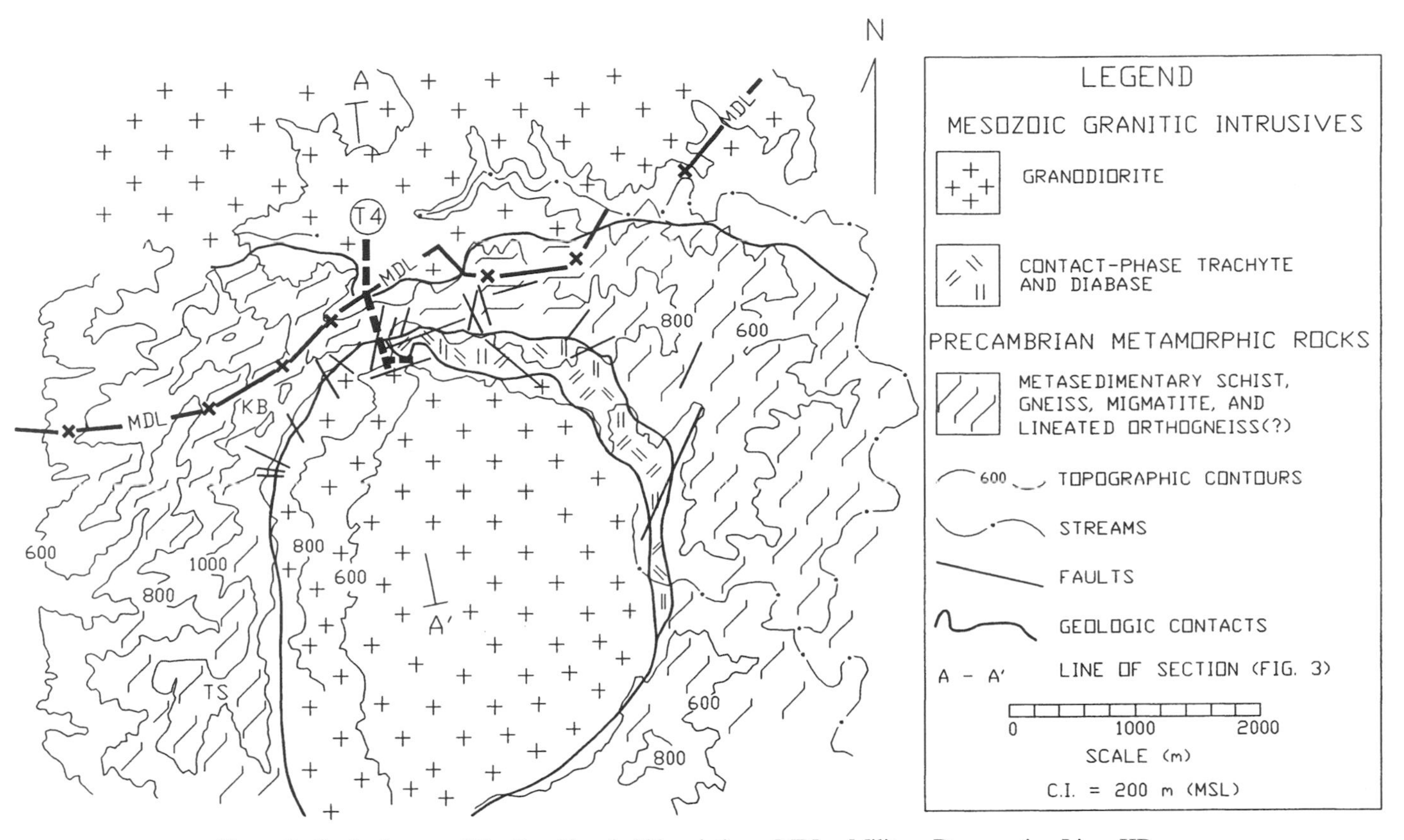

Figure 2. Geologic map of the Punchbowl. Abbreviations: MDL—Military Demarcation Line; KB—Kach'il Bong; TS—Taeu San; T4—Clandestine Tunnel-4.

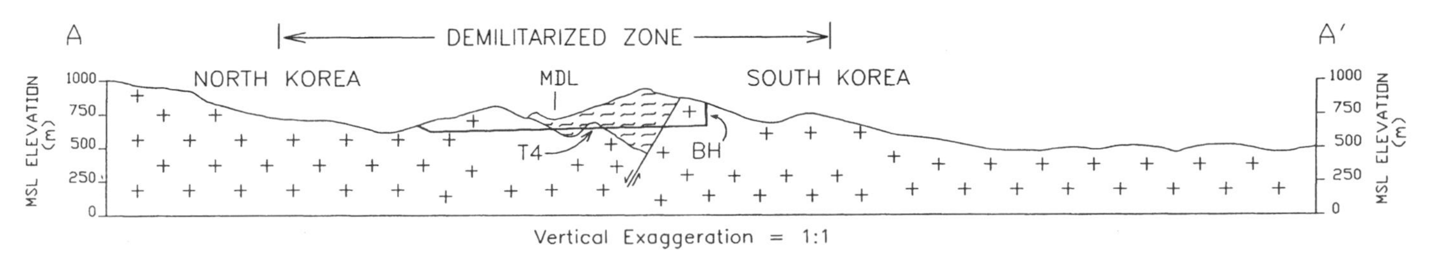

Figure 3. Regional geologic cross-section A–A′, northern Punchbowl region. Abbreviations: MDL—Military Demarcation Line; BH—Discovery borehole 80-14-4-6; T4—Clandestine Tunnel-4. Lithology symbols same as for Figure 2.

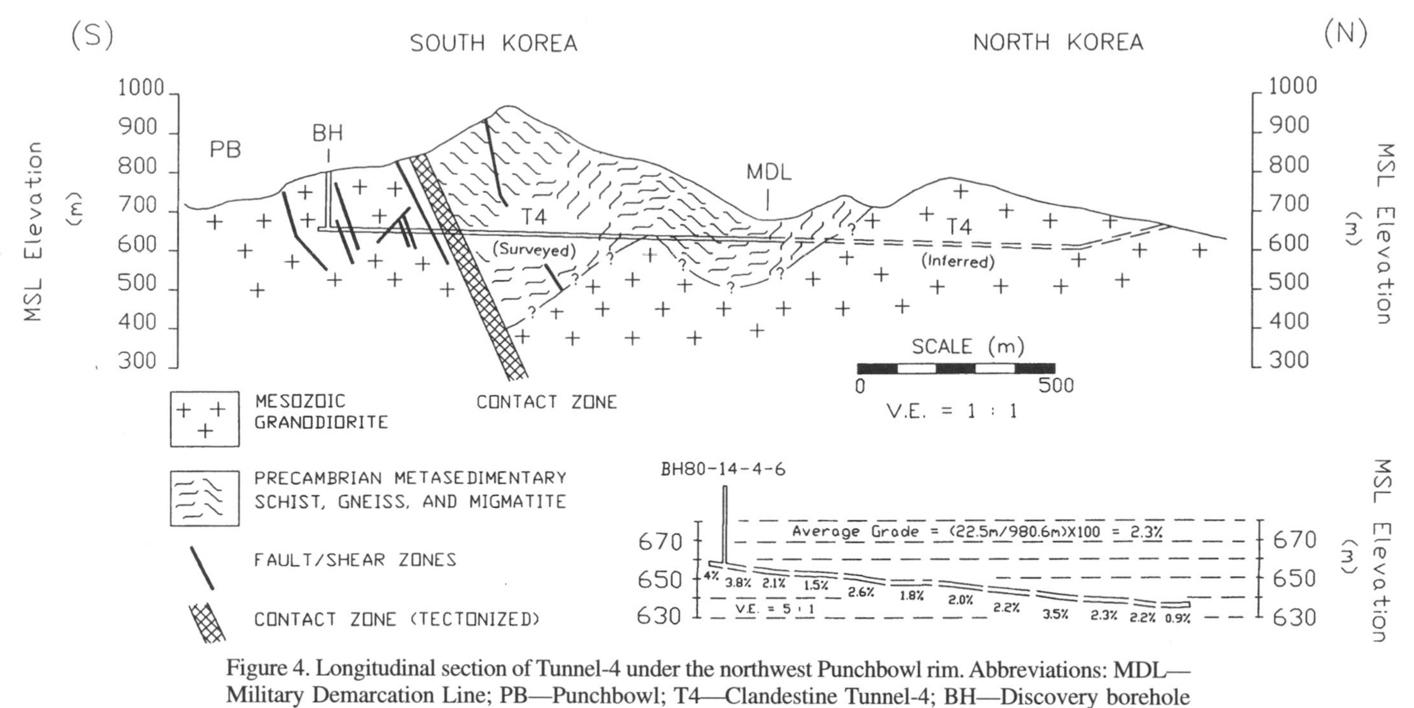

Figure 4. Longitudinal section of Tunnel-4 under the northwest Punchbowl rim. Abbreviations: MDL—Military Demarcation Line; PB—Punchbowl; T4—Clandestine Tunnel-4; BH—Discovery borehole 80-14-4-6.

dark hornblende-biotite-pyroxene diabase and diorite that appear to occur as sills and dikes marginal to the main intrusive phase.

Surface and subsurface mapping established that portions of the northwestern and northern margins of the Punchbowl Granite are fault controlled. Faults combined with erosion control the elevation and position of the Punchbowl Granite as it trends into North Korea through the topographic saddle that splits the northern rim. This is the only locale in the study area where outcrop continuity of the Kyônggi Gneiss Complex along the rim is broken by the Punchbowl Granite.

The age of the Punchbowl is problematic. This intrusive does not appear on the large-scale geologic maps of Korea produced by the Republic of Korea Institute of Energy and Resources (KIER), (Hyun et al., 1981), the Democratic Peoples Republic of Korea Geological Research Institute (GRI, 1993), or on earlier maps produced from Japanese records. This is a strange omission because the Punchbowl is scarcely a subtle feature and its granite floor could hardly be overlooked in any regional study. The large granitoid intrusive mass to the north, Kumgangsan Granite, was mapped as Cretaceous by Hyun et al. (1981) but is assigned to the Jurassic or Daebo Event by North Korean geologists on the basis of two whole-rock, K-Ar ages of 133 and 168 Ma (Paek et al., 1993). The Yang-gu Granite south of the Punchbowl is assigned a Late Cretaceous age (Bulguksa Event) on both the KIER and GRI maps; however, no radiometric ages have been reported for this rock mass in the literature. Whereas the Kumgangsan Granite supports steep slopes and elevated topography, the Yang-gu and the Punchbowl granites are softer than their surrounding terranes and form topographically subdued, bowl-shaped depressions. Radiometric-age dates, detailed petrography, and petrochemistry for these topographically subdued granite masses are needed to prove the genetic relationship suggested by their topographic expression.

Structure

Texture and fabric. The gneiss ranges in texture from very coarse to medium grained. An equally diverse suite of fabrics includes large-scale banded, lit-par-lit structure, augen porphyroblasts, boudinage created by extensional strain during folding, and ptygmatic folds and veins indicative of compressional strain during folding. Strong lineation characterizes the orthogneiss that forms the terrane along the eastern rim of the Punchbowl. Where affected by faults and shears, the gneiss is commonly phyllonitized, mylonitized, or cataclastized; that is, the rock mass has undergone smearing, granulation, crushing, mineral alteration, and recrystallization in zones of maximum stress and rock-mass displacement. Where dilation occurred along faults and shears, quartz veins are common.

Analysis of foliation orientations along the northern and northwestern Punchbowl rim revealed that decimeter-scale wavelength, low-amplitude folds along north-northeast–south-southwest– and north-south–plunging fold axes are common in the paragneiss. Foliation dips vary widely, from low to very steep, with pronounced steepening and even overturning of fold limbs in zones disrupted by faults. In these fault zones, the noses of the folds are commonly sheared out by axial-plane faults that trend east-northeast–south-southwest or are cut off by east-northeast faults. This structural style is very common along the steep and broken ridges between the Punchbowl and the Pukhan-Gang. At least one episode involving longitudinal strain in an uplifted and buckled layer of Kyônggi gneiss resulted in the development of concentric folds. In some areas, for example, the Hill 1220 area approximately 18 km west of the Punchbowl, field evidence shows that concentric folds were superimposed on earlier folds, but this is difficult to document in the Punchbowl area. The strong compressive strain perpendicular to layering predicted by this model is manifested in steep strike-joint sets, severe internal deformation attributed to flexural slip along foliation planes or layers, small subparallel tension gashes, and the presence of slaty cleavage near some fold hinges. Late-strain increments led to the development of small thrust faults whose planes may have been involved in the folding.

The intrusion of the Punchbowl Granite activated or reactivated high-angle reverse and normal faults that form a ring and radial pattern around the margin of the stock. Studies of cores drilled immediately northeast of Tunnel-4 reveal multiple episodes of displacement in rocks brecciated along these bounding faults. The breccia in the faulted, intrusive contact exposed 300–315 m north of the countertunnel intercept in Tunnel-4 is ruptured and sheared-out by fault reactivation, possibly during emplacement of the Punchbowl Granite to its present structural level (Fig. 5).

Where the gneiss is affected by faults, a gouge zone 0.3–8.0 m thick, in places accompanied by sheared leucocratic dike rocks or cataclastized gneiss, generally is present. Failure along foliation planes in the metamorphic rocks probably was aided by zones of graphite and chlorite whose low shear strength lubricated the fault plane. This situation also is characteristic of those fault zones that control the margin of intrusive Punchbowl Granite.

Joints and faults. Foliation, joint, and fault planes were measured at more than 100 surface outcrop stations in the northern Punchbowl and at intervals along Tunnel-4 (Fig. 6). Analysis of these data clearly indicates that the trends of faults, shears, veins, and dikes that transect the gneiss complex are generally aligned with the trend of metamorphic foliation. In most of the northern Punchbowl this trend is N5°–15°E with moderate to steep dips. This is not true of the intrusive Punchbowl Granite terrane. The predominant fault trend in the granite appears to parallel that of a major joint set that strikes N70°–90°E. This trend also is present in the metamorphic rock but to a lesser degree than in the granite.

CONSTRUCTION FEATURES AND ROCK MASS PROPERTIES

Mapping and surveying of Tunnel-4 revealed that the tunnel penetrated approximately 1.1 km into the Republic of Korea from the point where it crossed the MDL. Unlike previously discovered DMZ tunnels, Tunnel-4 was constructed in two distinct lithologies: Precambrian metamorphic rock and intrusive Mesozoic granodiorite.

Tunnel-4 slopes up to the south at an average grade of 2.3%, ranging from a low of 0.91% to a high of 3.5% (Fig. 4). The survey data suggest that the tunnel gradient may reverse near its suspected entrance in North Korea; however, this is uncertain. The elevation control at Borehole 80-14-4-6 is based on the U.S. Army Precision Azimuth Determination System (PADS) data. PADS is a vehicle-mounted automatic system used for artillery surveying. The PADS system uses a different geodetic datum than that on which the 1:25,000-scale map sheet of the area was contoured; this sheet was used to estimate the elevation of the suspected tunnel entrance in North Korea. Experience demonstrated that PADS elevations can differ by 20–30 m from those estimated on the contoured map sheets. Furthermore, the map sheets themselves may have elevation errors of 20–40 m. Discrepancies of this magnitude could, over the distance of the tunnel, significantly affect the longitudinal configuration and remove the necessity for a slope reversal near the suspected entrance.

Visual aspects of Tunnel-4 are shown in Figures 7–10. The average dimensions of the tunnel, 1.6 m × 1.6 m, are somewhat smaller than other discovered hard-rock tunnels in the DMZ. The tunnel was constructed by drill and blast methods both in the Mesozoic intrusive rock and Precambrian gneiss. These hard rocks are estimated to have high compressive strengths, probably 100–200 Mpa. Most rock discontinuities are joint planes that have rough, interlocking apertures filled or partially filled with clay and quartz-calcite veinlets. Joint spacings are moderately close (0.3–1 m) to wide (1–3 m) and, occasionally, very wide (>3 m), depending on the set, lithology, and proximity to faults and tectonized contacts. The ground-water condition generally is wet, that is, dripping and occasionally flowing at fault contacts or where discontinuity fillings are washed out, and is highly variable, dependent on local precipitation and seasonal variations.

The rock-mass rating, estimated using methods developed by Bieniawski (1979, 1989) and Gonzalez de Vallejo (1983), is fair-poor in the gneiss, significantly less than in the homogeneous granite, which is rated good-excellent. The rock quality and condition in the metamorphic rock, however, did not substantially impede tunnel construction, nor did the 50-m-wide tectonized zone of intrusion at the granite-gneiss contact. This zone was negotiated without remediation and timbering. A significant 8-m-wide fault severely disrupts the metamorphic section about 480 m along the tunnel. Although the rock mass is very altered and ostensibly weak, the zone was crossed and supported with a minimum of timbering and remediation (Fig. 7). Tunnel stability was aided by the fact that the fault zones are occupied by broken rocks embedded in a matrix of chlorite. Although soft and prone to failure by shear, the chlorite served to create a permeability barrier in the fault zones; thus water inflow is minimal except locally at fault contacts.

A steep, wide-spaced joint set that trends north-northwest follows the general bearing of the tunnel. As is shown on the tunnel profiles and in Figures 5, 8, and 9, these joints form even, unbroken walls in many places, a factor that lends stability to a tunnel of small dimension. Muck removal was accomplished using carts on narrow-gauge rails, an operation that was facilitated by the use of side-wall galleries 12–15 m long and as wide

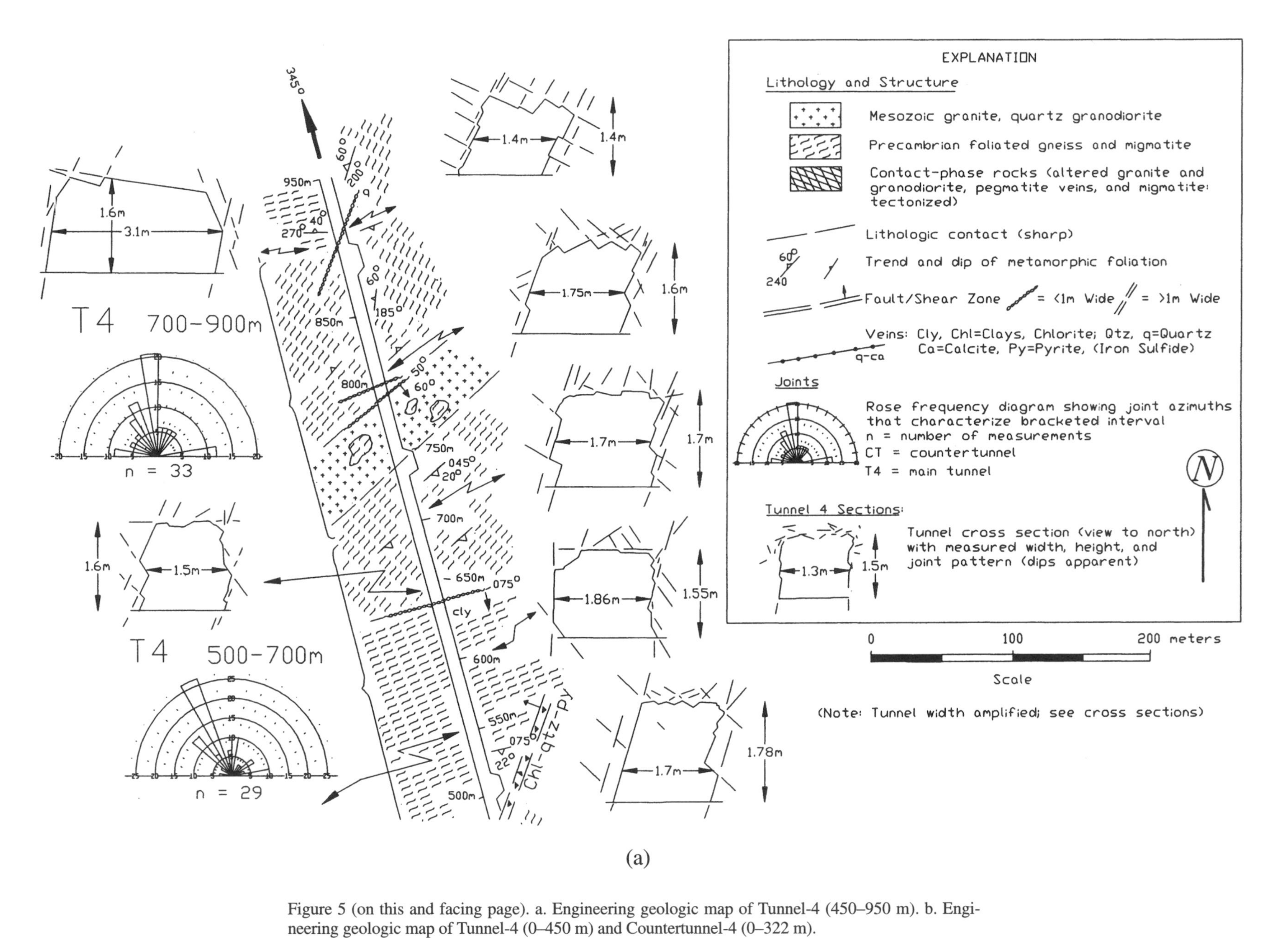

Figure 5 (on this and facing page). a. Engineering geologic map of Tunnel-4 (450–950 m). b. Engineering geologic map of Tunnel-4 (0–450 m) and Countertunnel-4 (0–322 m).

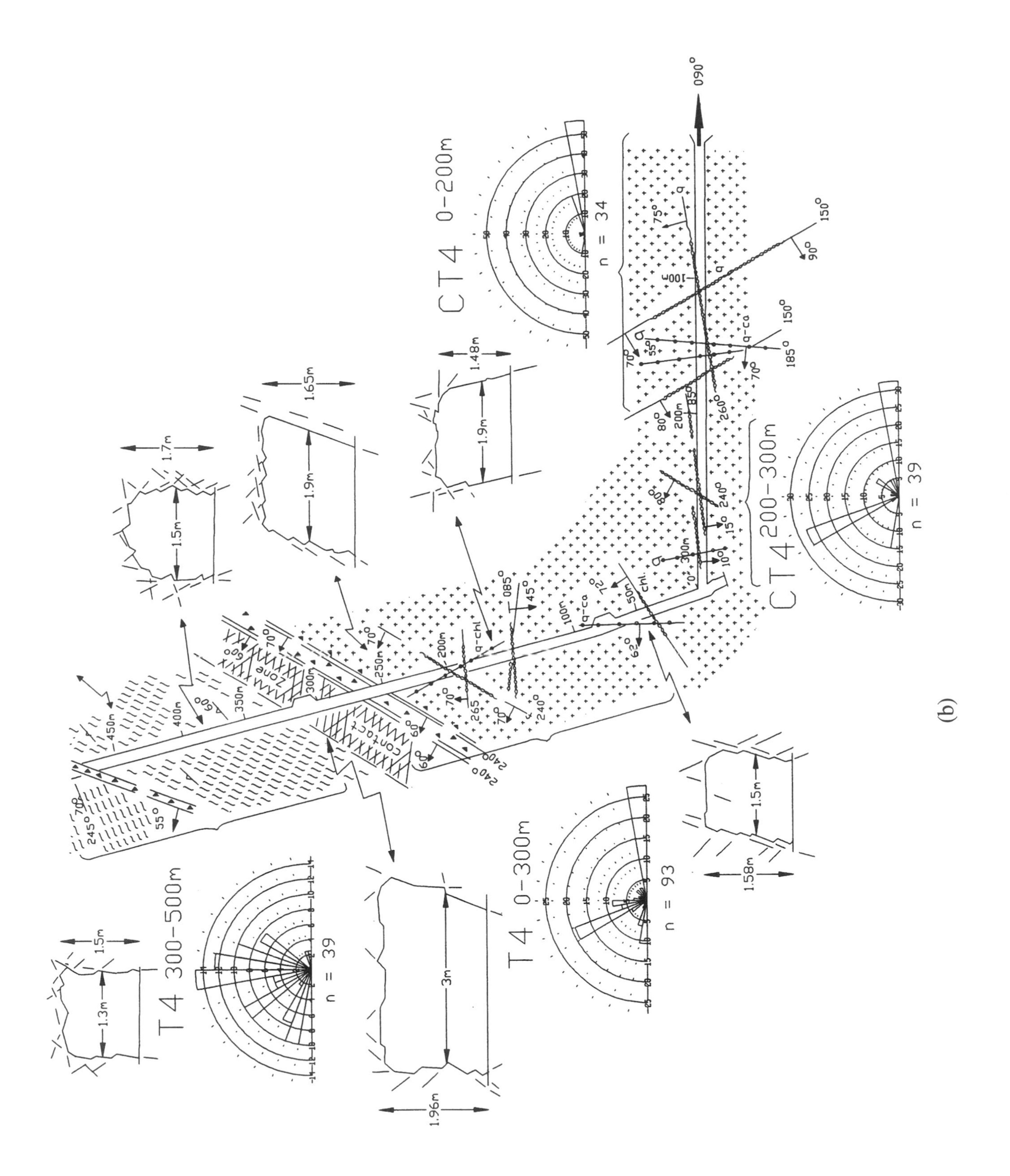

CT4 0-200m
n = 34
CT4 200-300m
n = 39
T4 0-300m
n = 93
T4 300-500m
n = 39
090°
contact
Zone
1.5m
1.7m
1.9m
1.65m
1.48m
1.3m
3m
1.96m
1.58m

(b)

as 3 m, located at intervals of 150–200 m along the east wall of the tunnel (Fig. 10). Loaded rail carts were sidetracked into the galleries, allowing empty carts access to the working face.

EXPLORATION CASE HISTORY

The potential for the Punchbowl terrain to host clandestine NK tunnels was identified by Allied intelligence personnel, who recorded observations of suspicious movements of ground personnel and pursued sounds typical of underground drill and blast operations in hard-rock terrains. Follow-up field operations commenced late in 1979, following the discovery of Tunnel-1 (1974), Tunnel-2 (1976), and Tunnel-3 (1978). The effectiveness of the efforts in neutralizing the tunnel threat on the northwest rim of the Punchbowl can be inferred from a study of the historic pattern of exploration drilling in the immediate area. The first boreholes drilled in the Punchbowl, K79-1, K79-2, and K79-3 were drilled on a narrow winding road that was in existence for many years as the only low-elevation, vehicular-access route on the northwest rim (Fig. 11). These boreholes effectively straddled the trend of the tunnel and drilling in 1980, filled in the pattern, and further narrowed the location of the suspected tunnel axis.

Stubbornly determined pursuit of objectives by the Republic of Korea Army Tunnel Detection Section (ROKA-TDS) and the Eighth U.S. Army Tunnel Neutralization Team (EUSA-TNT) resulted in a series of exploration programs that combined percussion drilling and cross-hole geophysics. The drilling program, begun in the spring of 1989, also was supported by engineering-site geologic mapping and rock-mass characterization, including core drilling and fracture analysis. The success of this program is now part of the historical record.

In December 1989, a coherent cross-hole electromagnetic (radar) anomaly with a signature indicative of an air-filled cavity was recorded by the Korean Advanced Institute for Science and Technology (KAIST) Continuous Wave System in new boreholes drilled by the ROKA-TDS. These new holes were drilled to follow up earlier intelligence studies and exploration programs. The anomaly, detected at a depth of 145 m, was confirmed as a probable air-filled cavity by the Pulsed Electromagnetic Search System (PEMSS) developed for the U.S. Army by Southwest Research, Inc. (Fig. 12).

Both the presence of massive, homogeneous granodiorite and the evidence provided by borehole water-level elevations

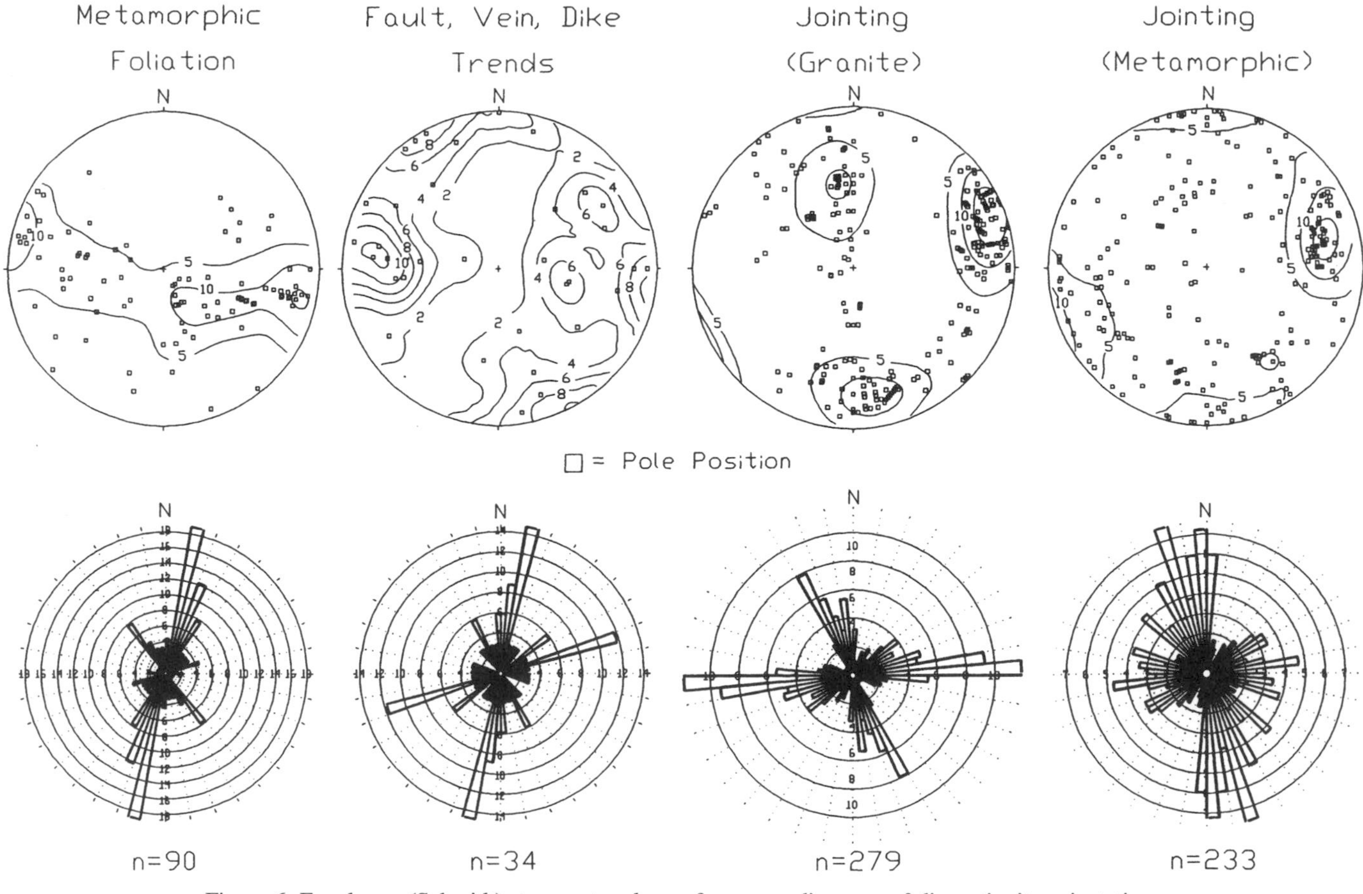

Figure 6. Equal area (Schmidt) stereonet and rose frequency diagrams of discontinuity orientations, northern Punchbowl area. Contours (in percent of total points) on the stereonets depict lines of equal point density generated by Gaussian density gridding with a search value of 5%. The plots combine measurements made on the surface and in Tunnel-4. Abbreviation: n—sample population.

precluded the possibility of a natural air-filled cavity at this depth. Target-evaluation drilling followed, and on December 24, 1989, at 0130 hours, a drill bit penetrated "Tunnel-4." EUSA-TNT deployed a borehole television camera into the tunnel that provided clear evidence of human construction activity, e.g., jackleg drill-rod scars on the tunnel walls, rail ties, and so on. Two and a half months later, Korean engineers, using a Wirth 3-m tunnel-boring machine (TBM), gained entrance to the tunnel and confirmed its nature and purpose. The borehole TV once again played a significant role in that it was able to provide continuous coverage of the tunnel as the TBM broke through. Inspection and clearing of the tunnel were accomplished by a ROKA reconnaissance team, which suffered the loss of its mine-sniffing dog when he triggered a mine hidden under shallow water pooled in the vicinity of the MDL.

Figure 7. The 8 m wide fault zone at about 480 m north of the countertunnel intercept was penetrated with a minimum of crude timbering and remediation. The overbreak owing to poor rock quality increased the height of the tunnel in the immediate area of the fault zone. Eighth U.S. Army personnel are holding a device for determining tunnel dimensions.

Figure 8. Steep west-dipping joints characterize foliated gneiss and migmatite 804 m north of the countertunnel intercept. The electrical cables, piping, and supporting brackets at the base of the west wall (right) were installed by ROKA engineers during postdiscovery operations. The four white insulators on the west wall supported electrical wires that were stripped from the tunnel by NK personnel at some time prior to discovery.

Discussion. The following facts should be taken into consideration before entertaining speculations regarding clandestine tunneling operations at Tunnel-4. In the first place, the narrow road traversing the lower elevations of the northwest Punchbowl rim was the only such access for large, truck-mounted drilling rigs during the late 1970s and mid-1980s. Sections of this route were under direct observation through the gap in the northern Punchbowl rim by the NKPA guardpost situated on Hill 1052 to the northeast. The terrain that flanks this access route is very steep, particularly upslope to the northwest. The ROKA-TDS drill teams had little choice but to drill where they did in 1979–1980. In a

Figure 9. West-dipping joint planes in the Punchbowl Granite form large slabs that control tunnel walls 50–100 m north of the countertunnel intercept.

Figure 10. A south facing view in Tunnel-4 showing the east-wall gallery 735–750 m north of the countertunnel intercept.

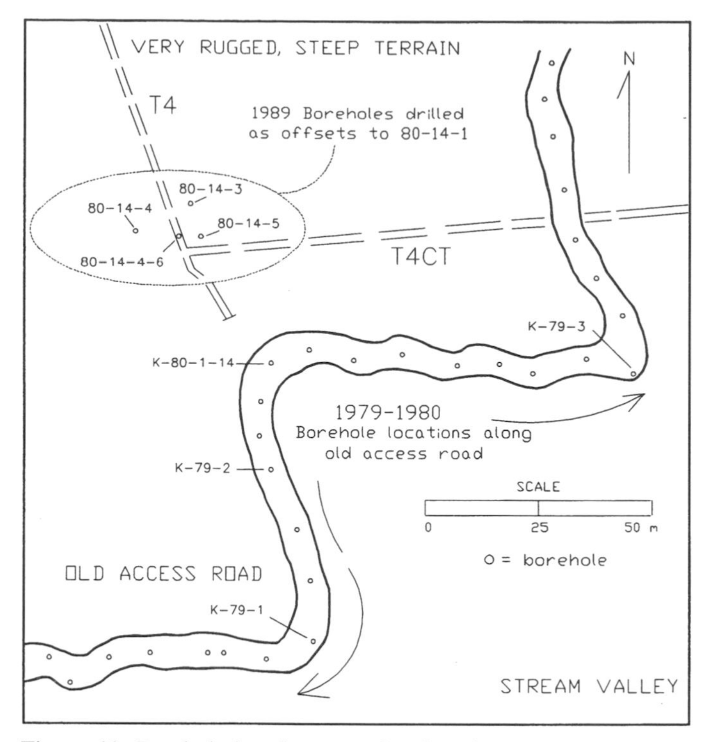

Figure 11. Borehole-location map showing the two generations of drilling (1979–80 and 1989) that neutralized and found Tunnel-4. Abbreviations: T4—Clandestine Tunnel-4; T4CT—Tunnel-4 Countertunnel.

classic near-miss exploration situation, K-1-14 (1980) was drilled directly on trend and only approximately 12–13 m southeast of the end of the tunnel (Fig. 9). Interestingly, the tunnel bearing shifts about 7° counterclockwise to a more easterly direction in its final 25 m, and its gradient steepens sharply to about 4%. Underground mapping established that these directional shifts were not occasioned by a change in rock quality or condition. From these observations the following scenarios can be postulated, albeit they all are speculative at this time:

(1) The NK command responsible for tunnel construction in this area gained knowledge of the ROKA borehole placement and drilling via direct observations of drilling rigs being ordered into position(s) directly "on-axis" with their tunnel.

(2) Alternatively, NK personnel engaged in underground construction actually heard the sound of nearby percussion drills searching for them.

(3) The capture of NK Tunnel-3 in 1978 resulted in a significant shift in NK thinking with respect to the advisability of further construction of tactical tunnels under the DMZ.

(4) Two possibilities are suggested by the directional shift of the tunnel in its last 25 m. The first is that the tunnel was swinging to its breakout position: probably about 300 m to the southeast, and commencing to develop two or more splayed branches. The second is that the NK were trying to avoid the straddling effect of the 1979 boreholes, indicating that the NK had the drill site under observation.

CONCLUSIONS

The discovery of Tunnel-4 demonstrated the effectiveness of the Continuous Wave (CW) and Pulsed Electromagnetic Search Systems (PEMSS) in locating and defining deep hard-rock tunnels. Confirmation by a borehole television camera of human construction activity was a resounding technical success, as was the ROKA TBM countertunneling operation.

Tunnel-4 extends 2.1 km from its suspected entrance. The tunnel slopes up to the south at an average grade of 2.3% (0.91%–3.5%) and may contain an invert near its entrance in North Korea. The average dimensions of Tunnel-4, 1.6 m × 1.6 m, is less than that of other discovered DMZ tunnels; however, Tunnel-4 contains more sidewall galleries, where the width doubles, to facilitate muck removal.

The tunnel was constructed through both granitic and metamorphic rock, unlike previously discovered DMZ tunnels that were constructed entirely in granite. It was demonstrated once again that "tunnels are most easily found where they are most easily made," that is, in hard, relatively homogeneous granite with minimal structural discontinuity.

The rock mass rating in the metamorphic rock, fair-poor, is significantly less than that of the granite, good-excellent. The relatively poor rock quality in the metamorphic rock did not significantly affect tunnel-construction activity that was conducted by drill and blast methods and with a minimum of timbering.

In a manner very analogous to the situation at Tunnel-2, the walls of Tunnel-4 are formed in many places by a steep, easterly dipping, north-northwesterly trending master-joint set. This structural feature, combined with the generally wide joint spacing, enhances rock-mass quality and substantially reduces support requirements, even in zones of otherwise poor rock quality. It is probably for this reason that rock-mass rating and classification schemes by Bieniawski (1979, 1989) and Gonzalez de Vallejo (1983) yield overly pessimistic results in the rock mass under consideration here, with respect to construction difficulty and support requirements for tunnels <2 m in diameter. Particularly with respect to support requirements, these methods must be considered in the context of the quantitative assessments by Houghton and Stacey (1980), which are also compiled in Bieniawski (1989, p. 134). These assessments show that factors of safety of 1.2 or greater can prevail for tunnels of small diameter, 1.5 m to 2 m, when the rock-mass rating is in the "fair" and "poor" range.

The Tunnel-4 military-geology case history strongly suggests that sound intelligence combined with early drilling efforts successfully neutralized this attempt at clandestine infiltration, although the tunnel was not discovered until a decade later. The events that led to these efforts also tend to reinforce the prescient observations of Brooks (1920, p. 107) who wrote of a hard-rock tunnel war at another time and in another place:

> One of the first principles of underground warfare is that operations should be carried on with as little sound as possible, so as not to reveal the position of the mines to the enemy. . . .

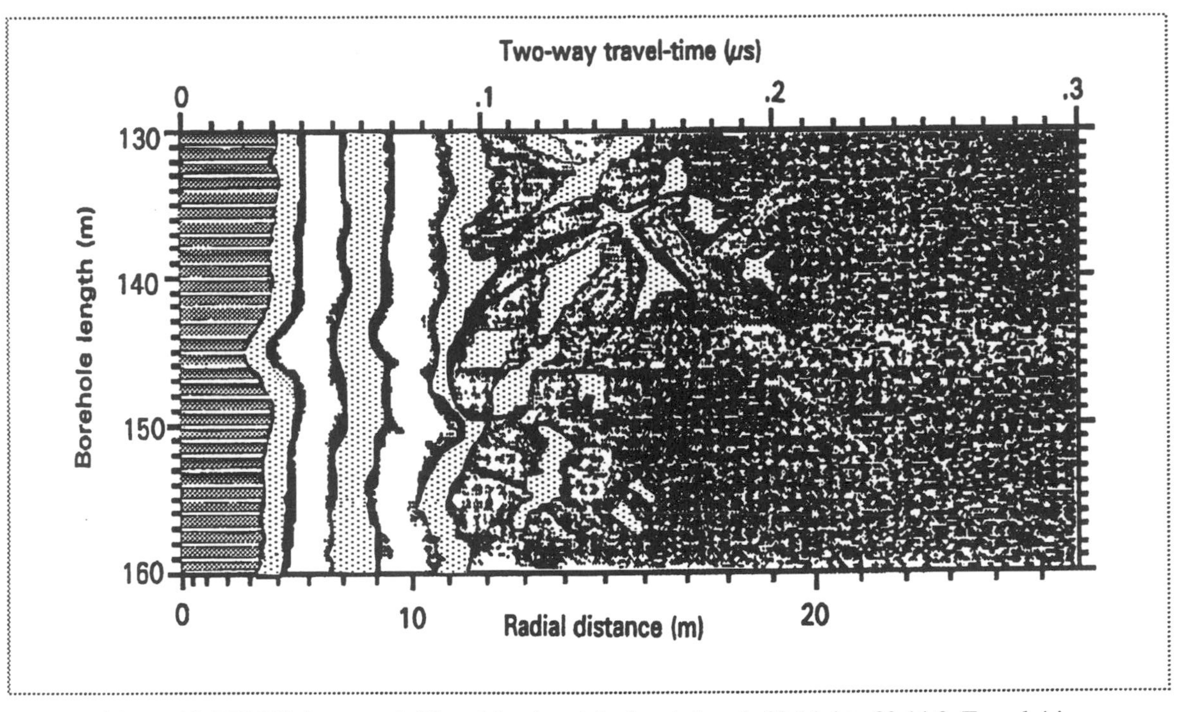

Figure 12. PEMSS data record (filtered-level run) for borehole pair 80-14-4 to 80-14-3. Tunnel-4 is clearly indicated by the strongly anomalous early arrival of the PEMSS signal at 145 m. Borehole separation is approximately 15 m at the tunnel depth. From Alleman et al. (1993).

The noise of the operations was so great, however, that both armies had fairly definite information about the operations of their opponents. In general, the plan was ill-conceived and led to no decisive results. . . .

ACKNOWLEDGMENTS

The geological results discussed herein formed part of the operations and research conducted under the Tunnel-Detection Program of the U.S. Army. Funding for this mission was provided by the U.S. Army Belvoir Research, Development, and Engineering Center (BRDEC). The author's participation was funded by a Broad Agency Grant Contract (DACA39-90-K-0029) issued by the U.S. Army Engineer Waterways Experiment Station (WES). Sincere appreciation is extended to Ray Dennis (BRDEC) and to Bob Ballard (WES) for their long-term support and encouragement during the course of this work. This work would not have been possible without the enthusiastic cooperation of the EUSA-TNT, the ROKA-TDS, and the ROKA 12th and 21st Divisions. Particular appreciation is extended to SPC Chris Hacker, U.S. Army, who assisted the author in all phases of the surface mapping and field work, and to SFC Mike Joos, Sgt. John Britton, and Sgt. John Rodgers, all of the U.S. Army, whose hard work in conducting the Tunnel-4 topographic survey was both inspiring and invaluable. The views expressed herein do not purport to reflect the views or the position of the U.S. Department of the Army or the U.S. Department of Defense.

REFERENCES CITED

Alleman, T. J., Cameron, C. P., and MacLean, H. D., 1993, PEMSS response of rock tunnels to "in-axis" and other nonperpendicular antennae orientations, *in* Proceedings, Fourth Tunnel Detection Symposium on Subsurface Exploration Technology: Fort Belvoir, Virginia, Combat Engineering Directorate, U.S. Army Belvoir Research, Development, and Engineering Center, p. 19–44.

Bieniawski, Z. T., 1979, Tunnel design by rock mass classifications: Vicksburg, Mississippi, Technical Report GL-79-19, U.S. Army Engineer Waterways Experiment Station, 71 p.

Bieniawski, Z. T., 1989, Engineering rock mass classifications: New York, John Wiley and Sons, 251 p.

Brooks, A. H., 1920, The use of geology on the Western Front: Washington, D.C., U.S. Geological Survey Professional Paper 128-D, p. 85–124.

Cameron, C. P., 1990, The geology of the northern Punchbowl: Seoul, Republic of Korea, Technical (Draft) Report, EUSA-TNT-J2, Yongsan Army Garrison, and Vicksburg, Mississippi, the U.S. Army Engineer Waterways Experiment Station, 40 p.

Cameron, C. P., 1993, Exploration case history and engineering geology of "Tunnel-4," northern Punchbowl sector, Korean Demilitarized Zone, *in* Proceedings, Fourth Tunnel Detection Symposium on Subsurface Exploration Technology: Fort Belvoir, Virginia, Combat Engineering Directorate, U.S. Army Belvoir Research, Development, and Engineering Center, p. 595–615.

Gonzalez de Vallejo, L. I., 1983, A new rock classification system for underground assessment using surface data, *in* Proceedings, International Symposium of Engineering Geology: Underground Construction, Laboratorio Nacional de Engenharia Civil, Lisbon, v. 1, p. 1185–1194.

Geological Research Institute, 1993, Geological map of Korea: Pyongyang, Democratic Peoples Republic of Korea, Foreign Languages Books Publishing House, scale 1:1,000,000, 1 sheet.

Houghton, D. A., and Stacy, T. R., 1980, Application of probability techniques to underground excavation, *in* Balkema, A. A., ed., Proceedings of the 7th Regional Conference for Africa on Soil Mechanics and Foundation Engineering: Accra, Ghana, v. 2, p. 879–883.

Hyun, B. K., Um, S. H., and Chun, H. Y., compilers, 1981, Geological map of Korea: Seoul, Republic of Korea, Korea Research Institute of Energy and Resources (KIER), scale 1:1,000,000), 1 sheet.

Lee, D.-S., editor, 1987, The geology of Korea, 1st ed., Seoul, Republic of Korea, The Kyohak-Sa Publishing Co., 514 p.

Paek, R. J., Kan, H. G., Jon, G. P., Kim, Y. M., and Kim, Y. H., editors, 1993, The geology of Korea, Appendix 1: "Isotope Geochronology": Pyongyang, Democratic Peoples Republic of Korea, Foreign Languages Books Publishing House, p. 545–560.

Manuscript Accepted by the Society October 29, 1997

Printed in U.S.A.

Geological Society of America
Reviews in Engineering Geology, Volume XIII
1998

Swords into plowshares: Military geology and national security projects

James T. Neal*
Sandia National Laboratories, Albuquerque, New Mexico 87185-0706

ABSTRACT

Military geology and national security projects are comparable, achieving their raison d'etre in support of national goals, military operations, and the systems that support them—all for vital national interests. The application of geoscience to these ends, especially engineering geology, has occurred from pole to pole and included every conceivable environment and natural condition. In the conduct of such projects, the geosciences have advanced, and vice versa.

Desert trafficability, most notably regarding playa surfaces, is temporary and variable and not a persistent condition as some early authors believed. Playas in Australia, Iran, and the United States show that saline efflorescence is removed following surface water dissolution and subsequent deflation, resulting in very hard crusts. Magadiite, a hydrous sodium silicate and possible precursor of bedded chert, was first discovered in North America at Alkali Lake, Oregon, during a military project. Pleistocene Lake Trinity, a small and mostly buried evaporite basin in the northern Jornada del Muerto, New Mexico, was discovered during exploratory drilling in support of a military test program.

The Strategic Petroleum Reserve (SPR), operated by the U.S. Department of Energy, has underground cavern storage of ~600 million barrels of crude oil in five Gulf Coast salt domes. The geologic characterization of the SPR sites is a major component of these comprehensive engineered works—unparalleled in modern times and on a comparable scale with the Panama Canal. Numerous studies of salt-stock heterogeneity, salt-karst features, and structural and physical attributes of salt deposits are broadening the database for use in the commercial storage industry. Geologists serving in military and national security endeavors are fully functioning members of the project technical teams and have made significant advances to the geosciences.

INTRODUCTION

They shall beat their swords into plowshares, and their spears into pruning hooks; nation shall not lift up sword against nation, neither shall they learn war anymore. . . .

Isaiah II, 4

The biblical vision to convert instruments of war into tools for peace may or may not be materializing as we approach the year 2000, but hope has not faded. In fact, that aspiration is even more urgent because of the advent of ever more deadly weapons and associated delivery systems.

The U.S. Plowshare Program of the 1950s and 1960s examined the use of nuclear explosives for peaceful applications, such as rapid excavation and tunneling. Although never applied commercially in peacetime, that technology is available for future generations (Hamburger, 1973). The Plowshare Program exemplifies the metaphor discussed in this paper—the beating of swords into plowshares or, literally, applying information and technology obtained for military purposes to peaceful endeavors.

The application of the geosciences to military strategy and

*Present address: 1911 Crestview Drive, Prescott, Arizona 86301.

Neal, J. T., 1998, Swords into plowshares: Military geology and national security projects, *in* Underwood, J. R., Jr., and Guth, P. L., eds., Military Geology in War and Peace: Boulder, Colorado, Geological Society of America Reviews in Engineering Geology, v. XIII.

tactics has been taught in the military academies of most countries in one form or other for many years and is hardly novel; it is practiced in combat theaters often. Examples of the use of engineering geology in wartime were described by Kiersch (1991), and by Kiersch and Underwood (this volume). Peacetime benefits from military technology are perhaps best known in regard to reconnaissance systems, especially remote sensing. The widespread availability of satellite images in their many manifestations for use in modern geoscience is remarkable when we consider its rudimentary beginnings in the 1960s (Williams and Carter, 1976). Likewise, current use of the military's Global Positioning System in geologic mapping, geodesy, and surveying is impressive, if not awe inspiring. Handheld receivers capable of providing virtually immediate ~50–75 m horizontal accuracy are as affordable today as the time-honored Brunton® compass. And, even more sophisticated receivers and methods achieving centimeter-range accuracy are employed regularly today in studies of regional and global tectonics.

Discoveries made during the course of U.S. military geology and national security projects over the past half-century are legion, and a few of them are described in this volume. This author has been personally associated with several: playa trafficability, the mineral magadiite, Pleistocene Lake Trinity, and salt-karst dissolution.

Figure 1. The extensive, hard playa crust on Rogers Dry Lake at Edwards Air Force Base, California, has been used routinely for the landing of air and spacecraft for some 50 years. The Space Shuttle of the 1990s now makes precision landings on concrete runways, but numerous playas similar to this provide emergency landing sites all over the world.

PLAYA-SURFACE TRAFFICABILITY

Playas attract attention because they are among the flattest of all landforms and, at their smoothest and most trafficable condition, they provide unimpeded mobility for heavy vehicles and aircraft. Playas have been used routinely for aircraft operations at Edwards Air Force Base and nearby Fort Irwin military reservations in California for more than 50 years with relatively little maintenance (Fig. 1). Each flooding by infrequent desert rains reconditions the upper few centimeters of the clayey soil, forming a hard surface crust upon drying.

At other playas such as Harper Lake, California (between Edwards Air Force Base and Ft. Irwin), evaporite efflorescence creates a friable surface with greatly reduced trafficability. These surfaces have been described as "puffy" or "self-rising" ground (Motts, 1972). Previous authors have implied that these conditions were persistent and attempted to classify playas on the basis of these surface characteristics. However, extended observations by Neal (1972), Krinsley (1976), and others have showed that playa surfaces change over time, responding to both short- and long-term influences in *both* surface and ground-water hydrology (Fig. 2). For example, jagged accumulations of thick evaporites on some salt pans are made smooth naturally by regular surface flooding. Automobile racing and speed trials at the Bonneville Salt Flats, Utah, would be impossible otherwise.

Figure 2. Automobile tracks show transition from an extremely hard playa surface to a soft, friable condition. Such changes are sometimes caused by slight topographic variations that influence surface-water movement, producing a flushing action and the removal of evaporite minerals by dissolution and deflation. Amargosa playa, Nevada, 1963.

Playas are very sensitive indicators of changing environment and show a wide range of climatic, hydrologic, and geomorphic responses (Neal and Motts, 1967; Neal et al., 1968; Neal, 1972; Krinsley et al., 1968). Large fissures in playas reveal long-term geohydrologic response and are one indicator of human activities or of climatic change. Some giant fissures are of tectonic origin (Fig. 3), although most result from deep-seated desiccation (see Neal, this volume).

Other peaceful applications of playa investigations include the widespread artificial recharge of ground water in arid and semi-arid regions around the globe, installations of solar and radiotelescope arrays that require very large and flat areas, and sites for automobile racing and other recreation.

Figure 3. Fissures of tectonic origin transect numerous playas, e.g., Yucca Lake, Nevada. This occurrence in 1963 drained the playa of more than 1×10^6 m^3 of water overnight. The fissure was initially more than 30 m deep and 1 km long but was quickly infilled with clayey sediment as eroded gullies formed adjacent to it. Similar fissures elsewhere, which form giant polygons 50 m across, are attributed to long-term desiccation.

MAGADIITE AT ALKALI LAKE

The mineral magadiite ($NaSi_7O_{13}[OH]_3$ $3H_2O$) was named for the locale of its discovery and identification at Lake Magadi, Kenya (Eugster, 1967). Eugster believed that magadiite could be a precipitate that might be subsequently converted to bedded chert. Such a mechanism could explain the previously enigmatic depositional environment for bedded chert. During a 1965 military evaluation of the playa surface at Alkali Lake, Oregon, the uncommon mineral occurrence was observed but not recognized as a new species (Fig. 4). Following the publication of Eugster's 1967 paper, the similiarity between Lake Magadi and Alkali Lake was noted and the first identification of magadiite in North America was announced (Rooney et al., 1969).

PLEISTOCENE LAKE TRINITY, NEW MEXICO

During the search for a test site that would represent a geologic environment similar to a potential military objective, a buried 200-km^2 Pleistocene evaporite basin was identified, and its equivalency was verified by several core holes. The basin was shown to be similar to Pleistocene Lake Otero in the adjacent Tularosa Basin of southcentral New Mexico, and yet it has distinctive sulphate hydrochemistry unlike any other lake in the western United States (Neal et al., 1983) (Figs 5, 6). The discovery of this basin has enhanced the overall understanding of Pleistocene and Recent pluvial climates, providing further evidence of widespread, perennially lacustrine conditions throughout the Great Basin of California, Nevada, Oregon, and Utah and extending into enclosed basins of Arizona, New Mexico, western Texas, and southern Colorado.

U.S. STRATEGIC PETROLEUM RESERVE

Following the oil embargoes during the winter of 1973–1974, the U.S. Congress created the Strategic Petroleum Reserve (SPR) in 1975 to provide a buffer against further interruption in imports of foreign oil. Six salt domes were identified for petroleum storage, and ~600 million barrels of crude oil are now stored in four remaining sites; one site, Sulphur Mines, Louisiana, was decommissioned in 1992 to increase system efficiency. During the Persian Gulf War of 1990–1991, the SPR was partially drawn down, largely to encourage market stability by showing the Reserve's immediate availability.

The site at Weeks Island, Louisiana, the only site created by room-and-pillar mining, was being decommissioned starting in November 1995 because a fracture in the salt had created a flowpath between ground water and the oil storage chamber, resulting in a sinkhole (Neal et al., 1997). The diagnostic investigations

Figure 4. Magadiite, a hydrous sodium silicate first discovered in Kenya, occurs in veins and forms rectilinear patterns at Alkali Lake, Oregon, a hyperalkaline playa.

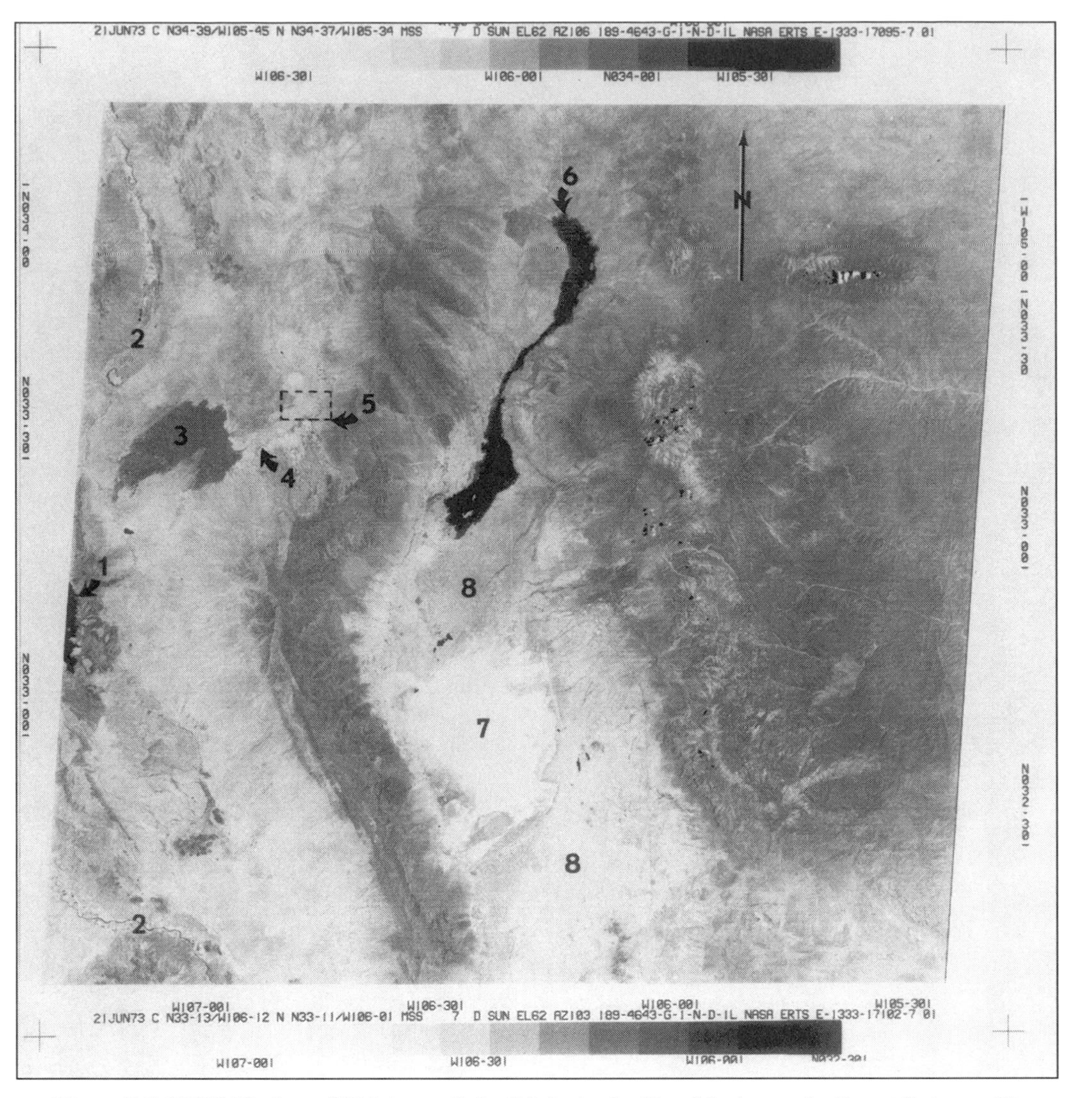

Figure 5. LANDSAT view of Pleistocene Lake Trinity basin, New Mexico, and adjacent features: (1) Elephant Butte Reservoir, (2) Rio Grande, (3) San Marcial lava flow, (4) Lake Trinity shoreline, (5) location of Figure 6, (6) lava of Tularosa basin, (7) White Sands (gypsum dunes), and (8) basin of Pleistocene Lake Otero (scale 1:1,500,000).

used in studying the sinkhole, and ground-water control and mitigation achieved by the injection of saturated brine into the sinkhole throat, are directly relevant to conventional salt mining (Neal and Magorian, 1997).

Brine has been extracted from caverns in salt during most of the 20th century, but the underground storage of liquefied petroleum gas (LPG) in salt did not become a widely used technology until after 1950. Geologic site characterization has been an essential element of studies required to support the SPR, and this methodology is directly applicable to cavern storage for other purposes (Neal and Magorian, 1997). The understanding of salt-dome geology has increased markedly during the past 20 years, partly because of SPR studies. During the last 50 years, more than 1,000 caverns have been constructed in domal salt, and they contain a variety of liquid and gaseous hydrocarbons. Salt-creep phenomena, which led to cavern closure and surface subsidence, have been modeled with remarkable accuracy (Hoffman, 1992; Ehgartner, 1992) and require attention in all salt-cavern applications. The recognition of inhomogeneities in domal salt in particular has led to the conceptualization and understanding of anomalous zones; these zones are believed to separate individual spines or lobes of salt that are involved during differential movement occurring in salt diapirism (Kupfer, 1990; Talbot and Jackson, 1987; Neal et al., 1993).

Although the SPR was established for National Security purposes, the same geotechnical data have been applied to industrial use, such as storing liquefied petroleum gas. Natural gas also is now stored, especially for seasonal peak demands, in 50 caverns in both domal and bedded salt. New applications are continually being considered; for example, compressed air energy storage

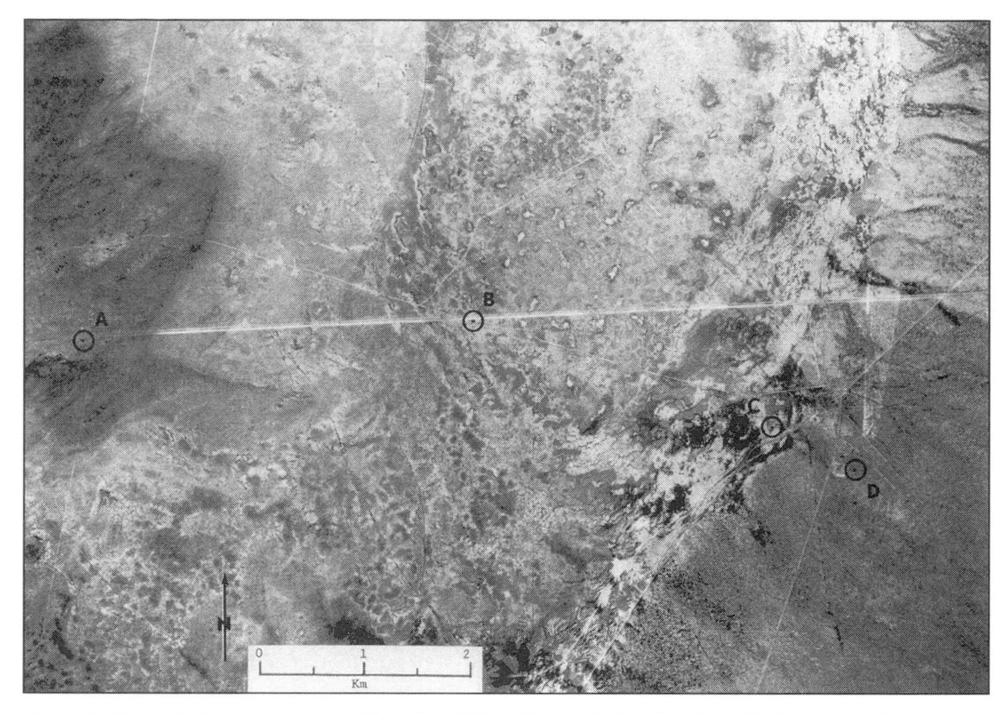

Figure 6. Expanded airphoto view of location "5" on Figure 5, showing four distinct facies from center to edge of Pleistocene Lake Trinity basin, verified in bore holes A, B, C, D: (A) windblown sand (Holocene), (B) sulfate, (C) shoreline, and (D) alluvial fan underlain by B at depth.

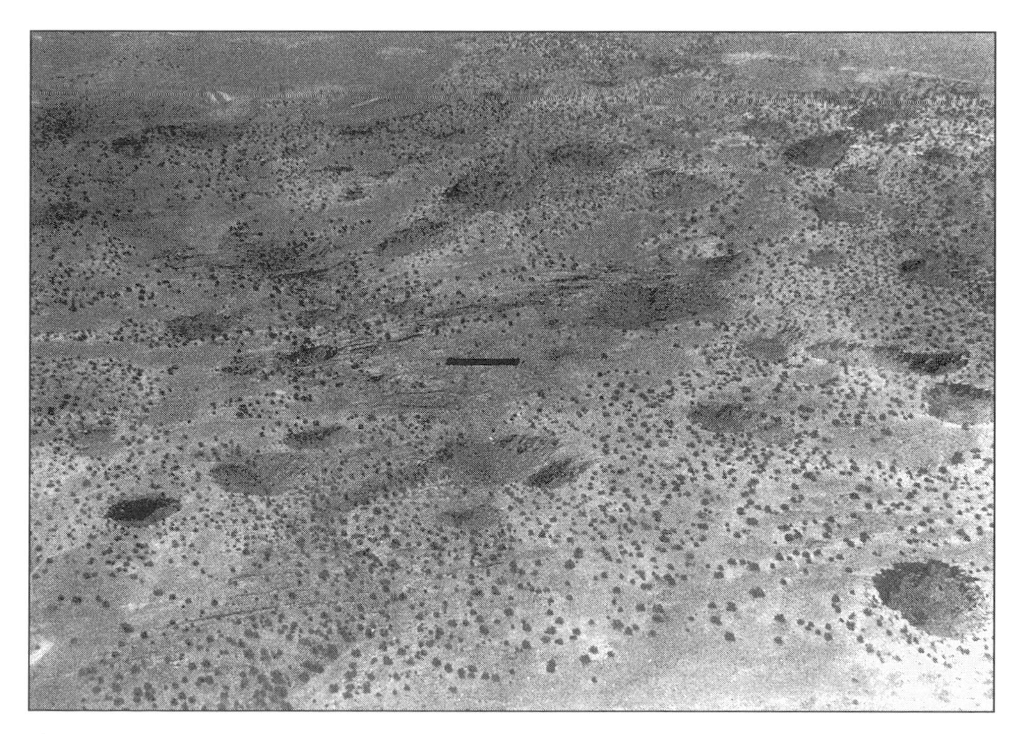

Figure 7. Salt-karst features overlying Permian evaporites in Holbrook basin, Arizona. Joints in surface Coconino Sandstone control development of sinkholes, infiltration, and the subsurface dissolution front coincident with the Holbrook anticline. Largest sinkholes are 200 m across and 50 m deep. Black scale bar in center = 75 m.

(CAES) has been tested but is not yet economically practical. The storage of other products, including grain and hazardous waste, has also been proposed (Bishop, 1993).

Salt karst is relatively poorly known in the geologic literature in comparison with limestone karst. During the last 35 years, much of the technical literature about salt karst has been produced under the sponsorship of national security projects—Projects Gnome and Salmon, Office of Nuclear Waste Isolation, Waste Isolation Pilot Plant, and the Strategic Petroleum Reserve—all managed by the U.S. Department of Energy and its predecessor agencies (Neal, 1994). As a result of these projects conducted in salt, the geologic literature is replete with estimates of rates of dissolution and the potential risks to underground facilities, and of ground-water travel times for storage projects that involve hazardous materials. It seems likely that without the intense effort invested in these projects, today's understanding of salt karst would be much less mature (Fig. 7).

CONCLUSIONS

The studies described here are a direct outgrowth of military and national security projects and are relevant to peacetime applications of the geosciences. Some of the results described here were not specifically sought, and their application to peacetime purposes was not immediately obvious. Many other projects also have resulted in peacetime applications. The geosciences can benefit in many ways from insights gained during military and national security projects; however, geologists must identify opportunities for the use of such insights *and must be vigorous and effective in disseminating it.*

ACKNOWLEDGMENTS

This work was performed at Sandia National Laboratories and supported by the U.S. Department of Energy under Contract DE-ACO4-94AL85000.

REFERENCES CITED

Bishop, W. M., 1993, Storage of grain in underground leached salt caverns: Deerfield, Illinois, Solution Mining Research Institute, Houston, Texas, Spring Meeting Proceedings, 12 p.

Ehgartner, B. L., 1992, Effects of cavern spacing and pressure on subsidence and storage losses for the U.S. Strategic Petroleum Reserve: Albuquerque, New Mexico, Sandia National Laboratories Report SAND91-2575, 47 p.

Eugster, H. P., 1967, Hydrous sodium silicates from Lake Magadi, Kenya: Precursors of bedded chert: Science, v. 157, p. 1177.

Hamburger, R., 1973, Geologic factors in rapid excavation with nuclear explosives: *in* Pincus, H., ed., Geological factors in rapid excavation: Geological Society of America, Engineering Geology Case History No. 9, p. 17–25.

Hoffman, E. L., 1992, Effects of cavern depth on surface subsidence and storage loss of oil-filled caverns: Albuquerque, New Mexico, Sandia National Laboratories Report SAND92-0053, 30 p.

Kiersch, G. A., 1991, The heritage of engineering geology, the first hundred years: Geological Society of America, Decade of North American Geology, Centennial Special Volume 3, 605 p.

Krinsley, D. B., 1976, Selection of a road alignment through the Great Kavir in Iran, *in* Williams, R. S., Jr., and Carter, W. D., eds., ERTS-1, A new window on our planet: Washington, D.C., U.S. Geological Survey Professional Paper 929, p. 296–302.

Krinsley, D. B., Woo, C. C., and Stoertz, G. E., 1968, Geological characteristics of seven Australian playas, *in* Playa surface morphology: Miscellaneous investigations: Bedford, Massachusetts, Air Force Cambridge Research Laboratories Environmental Research Paper 283 (AFCRL-68-0133), p. 59–103.

Kupfer, D. H., 1990, Anomalous features in the Five Islands salt stocks: Baton Rouge, Louisiana, Gulf Coast Association of Geological Societies Transactions, v. 40, p. 425–437.

Motts, W. S., 1972, Some hydrologic and geologic processes influencing playa development in western United States, *in* Reeves, C. C., Jr., ed., Playa lake symposium: Lubbock, Texas, International Center for Arid and Semiarid Land Studies, ICASALS Pub. 4, p. 89–106.

Neal, J. T., 1972, Playa surface features as indicators of environment, *in* Reeves, C. C., Jr., ed., Playa Lake Symposium: Lubbock, Texas, International Center for Arid and Semiarid Land Studies, ICASALS Publication No. 4, p. 107–132.

Neal, J. T., 1994, Surface features indicative of subsurface evaporite dissolution: implications for storage and mining: Deerfield, Illinois, Solution Mining Research Institute, Houston, Texas, Spring Meeting Proceedings, 18 p.

Neal, J. T., and Magorian, T. R., 1997, Geologic site characterization (GSC) principles derived from storage and mining projects in salt, with application to environmental surety: Berlin, Springer International, Environmental Geology, v. 29, p. 165–175.

Neal , J. T., and Motts, W. S., 1967, Recent geomorphic changes in playas of western United States: Journal of Geology, v. 75, no. 5, p. 511–525.

Neal, J. T., Langer, A. M., and Kerr, P. F., 1968, Giant desiccation polygons of Great Basin playas: Geological Society of America Bulletin, v. 79, p. 69–90.

Neal, J. T., Smith, R. E., and Jones, B. F., 1983, Pleistocene Lake Trinity, an evaporite basin in the northern Jornada del Muerto, New Mexico, *in* Chapin, C. E., ed.: Socorro, New Mexico, 34th New Mexico Geological Society Guidebook, Socorro Region II, p. 285–290.

Neal, J. T., Magorian, T. R., Thoms, R. L., Autin, W. J., and Harding, R. S., 1993, Anomalous zones in Gulf Coast salt domes with special reference to Big Hill, Texas, and Weeks Island, Louisiana: Albuquerque, New Mexico, Sandia National Laboratories Report SAND92-2283, 65 p.

Neal, J. T., Bauer, S. J., and Ehgartner, B. L., 1998, Mine-induced sinkholes over the U.S. Strategic Petroleum Reserve storage facility at Weeks Island, Louisiana: Geologic causes and effects, *in* Borchers, J., ed., Land subsidence: Current research and case histories. Proceedings of Dr. Joseph F. Poland Symposium on Land Subsidence: Sacramento, California, October 4–5, 1995, Association of Engineering Geologists Special Publication 8.

Rooney, T. P., Jones, B. F., and Neal, J. T., 1969, Magadiite from Alkali Lake, Oregon: American Mineralogist, v. 54, p. 1034–1043.

Talbot, C. J., and Jackson, M. P. A., 1987, Internal kinematics of salt diapirs: American Association of Petroleum Geologists Bulletin, v. 71, p. 1068–1093.

Williams, R. S., Jr., and Carter, W. D., editors, 1976, ERTS-1, A new window on our planet: Washington, D.C., U.S. Geological Survey Professional Paper 929, 362 p.

Manuscript Accepted by the Society October 29, 1997

Printed in U.S.A.

Geological Society of America
Reviews in Engineering Geology, Volume XIII
1998

Military geology and the Gulf War

Robert B. Knowles
U.S. Army Topographic Engineering Center, 7701 Telegraph Road, Alexandria, Virginia 22315-3864
William K. Wedge
Missouri Department of Natural Resources, Division of Geology and Land Survey, Rolla, Missouri 65401

ABSTRACT

Engineering geologists and hydrogeologists assigned to the 416th Engineer Command (ENCOM) supported the planning and execution of construction and tactical operations during the Gulf War. Military geology applications included locating potential quarry sites for sources of construction aggregate and fill, evaluating terrain features such as sabkhahs to assess cross-country mobility, and developing water sources.

Sources of construction aggregate were needed to support sustainment engineering requirements in building and maintaining roads, heliports, and aircraft parking aprons in Saudi Arabia. Technical advice and assistance were provided to host nation forces who supported the production and transportation of aggregate from the source to the stock pile.

Terrain analysis contributed to the success of the ground war. Obsolete or inaccurate maps were updated with new satellite images and field reconnaissance. Areas with inadequate terrain data were investigated to document natural as well as manmade obstacles. Coastal sabkhahs were evaluated and tested to determine their effect on mobility. Extensive surficial samples were collected for detailed geologic analysis, and field-expedient methods to improve trafficability were recommended.

Military hydrogeologists and engineers worked closely with the Saudi Ministry of Agriculture to design and site new water wells. Several water wells were drilled by military teams to support operations deep in the desert.

Satellite images, aerial photographs, maps, existing reports, and field reconnaissance were utilized to evaluate geologic conditions, thorough knowledge of which greatly contributed to the success of the ground war.

INTRODUCTION

Engineering geologists assigned to the 416th Engineer Command (ENCOM) supported the planning and execution of construction and tactical operations during the Gulf War. Military geology applications included evaluating terrain features such as sabkhahs to assess cross-country mobility, locating potential quarry sites for sources of construction aggregate, and developing water sources.

Physiography

Operations Desert Shield and Desert Storm occurred in northeastern Saudi Arabia, southeastern Iraq, and Kuwait from August 1990 to March 1991. This area is part of the eastern physiographic region of Saudi Arabia (Ministry of Agriculture, 1984). The terrain is generally flat lying and includes the As Summan Plateau, the Gulf Coastal Plain, and the Ad Dibdibah Plain (Fig. 1).

As Summan Plateau. Chapman (1978) described the As

Knowles, R. B., and Wedge, W. K., 1998, Military geology and the Gulf War, *in* Underwood, J. R., Jr., and Guth, P. L., eds., Military Geology in War and Peace: Boulder, Colorado, Geological Society of America Reviews in Engineering Geology, v. XIII.

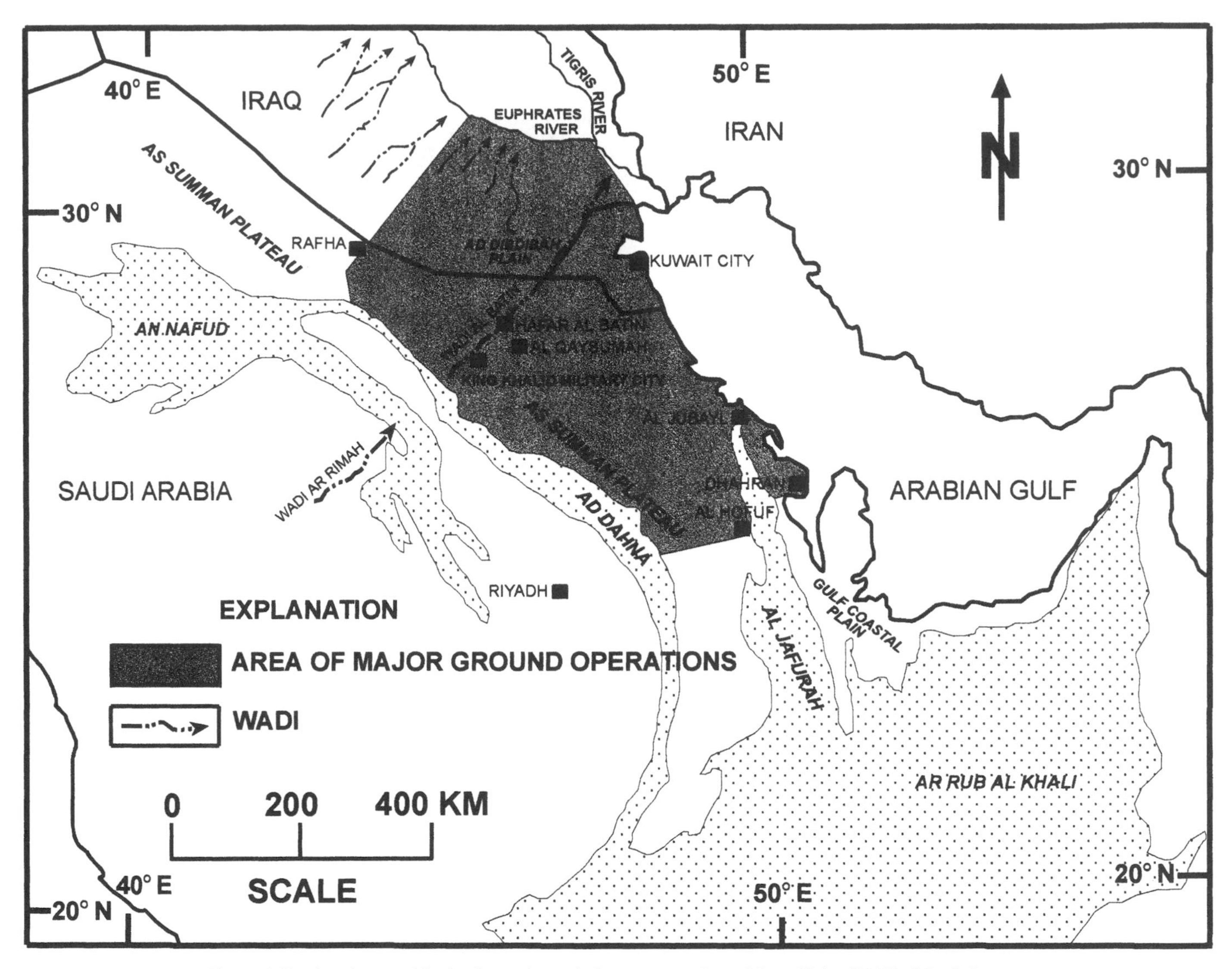

Figure 1. Regional map with physiography and place names adapted from Holm (1960). Stippled pattern represents major sand sheets. Area of major ground operations (shaded) extended over parts of the As Summan Plateau and Ad Dibdibah Plain.

Summan Plateau as a long, flat, barren, hard-rock plateau of varied width, trending north from a point about 300 km south of Al Hofuf northwestward across northeastern Saudi Arabia. The maximum width of the plateau is 250 km (Fig. 1). The As Summan Plateau is a prong of the larger Syrian Plateau. In Iraq, Syria, and northern Saudi Arabia the plateau is a vast desert of gravel and rock plains underlain by Cretaceous and Tertiary sedimentary rocks and Pliocene basalt. The plateau slopes eastward from an elevation of nearly 400 m at its western margin to about 245 m on its eastern edge. Bedrock is exposed in some parts of the plateau, but in many areas it is covered by surficial deposits that may be 30 m thick.

Gulf Coastal Plain. The Gulf Coastal Plain lies between the As Summan Plateau and the Arabian Gulf. It is an expanse of flat lowlands covered by sand and gravel that averages 160 km in width. It is composed of dunes, sabkhahs, and sandy plains that slope toward the Arabian Gulf. Low rolling plains are covered with a thin mantle of sand from Al Jubayl northward toward Kuwait City. The sand mantle averages 1 to 2 m thick but may be considerably thicker in dune areas. The roots of shrubs and grasses hold the sand in hummocks ranging in height up to 2 m and form an irregular, hummocky terrain called "dikakah." From Al Jubayl to Al Hofuf is a wide belt of drifting sand and dunes that merge with the Al Jafurah sand area and the Ar Rub Al Khali farther to the south. The shallow nearshore waters of the gulf may shift the coastline back and forth across a width of several kilometers. Large sabkhahs occur along the coast and are commonly saturated with brine and encrusted with salt. Barren-rock terrain developed on Eocene, Miocene, and Pliocene limestone is exposed in areas not covered with sand or sabkhahs (Chapman, 1978).

Ad Dibdibah Plain. The Ad Dibdibah Plain occurs in the area southwest of Kuwait. It is a large triangular-shaped gravel plain. With its apex near Al Qaysumah, it spreads northeastward almost to the Tigris and Euphrates valleys. It is composed of cobbles and pebbles of igneous and metamorphic rocks and limestones with finer sediments. The cobble and pebble sizes decrease toward the northeast. The Ad Dibdibah gravel plain is considered to be the delta of the Wadi Ar Rimah–Wadi Al Batin drainage system, which transported rock debris from the Arabian Shield area during one of the pluvial periods (Chapman, 1978). In many areas, the average thickness of surficial materials overlying a hardpan is less than 1 m. To the observer on the ground the plain appears to be perfectly flat, but actually it is an undulating surface with enough local relief, on the order of 20 m, to be tactically significant in many areas.

Geology

The Arabian Peninsula comprises the Arabian Shield and the Arabian Shelf. The Arabian Shield occupies the west-central one-third of Saudi Arabia and consists of igneous and metamorphic rocks of Precambrian age. These rocks are folded, faulted, and uplifted to form the Asir and other mountains along the Red Sea. The Arabian Shelf overlies the Arabian Shield over the eastern two-thirds of Saudi Arabia, southeastern Iraq, and Kuwait. The Arabian Shelf is a series of sedimentary formations ranging in age from Cambrian to Pliocene, dipping from the center of the Arabian Peninsula eastward toward the Arabian Gulf. In central Arabia the beds dip gently and uniformly northeast, east, and southeast from about one degree in older units to less than half a degree in the Upper Cretaceous and Eocene beds (Chapman, 1978). The sedimentary rocks consist mainly of limestone, sandstone, and shale with an aggregate thickness of strata of 5,500 m; they thin toward the west. The As Summan Plateau and Gulf Coastal Plain are part of the Arabian Shelf (Fig. 2).

TERRAIN ANALYSIS

Although the terrain of the As Summan Plateau and Gulf Coastal Plain is generally flat, areas of sand dunes and rounded hills, 5 to 10 m high, concerned military planners. Terrain analysts contributed significantly to the success of the ground war by delineating areas where cross-country movement could be adversely affected by sand areas, wadis, and sabkhahs. Obsolete or inaccurate maps were updated with new satellite images and field reconnaissance. Both Landsat and Systeme Probatoire d'Observation de la Terre (SPOT) satellite images were used in terrain analysis. Landsat Thematic Mapper system multispectral bands 2, 4, 7 (blue, green, red) with 30-m-resolution elements were used by Rinker et al. (1990) to produce rapid-response terrain analysis for the eastern physiographic region of Saudi Arabia, southern Iraq, and Kuwait. Areas with inadequate terrain data were investigated by ground reconnaissance teams to document natural as well as man-made obstacles.

Sand areas

Approximately one-third of the Arabian Peninsula is covered by mobile sand in one form or another. Three major sandseas

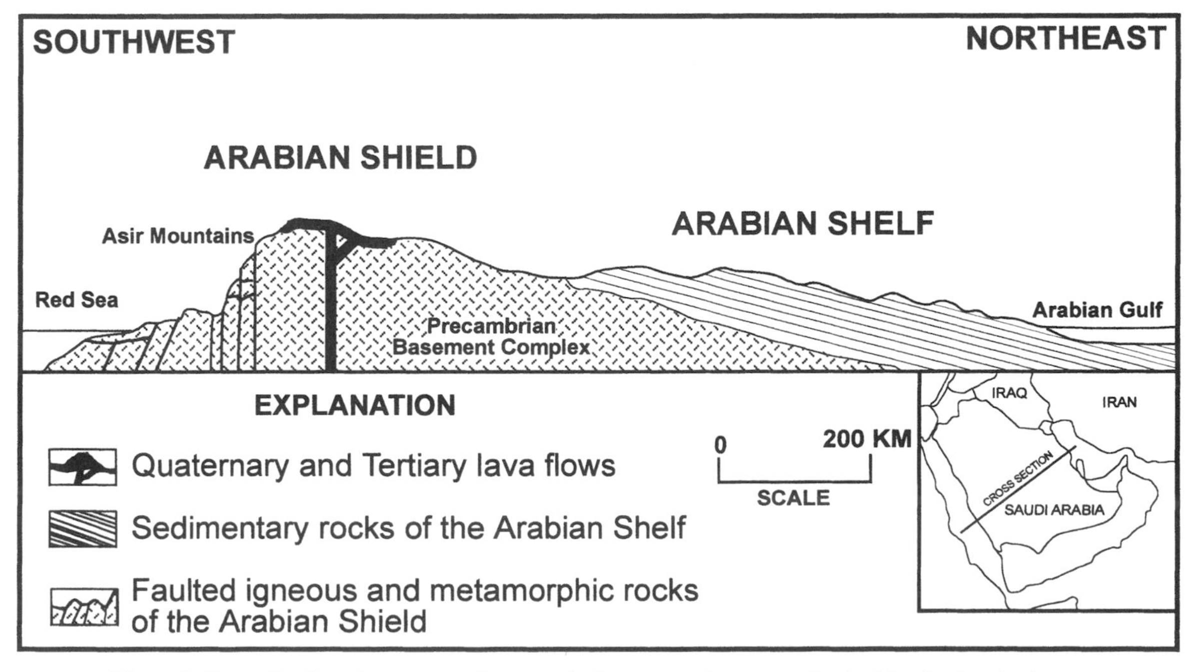

Figure 2. Generalized southwest to northeast geologic cross section across the Arabian Peninsula showing the Arabian Shield and Arabian Shelf, modified from Ministry of Water and Agriculture (1984).

occur: the An Nefud, the Ad Dahna, and the Ar Rub Al Khali. In the northwestern part of the peninsula, the An Nefud occupies a broad basin covering an area of approximately 57,000 km^2 with reddish sand and rolling dunes. The An Nefud is devoid of streams and oases and has sparse vegetation. The Ad Dahna is a long narrow belt of shifting sand and dunes extending nearly 1,300 km in a broad arc from the An Nefud in the north to Ar Rub Al Khali in the south. It lies between As Summan Plateau on the east and the cuesta region on the west, an area in central Arabia where upturned edges of sedimentary beds form west-facing escarpments along the eastern edge of the Arabian Shield. The sand of Ad Dahna is bright red-orange owing to an iron oxide coating of the grains. The Ar Rub Al Khali, or Empty Quarter, in southern Arabia is the largest continuous sand body in the world. The region covers a total area of 600,000 km^2; the sand is red-orange and medium to fine grained like that of the An Nefud and Ad Dahna (Chapman, 1978). Although Operations Desert Shield and Desert Storm avoided the major sand seas of the Arabian Peninsula, localized sand bodies and hummocky areas of dikakah presented trafficability problems, especially to large trucks operating off existing roads.

Trafficability most often depended on the degree to which the sand had been compacted by the wind. In sand-dune areas, trafficability depended on the types of dunes and their associated interdune areas. On dunes with asymmetric slopes, the gentler windward slope was wind compacted and could usually support foot and light vehicular traffic. The steep lee slope, or slip face, could not support either foot or vehicular traffic without avalanching (Rinker et al., 1991). Most of the sand-induced trafficability problems occurred in dikakah, areas of sand drift, and in loose sand in wadi channels. In general, these areas were only significant locally. Areas of dikakah were depicted on existing maps and were further delineated with satellite images.

The U.S. Army field manual *Desert Operations* (U.S. Army Armor School, 1977, p. 2–22) states that "dust and sand are probably the greatest danger to the efficient functioning of equipment in the desert." Helicopters in particular were affected because of their normally low operating altitudes. Sand abraded leading edges of rotor blades, rotor heads, and exposed flight control surfaces. Sand ingestion by engines and disorientation created by blowing dust and flying sand also were operational hazards (U.S. Army Armor School, 1977). Military construction in sandy areas was complicated by the excavation problems that resulted from lack of cohesion and the need to protect horizontal structures from drifting sand. Potential problems occurring in dikakah, loose drift sand, sand sheets, sand dunes, and loose sand in wadi channels were for the most part minimized by avoiding these areas. Where necessary, sand was stabilized with various types of mattings, grids, and dust pallatives. Dust pallatives cover the dust source areas with a thin protective layer, cement dust particles together, or cause dust particles to agglomerate into larger particles not readily airborne. Bituminous materials and oil were two common pallatives used to stabilize sand and control dust.

Wadis

For military terrain analysis in desert environments, Rinker et al. (1991) described wadis as channels that were dry or had only intermittent or ephemeral streamflow. They ranged in size from small gullies less than a few meters wide and deep, to large broad valleys several kilometers wide and tens of meters deep, to large, deep mountain canyons hundreds of meters wide and deep. They were classified as wadi washes that were dry, intermittent, or ephemeral drainage courses marked by deposits of alluvial material that were not confined to a specific channel. The channels were commonly braided and covered an area up to several kilometers wide. These broad deposits on open valley floors were, in most places, generally good surfaces for travel. In the Desert Shield/Desert Storm area most of the wadis either were small gullies a few meters wide and deep, or broad valleys several kilometers wide with very little relief. Many of these wadis became the axis of advance for the coalition forces attacking into the southern Iraqi desert.

The largest wadi in the Desert Shield/Desert Storm area is Wadi Al Batin. This wadi appeared clearly on satellite images and maps as one of the most distinct features in the area, which led to some initial misconceptions as to its military significance: "Wadi Al Batin appeared to be a geologic rift—a formidable obstacle—but on the ground it turned out to be passable for both tracked and wheeled vehicles" (Brinkerhoff et al., 1992, p. 60). Military planners initially believed that Wadi Al Batin could not be traversed, which would have significantly affected the battle plan for Operation Desert Storm.

A thorough analysis of maps and Landsat satellite images by Rinker et al. (1990, p. 5–6), which was confirmed by field reconnaissance during the war, determined that:

> From Hafar al Batin to north of the border with Kuwait, the wadi is about 12.5 kilometers wide. North of the border it broadens out and the southeastern rim becomes indistinct. Near Hafar al Batin the wadi depth is about 60 m (JOG sheet), 50 m near the border, and less than 25 m in the northern part. This means a cross section of gentle slopes. A drop of 60 m over about 6 kilometers (half the wadi width), or about 1 m per 100 m, a 1% slope. The main channel slope ranges from about 1% in the southwest to nearly level in the north.

Further analysis predicted that the wadi was trafficable in all directions except for local obstacles such as rocks or loose sand. South of Hafar al Batin, the wadi is more incised and is characterized by rocky sideslopes and steep walls. Between King Khalid Military City and the Ad Dahna sands, few trails lead into the wadi. It was critical for operations planners to understand that for military operations conducted north of Hafar al Batin the wadi was not an obstacle, but farther south it could be. The characteristics of Wadi Al Batin helped shape the overall strategy of outflanking the Iraqi army in Kuwait by rapidly shifting forces west of the wadi just prior to the attack into the southern Iraqi desert.

Sabkhahs

Sabkhahs are common along the coastal plain from Kuwait to the southern end of the Arabian Gulf. Johnson et al. (1978) defined sabkhahs as saline flats underlain by clay, silt, and sand, and often encrusted with salt that are equilibrium surfaces whose level is largely controlled by the local ground-water table. Rinker et al. (1991) described two types of sabkhahs on the Arabian coast. One is arenaceous, or sand filled, and the other is argillaceous, or clay filled. The arenaceous sabkhah is formed by the in-filling of embayments of the sea with eolian sand and is typical of sabkhahs along the coast of the eastern physiographic region of Saudi Arabia. During dry periods, capillary water in the sand evaporates and concentrates as a brine near the surface, producing a soft quicksand of low bearing strength. Evaporation of standing water forms a coating of salt crystals, which can thicken into a crust. Sabkhahs are a concern for cross-country mobility because vehicles crossing sabkhahs may break through the surface crust and become mired in the mud.

Operation planners for Operation Desert Shield were concerned about trafficability along the large coastal sabkhahs near the Kuwait border. One initial concern was whether Iraqi armored columns could cross the large coastal sabkhahs from Kuwait into the eastern physiographic region of Saudi Arabia. In addition, it was important to delineate the sabkhahs and determine which ones would impede movement. Sabkhah trafficability is a valid military concern as failed crossings have resulted in the loss of life. For example, in December 1940 the British army forces attempted to cross a sabkhah covered by enemy antitank guns while pursuing Italian forces west of Buqbuq, Egypt:

> Here the 3rd Hussars . . . anxious to close the range regardless of opposition, weaved aside to avoid the fire, drove hard into the salt pans and bogged to their belly-plates. Soon a line of wallowing Mark VIs were a row of burning hulks, the victims of an ill-considered attempt at charging an unshaken enemy across impossible ground (Macksey, 1971, p. 86).

Because the available maps indicated the presence of sabkhahs in almost every topographic low, some concern existed about their widespread occurrence. A terrain team with a heavily loaded five-ton dump truck and water trailer utilizing Geographic Positioning Systems (GPS), moved from map-designated "sabkhah" to "sabkhah" near the Kuwait border to establish the fact that most of those shown could be described more accurately as playas, and as long as they were dry were no impediment to cross-country mobility (Fig. 3).

Rinker et al. (1990, p. 9) encountered the same problem when comparing maps with satellite images: "The JOG sheet (NH 38-12) shows the area as wadis and sabkhahs. Note: no indication of such patterns [sabkhahs] in the image." As Landsat satellite image maps became available they were utilized to help distinguish sabkhahs from playas. Many sabkhahs on Landsat Thematic Mapper images featuring multispectral bands 2, 4, 7 (blue, green, red), have a distinct dark-brown tone

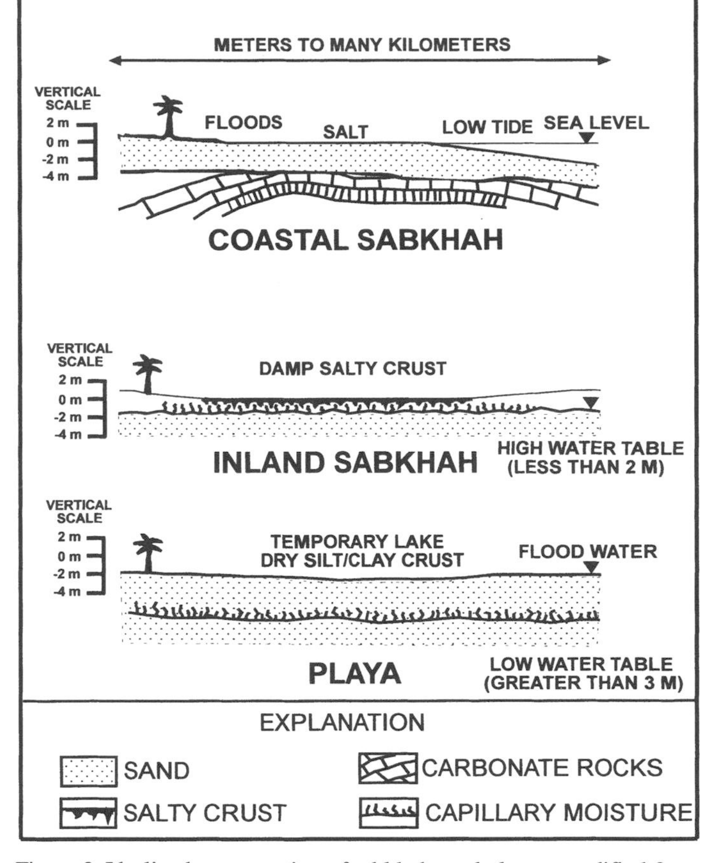

Figure 3. Idealized cross-section of sabkhahs and playas, modified from Fookes (1976). Coastal sabkhahs are influenced by tidal action and may be flooded during high tides. Inland sabkhahs are influenced by high water tables and the capillary moisture maintaining a damp salty crust. Playas are intermittent lakes with dry silt and clay crusts.

that can help distinguish them from playas (Fig. 4). But if sabkhahs are covered by dust or sand, their tone may be very similar to that of playas.

In general, playas may be described as enclosed shallow depressions in desert basins that contain deposits of evaporites from the impoundment of episodic stream flow. When dry they support foot and vehicular traffic and aircraft operations (Rinker et al., 1991). In Saudi Arabia, non-saline playas well above the water table composed of silt, fine sand, and clay had been labeled on military maps as sabkhahs. Although resembling playas, sabkhahs are characterized by the presence of salt, and the term "sabkhah" always refers to the saline, puffy, crust-surface flat basins that intersect the water table and cannot be assumed to be trafficable.

Johnson et al., (1978) described sabkhahs as either coastal or inland. Coastal sabkhahs are supratidal surfaces produced by depositional offlap of marine sediments, sometimes with the addition of eolian sediments. The associated brines are primarily from seawater. Inland sabkhahs may be many kilometers from the coast and at elevations as high as 150 m (Chapman, 1978). They represent

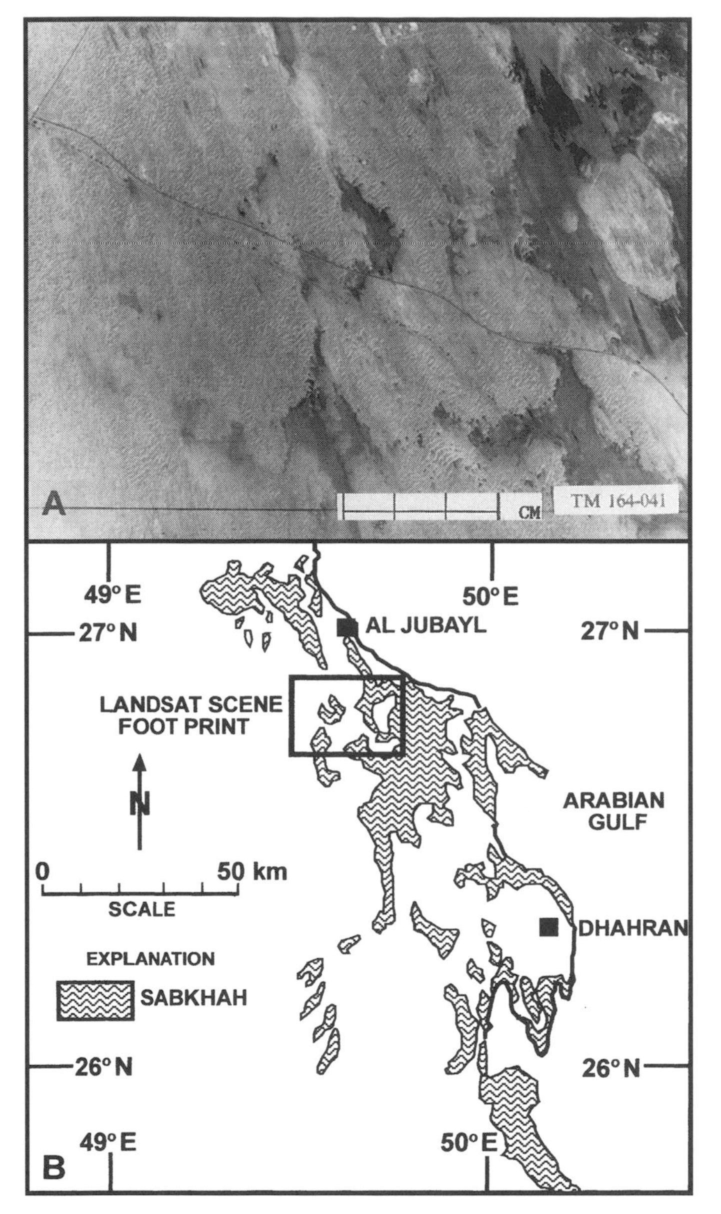

Figure 4. A. Sabkhahs and sand dunes on Landsat Scene 164-041 (09/07/87) in eastern Saudi Arabia. Sabkhahs appear as dark tone and smooth texture in contrast with the light tone and texture of surrounding sand sheets and dunes. B. Sketch map of area of Landsat scene 164-041 (09/07/87) on coastal region map modified from Johnson et al. (1978).

areas of equilibrium between eolian sedimentation and deflation, a condition controlled by the elevation of the local water table (Johnson et al., 1978). The base level of deflation is just above the capillary fringe in the sediments above the water table. Evaporation of ground water causes the formation of brine and the precipitation of evaporite minerals. The report of the U.N. Food and Agriculture Organization (1979) indicates that a discharge line of sabkhahs and springs is continuous from northwestern Iraq, through the Euphrates region into western Kuwait, and through the coastal region of Saudi Arabia and into the Ar Rub Al Khali.

Coastal and inland sabkhahs were evaluated in the field and tested to determine their effect on mobility. Heavy trucks and trailers were used to test sabkhah trafficability and pulled out where necessary. Field-expedient methods to improve trafficability were recommended; they included utilizing geotextile membranes and constructing causeways. Multispectral images can assist military planners in differentiating sabkhahs and playas by their texture and tone, but only direct observation can determine their potential trafficability.

MILITARY CONSTRUCTION

Sources of construction aggregate were needed to support sustainment engineering requirements in building and maintaining roads, heliports, and aircraft parking aprons in Saudi Arabia. Historically, construction has played a paramount role in providing the logistical support needed by U.S. combat units. Movement of these units and support units within a theater depends on lines of communication. Because of the limited number of existing roads and the difficulty of off-road mobility in the desert, considerable effort was required to construct and maintain roads forward to maneuver units. The harsh Middle East environment presented critical horizontal construction problems related to the lack of water, temperature extremes, dust, lack of construction materials, and soil conditions (Kao and Hadala, 1981). Difficult soil conditions included unstable soils (e.g., drifting sand), aggressive salty ground (e.g., sabkhahs), unsuitable construction materials (e.g., silt, fine sand, soft carbonate rock), and those soils subject to rapid erosion by wind and flash floods.

Fookes (1976) described engineering properties of various geographic units found in desert areas. He divided desert regions into four geographic zones: (I) mountain slopes, (II) the apron fan or bajada, (III) the alluvial plain, and (IV) the base plain, which includes sabkhahs, playas, salt playas, salinas, and sand dune areas. The eastern physiographic province of Saudi Arabia, the southeastern Iraqi desert, and Kuwait lie within the base plain (IV) engineering zone. This zone includes desert flats, playas and sabkhahs, and sand dunes. Fookes (1976) described this zone as the most widespread of all the zones and the one with the most engineering problems, including erratic behavior of load-bearing materials, migrating dunes, saline soil, sabkhahs, crusts, and a lack of coarse materials needed for military construction of roads and airfields.

Aggregate for military construction was not a logistical constraint in Saudi Arabia, although the quality of available materials was, in places, unsuitable for long design-life commercial projects. Limestone is the major construction aggregate in the Gulf and is extensively quarried for road metal, aggregate, and to a small extent for building stone (Johnson, 1978). Marl and marly shale are excavated for use as impermeable barriers by commercial interests, primarily for underlayments for garden areas around building sites and for military road construction in some areas. Although sand is plentiful, most of it is too fine grained, too badly graded, and too chemically impure to be used in any form of construction other than as fill. In Kuwait, "gatch," the local term for a variety of marine-deposited sandy soils, can be used in road bases. It is composed of grains of silica sand and clay materials (Evans, 1977).

Sand and desert gravel were abundantly available to support military construction but not in concentrated deposits that were easy to exploit. Military geologists conducted reconnaissance to locate potential quarry sites in the northern and eastern provinces of Saudi Arabia for sources of construction aggregate. Existing geologic maps and global positioning receivers were critical in this effort. Existing sources were identified and technical advice and assistance were provided to host-nation forces who supported the production and transportation of aggregate from the source to the stock pile. More than 500,000 m^3 of gravel were procured to repair and maintain 2,200 km of roads, construct two 48,000-person enemy prisoner-of-war camps, construct numerous helipads/aircraft bed-down sites, and assist in constructing theater logistic bases. In addition, 170,000 metric tons of asphalt, and 93,000 m^3 of ready-mix concrete were used to improve and repair existing facilities for expanded military use (Mulcahy, 1992).

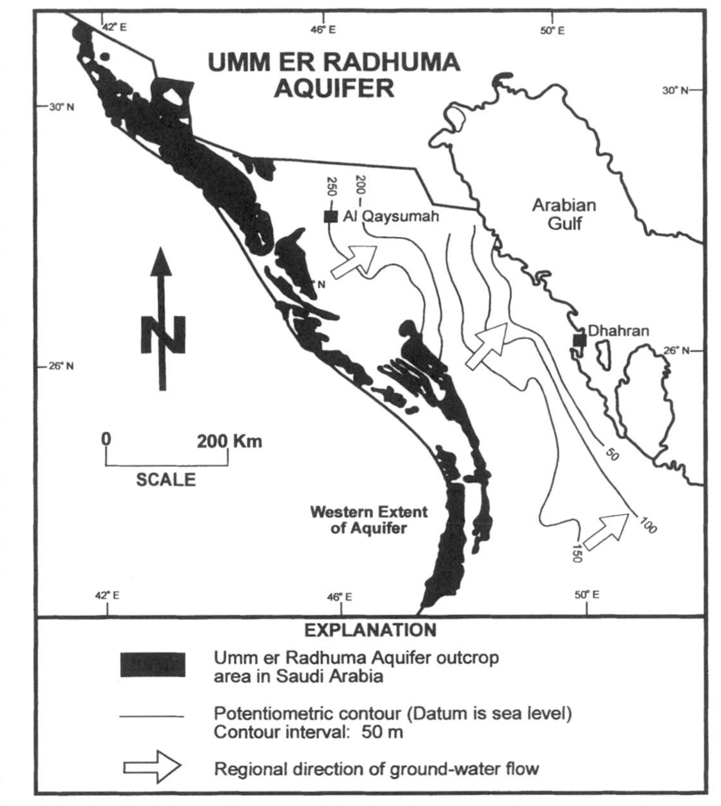

Figure 5. Outcrop area and extent of Umm er Radhuma aquifer in Saudi Arabia, modified from Ministry of Water and Agriculture (1984). Regional ground-water flow is toward the Arabian Gulf. Ground water is generally fresh in the outcrop zone and becomes more brackish toward the coast.

WATER SUPPLY

Water supply is one of the most important logistical concerns of military operations in a desert environment. In the Desert Storm theater of operations, the average rainfall is approximately 100 mm per year; no surface-water supplies exist in the area. The gently dipping sedimentary formations of the Arabian Shelf contain virtually all of the naturally occurring fresh water in the Arabian Peninsula. Aquifers of the Arabian Shelf are composed of sandstone, limestone, and dolomite, which have large areal extent and great volumes of stored water. Nine major aquifers are producing water in large volumes for industrial, agricultural, and public supply uses in Saudi Arabia (Ministry of Agriculture and Water, 1984).

Two of these aquifers, the Umm er Radhuma and Wasia aquifers, are the primary water producers in the Desert Shield area. The Umm er Radhuma provides water for most of the irrigated farms in the region from 10 to 400 m depths; the deeper Wasia aquifer provides drinking water from 70 to 800 m depths for large villages such as Hafar al Batin. The Umm er Radhuma aquifer provided water for numerous existing water wells that were used by U.S. forces as water-supply points. The Umm er Radhuma aquifer crops out along the border with Iraq and Jordan in a 50- to 100-km-wide band 1,200 km long (Ministry of Agriculture and Water, 1984). It crops out or is near the surface in much of the Desert Shield build-up area and has excellent water-bearing properties. The water quality is generally better near the outcrop and recharge areas and becomes more saline eastward (Fig. 5).

Hydrogeologic data were provided by the Saudi Ministry of Agriculture and Aramco, and military hydrogeologists and engineers worked closely with the Saudi Ministry of Agriculture to design and site new water wells. Several water wells were drilled by military teams to support rear-area operations deep in the desert, whereas others were drilled by commercial drilling companies under contract to support military operations. The depth of these wells averaged 300 m.

Satellite images and aerial reconnaissance photographs were used to identify existing wells, especially in denied areas of the southern Iraqi desert. In northern Saudi Arabia, center-pivot irrigation farms and the more traditional ridge and furrow irrigated farms appear on satellite images. Water wells or springs are associated with each irrigated area. Most of these irrigation wells tap the Umm er Radhuma aquifer, although many of the water wells shown on existing maps were either dry or unusable. All potential water sources had to be verified with ground reconnaissance.

The influx of several hundred thousand soldiers into the northern Saudi Arabian desert created a tremendous water demand in a largely uninhabited region. Fortunately, the region is the recharge and outcrop area of one of the most productive aquifers in the Arabian Peninsula where the ground-water resource is well understood and exploited. This allowed for the temporary support of a large population without disrupting existing municipal or agricultural supplies.

CONCLUSIONS

During the Gulf War military geologists contributed to the success of the planning and execution of both tactical and strategic operations in Saudi Arabia, Iraq, and Kuwait. Military geologists supported the terrain analysis that allowed operational planners to devise the decisive strategy that ultimately won the war. Military geologists supported the development of infrastructure to support the tremendous logistical effort required to sustain modern mobile warfare by identifying, developing, and managing sources of construction aggregate. In addition, military geologists identified the sources of subsurface water supply that insured that water never became a logistical constraint to ground forces operating deep in the desert.

REFERENCES CITED

Brinkerhoff, J. R., Silva T., and Seitz, J., 1992, United States Army Reserve in Operation Desert Storm—Engineer support at echelons above corps: Washington, D.C., The 416th Engineer Command: Department of the Army, p. 77.

Chapman, R. W., 1978, Geomorphology, *in* Al-Sayari, S. S., and Zolt, J. G., eds., Quaternary Period in Saudi Arabia: Vienna, Austria, Springer-Verlag, p. 19–25.

Evans, P. L., 1977, The Middle East—An outline of the geology and soil conditions in relation to construction problems: Watford, England, Building Research Station, 17 p.

Food and Agriculture Organization (FAO), 1979, Survey and evaluation of available data on shared water resources in the Gulf States and the Arabian Peninsula, Volume One: Rome, United Nations, p. 11.

Fookes, P. G., 1976, Road geotechnics in hot deserts, The Highway Engineer: Journal of the Institution of Highway Engineers, v. 33, no. 10, p. 11–23.

Holm, D. A., 1960, Desert geomorphology in the Arabian Peninsula: Science, v. 132, no. 3437, p. 1371.

Johnson, D. H., 1978, Gulf Coastal Region and its hinterland, general geology, *in* Al-Sayari, S. S., and Zolt, J. G., eds., Quaternary Period in Saudi Arabia: Vienna, Austria, Springer-Verlag, p. 45–50.

Johnson, D. H., Kamal, M. R., Pierson, G. O., and Ramsay, J. B., 1978, Sabkhahs of eastern Saudi Arabia, *in* Al-Sayari, S. S., and Zolt, J. G., eds., Quaternary Period in Saudi Arabia: Vienna, Austria, Springer-Verlag, p. 84–93.

Kao, A. M., and Hadala, P. F., 1981, Theater of operation construction in the desert: Handbook of Lessons Learned in the Middle East: Champaign, Illinois, U.S. Army Construction Engineering Research Laboratory, 164 p.

Macksey, K. J., 1971, Beda Fomm: The classic victory: New York, Ballantine Books Inc., p. 83–86.

Ministry of Water and Agriculture, 1984, Water Atlas of Saudi Arabia: Riyadh, Saudi Arabia, Saudi Arabian-United States Joint Commission on Economic Cooperation, 111 p.

Mulcahy, T. D., 1992, Engineer support in the COMMZ: Military Review, v. 72, no. 3, p. 14–21.

Rinker, J. N., Breed, C. S., McCauley, J. F., and Corl, P. A., 1991, Remote sensing field guide—desert: Fort Belvoir, Virginia, ETL-0588 U.S. Army Engineer Topographic Laboratories, 568 p.

Rinker, J. N., Corl, P. A., Breed, C. S., and McCauley, J. F., 1990, Remote sensing field guide—desert—Landsat Thematic Mapper 166-040: Quantico, Virginia, U.S. Marine Corps Operational Handbook No. 0-52D, U.S. Marine Corps, 16 p.

U.S. Army Armor School, 1977, Desert operations: Washington, D.C., U.S. Army Field Manual (FM) 90-3, p. 2-20–2-25.

MANUSCRIPT ACCEPTED BY THE SOCIETY OCTOBER 29, 1997

Printed in U.S.A.

Geological Society of America
Reviews in Engineering Geology, Volume XIII
1998

Background and recent applications of military geology in the British armed forces

Michael S. Rosenbaum
Faculty of Construction and Environment, Nottingham Trent University, Burton Street, Nottingham, NG1 4BU, United Kingdom

ABSTRACT

Military geology involves the application of geological science to the decision-making processes required by military command; hence the individual geologist needs to be professionally experienced in applied geology and trained in military staff work and doctrine.

The importance of establishing an adequate and relevant database of information is now widely recognized and the trend for its compilation has been toward digital recording in support of the existing paper library information. The provision of geological information together with its interpretation and the means of giving advice are now established components of decision support within headquarters at Corps and Division. Generally the tasks have to be dealt with in emergency situations and so time is very short by comparison with comparable civilian projects. What is primarily required is a rapid assessment of the ground conditions within the context of the prevailing military situation. For the advice to be useful, it has to be presented in a manner compatible with the standard military format and avoiding use of technical jargon.

Construction work is required in support of the battle: preparing defenses, supporting an advance and consolidating the new positions. Interaction of these works with the ground and the supply of natural materials, particularly water, requires characterization and management sensitive to the contemporary military operations. Local supplies, even if undamaged, are unlikely to be able to sustain the quantities required by the influx of large numbers of troops. Health risks from poor water, not only due to natural bacteria but also deliberately contaminated from terrorist sabotage or NBC attack, require that suitable supplies be established early in the campaign.

Recent actions in the Falkland Islands, the Persian Gulf, and mainland Europe demonstrate how military geology has been used, directly or indirectly, by the British armed forces. The trend throughout the 20th century has been of increased mobility during armed conflict, although the scale of operations has varied enormously.

THE NEED FOR MILITARY GEOLOGY

The British armed forces currently provide their military geological support from a number of individuals administered and trained by the Royal Engineers Specialist Advisory Team (RESAT) of the Territorial Army (TA). This team provides a worldwide consultancy service for the army, navy and air force in locations such as Cyprus, Germany, Belize, the Falkland Islands, and the United Kingdom. Such tasks include the geological aspects of terrain analysis, assessing the potential for abstracting ground water, establishing quarries, and conducting site investigations for various types of construction projects.

Geologists and soldiers share a common interest in the ground, and as geology has developed from a descriptive to a predictive science, so its military applications have become more apparent (Rose and Rosenbaum, 1993a). In Britain the term "civil" implies a military branch of the subject, widely recognized for engineering in public perception of castles and fortified towns. As used in this chapter, "military geology" is regarded as the application of geological sciences to the decision making processes required by mili-

Rosenbaum, M. S., 1998, Background and recent applications of military geology in the British armed forces, *in* Underwood, J. R., Jr., and Guth, P. L., eds., Military Geology in War and Peace: Boulder, Colorado, Geological Society of America Reviews in Engineering Geology, v. XIII.

tary commanders as opposed to geological activities that are in general support of the war effort. This latter type of work, notably the location of supplies of ground water, strategic minerals, and fossil fuels is generally undertaken for the United Kingdom by the British Geological Survey, within which the Overseas Directorate deals with information for areas outside the British Isles.

Geologically related problems faced by military commanders in an operational situation often need to be solved within a very short time, and the formulation of each problem is usually not straightforward in the absence of geological training. There is, therefore, a requirement for the expertise of geologists who are not only fully qualified and professional but also trained in military staff work and doctrine, which necessitates a level of training beyond the provision of academic or civilian institutions. These individuals are the military geologists who are provided for the British armed forces in the shape of Territorial Army (TA) officers and administered by the Royal Engineers (RE); they are situated in a unit known as RESAT (Royal Engineers Specialist Advisory Team). The history of how this team has been established has been described in detail elsewhere (Rose, 1978, 1988; Rose and Hughes 1993a, b, c). These officers have to be professionally qualified, practicing geologists, and they have to be commissioned through the TA.

The potential use of geology in military activities has long been recognized (Rose and Rosenbaum, this volume), but there was little direct application of the modern science of geology to military campaigns before the 1914–18 war when the first military geologist to be used as such was appointed by the British army in April 1915. The slow pace of this development was not so much the failure to organize geological staff or to give support to geological investigations as the failure to apply the results of the studies; this was because few officers in high command had any adequate perception of how to apply geology to either military or engineering problems. Even after the 1914–18 war, Brooks (1920) wrote that geology had recently been thought of as merely a speculative and abstract science rather than a practical subject. The military geologists, therefore, had to spend much time and energy educating such people about the possible applications. That the pattern of military conflict tends to repeat itself is perhaps a reflection of the fundamental control on movement and deployment exerted by the configuration and state of the ground (King, 1951).

PERSONNEL

Operational requirements

The essential need of the military is for a geologist who, on his own initiative within a headquarters, can identify tasks in which potential contributions would support the military problem that is to be addressed. The geological problem to be solved needs to be articulated, and the information that will be required within the available time frame needs to be identified. Data acquisition then follows, perhaps with the support of military geologists further back in the chain of command, the government survey, and the armed forces' own database. Analysis follows and the conclusions are drawn prior to briefing the commanders in the field.

Predicting likely ground conditions, "terrain analysis," to use the military terminology, is the principal operational requirement for geological advice, essentially the prediction of cross-country mobility of tracked and wheeled vehicles, and excavation for troop protection. With the ever-increasing mobility of modern warfare both in the air as well as on the ground, it is with the analysis of terrain that most military geological effort is being concentrated. Trafficability is mainly determined by ground strength, and this depends greatly on moisture content. Thus the soil mechanics characteristics effectively determine the ability of the ground to support vehicle movements in single and multiple passes. Low plasticity soils such as kaolinite clays will quickly lose their ability to sustain vehicle movements as they get wet, whereas high plasticity clays such those rich in smectite will be more resistant. However, the rate of recovery on drying out will be far slower with the more plastic materials. Sands and coarse silts will generally tend to be insensitive to changes in moisture; their ability to support vehicle movements will be determined by their compactness.

Thus it is the short-term behavior of the near-surface ground that is of greatest concern to the military geologist, just as it is to the environmental and engineering geologist. The investigation of bulk aggregates for construction and the consideration of hydrogeology in connection with water supply and waste disposal are similarly concerned with the near-surface ground conditions. The contribution of such applied geologists is therefore determined by an ability to communicate geological knowledge to the engineer in a form that is useful and usable. Thus the general education that a young geologist receives at university is not usually directly applicable to the needs of the military geologist, and a period of specialized training and acquisition of practical experience is generally necessary before the individual can be expected to be effective in this role.

Academic and professional training

The focusing of the basic skills of data recording, analysis, and presentation already acquired is provided in the United Kingdom by postgraduate Master of Science courses, building on the first degree and a number of years' experience previously acquired in industry. Professional training is provided by industry, and guidance regarding an acceptable program of work experience in a variety of situations—office and field—supported by relevant courses is being developed and checked by the Geological Society of London (de Freitas, 1994). This enables the individual geologist to take the Professional Examination and, if successful, to be awarded Chartered Geologist status, corresponding to the long-established route for engineers who become Chartered Engineers.

Geological Survey support

The British Geological Survey (BGS) may at first be thought of as the obvious source of manpower for providing geological advice in support of the armed forces as it was with the U.S. Geological Survey's Military Geology Unit as described by Hunt (1950). Indeed, the BGS has been actively involved with wartime support, notably in the production of the wartime pamphlets concerning water supply, aggregates, and fuel. The survey has also been requested to advise the armed forces as to the availability of geologists suitable for active service, and, as such, in 1915 the director recommended Bill King as the first full-time geological adviser to be appointed as such by any modern army (Rose and Rosenbaum, 1993a). Again in 1940, the director was approached to advise on the appointment of a suitable geologist prepared to "volunteer as a Staff Officer for water supply" (Rose and Rosenbaum, 1993b).

Military training

To operate effectively within a military command structure, the geologist must be militarily trained and exercised to an appropriate level, able to recognize when geological advice is required, and provide that advice in an appropriate way.

Unlike the medical field, where individuals require a similarly long academic and professional training as the geologist but generally work in a hospital environment and are not expected to bear arms, military geologists have to operate on their own within a military headquarters, regularly interacting with field and staff officers. Direct commissioning of military geologists has therefore not been favored since experience has shown that it is necessary to build up the expertise and confidence commensurate with rank over a period of years.

Traditionally, most British military geologists join RESAT between the ages of 25 and 35 and are members of university teaching staffs or are employed in engineering, environmental, petroleum, or mining companies or national or local government organizations. During the initial phases of training, concern is focused on developing leadership qualities. These include personnel management, decision making, and planning and execution, often under conditions of considerable physical and mental pressure. Practical skills are also taught in first aid, map reading, report writing, abstraction and delivery of orders, and in fieldcraft. Such tuition is equally useful in the civilian world, and British employers generally look favorably on their staff volunteering for such service in the Territorial Army. The essential difference between civil and military tasks is that for the latter one has to be trained for work in areas subject to insurrection or combat, where self-protection and protection of the troops under command are as much a part of the mission as being a professional geologist. The successful completion of such training depends as much upon the individual's strength of character and mental endurance as it does on physical stamina and military knowledge.

Within the terms of service for RESAT, at least 30 days of training have to be completed within each two-year period, one must pass basic military skills and be medically and physically fit. In practice, the military geologists contribute considerably more time than the minimum requirement, particularly in support of operations and in maintaining and developing the role of military geology as a part of terrain analysis in higher formation headquarters on exercise. In addition, they attend military training courses appropriate for the development of their military careers within the TA to equip them for roles in a potential future conflict. The technical aspects of military geology are regularly practiced through peacetime projects concerning the development of ground-water supplies, locating material for road construction, and evaluating the stability of excavations, both surface and underground.

The rank of captain is currently regarded by the army as the lowest acceptable rank for duty at Corps level when undertaking military geological tasks, promulgating the results of the analysis, and interacting effectively with the military command structure. This means a minimum of five years' military training after commissioning into the Royal Engineers, entry to which is likely to take two years if no prior military experience has been acquired.

The establishment post for the senior geologist is that of lieutenant-colonel, but promotion to this rank is dependent upon acquiring the appropriate military training and having satisfactorily served in the subordinate ranks of captain and major. Command and Staff courses have to be completed to provide the individual with the necessary knowledge and skills commensurate with field rank and the training necessary to operate effectively within a headquarters.

Training in command is not possible in a strictly military geological role, but the Royal Engineers Specialist Well Drilling Team provides one opportunity for individuals to gain appropriate experience as a commander of troops in the rank of lieutenant or captain. This is also an appropriate position from which to give geological advice in connection with well drilling and water development, the main roles of the Team. In general, each well drilling task needs a record search and preliminary site reconnaissance—perhaps drilling some small-diameter site investigation boreholes—before the well design can be finalized and the mobilization effected. The logistics of getting equipment and troops to the required location, often in militarily sensitive areas, is itself a nontrivial task but one that enables the young military geologist to gain valuable experience.

DATABASE PREPARATION

Data acquisition

A major factor concerning geological data is that it is frequently impossible, for reasons of time or safety, to undertake direct observations of the area for which advice is being sought. Ready access to preexisting database information is therefore of paramount importance, and training for contingencies governs the preparation for emergency situations. This requires the integration and analysis of a wide variety of information including geological maps, reports, logs, well records, company reports, and survey records. The British Geological Survey is the principal custodian for geological mapping in the United Kingdom and thus provides an essential role for the initial phase of most military geological projects. Aerial photography and satellite imagery are readily available and are also consulted.

However, the information, most of which is in paper report or map form, is scattered among libraries, universities, and diverse government departments, which makes its access and compilation a time-consuming task; procurement from local sources in hostile regions is impractical. Some familiarity with the terrain is therefore necessary as is rapid access to pertinent information. This has implications both for training and for database compilation.

Some record keeping can be undertaken during a conflict, indeed the water-supply work undertaken by the first military geologist in northern France included the preparation of a card index of boreholes that was maintained and updated right until the end of the 1914–18 war (Anonymous, 1922). This comprised cards arranged alphabetically by place name. Each card recorded the geological succession, depth, elevation, and details of water levels, yield, and pumping equipment.

However, the volume of information that has to be dealt with and the limited time available in preparation for most modern military tasks necessitates forward planning so that interpretations are already prepared for the areas and scenarios envisaged. The well-defined areas of operation that were the responsibility of the British 1 (BR) Corps in northern Germany during the cold war years provided the setting for much military geological activity during the 1960s, 70s and 80s (Rose and Rosenbaum, 1993b).

Information technology

With the development of information technology has come the ability to use computers to integrate different influential variables that are known to affect ground behavior, forming what is known as the Tactical Terrain Analysis Data Base (TTADB). An early example has been to model the off-route mobility of military vehicles, and thus Wallenhorst (1969) has been able to describe how soil type, slope angle, vegetation, and hydrology have been combined to predict the performance of five different classes of vehicles ranging from tanks to armored personnel carriers to wheeled trucks. He categorized surface composition into 10 classes based on soil strength using the cone test; surface geometry into 40 classes taking into account slope angle and the shape and spacing of obstacles; vegetation into 18 classes using the average diameter and spacing of trees; and hydrology based on 60 classes dependent upon current flows at high and low water. The resultant "Go/No Go" map has immediate appeal to commanders since a quick decision is facilitated, but of course the relevance of the classes and the boundary limits are of paramount importance to the effectiveness of any decisions that are made. A particular problem arising from such analyses concerns the general request, on maneuvers, for the off-road trafficability of a main battle tank to be assessed. The initial success of an armored operation can only be maintained in practice if the support vehicles can also follow through, in particular facilitating ammunition supply, fuel, and maintenance. Such support vehicles tend to be wheeled and are far less mobile off-road than the tracked and armored vehicles.

Military geologists would still rather have the original information available for interpretation than have to depend on a categorized product prepared in advance by others. Thus the TTADB approach, though it facilitates rapid deployment in the field, is somewhat limited when changed conditions require a reevaluation of the ground. In particular, funding constraints rarely permit a thorough ground-truth evaluation of a model once it has been developed. Feedback of information from field performance is therefore essential once operations begin in a theater in order that the mobility model be amended in the light of experience.

DECISION SUPPORT

General

The key questions in decision support are first So what? and then What next? These questions control the acquisition and dissemination of information.

Ten years after the Falkland Islands conflict, the provision of geological information and advice as a component of decision support within headquarters has become established for the British army at Corps and Division level. The organization of such advice is within the framework of the Geographic Branch (Geo) of the Royal Engineers, and its execution can be illustrated by what happened during Operation Granby, the preparation phase for the British contribution to the U.N. Coalition force assembled to liberate Kuwait during the Gulf War of 1990–91. The operational support was provided to military formations through Geo staff in headquarters down to the level of 1 (UK) Armoured Division (Armd Div) (Anonymous, 1991a).

The preparation of maps in direct support of an operation are produced in theater by Geo, but information for contingency plans and training requirements in peace are principally undertaken by units staffed by civilians. Generally the operational tasks have to be dealt with in emergency situations. Time is therefore very short by comparison with civilian projects of a similar magnitude, and the requirements for technical support tend to be ill-defined, highlighting the importance of geological advice being guided by officers who are professional geologists and who also understand the military situation. The tasks require analysis to identify the key military objectives and the geological factors that will influence them, and then the military geologist needs to set about defining these factors within the available time frame. The pertinent geological information must therefore already be available, and what is primarily required is a rapid assessment of the ground conditions within the context of the prevailing military situation, both with respect to friendly and enemy forces. It is generally not feasible to compile a standardized database, so the existing information needs to be available for wartime use in a form tailored to the perceived actions. The development of an appropriate methodology is therefore of more importance than just having the raw data held on file.

Thick documents of a detailed nature tend not to get distributed widely, so information is "lost" unless individuals communicate and pass along knowledge of its availability. Such dissemination is essentially a combination of the raw data and an initial appraisal of its significance. For the advice to be useful, it has to be presented in a manner compatible with standard military formats and communicated without technical jargon. This in order that the advice can be incorporated with the information gathering process being undertaken by the headquarters staff.

Falkland Islands, 1982

During the Falkland Islands conflict in 1982, geological advice was provided for planning the beach landings and subsequent advance across East Falkland. This was provided in the form of a briefing map, printed by the Royal Engineers (EinC[A], 1982), as the product of a terrain evaluation to present to formation commanders the presence of obstacles and ground that could significantly impede troop movements. The tectonic stability of the islands since the early Mesozoic has enabled considerable terrestrial weathering to develop; thus the lithology closely controls the local topography as discussed in more detail by Rose and Rosenbaum (1993b). More recent periglacial processes during the Quaternary have created a number of tracts of peat that impede vehicle movement and extensive stone-runs that make crossing on foot somewhat hazardous, particularly at night.

Only four days were available between the initial order and the requirement for producing the first edition of the briefing map. Even in such a short time the generally inhospitable nature of the western part of East Falkland could be recognized, from which the Argentine command seemed to have concluded that a landward advance by the British from the west would be unlikely. However, compared to the strong quartzites and quartzitic sandstones of the lower Carboniferous Port Stanley Beds, the weaker micaceous sandstones and slaty shales of the Middle Devonian Fox Bay Beds had been eroded to form narrow topographic depressions that could be seen striking east to west, both north and within the central mountain range. Along these depressions troops on foot could advance so the corridors could be exploited to advantage by the British. Even so, lines of supply were severely stretched because only light-tracked vehicles could negotiate the boggy and rocky ground, and therefore helicopter support was essential.

Persian Gulf, 1990–91

Tactical terrain analysis. For Operation Granby, the two main components of terrain analysis concerned the ground and the infrastructure. Geology fundamentally controls the ground but also influences the general infrastructure. The 7 Armoured Brigade (Armd Bde) was deployed as the spearhead for 1 (UK) Armd Div under 7 (US) Corps (described by Moore-Bick in 1991). Their initial concern was the likely effects of the ground on cross-country mobility. This led to the early production and printing of "going" maps at a scale of 1:50,000 that were used as the basis for the deployment plan. Once the Division had become established in theater, the material requirements could be more readily defined and advice could then be provided on aspects of specific site selection, water and aggregate supply, and terrain and excavation problems.

The lack of a comprehensive database for the area led in the early stages of deployment to terrain reconnaissance being undertaken by Military Survey personnel, particularly along the Kuwaiti border (Elder, 1991), but such work is fraught with subjective judgment and tends to be time consuming and dangerous when approaching enemy lines. Nevertheless, such reconnaissance yielded the most comprehensive information regarding the ground for the commanders and gave the terrain analysis team a basis for interpreting other data that had been acquired concurrently such as published reports on the recent surface deposits (Khalaf et al., 1984).

It soon became clear that the main terrain analysis concern during Operation Granby would be the assessment of areas of poor trafficability. These were associated inland with extensive wadi development and desert pavement. The wadis exert a strong directional control on the speed of cross-country mobility since they typically have steep 5 to 10 m high cliffs to either side of a predominantly sandy floor. These sands often contain boulders that make the going slow for wheeled vehicles, although cover from view is good. Movement across the gravel-capped desert pavement is, by contrast, excellent in all directions.

The surface material covering much of the inland area represents an inversion of recent geological processes (Khalaf et al., 1984) because the Pleistocene gravels were deposited as northeast to east trending sinuous channel deposits by rivers during wetter climatic conditions. During the subsequent more arid conditions, deflation of sand, silt, and clay has left these early channels as high ground and deposited the finer sediments downwind as a blanket of loess-like material, partly resorted by periodic flash floods. Where the old drainage system has become dammed by the deposition of loess, extensive playas have developed, within which ephemeral lakes occur, notably in northern Kuwait. Periodic flooding and subsequent settling of the suspended particles has left a surface covering of clay and fine silt over most playa floors. The perfectly flat surface gives such relatively small areas considerable strategic importance as aircraft landing sites and helicopter operating bases. The deflation has been unable to remove the gravel-sized components of the original sediment, and these have remained and settled to develop an extensive surface concentration of gravel forming the desert pavement. The deflation process would ultimately be stopped by the water table, but because this lies at some depth below the wadi floors, pavement development still continues slowly. Although most of this terrain is passable in dry conditions, after rain much of the ground covered in silt and fine sand becomes impassable away from the made roads. Particularly troublesome is sheet flooding, which arises when rain falls on dry sand. The high surface tension of the water is unable to wet the sand so the rain drops pick up a coating of dust that armors them and allows the drops to aggregate, only mixing when their weight breaks down the coating; a flood ensues.

In the coastal regions, poor mobility is associated with sand dunes and sabkha deposits, evaporite-rich clastic sediments generally consisting of silt and clay formed by prolonged intertidal flat sedimentation and associated with a very high water table (Shehata et al., 1990). The ground water is fed from the sea as well as from ground water inland and is highly saline in character.

Therefore the sabkha profile is capped by a salt-cemented crust approximately 0.6 m thick that can generally support wheeled traffic when dry, but it loses much of its strength when wet. Beneath there is a zone 1 to 10 m thick within which the ground water fluctuates and consists of soft or loose sand with some silt and clay; only at greater depths is dense, compact material found. The sabkha becomes very weak on wetting (Akili and Ahmad, 1983) and where multiple vehicle passes are attempted movement is significantly impeded, effectively because of the cement breaking up and subsequent remolding of the uppermost 0.1 m of the ground. This upper zone is particularly vulnerable because of the high degree of saturation but low rate of percolation for water.

These geotechnical characteristics clearly show the constrictions that the sabkha salt marshes place on cross-country movement in Kuwait and neighboring regions, graphically illustrated by the bogging of seven Iraqi tanks during their incursion into Khafji at the end of January 1991. In order to reopen the Main Supply Routes (MSRs), several kilometers of matting and thousands of individual sand channels were required by the Coalition forces.

Strategic terrain analysis. On a wider scale, the distribution of surface materials in the area of Kuwait and southern Iraq played a significant role in developing the Coalition forces' strategy for removing the Iraqi invasion force. The widespread distribution inland of aeolian deposits and alluvial fans cut by wadis had been summarized in the maps published by Fookes and Higginbottom in 1980. These are bounded along the coast to the east by variable littoral deposits and weak sabkha. To the north, the Euphrates river, which flows across the Lower Mesopotamian Plain, has deposited extensive silt and clay alluvium both in Pleistocene terraces and across the Holocene flood plain. The Euphrates fluviatile sediments have a fine-grain size and are generally normally consolidated (although desiccation locally improves the near-surface strength [Saeedy and Mollah, 1990]), which, together with their high organic content, generally creates very soft soils that are very poor for off-road movement.

This distribution of surface materials forms an extensive tract of country across which off-road mobility tends to be good in dry conditions. It is bounded along the coast and to the north by extensive regions where mobility tends to be severely impeded by soft ground, not to mention the strong defenses and extensive minefields installed by the invaders. This favored the development of a battle plan that envisaged a rapid mobile offensive inland, outflanking the invaders to the west and forcing their movement northward where they would be seriously slowed down in the soft Lower Mesopotamian Plain mud. This is reminiscent of Hindenburg's victory in the Mazurian Lake region of East Prussia where the Russian army was maneuvered into soft organic-rich swamps. The German army kept to those swamps known from detailed geologic mapping to have firm sand at shallow depth, and they were aided by maps of the flora distinctive of those coarser bottom sediments (Cross, 1919). Once the land battle commenced, Iraqi troop movements maneuvered toward the city of Basra where the spread of alluvium is comparatively narrow and where the few surfaced roads converge onto a bridge across the Euphrates just downstream of its confluence with the Tigris. This created a choke point and prevented effective regrouping and counterattack: *Defeat followed swiftly.*

As feedback was received from experience in the field, so the model for assessing mobility was refined and the assessments of "Go/No Go" improved. A single vehicle given sufficient time will always be able to get across an area, but what is needed is an assessment of how brigades, especially if armored, will manage under the tactical constraints of battle. It became clear that the successful implementation of the battle plan required geological and meteorological input combined with knowledge of soil-vehicle mechanical interaction. The experience gained by field observations overcame the sometimes disappointing initial appreciations that had to be based on a somewhat over-cautious interpretation using data sets compiled on a regional, rather than local, basis. For instance, three days of persistent rain a few weeks before the land offensive commenced made movement very difficult for wheeled vehicles, aggravated by slow rates of water percolation and multiple vehicle passes on highly saturated ground that churned up the near-surface material. Development of computer hardware to store the vast quantities of information will improve this situation in the future, helped by the development of Geographical Information Systems (GIS) software to manage the data and enable realistic models to be generated from it. However, high technology computers using information technology are vulnerable in a battle situation, and reliance on fixed links, particularly to electric power, means that for some time to come, information technology will need to be considered as a support facility rather than as the principal source of information within the headquarters.

CONSTRUCTION IN SUPPORT OF BATTLE

Construction work is required in support of battle, preparing defenses and supporting the advance. Consolidation of the new positions then requires rapid excavation, preparing new defenses and establishing the lines of communication and supply.

Tunneling

Tunneling has long been used to support an advance by destroying enemy defenses. Two historical British examples are provided by tunneling during the 1914–18 war in northwest Europe along the western front. The first was in preparation for the destruction of Hill 60 in the Messines Ridge at the start of the third battle of Ypres (MacLeod, 1988). The second was at Vimy Ridge (Rosenbaum, 1989) where offensive tunneling was carried out in close proximity to the enemy in conditions where there were significant hazards from artillery fire, infantry attack, and logistical support, and thus excavation and support techniques needed to be both simple and sound. The situation at Vimy Ridge gave the opportunity to exploit the geologic structure to undermine the enemy stronghold at the top of the ridge. The detailed structure was deduced from field mapping, including new exposures in trenches and artillery shell craters, verified by small boreholes. The success of the Vimy attack, in which the ridge was captured, testified to the need for such careful planning prior to the onset of military mining operations.

Concealment

When excavating defensive positions or mines, disposal or dispersal to conceal the considerable volumes of waste can be a significant logistical problem. Even the color of waste can be important because it can give away the extent or the depth of the operations. However, the scope for deceit is tremendous: Careful placing of materials can be designed to mislead the enemy intelligence while the application of appropriate camouflage hides the real piles of spoil. However, this can be a rather difficult problem in some areas, as in the chalk ground of northwestern Europe. There the soil is typically of the order of 0.3 m thick and its dark brown color and characteristic green turf contrast so markedly with the white of the chalk rock beneath. Similar contrast occurs in the Middle East where dark brown silica varnish has tended to develop on the uppermost side of the gravel on the desert pavement as a result of solution by dew or wind abrasion. The clasts can easily be disturbed, thus revealing the lighter brown color more characteristic of the gravel and sand beneath.

Dust suppression is another concern in arid regions. Dust clouds reveal the location of vehicle movements, but for aircraft, particularly helicopters, the major hazard is the rapid wear and even failure that the engines can suffer by ingestion of abrasive dust. Silt-sized quartz is particularly abrasive because of its hardness (greater than steel) and fine-grain size. The use of sprayed tar or water to bind the dust on helicopter operating bases (HOBs) temporarily reduced the problem for the British HOBs during Operation Granby, but early consideration of the distribution of surface geologic materials might be used to advantage so as to avoid siting such landing facilities in areas likely to be troublesome.

Maintenance of the main supply routes

Consolidation of new positions requires that the troops be adequately provided with supplies, particularly munitions, and food and water. This presents a considerable logistical problem, and wheeled vehicles are generally used for transport, although when fully laden these place a severe additional load on the existing road network. The principal need is to keep such main supply routes (MSRs) open, but this means the provision of aggregates suitable for road construction or repair, a recent example being the repair of the border patrol tracks in Cyprus (Nathanail, 1992). Similarly, new aircraft landing sites and HOBs will be required, also needing aggregate such as for the resurfacing of the airfield runway in Ascension Island (Rosenbaum and Skene, 1995).

The provision of aggregates for road repair was a major task for the plant troop of the Royal Engineer Logistic Support Squadron during Operation Granby and more than one million cubic meters of rock had to be quarried in support of 1 (UK) Division alone. This was supplied by opening small quarries in the strongest local materials—limestone and sandstone—taking advantage of where these rocks were exposed in wadi sides if there were no existing quarries, or using the calcrete and dolocrete duricrusts produced during the development of old weathered profiles. It was realized that there would be few local sources of high-quality rock aggregate in Kuwait or neighboring areas. Fookes and Higginbottom wrote in 1980 that "along the Gulf coast . . . relatively young and weak sedimentary rocks predominate. From south of Dammam to Kuwait, scarcely any useful aggregate materials can be obtained from the solid formations, but near Dammam, in Bahrain and in Qatar, restricted areas of Palaeogene limestones form virtually the only local sources of coarse aggregates" (p. 40). Interpretation of stereo monochrome aerial photographs greatly facilitated the location of suitable sites.

Extensive use was also made during Operation Granby of the widely developed "gatch," a duricrust produced by strong upward leaching and evaporation of ground water. This consists essentially of calcite- or dolomite-cemented quartz sand. Gatch has been used extensively as a subbase material for road construction in Kuwait (Al-Sulaimi et al., 1990). However, its distribution and thickness are rather variable and depend largely on local environmental conditions that control the rate of evaporation and salt content (Saeedy and Mollah, 1990). Such duricrust is generally found beneath a cover of aeolian silt or sand that may be as much as 2 m thick, and this can overlie as much as 3 m of aeolian or alluvial sand and therefore can be excavated using dozers and back-actor excavators. However, gatch has a high shrinkage limit (averaging 17%) that makes it likely to swell when saturated and crack when dry; thus it requires suitable drainage to maintain long-term durability. Nevertheless, when the moisture content is controlled, high california bearing ratio (CBR) values can be obtained (between 56 and 93% are reported by Al-Sulaimi et al., 1990); thus it can be employed as a base course for military pavements.

Locating battlefield landing sites

Support for front-line troops generally requires air support with the concomitant need for aircraft landing sites and HOBs in forward locations. At the end of the Falklands campaign in 1982, although rapid replenishment of supplies could be undertaken only by air, the existing runway at Stanley was incapable of accepting jet aircraft; indeed, the C130 Hercules transport planes could land only if arrangements for inflight refueling were employed. The most urgent requirement was to repair the cratered runway so it could accept the C130 aircraft and then to extend the runway 650 m to enable the operation of jet fighters such as Phantoms. This required rock aggregate to make a construction base for laying AM2 matting, a portable aluminum planking material that could be keyed together to form a pavement surface and therefore be suitable for rapid deployment in the field. The relatively flat terrain and absence of existing quarries with suitable access led to the need to develop new sources of stone. Although a small quarry at Mary Hill, near the airfield, had been established to provide the well-cemented sedimentary quartzite used for the initial runway construction, its extension quickly led to difficulties because of a cover of highly plastic marine clay (Rosenbaum, 1985). This had to be removed prior to the main rock blasting. Difficulties also arose from the abrasiveness of the quartzite rock. Jaw crushers were wearing down in a tenth of the time that had been expected. Extending the runway required assessment of the bearing capacity of the ground, hence the need for the AM2 matting once the wide extent and weak strength of the near-surface peat and clay had been confirmed.

Siting amphibious crossings

Major rivers pose significant obstacles to military maneuvers within an operational theater. Existing bridges are prime targets for demolition; therefore heavy reliance has to be placed on the provision of portable bridging and ferry equipment. The principal geological task is to assess the strength and deformability of bank and pier positions, both with regard to direct loading and to the shear strength and slope of the ground in the approaches. The need to anchor the equipment requires that stakes appropriate to the ground conditions be available. The equipment for sandy soils is rather different from that suitable for clays.

An even greater obstacle is presented by the sea. Major seaborne offensives require a great deal of careful, coordinated planning, the geological aspects of which largely concern the ease with which vehicles will be able to move over the beach and across the ground immediately inland. In particular, any possible changes in beach profile or sediment grading with time require defining (e.g., the drawdown of the beach sand during winter storms). In support of an advance, the Normandy beach landings of Operation Overlord in 1944 may be cited as an example and are described more fully Rose and Pareyn (this volume).

Locating water supply

Water supply is a vital resource that requires careful management during and following military operations. The local supplies, even if undamaged, are unlikely to be able to sustain the quantities required by the influx of large numbers of soldiers, particularly where troop camps and field hospitals are to be established. Risk of contamination, not only from waste water but also from toxins deliberately introduced by terrorist sabotage or nuclear, biological, and chemical attack, requires that suitable supplies be established early in the campaign. The quickest option is to develop surface water supplies in lakes or rivers, but these are difficult to secure, and ground water frequently has to be used. Historically, the supply of water from underground sources has been of fundamental importance to the success of a campaign. Three examples concerning the British army are: (1) behind the western front during the 1914–18 world war when horses provided the main source of locomotion, requiring the first military geologist as such to be appointed (Anonymous, 1922); (2) in North Africa during the 1939–45 world war when the large-scale maneuvers of the belligerents required careful planning of supplies (Shotton, 1946); and, (3) most recently, in Bosnia for support of base camps as described in detail by Wye (1994) and Nathanail (this volume).

During the early stages of Operation Granby in the Persian Gulf, it was clear that water supply and replenishment would be a major limiting factor for troop support because only a few days' supply could reasonably be carried by each fighting unit (Walton-Knight, 1994). The hydrogeology is controlled by the structurally simple Arabian Shelf sediments, consisting mostly of sandstone and carbonate formations overlying the metamorphic Arabian Shield at considerable depth (Powers et al., 1966). In southern Iraq these sediments date back as far as the Triassic, but in Kuwait they range from the Eocene to the Holocene. The best aquifers would be expected to have developed in the sandstone formations, but their storage is considerably reduced by the development of diagenetic evaporite mineral cement. Existing supplies had to be assessed, but there was particular concern about the suitability of the ground water for human consumption. This was exacerbated by the presence of shallow aquifers that are associated with aeolian sands and wadi sands and gravels because these are particularly prone to contamination, and the salinity is known to be high. The regional ground water flows from west to east, but heavy pumping in coastal regions easily leads to saline intrusion from the Persian Gulf. The potability of the ground water was uncertain because of the likely high salt content, so reverse osmosis water production equipment was deployed (Anonymous, 1991b). This extracted water from a network of existing wells bored into the Tertiary limestone aquifer and therefore was able to accept the partial recharge from alluvium in nearby wadis.

The sinking of boreholes into an aquifer demonstrates whether ground water is present; pump testing will produce quantitative figures for the yield and storage. However, it takes considerable time to set up and drill such wells, and therefore geologic mapping supplemented by geophysics (notably electrical resistivity and seismic refraction) provides much more efficient methods for hydrogeologic mapping. Such an approach has been employed for military water development for some time, for example, by Moseley (1966) based on his experiences in Aden.

The alternative technique of water divining for locating ground water still has its proponents. In the early 1940s this approach was preferred in the higher echelons of some headquarters to such an extent that the fate of the British 8th Army in the Western Desert of North Africa was almost at the mercy of dedicated dowsing (Moseley, 1973). Indeed, it was not until a War Office directive that prohibited divining was issued that geological methods prevailed.

In 1946 Shotton wrote a series of articles describing how hydrogeological work had been carried out during the North African campaigns and explaining how the supply of water had been investigated and established. Geologically the area is reasonably simple, consisting of almost horizontal Miocene limestone with subordinate sands and clays. There are occasional faults and a general distribution of stress relief fractures that provide the main storage for ground water. There was, as in the Persian Gulf, a significant risk of saline intrusion should the water be pumped for any period of time, particularly in the near-coastal regions. However, because the supplies were needed for relatively short periods, it was only where there was a significant gypsum content that significantly high-salinity levels arose. The discovery of perched water tables helped supplement the groundwater supply, and here the work of the 42nd Geological Section of the South African Engineer Corps led to the innovative use of the electrical resistivity method. This was particularly successful for locating the depth of shallow limestone-clay contacts, the most favorable trap for perched water. Larger tectonic structures were also able to provide effective ground-water traps. A historic example is the Fuka Basin consisting of a syncline that could deliver 19 m^3 (5,000 gal) an hour from boreholes just 30 m deep (Shotton, 1944).

Providing garrison support

Longer term support of military garrisons is necessary once hostilities cease. Until recently such support has been provided for the British armed forces by the Property Services Agency (PSA), but now garrisons are responsible for their own budgets and the trend has been toward awarding individual contracts, often to local companies. Nevertheless, construction for defense has grown to become the largest sector of British domestic public work, accounting for one-third of the entire public nonhousing turnover of U.K. contractors and almost twice the size of the road-building program (Anonymous, 1987).

The PSA has tended to act in a project management role, as illustrated by the construction in the mid-1980s of a new airfield at Mount Pleasant in the Falkland Islands, 30 miles distant from the old airfield at Stanley. At Mount Pleasant a consortium of three major contractors (Laing, Mowlem, and Amey Roadstone) carried out the construction under a range of consultants. At the stage of letting the contracts, geological advice was provided by the army and by the PSA, but the responsibility for this project for the main ground investigation lay with the contractors. Indeed, the siting of the airfield was in large part dictated by the availability of sufficient rock material in close proximity and the geology determined the topography. The principal geologic units at Mount Pleasant are lower Carboniferous quartzite overlain by upper Carboniferous tillite and by a thick series of Permian sandstone interbedded with subordinate slaty shale. The tillite, which outcropped at the site of the proposed runway, behaves as a siltstone, and the contractors initially envisaged its use for most of the construction work requiring rock. However, problems arose because of its unduly high clay content and resulting low shear strength, caused by extensive weathering that had attacked this material during the early Mesozoic. The specifications that needed to be met by the best quality rock, for instance, when used as aggregate for pavement quality concrete, could therefore not be satisfied by the tillite in practice. Alternative quarries then had to be opened in the quartzite, the nearest outcrop being 16 km distant, which required haulage by road and added considerably to the cost as well as adversely affecting the schedule for the whole operation.

A second example of geology influencing garrison support may be taken from Cyprus where the provision of water supplies to the garrison have, since 1953, been from a spring located 25 km to the northwest. During the depleted supplies of a dry summer, this water has had to be supplemented by pumping from operationally secure boreholes in the garrison. However, these boreholes are fed by water flowing underground from the northeast where the increased use of nitrate fertilizers and the building of the Kouris Dam has led to a gradual reduction in both the quality and the quantity of the ground water available (Swanson, 1991). Location of a dam at Symvoulos within the garrison area has been accepted as being the most effective means of storing water, but the limestone bedrock and the suspected presence of tectonic fractures detected from remotely sensed imagery made siting of this dam difficult. The involvement of military geologists was necessary for establishing the nature of the groundwater supply and assessing the likely effects on abstraction of the increasing local population. It quickly became evident that equally important controls would be exerted by the construction of works such as the Kouris Dam and its feeder, the underground aqueduct then being proposed to form part of the Southern Conveyor Project supplying water from the hills to the west into the Kouris Dam and from there to supply southeast Cyprus.

SUMMARY

Examples from recent actions demonstrate how military geology has been used, directly or indirectly, by the British armed forces. The trend throughout the 20th century has been of increased mobility during armed conflict, demonstrated by just 97 hours required for the land battle in the Persian Gulf at the end of February 1991 when the front moved some 350 km from Saudi Arabia through southern Iraq to the edge of Kuwait City. However, even though the scale of operations involving the British army has varied enormously, the frequency with which it is called into action remains high. In support of this activity, the importance of establishing an adequate and relevant database of information is now widely recognized, and the trend for its compilation has been toward electronic cataloguing of the existing library holdings.

The application of geological knowledge to operational situations has been shown to require the skills of professional geologists familiar with the management structure of a military headquarters and conversant with staff procedures. Tasks requiring a geological input tend to be sporadic, but when they arise they require urgent action and an intense effort to achieve the desired results in time. This often necessitates a number of specialists working concurrently on a particular task in order to solve the problem. The provision of advice by civilians in a battlefield situation has not been favored by the British armed forces because of the length of time required to mobilize the appropriate personnel. Instead, the British army has on hand reserve officers who possess the necessary combination of geological and military skills and who have been trained to operate in a headquarters environment. An additional advantage of utilizing reserve officers has been to enable them to develop their professional skills in the wider geological community so that their knowledge is as up to date as possible and their personal career development encouraged to progress in a specialist technical role, unlikely to have been feasible within the framework of the regular army. Nevertheless, the annual training obligation maintains the individual's military awareness and provides sufficient time for regular training in personal military skills and for staff training, building upon the foundation of basic military training and Engineer skills that form the basis of the young officer's initial training.

ACKNOWLEDGMENTS

The author would like to thank the members of RESAT, in particular the military geologists past and present who have contributed their specialist expertise to the British armed forces in recent years, for their contributions to the operations and tasks that have been described in this chapter, some of which have not yet been published.

REFERENCES CITED

Akili, W., and Ahmad, N., 1983, The sabkhas of Eastern Saudi Arabia—Geotechnical considerations, *in* Proceedings of the 1st Saudi Engineering Conference: Saudi Arabia, Dahran, v. 2, p. 300–322.

Al-Sulaimi, J. S., Mollah, M. A., and Matti, M. A., 1990, Geotechnical properties of calcrete soil (gatch) in Kuwait: Engineering Geology, v. 28, p. 191–204.

Anonymous, 1922, The work of the Royal Engineers in the European War, 1914–19: Geological Work on the Western Front: Chatham, United Kingdom, Institution of Royal Engineers, 71 p.

Anonymous, 1987, Defence: Unsung civils achievement: Construction for Defence: London, Thomas Telford, p. 5–8.

Anonymous, 1991a, Report by Director General of Military Survey: Royal Engineers Journal, v. 105, p. 111–113.

Anonymous, 1991b, Report by Engineer in Chief: Royal Engineers Journal, v. 105, p. 102–111.

Brooks, A. H., 1920, The use of geology on the Western Front: U.S. Geological Survey, Professional Paper 128, p. 85–124.

Cross, W., 1919, Geology in the World War and after: Geological Society of America Bulletin, v. 30, p. 165–188.

de Freitas, M. H., 1994, Keynote lecture: Teaching and training in engineering geology: Professional practice and registration, *in* Proceedings of the 7th Congress of the International Association of Engineering Geology, Lisbon: Amsterdam, Balkema, September 5–9; v. 6, p. LVII–LXXV.

EinC(A), 1982, Falkland Islands briefing map: Feltham, United Kingdom, Directorate of Military Survey, Series GSGS 5453 Edition 3—GSGS, Scale 1:250 000.

Elder, J. P., 1991, Survey operations: Royal Engineers Journal, v. 105, p. 125–131.

Fookes, P. G., and Higginbottom, I. E., 1980, Some problems of construction aggregates in desert areas, with particular reference to the Arabian peninsula. 1: Occurrence and special characteristics, *in* Proceedings of the Institution of Civil Engineers Part 1: London, Institution of Civil Engineers, v. 68, p. 39–67.

Hunt, C. B., 1950, Military geology, *in* (Ed. S. Paige) Application of geology to engineering practice: Geological Society of America, Berkey Volume, p. 295–327.

Khalaf, F. I., Gharib, I. M., and Al-Hashash, M. Z., 1984, Types and characteristics of the recent surface deposits of Kuwait, Arabian Gulf: Journal of Arid Environments, v. 7, p. 9–33.

King, W. B. R., 1951, The influence of geology on military operations in northwest Europe: Presidential Address to Section C of the British Association for the Advancement of Science: Edinburgh, British Association for the Advancement of Science, v. 8, p. 131–137.

MacLeod, R., 1988, Phantom soldiers: Australian tunnellers on the Western Front, 1916–18: Journal of the Australian War Memorial, no. 13, p. 31–43.

Moore-Bick, J. D., 1991, Operation Granby: Preparation and deployment for war: Royal Engineers Journal, v. 105, p. 260–267.

Moseley, F., 1966, Exploration for water in the Aden Protectorate: Royal Engineers Journal, v. 80, p. 124–142.

Moseley, F., 1973, Desert waters of the Middle East and the role of the Royal Engineers: Royal Engineers Journal, v. 87, p. 175–186.

Nathanail, C. P., 1992, Blastability assessment in the Troodos Massif, *in* Proceedings of the 28th Annual Conference of the Engineering Group of the Geological Society, held in Manchester: London, Geological Society Engineering Geology Special Publication, p. 355–359.

Powers, R. W., Ramires, L. F., Redmond, C. D., and Elberg, E. L., 1966, Geology of the Arabian Peninsula—sedimentary geology of Saudi Arabia: U.S. Geological Survey Paper 560-D, 147 p.

Rose, E. P. F., 1978, Engineering geology and the Royal Engineers: Royal Engineers Journal, v. 92, p. 38–44.

Rose, E. P. F., 1988, The Royal Engineers Specialist Advisory Team (V): Royal Engineers Journal, v. 102, p. 291–292.

Rose, E. P. F., and Hughes, N. F., 1993a, Sapper Geology: Part 1—Lessons learnt from world war: Royal Engineers Journal, v. 107, p. 27–33.

Rose, E. P. F., and Hughes, N. F., 1993b, Sapper Geology: Part 2—Geologist Pools in the Reserve Army: Royal Engineers Journal, v. 107, p. 173–181.

Rose, E. P. F., and Hughes, N. F., 1993c, Sapper Geology: Part 3—Engineer Specialist Pool Geologists: Royal Engineers Journal, v. 107, p. 306–316.

Rose, E. P. F., and Rosenbaum, M. S., 1993a, British Military Geologists: The formative years to the end of the First World War: Proceedings of the Geologists' Association, v. 104, p. 41–49.

Rose, E. P. F., and Rosenbaum, M. S., 1993b, British Military Geologists: Through the Second World War to the end of the Cold War: Proceedings of the Geologists' Association, v. 104, p. 95–108.

Rosenbaum, M. S., 1985, Engineering geology related to quarrying at Port Stanley, Falkland Islands: Quarterly Journal of Engineering Geology, v. 18, p. 253–260.

Rosenbaum, M. S., 1989, Geological influence of tunnelling under the Western Front at Vimy Ridge: Proceedings of the Geologists' Association, v. 100, p. 135–140.

Rosenbaum, M. S., and Skene, G., 1995, Airfield pavement construction using basalt aggregate: Bulletin of the International Association of Engineering Geology, v. 51, p. 45–53.

Saeedy, H. S., and Mollah, M. A., 1990, Geotechnical study of the North and Northwest Coast of the Arabian Gulf: Engineering Geology, v. 28, p. 27–40.

Shehata, W. M., Al-Saafin, A. K., Harari, Z. Y., and Bader, T. A., 1990, Potential sabkha hazards in Saudi Arabia, *in* Proceedings of the 6th Congress of the International Association of Engineering Geology, Amsterdam, August 6–10, v. 4, p. 2003–2010.

Shotton, F. W., 1944, The Fuka Basin: Royal Engineers Journal, v. 58, p. 107–109.

Shotton, F. W., 1946, Water supply in the Middle East campaigns: Water and Water Engineering, v. 49, p. 218–226, 257–263, 427–436, 477–486, 529–540.

Swanson, R. C., 1991, The Cyprus Connection August 1990 to August 1991: Royal Engineers Journal, v. 105, p. 280–285.

Wallenhorst, R., 1969, Improving off-road mobility: Buffalo, New York, Research Trends, Cornell Aeronautical Lab, v. 17, p. 3–8.

Walton-Knight, M. P., 1994, Supplying water to the British Army during the Gulf War: Royal Engineers Journal, v. 108, p. 154–159.

Wye, T. W., 1994, Well drilling in Bosnia: Royal Engineers Journal, v. 108, p. 149–153.

MANUSCRIPT ACCEPTED BY THE SOCIETY OCTOBER 29, 1997

Published in U.S.A.

Geological Society of America
Reviews in Engineering Geology, Volume XIII
1998

Geology education in the U.S. Army

John C. Jens
U.S. Army Topographic Engineering Center, 7701 Telegraph Road, Alexandria, Virginia 22315-3864
M. Merrill Stevens*
U.S. Army Engineer Center, Fort Leonard Wood, Missouri 65473-6650

ABSTRACT

The U.S. Army trains soldiers in geology under the guise of related topics. Thus, instructional responsibilities are disjointed.

The army considers geologic matters to be the responsibility of Engineer officers. Familiarization of topographic requirements and construction practices, both military and civil, is given in the Engineer Officer Basic Course, with specific concepts included in the Engineer Officer Advanced Course. Officers wanting more depth in topographic matters take the Mapping, Charting, and Geodesy Officers Course that emphasizes terrain analysis.

Enlisted personnel have more direct contact with geology. Construction materials specialists and quarry operations specialists receive instruction in rock identification and properties about quarrying and construction. On the topographic side, Basic Terrain Analysis teaches mineral and rock identification—weathering and soils are the targeted tasks. Sergeants learn more advanced geology concepts within an imagery analysis and interpretation block of classes in the Advanced Terrain Analysis Course.

Warrant officers, normally selected from the enlisted ranks, provide technical expertise in terrain analysis. Although they receive no formal training, the competitive selection process generally chooses those with the most extensive computer knowledge and field experience in terrain analysis. Geology skills per se are not a criterion.

The Topographic Engineering Center provides, when needed, a two-week intensive interdisciplinary imagery analysis and interpretation short course, which is primarily for federal agencies.

INTRODUCTION

Geology training can be found in U.S. Army instruction, but only under the guise of other related topics. The related discipline of topography encompasses general military consideration of geologic knowledge because geology is the underlying influence for the expression of the surface upon which battles are fought.

Army Regulation 115-11, Army Topography, currently dated 30 November 1993, establishes policies, responsibilities, and procedures for army topography (U.S. Army, 1993c, p. i, summary). The Commanding General, U.S. Army Training and Doctrine Command (TRADOC), provides "topographic doctrine, concepts, and training requirements and proponency through the U.S. Army Engineer School" (U.S. Army, 1993c, p. 2, para. 1–4j[1]). TRADOC, through its schools and associated Department of Defense schools, provides "training in the use of the terrain in combat and in acquiring and using topographic support." The Defense Mapping School at Fort Belvoir, Virginia, and the U.S. Army Engineer School at Fort Leonard Wood, Missouri, train topographic personnel through courses described in the schools' catalogs (U.S. Army, 1993c, p. 3, para. 2–4b). Under Department of Defense Directive 5105.40, Mapping, Charting, and Geodesy (MC&G), the Defense Mapping Agency is responsible for train-

*Present address: Department of Geological Engineering, 129 McNutt Hall, University of Missouri, Rolla, Missouri 65401-0249.

Jens, J. C., and Stevens, M. M., 1998, Geology education in the U.S. Army, *in* Underwood, J. R., Jr., and Guth, P. L., eds., Military Geology in War and Peace: Boulder, Colorado, Geological Society of America Reviews in Engineering Geology, v. XIII.

ing in topographic skills at the Defense Mapping School (U.S. Army, 1993c, p. 1, para. 1–4a[5]). Also under AR 115-11, the chief of engineers through the U.S. Army Topographic Engineering Center provides "instruction in the techniques of imagery analysis to derive and process information about the terrain and environment necessary to the solution of civil and military problems" (U.S. Army, 1993c, p. 2, para. 1–4h[12]). Even though the Army Engineer Center has responsibility for training in these matters, the actual instruction is accomplished by several institutions.

Training is based upon a system that is task oriented as opposed to knowledge oriented. Knowledge, if taught, must lead to the accomplishment of a task. For instance, in order to accomplish the task of extracting "essential elements of terrain information from all sources (unexploited and construction resources)," one must have some background knowledge of rock identification, the formation of that rock, rock structure, and engineering properties of rocks appropriate for quarrying (U.S. Army, 1993d, p. 12).

Army doctrine has officers in leadership and management roles to orchestrate the formulation and the execution of military operations. The actual accomplishment of work is performed by enlisted personnel and warrant officers. Due to these functional differences there are three distinct education systems.

OFFICER TRAINING

Officers in the army responsible for geologic matters wear the castle insignia of the Corps of Engineers. As stated in Army Field Manual 100-5, Operations, "in all types of operations, engineers advise the maneuver commander on the effective use of the terrain," (U.S. Army, 1993b, p. 2–24). Army Field Manual 5-105, Topographic Operations, further develops how this will be done by the topographic engineers (U.S. Army, 1993a, p. i). Familiarization of topographic responsibilities and construction practices, both military and civil, is given to second and first lieutenants in the Engineer Officer Basic Course in a one-hour lesson entitled "Terrain Analysis." By their sixth year captains have attended the Engineer Officer Advanced Course where more specific concepts are found in a three-hour lesson, "Topographic Support," and a six-hour elective, "Advanced Terrain Analysis." The emphasis in these lessons is the management aspect of where to find support and what can be provided. Familiarization includes the digital aspects of topographic information available. Another mandatory lesson is the "Construction Applications of Geology." This class is four hours long plus two hours of computer-aided instruction. In that time the officer learns rock and mineral identification, weathering and erosion, mass movement, soils development, structural geology, engineering properties of rocks, occurrence of natural construction materials, developing sources of construction materials, and locating and developing subsurface water supplies. Officers wanting more depth in topographic matters or choosing to pursue the military career path of a topographic engineer (21C) take the Mapping, Charting, and Geodesy Officers Course (MCGOC) at the Defense Mapping School where the army track emphasizes terrain analysis. Recently, MCGOC was reduced from eight to six weeks. The two-week segment dealing with terrain analysis provides hands-on practical terrain analysis skills experience rather than traditional geology training. Generally, however, the officer brings with him or her the geology training received in college. Officers with geology degrees are numerous, but the use of that knowledge is not rigorous unless individuals make a concerted effort to seek out the ways in which they can put their knowledge to use solving military problems. It is planned to eliminate the separate Topographic Engineer Career Field (21C). The concept is that all Corps of Engineers branch officers should be able to perform in any engineer assignment regardless of position, whether combat engineer, civil works, or topographic engineer.

ENLISTED TRAINING

Combat construction engineers

Enlisted personnel have more direct contact with geology than officers. In the Combat Construction Engineering Career Management Field (CMF) 51, Construction Materials and Quarry Operations Specialists (51G) and Engineer Technicians (51T) receive cursory instruction about quarrying, construction, soils analysis, construction surveying, and drafting.

Topographic engineers

On the topographic side, CMF 81, the Basic Terrain Analysis Course teaches young privates fresh from their basic training rock and mineral identification, engineering properties of rocks, weathering and erosion, mass movement, soils development, structural geology, occurrence of natural construction materials, developing sources of construction materials, and locating and developing subsurface water supplies. This section, entitled "Physical Geography," consists of approximately 32 hours of classroom lecture and laboratory practical exercise plus a one-day geology field trip. A basic understanding of the earth is established as a framework within which the encountered challenges of military operations are set. An additional field trip of the Smithsonian's Rock and Mineral display at the Natural History Museum is a personalized tour conducted by a curator. Soils are addressed separately through 20 hours of course work called "Surface Materials." Vegetation and surface configuration, or slope, round out the classroom instruction. The key elements in a terrain analyst's database are vegetation, slope, soils, transportation, obstacles, surface drainage, and water resources.

A terrain walk of the Manassas Civil War Battlefield sites early in the course provides an insight into the use of terrain during military operations. The course is currently being reviewed with the intent to revive several other field trips that would provide the students hands-on experience with geology, hydrology, soils, and vegetation.

At some time after their sixth year of service, sergeants learn more advanced geology concepts within a one-week "Interdisciplinary Imagery Analysis and Interpretation" seminar in the Advanced Terrain Analysis Course. Through the use of stereo imagery, the students develop a very real appreciation for detailed analysis and an interdisciplinary approach to problem solving through individual and team problems and a related field trip. Although they complain that they need more geology training to understand the problems, they find that by the end of the section they have a good appreciation of the earth system.

WARRANT OFFICER TRAINING

Warrant officers, normally selected from the enlisted ranks, provide technical expertise in construction and terrain analysis. The utility, operations and maintenance, technician (210A) warrant officers have a program of instruction with basic and advanced courses for certification. Their primary area of earth science emphasis is instruction in soils properties; in an area dominated by construction skills and applications, any experience they have with geoscience subjects is a bonus.

The competitive selection process is intended to provide the army with the most qualified terrain analysis technicians (215D). Generally, those with the most extensive computer skills and topographic-related field experience are selected. Geology skills per se are not a criterion, except for their relationship to terrain analysis. There is no formal training program. Currently, only a certification course conducted by senior terrain warrants provides formal tailored instruction for those technical areas in which the new warrant officer may be weak. In the future this will be formalized as a basic course. Also, an Advanced Terrain Warrant Officer Course is currently under development by the Defense Mapping School to replace warrant officer attendance at the Mapping, Charting, and Geodesy Officer Course. Emphasis will not be to hone terrain analysis skills but to provide production management training culminating in an intensive terrain analysis production-based command post exercise.

NONMILITARY OCCUPATIONAL SPECIALTY TRAINING

As directed by AR 115-11, the chief of engineers, through the U.S. Army Topographic Engineering Center, provides "instruction in imagery analysis techniques to derive and process information about the terrain and environment necessary to the solution of civil and military problems" (U.S. Army, 1993c, p. 2, para. 1–4h[12]). To fulfill this responsibility, a two-week intensive interdisciplinary imagery analysis and interpretation course, affectionately called the "short course," was offered. Until his retirement, Robert E. Frost, a civil engineer with a strong appreciation for geology, taught it on call. Rinker (1994, this volume) and Ehlen (1994, this volume) provide excellent examples of the applications of this interdisciplinary imagery analysis method. The only existing elements of this course in formal military instruction is the seminar in the Advanced Terrain Analysis Course for sergeants. The Topographic Engineering Center has revived this short course with one of this chapter's authors (Jens) providing the instruction upon request.

SUMMARY

Geology instruction can be found in various army courses. However, it is not as evident as it once was and often takes the disguise of geography or other related disciplines such as soils analysis, quarry operations, or hydrology, where the instruction emphasis is on specific construction applications. Army engineer officers, as the overall managers of MC&G-related information, bring geologic knowledge with them to the army rather than receive more formal geology training. Enlisted personnel receive minimal introductory training as necessary for application to their specific military occupational specialty. Warrant officers widen their base of geologic knowledge through their own educational development where, again, intent to apply the knowledge to a specific problem is the reason to acquire it.

Even though the terrain analyst relies upon a prepared database, an understanding of the earth system is essential to understanding the effects of terrain and weather on military operations. It is also very likely that a prepared database is not available for a contingency operation area in which case the terrain analyst must develop the basic information himself from imagery. An understanding of the earth system is then crucial. During Operations Desert Shield and Desert Storm the terrain analysts who had interdisciplinary imagery analysis training provided more accurate predictions of what lay ahead for the coalition ground forces (personal communications with terrain analysis warrant officers, 1991; unpublished notes, topographic executive officer after action meeting, 1991).

It is clear that these situations need to be reversed with more in-depth instruction in geology and its application to military operations if the engineers are, indeed, to advise their commanders on the effective use of the terrain.

ACKNOWLEDGMENTS

Opinions expressed in this paper are those of the authors and not necessarily those of the U.S. Department of the Army or the U.S. Department of Defense.

REFERENCES CITED

Ehlen, J., 1994, Analysis of fracture patterns using remote sensing: Geological Society of America Abstracts with Programs, v. 26, no. 7, p. A-346.

Rinker, J. N., 1994, Remote sensing, terrain analysis, and military operations: Geological Society of America Abstracts with Programs, v. 26, no. 7, p. A-346.

U.S. Army, 1993a, Field Manual 5-105, Topographic Operations, Washington, D.C., U.S. Department of the Army, 60 p.

U.S. Army, 1993b, Field Manual 100-5, Operations, Washington, D.C., U.S. Department of the Army, 175 p.

U.S. Army, 1993c, Regulation 115-11, Army Topography, Washington, D.C., U.S. Department of the Army, 15 p.

U.S. Army, 1993d, Pamphlet No. 81Q2093(AC&RC)N, Notice for MOSC 81Q2/3/4, Washington, D.C., U.S. Department of the Army, 14 p.

NOTE ADDED IN PROOF

Since the completion of this manuscript in 1995, the Department of the Army has combined the enlisted Terrain Analyst (81Q) and the Cartographer (81C) military occupational specialties into a single career field, Topographer (81T). The attendant courses from this melding of two distinct but related specialties cut critical earth science instruction (SFC J. Jackson, Terrain Analysis Branch, Defense Mapping School, personal communication, 1997). Noting this deficiency, the Defense Mapping School currently is considering reestablishing the interdisciplinary imagery analysis and interpretation seminar in the Advanced Topographer Course (Major W. Piek, chief, Topography Team, Defense Mapping School, personal communication, 1998). (Jens)

MANUSCRIPT ACCEPTED BY THE SOCIETY OCTOBER 29, 1997

Printed in U.S.A.

Geological Society of America
Reviews in Engineering Geology, Volume XIII
1998

Recent activities in military geology at the U.S. Geological Survey

William Leith and John Rodney Matzko
U.S. Geological Survey, 920 National Center, Reston, Virginia 20192

ABSTRACT

Over the past 25 years, the focus of military geology at the U.S. Geological Survey has moved away from classical "terrain intelligence" traditionally used by ground forces (e.g., surface construction, trafficability), and toward the strategic assessment of very small areas of the Earth's surface—such as a single underground nuclear test site or foreign underground facility. Along with this change in focus has come a demand for great precision in the remote assessment of rock properties, as the users of geologic data assess the detailed effects of the rock environment on both weapons testing and targeting. The recent opening of the former Soviet Union has provided sources of "ground truth" for evaluating remote assessments of geology made over three decades. These sources have proved the efficacy of making reliable and detailed estimates of foreign geologic environments primarily through the use of published reports and satellite image data. Current U.S. Geological Survey efforts are in two broad fields: geologic support for monitoring treaties limiting nuclear testing and geologic assessments at the sites of underground structures. Nuclear test monitoring at low seismic magnitudes, where many seismic events may occur per day in countries in which nuclear proliferation is a concern, will require the construction of digital geologic data sets that can be used in an operational mode. Recent work in this area includes the assessment of mines and mine blasts as sources of seismicity and the feasibility of constructing large cavities for clandestine testing. Evaluation of underground facilities is also a subject of increasing interest. Without a "nuclear option," a nation's targeting strategies must rely on a combination of "smart" penetrating weapons and a detailed knowledge of the penetrability and shock-wave propagation of the near-surface materials. Geologic data are also required for many other types of facility monitoring, both remote and on-site.

INTRODUCTION

For more than half a century, the U.S. Geological Survey has provided information and expertise on all aspects of earth science to the defense and intelligence agencies of the U.S. government. This work has ranged from mapping all of the Earth's land areas to the detailed assessment of more than 1,000 foreign sites. This long-term effort continues at present in the Special Geologic Studies Group of the Geologic Division, which draws upon the full range of geologic expertise available at the U.S. Geological Survey to provide accurate, authoritative, and timely information.

History

The present-day Special Geologic Studies Group began as the Military Geology Unit (MGU) of the U.S. Geological Survey on June 24, 1942, when it was formally established at the request of the U.S. Army Corps of Engineers (see Terman, Chapter 5, this volume). Prior to this time, military-related geologic work was conducted on an informal basis by members of the U.S. Geological Survey staff. The MGU was organized to compile the terrain intelligence studies necessary to assist the Corps and other defense agencies in the war effort. By 1945, the original staff of 10 had grown to about 160 personnel, including 88 geologists, 11

Leith, W., and Matzko, J. R., 1998, Recent activities in military geology at the U.S. Geological Survey, *in* Underwood, J. R., Jr., and Guth, P. L., eds., Military Geology in War and Peace: Boulder, Colorado, Geological Society of America Reviews in Engineering Geology, v. XIII.

soil scientists, 15 other specialists, and 43 staff assistants, such as illustrators, typists, and photographers. After the war, the MGU became the Military Geology Branch, which continued terrain intelligence studies in all areas of the world. Later, the branch was downsized and identified as the Military Geology Project until 1993, when the name was changed to the Special Geologic Studies Group to reflect its efforts in areas not traditionally considered military geology (see Hunt, 1950).

Activities of the Military Geology Unit

The efforts of the wartime Military Geology Unit focused on terrain analysis and intelligence studies, soil trafficability, water supply, fuel and mineral resources, construction materials and problems, and airfield siting (see GSA, 1945). Special reports on natural resources and conditions that affected military engineering and deployment operations were also compiled. During this period of its history, the MGU completed many studies, some of which were used to assist in planning military operations in North Africa and the invasion of Sicily. An unclassified bibliography of the publications of the MGU from 1942 to 1975 was compiled by Selma Bonham in 1981 and published in Leith (1997).

In recent years, much of the terrain intelligence work was transferred to defense agencies such as the Corps of Engineers. The work of military geology in the U.S. Geological Survey turned to emerging issues of strategic and intelligence concerns, most notably the foreign nuclear weapons test programs and the regional geologic and geophysical assessments needed to support seismic monitoring. Further, many recent studies have focused on the geologic characterization of very small areas, such as a single underground nuclear test site or an underground military facility. Of increasing importance in these characterizations is the remote assessment of detailed rock properties at specific sites, which enables the users of these data to evaluate the influence of the local geologic environment on both weapons testing and target vulnerability.

DATA SOURCES

The use of data collected from all available sources, both classified and unclassified, is essential in compiling reports that

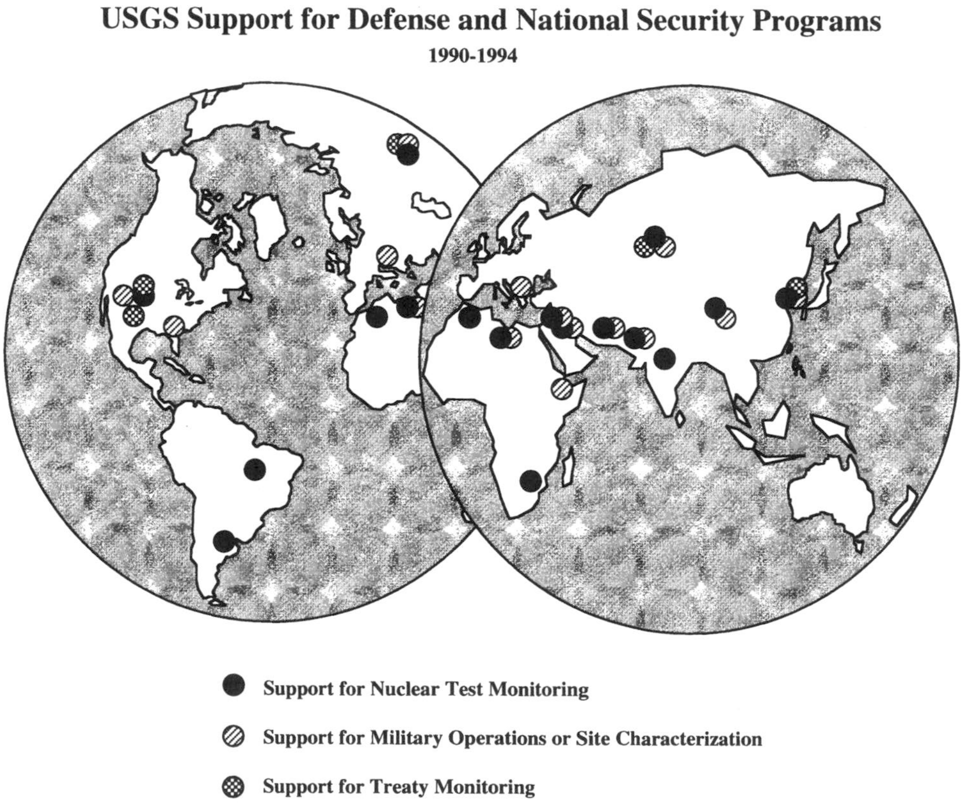

USGS Support for Defense and National Security Programs

1990-1994

Support for Nuclear Test Monitoring

Support for Military Operations or Site Characterization

Support for Treaty Monitoring

Figure 1. Regions of the World for which the U.S. Geological Survey has recently provided geologic information in support of U.S. defense and national security programs (from G. Eaton, Statement for the Record, House Subcommittee on Energy and Natural Resources, Feb. 23, 1995). Three categories of support are distinguished: support for nuclear test monitoring (such as the geophysical characterization of a country or region for evaluating seismic wave propagation); support for military operation or site characterization (such as the Yongbyon nuclear research facility in North Korea, see Fig. 3); and support for treaty monitoring (such as the evaluation of geotechnical factors affecting the verifiability of treaties calling for missile dismantlement at certain sites).

cover many geologic topics in diverse areas of the globe. Literature searches identify pertinent materials available in the U.S. Geological Survey library or other libraries where the material may be acquired. One of the requirements for work in the MGU was an ability to read at least two foreign languages; many in the early MGU could work in six or seven languages. This flexibility enabled geologists to use original language materials in compiling reports. A reading knowledge of scientific Russian was still required in the military geology group until the breakup of the Soviet Union.

To complete its all-source assessments, intelligence reports, air and ground photographs, civilian satellite imagery, and information derived from the National Technical Means are used where available and appropriate. When required, specialists among the staff of the U.S. Geological Survey are consulted for their expertise in a particular area of the earth sciences.

Cooperation within data exchange programs between the United States and the former Soviet Union has provided some "ground truth" (field corroboration) with which to evaluate the geologic assessments made by the U.S. Geological Survey over the past three decades. This means of authentication has proven the reliability of the methods of remote assessment that has been employed. For example, U.S. Geological Survey personnel were included on teams making the first on-site visits to the Soviet nuclear test sites in Kazakhstan and on the arctic island of Novaya Zemlya.

RECENT EFFORTS OF THE SPECIAL GEOLOGIC STUDIES GROUP

Figure 1 shows and Table 1 describes regions of the world for which the U.S. Geological Survey has provided geologic information in support of U.S. defense and national security programs. Many of these areas relate to U.S. efforts to monitor

TABLE 1. U.S. GEOLOGICAL SURVEY PROJECTS IN SUPPORT OF DEFENSE AND NATIONAL SECURITY PROGRAMS

Project	Program Supported	Description
Seismic Networks (foreign)	National security program in nuclear test detection	U.S.G.S. installs, operates, and maintains global seismic stations in support of the nuclear monitoring program of the Department of Defense (DoD). U.S.G.S. also provides data and analysis supporting seismic monitoring, such as the evaluation of seismic wave propagation on global and regional scales, and identifies areas having a potential for evasive nuclear testing.
(domestic)	Treaty verification	U.S.G.S. operates domestic seismic stations that the DoD will use to meet its obligations for seismic event reporting under nuclear test limitation treaties.
Nevada Test Site	Weapons testing	U.S.G.S., with 38 years of continuous service, has been the principal agent providing geological site characterizations in support of the DoD and Department of Energy (DoE) nuclear and conventional weapons testing operations.
	Remediation	U.S.G.S. is supporting DoE efforts to assess the impact of underground nuclear testing on the ground-water system.
Site Characterization (foreign)	Target assessment	U.S.G.S. provides detailed characterizations of foreign sites in support of DoD analyses of weapons effectiveness and target vulnerability.
(domestic)	Weapons development	U.S.G.S. provides detailed characterizations of domestic sites in support of DoD analyses of current and future weapons effectiveness.
	Treaty verification	U.S.G.S. has characterized foreign and domestic sites in support of DoD and DoE research on START and other nuclear treaty monitoring.
On-site Inspection	Treaty verification	U.S.G.S. scientists are included on on-site inspection teams for nuclear test verification and have participated in on-site visits under nuclear treaty protocols with the former Soviet Union.
Military Geology	Military operations	U.S.G.S. provides information and analysis products in support of military operations; recent examples include support for operations in Iraq, Somalia, Yugoslavia, and North Korea.
Classified Program	Intelligence interests	U.S.G.S. provides information and analysis products in response to queries from the intelligence community on any subject relating to geology, geophysics, hydrology, climate, oceanography, remote sensing, and other earth science disciplines.
Hazard Assessment	Overseas military bases and U.S. diplomatic missions	U.S.G.S. has provided assessments to establish the hazard from natural events to U.S. personnel and property at overseas military bases and diplomatic missions.

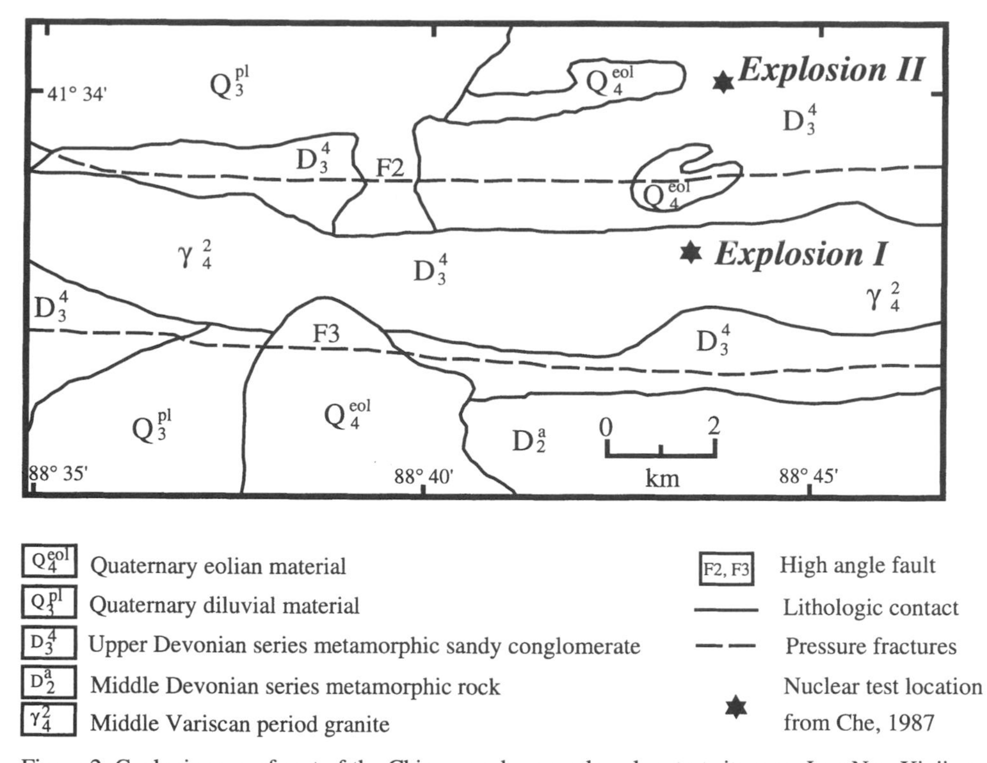

Figure 2. Geologic map of part of the Chinese underground nuclear test site, near Lop Nor, Xinjiang province, China. The map was compiled from Chinese published sources and is an example of the types of data available for remote and inaccessible areas. The two locations identified as Explosion I and Explosion II are sites for which subsurface data were published. Combined with image interpretation and other data, these data allowed a detailed assessment of the subsurface environment at the nuclear test site.

Figure 3. Digital geographic model of the area of the Yongbyon nuclear research facility, North Korea. Research on this site has included the evaluation of terrain, meteorology, hydrology, soils, geology, engineering geology, and crustal geophysics in support of the potential monitoring of the nuclear complex under international agreements. Contour interval is variable (contours at 18, 46, 69, 91, 113, 137, 165, 203, and 489 m).

nuclear testing and to plan verification measures for a comprehensive nuclear test ban. One of the technical challenges of such a ban is the detection and identification of nuclear tests at low seismic magnitudes, as there may be many seismic events per day in some countries in which nuclear proliferation is a concern. Recent work in this area includes the geologic assessment of the Russian and Chinese nuclear test sites (Matzko, 1994; see Fig. 2), the assessment of mines and mine blasts as sources of seismicity, and the feasibility of constructing large cavities that may be used for clandestine testing (see Leith and Glover, 1993).

Another area of current interest is geologically assessing the sites of underground, military-related structures. The overall characterization of underground structures has recently been given a high priority among areas for increased emphasis within the Department of Defense (see the recommendations of the "Deutsch Report," summarized in Taubes, 1995). Knowledge of the physical properties of near-surface materials can be used to predict both their penetrability and the shock-wave propagation of weapons detonated against potential targets buried in them. Such data are even more critical in the absence of a nuclear weapons option, since targeting strategies must rely on a precise knowledge of the near-surface geologic materials in order to utilize effectively the smaller charges available in the conventional weapons inventory.

For example, a region of interest for detailed site characterization is the Yongbyon nuclear research complex in North Korea. Figure 3 shows a digital terrain model of the region of the nuclear research complex. This model was constructed to evaluate the feasibility of monitoring the site for compliance with agreements limiting the production of nuclear materials at this site. The site study included analyses of the bedrock geology, soils and engineering geology, hydrology, terrain, and weather. The terrain and geologic models that were constructed allow different monitoring proposals for this site to be evaluated by computer simulation.

In summary, the Special Geologic Studies Group at the U.S. Geological Survey is a unique resource that provides geological and geophysical data for inaccessible areas by using a proven technique of augmenting published data with other information sources. This group has a long history of service to the nation and has evolved in character as both politics and military technologies have evolved.

ACKNOWLEDGMENTS

The manuscript benefited from reviews and material contributed by M. Terman, D. Percious, and J. Unger. Figure 3 was contributed by J. Unger. Thanks to J. Underwood and M. Terman for their continuing encouragement. U.S. Geological Survey publication no. R95-0367.

REFERENCES CITED

Che, Yongtai, 1987, Response of ground water levels in wells to underground nuclear explosions: Shuiwendizhi Gongchengdizhe (Hydrology and Engineering Geology), v. 4, p. 7–12 (in Chinese).

GSA (Geological Society of America), 1945, The Military Geology Unit: Pamphlet distributed at the Pittsburg meeting of the Geological Society of America, December 1945, 22 p.

Hunt, C. B., 1950, Military Geology: Geological Society of America, Berkey Volume, p. 295–327.

Leith, W., editor, 1997, Reports and Maps of the Military Geology Unit, 1942–1975, U.S. Geological Survey Open-File Report 97-175, 130 p.

Leith, W., and Glover, D., 1993, Cavity construction achievements and decoupling opportunities, worldwide, *in* Proceedings, ARPA/PL Seismic Research Symposium: Phillips Laboratory, September 1993.

Matzko, J. R., 1994, Geology of the Chinese nuclear test site near Lop Nor, Xinjiang Uyygur Autonomous Region, China: Engineering Geology, v. 36, p. 173–181.

Taubes, G., 1995, The defense initiative of the 1990s: Science, v. 267, p. 1096–1100.

Manuscript Accepted by the Society October 29, 1997

Geological Society of America
Reviews in Engineering Geology, Volume XIII
1998

Remote sensing, terrain analysis, and military operations

Jack N. Rinker
U.S. Army Topographic Engineering Center, 7701 Telegraph Road, Alexandria, Virginia 22315-3864

ABSTRACT

The army moves over, digs in, hides in, and builds on the land. Success in these endeavors relies on information about landform, structure, composition (rock and soil types), nature of the surface (sticky, dusty, hard, soft, etc.), and an evaluation of obstacles, engineering materials, water sources, and potential sites for ambush, defilade, and cover and concealment. Geology looms large. For many world areas, such information is not in the databases nor on maps; yet it is sometimes needed on short notice. The information can be derived from image analysis, and available imagery covers most of the world. Examples of such applications include Thule Air Base, Icecap access routes, Project Sanguine, Southeast Asia trafficability studies, and Operations Desert Shield/Storm. The *Remote Sensing Field Guide—Desert*, developed by a joint effort between the U.S. Army Topographic Engineering Center (TEC) and the U.S. Geological Survey, was used extensively in Operations Desert Shield/Storm in support of military operations. These materials plus spectral reflectance data are being blended into a hypermedia terrain database to support interactive image analysis between army elements.

INTRODUCTION

Predicting terrain characteristics for military operations relies on evaluating the geologic setting in terms of landform, structure, fractures/faults, and the composition, properties, and conditions of the resident rocks and soils. Military analysts are tasked to provide such information, on a rapid-response basis. For much of the world, databases and maps with the requisite detail of geology, topography, drainage, and so on, are sparse or do not exist. Most requirements, however, can be met by information derived from the manual or "eyeball" analysis of image patterns, and airborne and satellite imagery are available for most of the world. Geology, remote sensing, and image analysis support much of the strategic and tactical planning and execution of military operations.

Evaluation of the terrain for cross-country movement is a typical application of remote sensing and image analysis. Information needs include: Will the surface support tracked and wheeled vehicles? Is it flat, rolling, hilly? Are the slopes steep, gentle, smooth, stepped (thin or massive beds)? Are there restrictions due to surface roughness, obstacles, thorns, slopes, crusts? Is the surface dissected (drainageways as indicators of composition, as obstacles, as site for ambush or concealment)? Are there areas to support fixed-wing and rotary-wing aircraft operations? What is the potential for dust? Are there sites and approaches for cover and concealment or for ambush and defilade? Are there unstable materials and conditions? What are the water sources? The answers come from the geologic setting.

The three-dimensional shape of landscape elements is directly related to composition, structure, physical properties, and conditions and climatic regime of the resident materials. A derived corollary is: Like patterns, wherever they occur, indicate like materials and/or conditions, and unlike patterns, wherever they occur, indicate unlike materials and/or conditions. Image patterns can be classified as landform, drainage (plan and elevation), erosion, deposition, vegetation, tone and texture, cultural, and special (which includes such features as joints, faults, slumps, tears, mudflows, trim lines, and so on). Landform identifies terrain elements as plains, valleys, hills, mountains, fans, and terraces, and their specific shapes provide clues about composition and physical properties. The shape of a drainageway in plan and elevation is related to soil type, soil texture, mantle thickness, homogeneity, and structural control. The finer, or more closely

Rinker, J. N., 1998, Remote sensing, terrain analysis, and military operations, *in* Underwood, J. R., Jr., and Guth, P. L., eds., Military Geology in War and Peace: Boulder, Colorado, Geological Society of America Reviews in Engineering Geology, v. XIII.

spaced the drainage net, the smaller the grain size of the material. Short V-shaped gullies are indicators of granular material. As the fraction of the plastic and nonplastic fines increases, the gully lengthens and the cross section becomes more saucer shaped. Erosion patterns are also linked to the hardness or softness of the material. Depositional patterns are indicative of origin (eolian, fluvial, gravity), as well as of composition, e.g., the higher the arch of an alluvial fan, the coarser the material. Tone and texture patterns provide information about surface structure and composition. For example, the tone patterns of glacial till from the Wisconsin ice age differ from those of Illinoian glacial till. By evaluating these patterns and checking each against the other for consistency of logic, the analyst can arrive at identities such as igneous, metamorphic, sedimentary (sandstone, limestone, shale, massive beds, thin beds, interbedded), soil types, and properties such as hard, soft, dusty, sticky, quick, or unstable. With this database in hand, predictions can be made about terrain characteristics in support of cross-country movement, engineering site selection, and environmental impact.

Three-dimensional shape is the critical factor in deriving terrain information, and, for this task, the manual analysis of stereo imagery is still state of the art. Monoscopic images provide information but of less quality and quantity. At present, digital techniques contribute little in evaluating three-dimensional shapes of terrain elements. Coupled to multispectral and hyperspectral imagery, however, digital techniques such as enhancement, band ratios, and so on, can improve pattern boundaries for visual examination, and, coupled to spectral databases, they provide the best means for targeting applications. The basics of the manual analysis procedure for terrain information can be found in Belcher (1944), Frost et al. (1953), Rinker and Corl (1984), and Philipson (1997).

Table 1 lists some army applications of image analysis. The three italicized entries selected for summary presentations relied heavily on an evaluation of the geologic setting and directly supported military operations. They are taken from Rinker (unpublished data).

CASE STUDIES

Thule Air Base, 1951 (air photo analysis—Project Blue Jay [Thule Air Base]). This contract report by R. E. Frost was submitted to Metcalf and Eddy, for the U.S. Army Corps of Engineers who supervised the project for the U.S. Air Force. Although done in 1951, this study still ranks as one of the larger engineering applications of air photo analysis to date. When it was decided to build an air base at Thule, on the northwest coast of Greenland, the engineering information had to come from air photos—detailed soils and geologic maps did not exist. Complicating the issue was the fact that permafrost conditions existed throughout the area, and such conditions had caused severe construction problems in Alaska and elsewhere. Photo analysis identified soil characteristics, marked critical permafrost areas, designated overburden or pad specifications for preventing thermal transfer from heated structures to the permafrost, located engineering materials, and laid out the runway with specifications for handling the underlying soil problems. Construction followed as soon as the ice broke up and ships could convey construction equipment, crews, and support facilities to Thule, Greenland. The successful project proved the worth of air photo analysis for obtaining rapid and accurate terrain information in support of engineering construction in an isolated, unmapped, and problem-beset region.

TABLE 1. SOME EXAMPLES OF MILITARY AND ENGINEERING APPLICATIONS OF AIR PHOTO AND IMAGE ANALYSIS COMPLETED WITHIN ARMY AGENCIES OR UNDER ARMY AUSPICES

Trafficability studies — Alaska — Corps of Engineers/Waterways Experiment Station — 1940s
Airfield site selection and layout — 1940s
Forty Civil Aeronautics Administration and Army Air Corps Training Fields
Six Alaska — Corps of Engineers/U.S. Air Force
Thule Airbase — layout and materials locations — 1951
Desert trafficability — Corps of Engineers/Waterways Experiment Station
Dewline site locations (8) — Alaska — 1954–58
Greenland icecap access routes — 1956–58
Anti-Ballistic Missile site selection (3) — late 1950s
Nike site selection and evaluation — mid-1950s
Ballastic Missile Early Warning System site selection
Trafficability — Mobility Environmental Research Study (Project MERS) — Southeast Asia
Environmental database — two areas in Vietnam
Environmental database — two areas in Puerto Rico
Special terrain evaluation — Southeast Asia — 1967
Project Sanguine — Navy — 1967–68
Military Geographic Intelligence and geology data base — Puerto Rico — 1970s
Environmental database — Panama — 1970s
Environmental database — Fort Bliss, Texas — 1980s
Disaster evaluation — Gander airplane crash 1985–86
Desert Operations — 1990–91

Project Sanguine, 1969 (air photo analysis of a portion of Wisconsin and Michigan). This analysis, done for the U.S. Navy, is probably the largest air photo study completed to date, as well as the largest air photo analysis report. A single copy (text, maps, ground photos, photomosaics, overlays) weighed about 250 pounds. Unlike Thule Air Base, the construction phase never started. In the mid-1960s the U.S. Navy considered installing a large antenna as a communication link to submarines (large meaning something more than 100 × 100 miles). The site was in the northern part of Wisconsin and the lower part of upper Michigan. In order to bid on construction, contractors had to have a common package of specifications that described the area in terms of landform, drainage, land cover–land use; surface materials in terms of rock and soil types; soil mantle characteristics in terms of homogeneity, presence of rock fragments, and depth of soil mantle; bog and wetland characteristics and descriptions of critical areas in relation to engineering problems and environmental impact. There was a paucity of map information. The only way to supply the required information within the imposed time constraints was by air photo analysis. The Photo Interpretation Research Division (PIRD) of the U.S. Army Cold Regions Research and Engineering Laboratory was charged with the responsibility for the analysis and the transfer of the information to 1:50,000 scale sheets. To augment the PIRD capabilities, people were pulled in from the U.S. Army Natick Laboratories, U.S. Army Waterways Experiment Station, the Huntington and Kansas City Districts of the Corps of Engineers, and the 30th Engineer Battalion at Fort Belvoir, Virginia. The analysis was completed in 1968 and field checked. The report, which was an up-to-date environmental database as well as an engineering evaluation, was finished in 1969 and delivered to the Navy and to the states of Wisconsin and Michigan.

Desert operations

Landsat Thematic Mapper (TM) scenes were used extensively during the Persian Gulf war. Although the handicaps of monoscopic format, small scale, and a 30 m ground resolution preclude its effective use for tactical applications, a 1:250,000 scale color composite TM image is ideally suited for a rapid regional analysis of an area in terms of general terrain characteristics in support of military operations. Details can be filled in by reconnaissance, stereo aerial photography, or other means. The sorts of information that can be gleaned from such an image include rock-soil types, restrictions on foot-vehicular traffic, probable obstacles and deleterious factors, potential for dust generation, navigation points, sites for observation, ambush, and defilade, and potential for cover and concealment. A first review of such images showed that the proposed operating area for Desert Storm was mostly on the surface of fan deposits in the form of an old lobate delta that extended from Saudi Arabia into Kuwait and Iraq. The terrain teams were informed early on that this undulating sand-gravel plain provided a good surface for wheeled and tracked vehicles, as well as potential for cover and concealment along the grain and exposure across the grain.

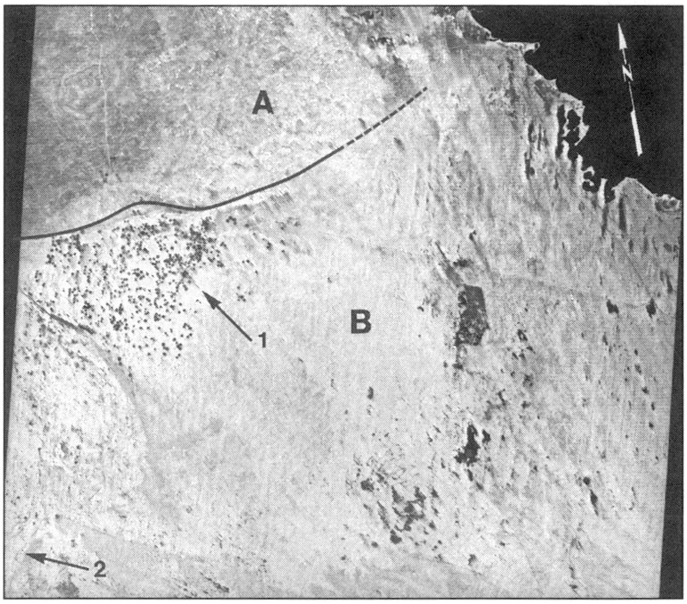

Figure 1. Landsat TM image 165-041 of August 31, 1990, over Saudi Arabia just south of Kuwait. Approximate scale is 1:2,200,000. This is a black and white copy of the 1:250,000 scale color composite (bands 2, 4, 7 coupled to blue, green, and red, respectively). The overlay is greatly simplified from the original. Area A is a portion of the large fan development, or delta, extending from Saudi Arabia into Iraq and Kuwait. Area B is a sand-gravel plain. Arrow 1 marks one of the exposed areas of gypsiferous hardpan that had a cyan color in the original color composite image. An enlargement of the area is in Figure 2. Arrow 2 marks the location of several inverted wadis. An enlargement of these features is in Figure 3.

Landsat TM image 165-041 of August 31, 1990, which covers a coastal desert region along the northeastern coast of Saudi Arabia, south of Kuwait, has interesting terrain patterns and can also serve to illustrate the coupling of spectral data to terrain information to gain a better understanding of surface materials. The analysis was done on a 1:250,000 scale hard copy color composite of bands 2, 4, and 7 coupled to blue, green, and red, respectively. Figure 1 is a black and white copy of the color image at a scale of about 1:2,200,000. Aside from loss of detail due to scale change, the patterns and pattern boundaries are not as discernible as on the color composite image. Also, airborne dust patterns in parts of the image cover up some of the ground detail. The original overlay of landform units (nine plus subsets) was too detailed to withstand scale reduction and was replaced with the simple version shown in Figure 1. The western part of area A shows a small portion of the deltaic landform that extends into Kuwait and Iraq. On the western edge of this area is a pattern of concentric circular rings, indicative of a truncated dome—not an uncommon feature in these regions. Area B is a sand-gravel plain showing bedrock control and center-point irrigation development in its northwestern portion. The crenate pattern along the coast and on some of the mesa edges suggested a calcareous sandy unit, which indeed it was. The following comments on terrain characteristics are confined to Unit B.

From landform unit descriptions. The area is a plain with isolated patterns of mesas, rock outcrops, mounds, dunes, ridges, yardangs, and similar variations in local relief. Escarpment patterns suggest that the beds dip gently to the east or southeast. The parallel pattern of the near-shore sabkhas is reinforced by parallel ridge-like features (highlight-shadow) that trend north-northeast—indicative of surface or near-surface bedrock. The ridge-like units are more massive to the east, and dwindle to barely perceptible trains of ever smaller hills and residual nubbins in the west. Most of the area between the ridges and mesas is a sand-gravel plain—residual material from the dissolved limestone forming a mantle of varying thickness over the underlying bedrock and hardpan that develops in such deposits. The surface, underlain by hard units, will support high-speed cross-country movement as long as that subsurface layer is continuous. Light blue (cyan) tones scattered through the image suggested that gypsum evaporites formed the hardpan throughout much of the area. Subsurface layers cemented by gypsum usually contain solution pits filled with soft sand. The prediction that the duricrust layer was most likely gypcrete was based, in part, on the cyan tones scattered throughout the image (Arrow 1 of Fig. 1), which shows how spectral reflectance data can sometimes be used to help identify surface composition (refer to Fig. 2).

Starting in the southwest corner and trending northeasterly through the various landform units are isolated and clustered patterns of discontinuous lineal, arcuate, and sinuous streaks whose shadow-highlight tones indicate they are ridges. Some of these are indicated by Arrow 2 in Figure 1 and are in the ground photograph in Figure 3. These interconnecting curvilineals are gravel or duricrust capped ridges—most likely inverted wadis—

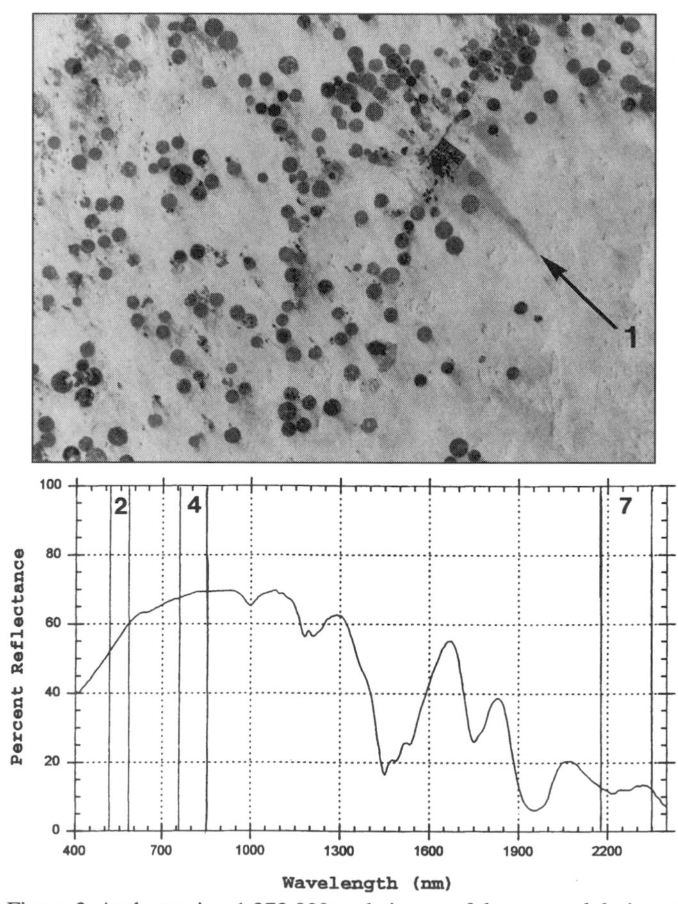

Figure 2. At the top is a 1:373,000 scale image of the exposed duricrust indicated by Arrow 1 in Figure 1. Below it is a graph of reflectance properties of gypsum. The TM bandpasses are marked on the graph. The TM color composite was compiled from bands 2, 4, and 7 coupled to blue, green, and red, respectively, and in it the indicated area was cyan. The graph shows about 60% reflectance for band 2 (blue), 70% for band 4 (green), and 12% for band 7 (red). Thus, if much gypsum were present, the image tones would be created predominantly by blue and green to give cyan, which is the tone evident in the image. Thus the prediction that the hardpan was probably gypsiferous.

Figure 3. At the top is a 1:316,000 scale image of the inverted wadis indicated by Arrow 2 in Figure 1. Photograph at the bottom is a typical ground view of inverted wadis in this area. They are about 10 m high. For the most part, the slopes are not too steep for four-wheel-drive vehicles. In some places the slope blends with the flat tops. In other places, there can be vertical faces as high as 1 m. Ground photograph by J. N. Rinker (P16-31, April 4, 1991).

that are remnants of large alluvial fans deposited by ancient river systems or channels of ancient river systems. Since their formation, a change in climate brought less moisture, higher temperature, and higher winds. Subsequent erosion has removed the associated fines, leaving behind the coarsest parts of the old channels and lowering the surrounding surfaces into which they were once inset.

From the regional summary. Throughout the lower two-thirds of Area B are mesas, buttes, mounds, knob-like hills, ridges, yardangs, clastic dikes, and inverted wadis—all of which can hinder cross-country movement and line of sight. On the positive side, many of these features can serve as observation sites, wind shelters, and navigation points, and they can provide opportunities for cover and concealment, defilading artillery, and conducting ambush and flanking maneuvers. Movement along the northerly trend will have fewer directional constraints to avoid obstacles (ridges, knobs, and so on) than in a westerly direction. Vehicles can cross ridges where they are low and smooth. Some ridge sections might have caps with a vertical edge high enough to impede crossing. Cross-country movement will be easier in the western and northwestern areas. The mesas and buttes are local highs, and they can provide key sites for

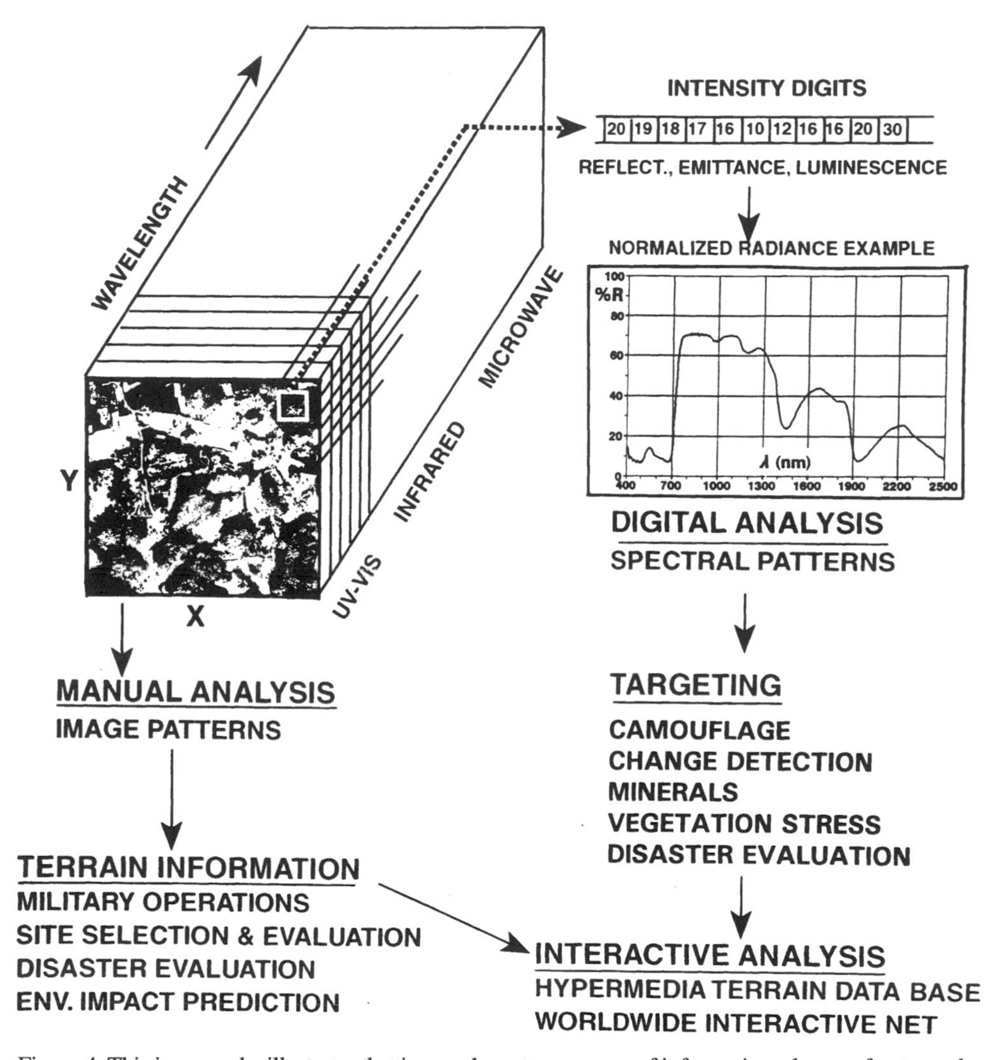

Figure 4. This image cube illustrates that images have two sources of information: shapes of pattern elements (best for terrain information) and brightness, or spectral characteristics of their surfaces (best for targeting). The X-Y axes set the boundaries of an image plane. The wavelength axis specifies regions of the electromagnetic spectrum that formed the image or series of images and extends from the ultraviolet to at least the radar P-Band frequencies. A panchromatic photograph would show one image plane (400–700 nm). A Landsat TM scene would show seven image planes: 1, 2, and 3 in the visible; 4, 5, and 7 in reflected infrared; and band 6 in the thermal infrared. A hyperspectral system such as the Airborne Visible Infrared Imaging Spectrometer (AVIRIS) would have over 200 narrow bandpass (10 nm) image files over the 0.4–2.5 micra range. For such files in a computer database, the cursor can be placed over a given area, and a string of intensity values will be displayed as line spectra or as an intensity versus wavelength plot. From Rinker (1994).

observation, radio communication (out of the sand stream), and so on. Some tops might be accessible by foot and vehicle via sand ramps (climbing dunes). The approach at the base will be rough and rocky. Because of the possibility of large cracks and cave-like features, there is potential for cover and concealment. The terrain between the rock ridges, hills, and mesas provides a good sand-gravel surface for vehicular traffic. Vehicles should not track each other because this increases the chance of following vehicles breaking through the firm surface into underlying soft sand, particularly over solution pits in the hardpan. Vehicles passing over them can bog down, while their partners on either side continue on their way. Where sand sheets obscure underlying sabkhas and wet materials, problems will be encountered such as stickiness, softness, or quick conditions.

CONCLUSION

To help image analysts relate image patterns to terrain characteristics, the TEC/USGS Desert Processes Working Group developed the *Remote Sensing Field Guide—Desert* (Rinker et al., 1991). It contains a classification of desert features and a series of single sheet entries that discuss each feature in terms of description, origin, and significance to engineering and military operations. As imagery becomes available, each feature will be illustrated in Landsat, radar, stereo aerial photos, aerial oblique photos, and ground photos. To assist in targeting tasks and the exploitation of hyperspectral imagery, spectral data sheets show the spectral reflectance characteristics of various desert surfaces and other materials, over the 0.4–2.5 micrometer wavelength band of the electromagnetic spectrum. Targeting refers to the detection, and sometimes identification, of specific features and conditions embedded in the landscape. To be indistinguishable, the item or condition must have the same reflectance, luminescence, and emittance characteristics as the background. This is seldom the case: someplace there will be a mismatch in the reflected (visual, infrared, radar) bands, thermal bands, or luminesced bands. In these applications, targeting is interpreted in its broadest sense, and includes diseased and stressed vegetation, flood boundaries, thermal springs, thermal plumes, hot spots, alteration zones, camouflaged sites, and so on. Figure 4 shows the concept of the image cube, its domains of image patterns and spectral patterns, and the separate yet interlocking roles they play in terrain analysis for military operations.

REFERENCES CITED

Belcher, D. J., 1944, The engineering significance of soil patterns: Washington, D.C., Joint Highway Research Project, Highway research reprints, v. 13, p. 569–598.

Frost, R. E., Johnstone, J. G., Mintzer, O. W., Parvis, M., Mantano, P., Miles, R. D., and Sheppard, J. R., 1953, A manual on the airphoto interpretation of soils and rocks for engineering purposes: West Lafayette, Indiana, School of Civil Engineering and Engineering Mechanics, Purdue University, 213 p.

Philipson, W. R., editor, 1997, The manual of photographic interpretation, 2nd ed.: Bethesda, Maryland, American Society of Photogrammetry and Remote Sensing, 689 p.

Rinker, J. N., 1994, Tutorial I—Introduction to spectral remote sensing, *in* Proceedings, International Symposium on Spectral Sensing Research, San Diego, July 1994, Volume I: Washington, D.C., U.S. Government Printing Office (U.S. Army Topographic Engineering Center, Alexandria, Virginia), p. 5–43.

Rinker, J. N., 1997, Chapter 18, Army, *in* Philipson, W. R., ed., The manual of photographic interpretation, 2nd ed.: Bethesda, Maryland, American Society of Photogrammetry and Remote Sensing, p. 613–639.

Rinker, J. N., and Corl, P. A., 1984, Airphoto analysis, photo interpretation logic, and feature extraction: Alexandria, Virginia, U.S. Army Topographic Engineering Center Report ETL-0329, 374 p.

Rinker, J. N., Breed, C. S., McCauley, J. F., and Corl, P. A., 1991, Remote sensing field guide—desert: Alexandria, Virginia, U.S. Army Topographic Engineering Center Report ETL-0588, 568 p.

Manuscript Accepted by the Society October 29, 1997

Printed in U.S.A.

Geological Society of America
Reviews in Engineering Geology, Volume XIII
1998

A proposed method for characterizing fracture patterns in denied areas

Judy Ehlen
U.S. Army Topographic Engineering Center, 7701 Telegraph Road, Alexandria, Virginia 22315-3864

ABSTRACT

Knowledge of three-dimensional, subsurface fracture patterns is necessary to solve many military and engineering problems. An understanding of subsurface fracture patterns is also essential for the field army with respect to penetrability and weapons effects. Many areas of the world in which the army has an interest are inaccessible (denied areas), and analysis of remotely sensed imagery provides a way to get needed information. This chapter describes a method that provides the basis for three-dimensional characterization of fracture patterns using remotely sensed imagery. Analysis of fracture patterns in the Dartmoor granite of southwest England shows that lineations delineated on imagery are very long, widely spaced joints that are members of joint sets found in outcrop. There are no statistically significant differences in orientation between joints and lineations; the smaller image scale allows the more widely spaced and longer members of a given joint set to be seen. With these relations established, fractal analysis of joint and lineation patterns on Dartmoor granite and in the East Pioneer Mountains, Montana, was done to determine whether fractal geometry could be used to predict subsurface fracture patterns. On Dartmoor, mean fractal dimension for vertical joint sets from outcrop and lineation patterns from imagery were comparable, but this was not the case for the Pioneer granites. Further research must address the problem of predicting three-dimensional, subsurface fracture patterns using remotely sensed imagery prior to attempting to meet the needs of the field army in denied areas.

INTRODUCTION

Many areas of the world in which the army has an interest are for various reasons inaccessible. Analysis of remotely sensed imagery provides a way to obtain information about such areas, called denied areas, in a timely, cost-effective manner. One uses analogues—areas similar to the denied area with respect to the subjects of interest—to do this. If information is needed about fracture patterns in a denied area, for instance, the fracture patterns in an area of similar geology are studied to predict the likely pattern in the denied area. If imagery were available over both areas, one could use the lineation data correlated to surface and/or subsurface data in the analogue area as the model for the denied area and, coupled with lineation data from the denied area, a more accurate prediction could be made. Development of this procedure is the goal of this effort. It is being done in two phases. The first phase, which addresses the relationships between field and imagery-derived data and the potential of fractal geometry to predict subsurface fracture patterns, is addressed in this chapter. The second phase, the modeling and prediction of subsurface fracture patterns using surface and image data, will be accomplished in the future as appropriate field areas become available.

Although local fracture patterns are best investigated at the outcrop scale, the scales of remotely sensed imagery are more appropriate for analysis of regional fracture patterns. Knowledge of fracture patterns at both local and regional scales is necessary for solving many problems that face the army. These include site characterization of terrain in denied areas, penetrability, and weapons effects, as well as more environmental problems such as the siting of repositories for nuclear, chemical, and toxic waste; identifying potential routes for contamination and pollution from leaking repositories, ammunition dumps, and so on;

Ehlen, J., 1998, A proposed method for characterizing fracture patterns in denied areas, *in* Underwood, J. R., Jr., and Guth, P. L., eds., Military Geology in War and Peace: Boulder, Colorado, Geological Society of America Reviews in Engineering Geology, v. XIII.

locating ground-water sources; and selecting sites for engineering structures, such as dams. Many of the latter problems will be addressed in studies that must be done in association with the Base Closure program.

The common thread linking these problems, in addition to the fact that they all involve an extensive understanding of both surface and subsurface fracture patterns, is that they are all three-dimensional. Unfortunately, there is no effective way to address these issues three-dimensionally at present. Most subsurface data come from boreholes that, by virtue of their vertical to subvertical orientation, provide very selective and limited information about fracture patterns. In addition, it is rare that borehole data are available in denied areas. Furthermore, there are no proven techniques for utilizing surface or remotely sensed data for predicting or modeling subsurface, three-dimensional fracture patterns. Study of relations between surface patterns and patterns identifiable on imagery may, nonetheless, provide guidelines to understanding subsurface patterns. Like borehole data, remotely sensed imagery provides limited information about fracture patterns because only vertical or steeply dipping lineations are easily seen. Coupled with fracture data from outcrops, however, a three-dimensional perspective can be obtained.

The relation between fracture characteristics measured in the field and lineations delineated on imagery must be determined prior to attempting to link surface and image patterns to subsurface patterns. Once these relations are established, the next step is to determine whether these one- and two-dimensional data sets can be used to predict and/or model three-dimensional, subsurface fracture patterns. One possible approach for addressing this problem is fractal geometry (Barton et al., 1991). Theoretically, fractal geometry provides a means to predict patterns of one dimension from data collected and analyzed in another dimension, assuming, of course, that the objects are fractal. Other approaches to these problems, including geostatistics and group renormalization procedures, are being investigated by others (Long et al., 1987; LaPointe, 1980; Hestir et al., 1987).

BACKGROUND

Evidence of relations between lineations on imagery and joint patterns in outcrop is contradictory, and none is quantitative. Except in the case of fracture-trace analysis used to locate productive water wells, little work has been done to study these relations. Lattman and Parizek (1964) found relations in some cases between lineations and fractures in limestone bedrock in Pennsylvania. Rinker (1974), in his endeavor to locate subterranean caves in Puerto Rican limestones, showed that subsurface fracture patterns can be related to surface fracture patterns delineated on air photos. Segall and Pollard (1983) reported that the same fracture trends occur on air photos and in outcrop in Sierra Nevada granites. Thorp (1967a,b), on the other hand, working in Nigerian granites, found the opposite relation, as did Mohammad (1987), who worked in limestones in Saudi Arabia.

A few studies have successfully attempted to relate subsurface fracture data to joint data measured in outcrop. Whittle and McCartney (1989) describe one such study in which surface and subsurface joint patterns in the Carnmenellis granite in southwest England are compared using a combination of field measurements, air photo and satellite data, and geophysical procedures. Other successful studies using a combination of geophysical methods, remotely sensed imagery, and ground data include those by Moore and Stewart (1980) on carbonates in Florida, Soonwala and Dence (1981) on Canadian granites, and Jammallo (1984) on metamorphic rocks in Vermont. In addition, the U.S. Geological Survey is currently investigating similar relationships at Mirror Lake, New Hampshire.

Fracture patterns in granites were selected for investigation because (1) granites are relatively common; (2) the photo patterns of granitic rocks are fairly well understood and are well documented (Belcher et al., 1951; Rinker and Corl, 1984; Gerrard, 1988); (3) the fracture patterns are highly distinctive and readily quantifiable, whether in the field or on imagery; and (4) there is a small body of published, quantitative data available on subsurface fracture patterns in granite (Thorpe, 1979; Whittle and McCartney, 1989).

Granitic rocks on Dartmoor, part of the Cornubian batholith in southwest England (Fig. 1), and in the Pioneer Mountains in southwest Montana were chosen for study (Fig. 2). The Dart-

Figure 1. Map showing location of Dartmoor in Southwest England.

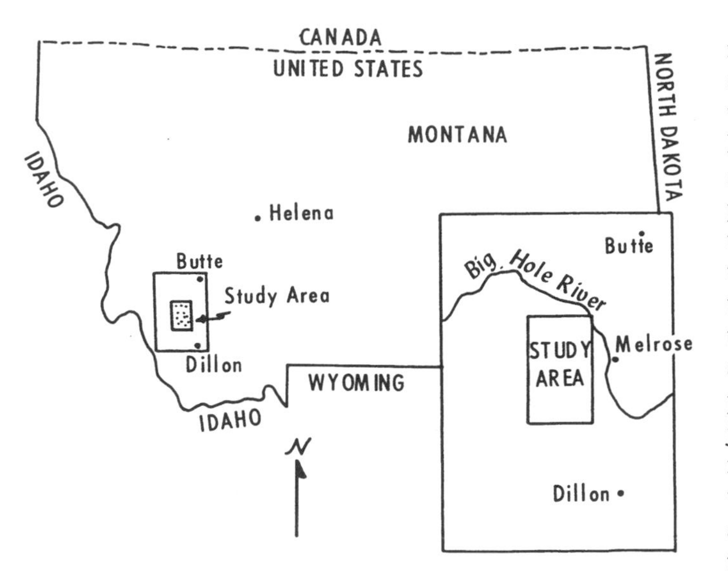

Figure 2. Map showing location of the Pioneer Mountains, Montana.

moor granite forms the easternmost exposure of the Cornubian batholith. Dartmoor is the most significant highland in southern Britain, with elevations ranging from 150 to 600 m. The rocks, which are late Carboniferous or early Permian in age, are coarse grained, strongly megacrystic, peraluminous biotite granites. Composition is relatively homogeneous throughout the pluton (Hawkes and Dangerfield, 1978; Ehlen, 1989). Extensive alteration and mineralization are present. The most comprehensive geologic mapping of Dartmoor was done by Reid et al. (1912), Brammall and Harwood (1923, 1932), and Brammall (1926).

The rocks of the Late Cretaceous to Early Tertiary Pioneer batholith in southwest Montana are predominantly medium to coarse grained, meta- to weakly peraluminous hornblende-biotite granites and granodiorites. Small quartz diorite and tonalite plutons are also present. Elevations range from 1,500 m to more than 3,000 m; several peaks rise above 3,200 m. The high areas of the batholith in the west have undergone multiple episodes of icefield and valley glaciation during the Pleistocene. The region has been mapped most recently by Zen (1988) in the north and by Snee (1978) in the south.

PROCEDURES

Field methods

The method most often used to characterize joint patterns in the field is the line survey method (Robertson, 1970; Hudson and Priest, 1979; LaPointe and Hudson, 1985). The orientation (dip and strike) of each joint along the line is determined and the distances between individual joints are measured continuously along the traverse. True orientations and spacings are then calculated from these data using the Terzaghi correction (Terzaghi, 1965). Once this is done, joint sets are identified using equal area projections.

A modified version of the line survey method was used to measure about 4,000 vertical joint spacings and orientations at 58 sample sites in the Dartmoor granite, and approximately 2,600 vertical joint spacings in 13 granite outcrops in the east Pioneer Mountains. Joint sets were identified in each exposure, and the spacings between individual members in each set were measured perpendicular to strike for those joint sets that appeared to characterize the outcrop. Most sample sites consisted of free-standing, three-dimensional outcrops, so the joint sets were measured in three dimensions (i.e., measured along two sides and/or on top of the exposure). Each sample site was selected because of its size, and only the largest exposures of continuous outcrop were sampled. Strike and dip were measured to the nearest 5° on several joints in each set, and an average orientation for the set was determined. This method was used rather than the traditional line survey because: (1) the field measurements are "true" measurements and no back calculation is required; (2) it is less time-consuming than the traditional line survey method; and (3) the different kinds of joints—primary and secondary (see below) or horizontal and sheeting—can be readily separated.

Spacings between all joints in each set were measured, regardless of aperture or length. Although it was rare that joint spacings were less than 1 cm, the minimum spacing measured was 0.5 cm. The joints within a set were separated into two categories: primary joints and secondary joints. A primary joint is a long, usually open, outcrop-shape–controlling joint that cuts across other joint traces. These joints are more likely to by hydrologically active. A secondary joint is a shorter joint, local in extent, that usually does not cross other joint traces. All field measurements were made between secondary joints. When a primary joint was encountered, it was noted as such, and primary joint spacings were determined by summing the secondary spacings between them. The field data and sample site descriptions are presented in Ehlen and Zen (1990).

Rose diagrams and frequency histograms showing the joint orientations in 5° classes were analyzed visually. Frequency histograms in which the data were grouped in 25 cm classes and descriptive statistics were used to analyze and compare joint spacings, and mean and modal spacings were calculated. The joint spacing frequency distributions were then compared using the chi square test.

Air photo methods

Lineations are natural linear features on imagery representing zones of joints or small faults. Lineations are comparable to the long, usually open, vertical or steeply dipping joints (primary joints) that control outcrop shape. The more abundant, shorter, secondary joints that occur at the scale of an outcrop and represent the bulk of the joints measured in the field are typically not visible at the scales of remotely sensed imagery: They are simply too short and too closely spaced. Lineation patterns were delineated on two black and white uncontrolled air photo mosaics in each area. In Montana, 1:40,000 and 1:80,000 scale photos were used, and on Dartmoor, air photo scales were 1:24,000 and 1:50,000. Figure 3

Figure 3. 1:24,000 scale air photo mosaic with lineation overlay, Dartmoor.

shows the lineation overlay on the 1:24,000 scale Dartmoor photo mosaic as an example.

The lineation overlays were done monoscopically but were checked using a mirror stereoscope. Many lineations are very long, extending beyond the stereoscope's field of view, and it is often easier to see them without stereo. All patterns were evaluated meticulously from different angles and with different kinds and angles of illumination. There are few to no shadows on either set of photos, so there was little to no bias with respect to photo illumination angle versus lineation orientation. To be included on the photo mosaic, a lineation had to be relatively straight and it had to parallel nearby lineations, forming a set. A series of straight, en echelon stream segments is a good example, as are linear vegetation patterns. In some cases, the lineations could be seen as dark lines cutting through light-toned rock. Very long, curved lineations were occasionally present as well.

Because the lineation overlays of the Pioneer Mountains have not yet been digitized so that spacings between lineations can be determined, evaluation of the relations between lineation spacing and orientation and analogous joint measurements must necessarily be restricted to the Dartmoor granite. The individual lineations on the Dartmoor overlays were digitized using a GIS (geographic information system) called CAPIR (Computer-Assisted Photo Interpretation Research). Both mensuration and classification were accomplished using CAPIR. Prior to digitizing, the overlays were rubber sheeted to topographic maps using multiple control points. The individual lineations were digitized, and the strike of each was automatically calculated from digitized coordinates. There are about 1,500 lineations on the 1:50,000 scale mosaic and 2,900 on the 1:24,000 scale mosaic. The lineations were grouped in 5° classes according to strike to match measurements made in the field. Spacings were then measured interactively between adjacent lineations along a line normal to strike, just as they were measured in the field. Spacings between about 1,000 lineations were measured on the 1:50,000 scale mosaic, and between about 1,700 on the 1:24,000 scale mosaic.

Rose diagrams and frequency histograms with strikes grouped in 5° classes and frequency histograms, in which the spacings were grouped in 75 m classes for the 1:24,000 scale lineations and 130 m classes for the 1:50,000 scale lineations, were used to analyze the lineation data in the same manner as described above for the field data. The chi square test was used to compare distributions.

Fractal geometry

Fractal geometry is a relatively young branch of mathematics. Basically, a fractal is a geometrical object whose fractal dimension is not an integer and is greater than the object's topological dimension (Barton et al., 1991). An object or feature is fractal if it is self-similar or scale invariant and if it follows a power-law distribution.

An object is self-similar if the parts resemble the whole. Most fractal objects in nature are statistically self-similar: The classic example is a coastline. A coastline looks very much the same if it is 1 km long or 100 km long. The two segments will not be identical, but statistically they will be the same. This phenomena is also known as scale invariance and is determined by measuring the object at different scales. These measurements are plotted in log-log space, and if a straight line passes through the data points using a least-squares fit, the distribution follows a power-law. The fractal dimension, or D, is defined as the slope of that line. Scale invariance allows predictions to be made from one dimension to another. For instance, if the fractal dimension (D) of a two-dimensional fracture network, such as a lineation overlay, is 1.7 then, theoretically, D for one-dimensional line survey data would be 0.7, and for a comparable three-dimensional pattern, that is, a subsurface fracture network, D would be 2.7.

Arguments have raged in the literature for years regarding the statistical distributions of joint characteristics. Joint spacing has been described most often as having either a lognormal or an exponential distribution (e.g., Baecher, 1983; Huang and Angelier, 1989). Reported distributions for other fracture characteristics, such as trace length and aperture, include lognormal and exponential, gamma or linear (e.g., Miller and Borgman, 1985). More recent work, however, has shown that most fracture characteristics, including joint spacing and joint length, exhibit power-law distributions and are fractal (e.g., Barton et al., 1988; LaPointe, 1988).

The fractal dimensions for the lineation overlays from the air photo mosaics were determined using the grid or box-counting method, the same procedure used by others for fracture-trace maps (e.g., Barton and Larsen, 1985; Barton et al., 1991). Grids of different sizes, which represent the different scales, were laid over the lineation overlays and the number of filled boxes containing segments of lineations was counted. These data were plotted against box size in log-log space to determine the fractal dimension using the following equation where N = number of filled boxes; δ = box size; and D = fractal dimension:

$$N(\delta) = 1/\delta^D$$

The joint spacings from the line surveys were plotted on strips of graph paper at a scale of 1:10. These data are Cantor's Dusts, and fractal dimension was determined as described by Velde et al. (1990). The grids were laid over the graph paper strips, and the proportion of boxes containing joints was determined for each grid. The fractal dimensions for these data were determined using the following equation where P = proportion of filled boxes (n/N); δ = box size; and D = fractal dimension:

$$P = \delta^{1-D}$$

JOINTS AND LINEATIONS

Orientations

Frequency histograms for the joint and lineation orientations on Dartmoor are shown in Figure 4. The mean vector for the joint

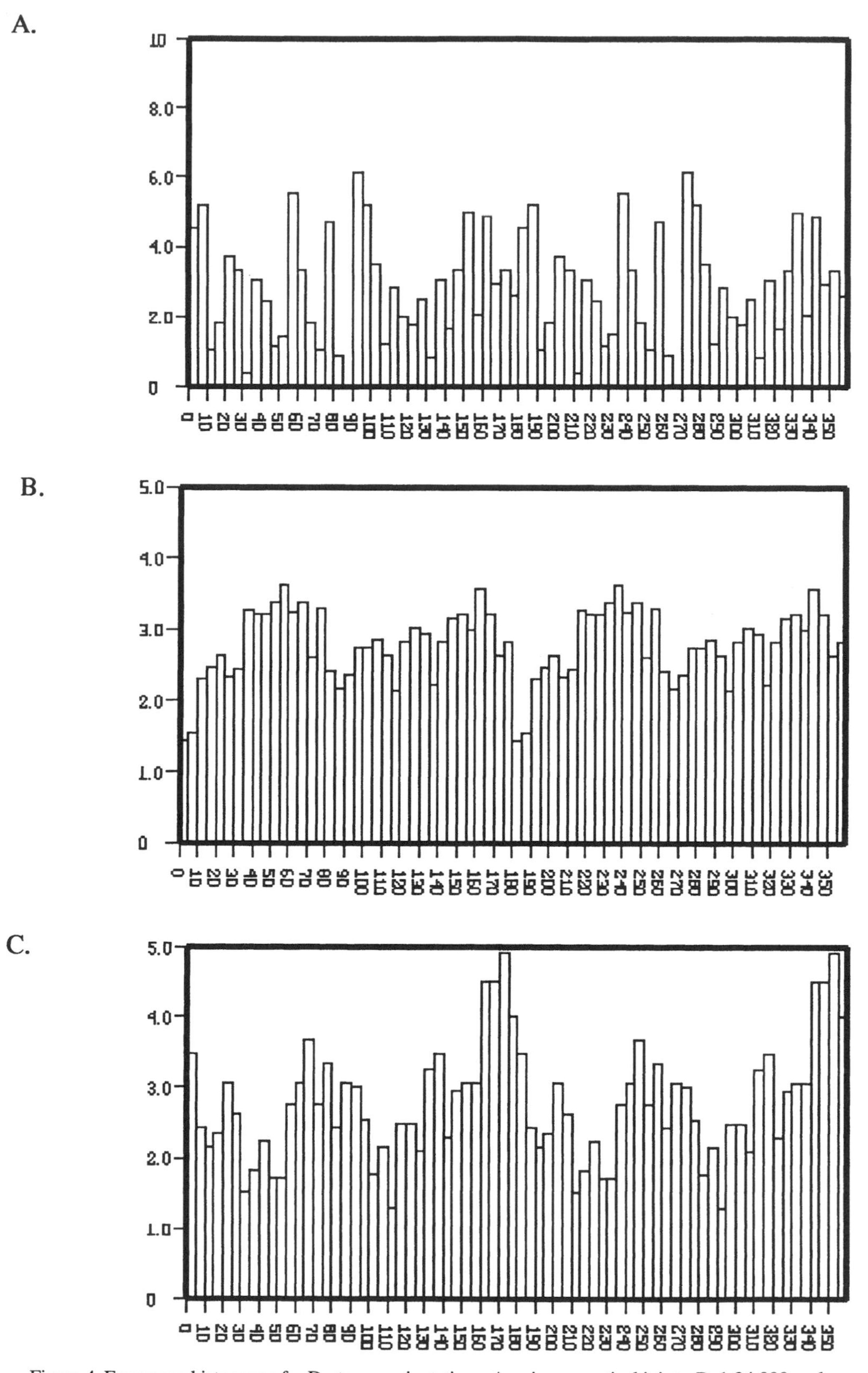

Figure 4. Frequency histograms for Dartmoor orientations: A, primary vertical joints; B, 1:24,000 scale lineations; and C, 1:50,000 scale lineations. Note each orientation is plotted twice (the angle and its supplement).

orientations measured in the field is 345°. The distribution is unimodal with 6.1% of the orientations in the modal category, 270–275°. The mean vector for the 1:24,000 scale lineations is 72°, and the modes in this bimodal distribution are 340–345° and 55–60°. Of the total number of orientations, 3.6% are in each of these modal classes. The mean vector for the 1:50,000 scale lineations is 341°, and the distribution is unimodal. The modal class, 350–355°, contains 4.9% of the lineations.

Analysis of the rose diagrams (Fig. 5) allows further analysis of the orientations, as well as comparisons to previous work. Table 1 summarizes the dominant and secondary trends in orientations for the three Dartmoor data sets. The most prominent trend for the joints is west-northwest (Fig. 5A) and for both scales of lineations is north-northwest (Fig. 5B and C). However, the weaker 330–340° joint trend is very similar to the prominent north-northwest lineation trend (330–360°), and the weak east-northeast joint trend (40–80°) includes the weak lineation trends, that is, 60–90° on the 1:50,000-scale mosaic and 45–80° on the 1:24,000-scale mosaic. The dominant west-northwest joint trend is present on the lineation rose diagrams, but it is less important than the north-northwest, east-northeast, and northwest trends.

Orientation data have been reported by a number of workers, and these data can be used to verify the joint orientations measured in the field and on the lineation overlays presented here. The prominent west-northwest joint trend was identified by Ormerod (1869), Worth (1930), and Gerrard (1978), and the northwest joint trend was noted by Brammall and Harwood (1925) and by Blyth (1962).

More often than not, lineations represent fracture zones, so no attempt was made to identify individual lineations in the field, and other than the results reported here, there are no studies of Dartmoor referring to lineations. The north-northwest lineation trend, however, is in agreement with joint orientations reported by de la Beche (1839), Brammall (1926), and Exley and Stone (1964), among others. In addition, de la Beche (1839) and Blyth (1962) identified the east-northeast lineation trend in their field data, and Brammall (1926) reported the east-northeast and west-northwest trends. These published data sets are much smaller than the ones used in this study, so the correlations are not exact; they do, however, support the lineation trends determined from the air photos.

Spacing

The frequency histograms for the three Dartmoor data sets are shown in Figure 6. Note that the x- and y-axes for the three histogram are different. Each distribution was maximized by plotting only the full range for that distribution. One hundred classes were used for each distribution, however, allowing comparisons to be more easily made between distributions. Fewer lineation spacings are shown in Figures 6A and B than lineation orientations in Figures 5B and C because, like the joint spacings, lineation spacings were measured between members of the same sets along a line, so that not all lineations in each set were included. Descriptive statistics for the three distributions are shown in Table 2.

Mean spacing for the primary vertical joints is 2.6 m; for the 1:24,000 scale lineations it is 961.4 m, and for the 1:50,000 scale lineations it is 1,487.3 m. The increase in spacing with the decrease in scale was expected because of the decrease in resolution with decreasing scale. In addition, the same sized pen was used to delineate the lineations on both photo mosaics, and although it had a narrow point, because of the scale differences, delineating closely spaced lineations at the 1:50,000 scale was more difficult. This in part accounts for the smaller number of lineations delineated on this mosaic. The three distributions were compared statistically using the chi square test: They are significantly different from each other. This was expected because of the large difference in scale between the three data sets. Even if the joints and lineations occur in the same sets at each scale, different individuals would be delineated at each scale and the spacings between them would be different.

Discussion

Six of the nine orientation trends shown in Table 1 are common to the field data and to both sets of lineations, and two additional trends are similar in two of the data sets. The prominent trends common to all three scales of data identify two regional fracture orientations: east-northeast (40–90°) and north-northwest (330–360°). The north-northwest trend parallels the major Lustleigh-Sticklepath fault on the east side of Dartmoor, and the prominent west-northwest joint trend (275–295°) occurs throughout the Cornubian batholith (the Armorican trend; Brammall, 1926). Lineations with the Armorican trend are present at both photo scales, but the trend is not prominent. Because the northwest trend (305–320°) is common to both lineation data sets but absent in the field data, it is likely that these lineations are too widely spaced to be identified as joint sets at the outcrop scale. The spacings between lineations are so wide that there were too few joints in the set in individual outcrops for spacing to be measured in the field. It is also likely that the north-northwest trend (330–360°), so strong on the air photos but less so in the outcrop data, also consists of very long, very widely spaced fractures that were not identified in outcrop because they were too sparse.

Although joint and lineation spacings are not statistically similar, the distributions for the three data sets are not dissimilar. None exhibit a normal distribution, and it is likely, as discussed above, that all three exhibit power-law distributions. In addition, few lineations coincide with the faults shown on the geologic maps of Dartmoor, providing evidence that the lineations are joints, not faults. It is unlikely, after more than 150 years of geologic mapping in this area, that the surface expressions of faults would remain unknown. The lineations are thus the longest members of any given joint set as well as the most widely spaced. Different members of the same joint sets are apparent at the different scales; decreasing scale merely allows these otherwise "invisible," very long, widely spaced joints to be seen.

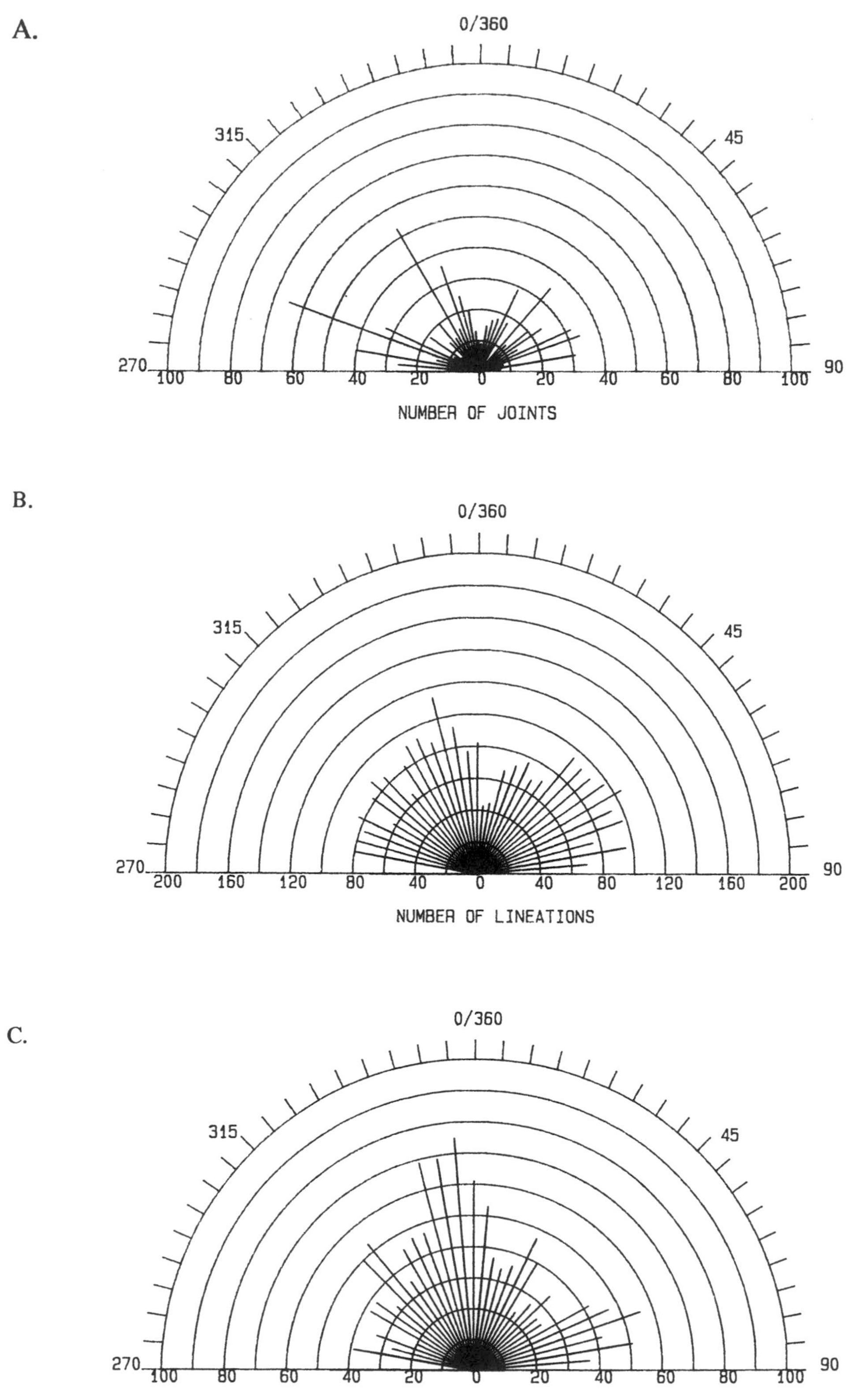

Figure 5. Rose diagrams showing: A, primary vertical joint strikes for 819 joints, Dartmoor; B, strikes for 2,916 lineations at 1:24,000 scale, Dartmoor; and C, strikes for 1,530 lineations at 1:50,000 scale, Dartmoor.

TABLE 1. PROMINENT TRENDS FOR JOINTS AND LINEATIONS, DARTMOOR GRANITE

	Ground Data	1:24,000 Scale Air Photo Data	1:50,000 Scale Air Photo Data
Dominant trends	275–295°	330–350°	345–360°
Secondary trends	40–80°	45–80°	60–90°
	330–340°	305–315°	315–320°

Figure 7 shows the log-log plot of mean primary vertical joint and lineation spacings for the three scales of data on Dartmoor. The slope of this line is 0.7 and the R^2 value is 100%. The regression line indicates a power-law distribution, strongly suggesting that lineation and primary vertical joint spacings are scale invariant. This means that the same joints and lineations are not seen at the three different scales, which was assumed, but that the spacing patterns are statistically self-similar. The distributions themselves are statistically significantly different, but then this was expected because of the large differences in scale, as noted above. Admittedly, the two points for the lineation spacings are very close together in log-log space, and both plot far from the point for the vertical joints, but regardless, fractal analysis of these patterns appears justified.

FRACTAL DIMENSIONS

Fractal dimensions (D) were calculated for the one-dimensional joint spacings for both the Dartmoor and the Pioneer granites. D was also determined for the two-dimensional fracture networks on the two lineation overlays for each area. The results are shown in Table 3.

As reported by Ehlen (1993), the mean fractal dimension for the vertical joint sets on Dartmoor is 0.77, and the range is 0.70 to 0.84. In the Pioneer granites, mean D is 0.55, and the range is 0.35 to 0.68. Vertical joints are much more widely spaced on Dartmoor than in the Pioneer granite—2.6 m and 0.9 m, respectively—so, at least in this case, wider joint spacing is associated with higher fractal dimension. The higher D for the Dartmoor granite may also reflect relative differences in joint spacing. The vertical joints on Dartmoor are more widely spaced than those in the Pioneer granite. Furthermore, the vertical joints on Dartmoor are much more widely spaced than either inclined or horizontal joints, whereas the vertical joints in the Pioneer granite are more closely spaced than either inclined or horizontal joints.

The fractal dimensions for the lineations in the Dartmoor granite are the same at both image scales, and they are higher than D for the lineations in the Pioneer granite (Table 3). If a lower fractal dimension corresponds to more clustered data (Velde et al., 1991; Snow, 1992), one would expect the lineation patterns to be more clustered in the Pioneer batholith than on Dartmoor, but this is not the case. In fact, the Dartmoor lineations at both scales appear more clustered—and more dense—than the Pioneer Mountains lineations. This apparent discrepancy may be due to glaciation in the Pioneer Mountains. Slopes are much steeper in the Pioneer Mountains than they are on Dartmoor, making it much more difficult to see lineations on valley sides in the photos, and the drainage network, which typically provides most lineation information, has been significantly altered by deposition of glacial materials, making joint control of the drainage pattern much less apparent.

Because drainage patterns in granitic rocks are typically fracture controlled, fractal dimensions were determined for the drainage patterns on the 1:50,000 scale Dartmoor mosaic and for both mosaics of the Pioneer Mountains (Table 3). The area over which the box-counting was done in the Pioneer Mountains, however, includes granodiorites, quartz diorites, and tonalites as well as granite, so these dimensions may not be directly comparable to the fractal dimension for the Dartmoor drainage pattern, which includes only granite. The mean D for the Pioneer drainage patterns, however, is 1.65, which compares favorably to the dimension of 1.70 for the Dartmoor granites. The lower fractal dimensions for the Pioneer Mountains may in part result from the Pleistocene valley glaciation, as noted above. Regardless, fractal dimensions for the drainage patterns for both areas are higher than for the lineations, indicating greater regularity in the former.

The fractal dimensions for the drainage patterns are greater than those for the lineations in each area but are lower than mean D for the outcrop data on Dartmoor and are higher than mean D for the outcrop data in the Pioneer granites. The higher dimensions for the drainage patterns relative to the lineation patterns were expected because there is no joint control in the first order tributaries. One would also expect the dimensions for the lineations to be lower than those for the outcrop data, as they are for the Dartmoor data, because the lineations are comparable to the less dense primary joints whereas the field data comprise secondary joints. The dimensions for lineations in the Pioneer granites, however, are higher than those for the joints. The close similarity between D for the 1:50,000 scale drainage overlay for Dartmoor (1.69) and the slope of the line on Figure 7 (0.70) suggests that a fractal dimension determined from a drainage overlay may be a better estimation of the true fractal dimension of the fracture pattern than dimensions determined from either line survey or lineation data.

CONCLUSIONS

Analysis of fracture patterns in the Dartmoor granite shows that there are important similarities between one-dimensional joint spacings measured in the field and two-dimensional lin-

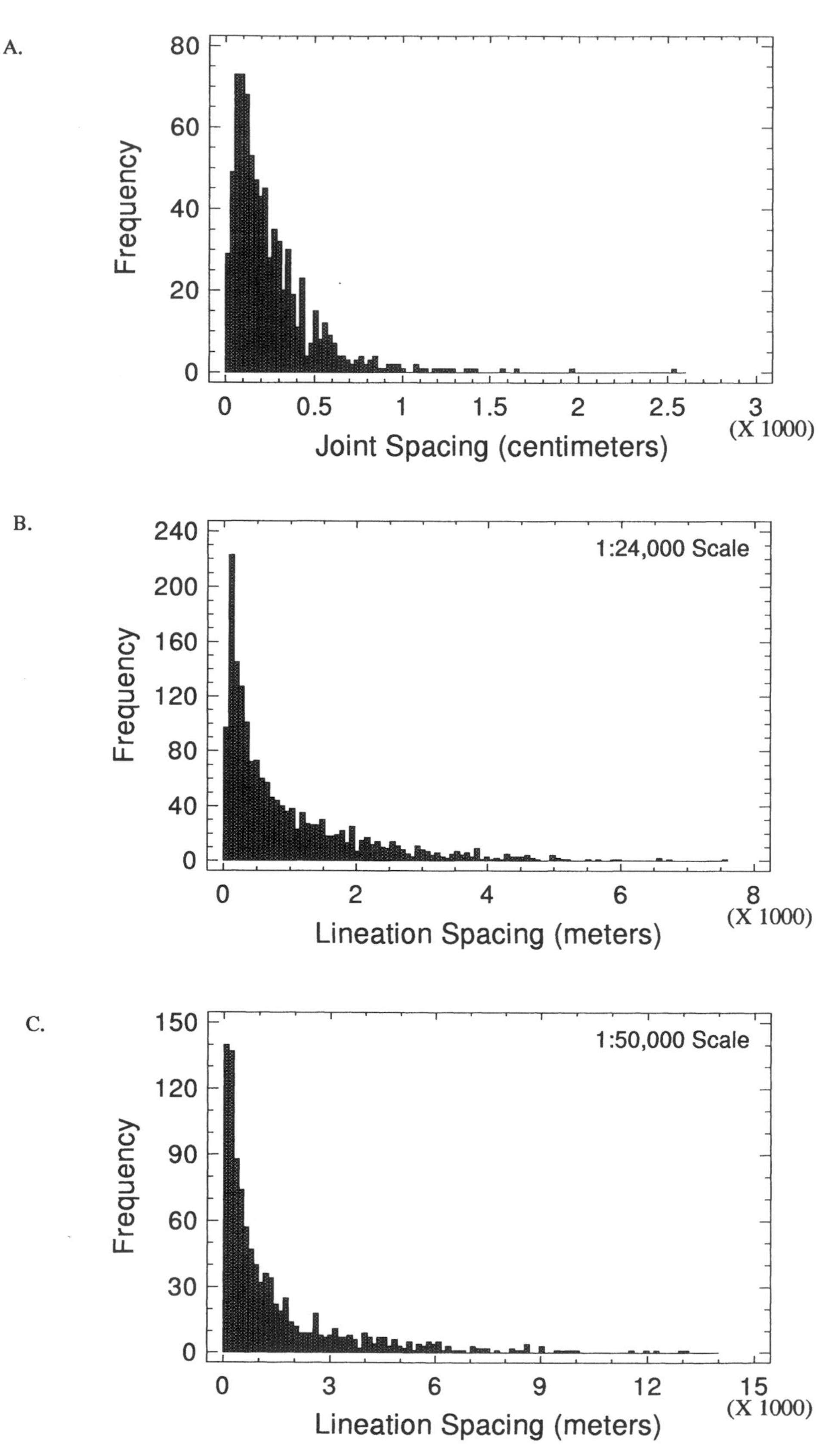

Figure 6. Frequency histograms showing A, primary vertical joint spacings for Dartmoor (819 joints); B, 1:24,000-scale lineation spacings for Dartmoor (1,698 lineations); and C, 1:50,000-scale lineation spacings for Dartmoor (999 lineations).

TABLE 2. SUMMARY STATISTICS FOR JOINT AND LINEATION SPACINGS, DARTMOOR GRANITE

Data Set	Number	Mean Spacing	Modal Spacing	Range	Class Size
Joints	800	260 cm	75–100 cm	4–2,500 cm	25 cm
1:24,000 scale lineations	1700	961 m	100–125 m	20–7,600 m	75 m
1:50,000 scale lineations	1000	1487 m	100 m	7–13,000 m	130 m

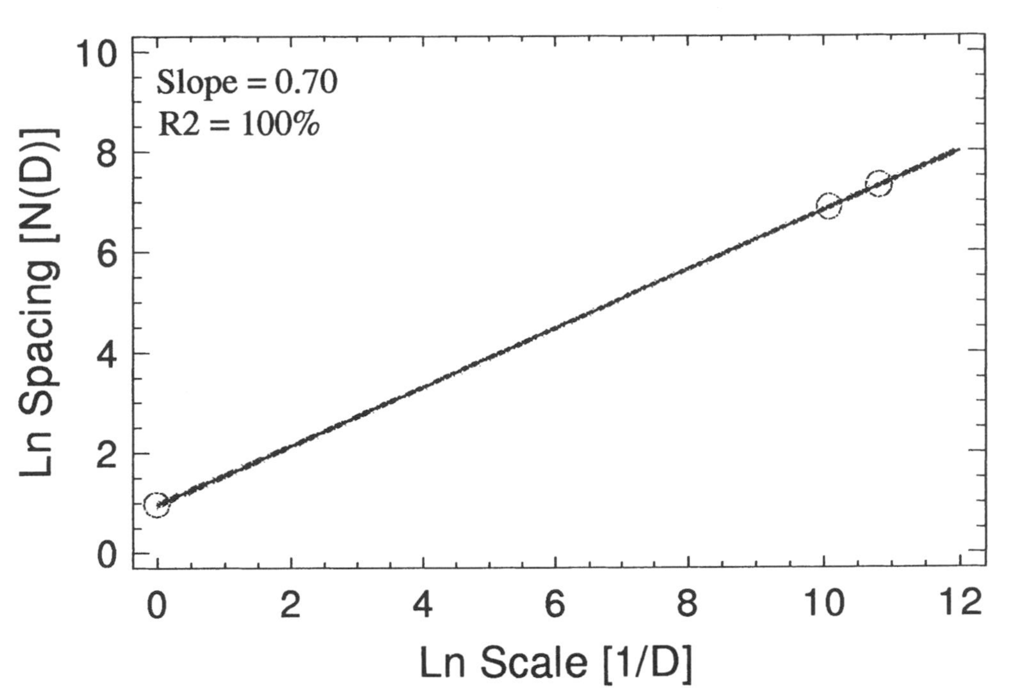

Figure 7. Log-log plot showing power-law distribution of joint and lineation spacings for Dartmoor.

eation patterns delineated on air photos. These results suggest that the lineations are joints and that the joint sets visible in outcrop are also apparent on remotely sensed imagery. The joints distinguishable at the different scales—outcrop, 1:24,000 and 1:50,000—are different individuals, with the longer and more widely spaced members of each set becoming visible as scale decreases. The scale-invariant nature of spacing relationships between the three Dartmoor data sets suggests that the use of fractal analysis to evaluate the possibility of predicting three-dimensional, subsurface fracture patterns from field and remotely sensed data is a reasonable approach.

There are differences between fractal dimensions for the joint and lineation data for Dartmoor, but if one can combine the dimensions determined using different procedures for the different types of data, a fair estimate of fractal dimension for vertical joints (disregarding dimensionality) is 1.7. The slope of the regression line on the log-log plot of mean joint and lineation spacings (0.70) strongly supports this, as does the fractal dimension for the drainage pattern (1.69).

Comparable data sets are not yet available for the Pioneer granite, but if the above assumptions can be accepted, a fair estimate of the fractal dimension for the Pioneer granites is likely to be near 1.5. The differences between the fractal dimensions for the outcrop data (mean: 0.55) and the lineations (mean: 1.21) are, however, much greater than for the Dartmoor granite, and the fractal dimensions for the Pioneer Mountains drainage patterns (mean: 1.65) bear little resemblance to the joint and lineation dimensions. This value of 1.5, however, is roughly in agreement with fractal dimensions reported by Barton et al. (1991) from fracture-trace maps of quartz monzonite and granodiorite, which are 1.50 and 1.52, respectively.

In conclusion, the determination that relationships exist between joint characteristics measured in the field and lineation patterns determined from imagery and the similarity in fractal dimensions for the different types of data in each of the two study areas suggest that fractal geometry is a reasonable approach for predicting and/or modeling three-dimensional, subsurface fracture patterns. Procedures must, however, be developed to include horizontal joints so that the three-dimensional character of fracture patterns is properly addressed. These results indicate that further research is needed to validate the relationships identified in this study prior to beginning the second phase of this effort—

TABLE 3. JOINT AND LINEATION FRACTAL DIMENSIONS, DARTMOOR AND PIONEER MOUNTAINS GRANITES

			D	R^2
Dartmoor				
	Outcrops:	Great Mis Tor	0.78	99.1%
		Haytor	0.81	99.1%
		Haytor	0.70	99.1%
		Lower Dunna Goat	0.84	98.4%
		Haytor	0.76	99.8%
		Emsworthy Rocks	0.71	99.7%
	Lineations:	1:24,000 scale	1.64	99.9%
		1:50,000 scale	1.64	99.4%
	Drainage:	1:50,000 scale	1.69	99.6%
Pioneer Mountains				
	Outcrops:	Browns Lake	0.69	97.2%
		Dinner Station	0.58	95.9%
		Granite Mountain Cirque	0.35	95.0%
		Crescent Lake	0.53	97.4%
		Browns Peak	0.48	97.6%
		Bond Lake	0.68	98.0%
	Lineations:	1:40,000 scale	1.25	99.5%
		1:80,000 scale	1.17	98.9%
	Drainage:	1:40,000 scale	1.74	99.8%
		1:80,000 scale	1.56	99.5%

modeling and prediction of subsurface fracture patterns in denied areas. Once this is accomplished, however, the analogue method should prove fruitful for site characterization in inaccessible areas of interest to the field army.

REFERENCES CITED

Baecher, G. B., 1983, Statistical analysis of rock mass fracturing: Mathematical Geology, v. 15, p. 329–348.

Barton, C. C., and Larsen, E., 1985, Fractal geometry of two-dimensional fracture networks at Yucca Mountain, southwestern Nevada, *in* Stephansson, O., ed., Proceedings of the International Symposium on Fundamentals of Rock Joints, Björkliden: Luleå, Sweden, Centek Publishers, p. 77–84.

Barton, C. C., Samuel, J. K., and Page, W. R., 1988, Fractal scaling of fracture networks, trace lengths, and apertures: Geological Society of America Abstracts with Programs, v. 20, p. A299.

Barton, C. C., LaPointe, P. R., and Malinverno, A., 1991, Fractals and their use in earth sciences: Short course manual, Geological Society of America Annual Meeting, San Diego, California, 312 p.

Belcher, D. J., Ta Liang, Fallon, G. J., Costello, R. B., Hodge, R. J., Ladenheim, H. C., Lueder, D. R., and Mollard, J. D., 1951, A photo analysis key for the determination of ground conditions, v. 3, Igneous and metamorphic rocks: Ithaca, New York, Cornell University School of Engineering, for the Amphibious Branch, Office of Naval Research, U.S. Naval Photographic Interpretation Center, 135 p.

Blyth, F. G. H., 1962, The structure of the north-eastern tract of the Dartmoor granite: Quarterly Journal of the Geological Society of London, v. 118, p. 435–453.

Brammall, A., 1926, The Dartmoor granite: Proceedings of the Geologist's Association, v. 37, p. 251–282.

Brammall, A., and Harwood, H. F., 1923, The Dartmoor granite: Its mineralogy, structure, and petrology: Mineralogical Magazine, v. 20, p. 39–53.

Brammall, A., and Harwood, H. F., 1925, Tourmalinization in the Dartmoor granite: Mineralogical Magazine, v. 20, p. 319–330.

Brammall, A., and Harwood, H. F., 1932, The Dartmoor granites: Their genetic relationships: Quarterly Journal of the Geological Society of London, v. 88, p. 171–237.

de la Beche, H. T., 1839, Report on the geology of Cornwall, Devon, and West Somerset: London, Longmans, p. 156–192, 270–282.

Ehlen, J., 1989, Geomorphic, Petrographic and Structural Relations in the Dartmoor Granite, Southwest England [Ph.D. thesis]: Birmingham, England, University of Birmingham, 2 volumes, 408 p.

Ehlen, J., 1993, Variations in fractal dimension for joint patterns in granitic rocks: Geological Society of America Abstracts with Programs, v. 25, p. A33.

Ehlen, J., and Zen, E-an, 1990, Joint spacings, mineral modes and grain size measurements for selected granitic rocks in the Northern Rockies and in Southwest England: U.S. Geological Survey Open-File Report 90-48, 102 p.

Exley, C. S., and Stone, M., 1964, The granitic rocks of southwest England, *in* Hosking, K. F. G., and Shrimpton, G. J., eds., Present Views of Some Aspects of the Geology of Cornwall and Devon: Penzance, Royal Geological Society of Cornwall, p. 131–184.

Gerrard, A. J., 1978, Tors and granite landforms of Dartmoor and eastern Bodmin Moor: Proceedings of the Ussher Society, v. 4, p. 204–210.

Gerrard, A. J., 1988, Rocks and landforms: London, Unwin Hyman, 319 p.

Hawkes, J. R., and Dangerfield, J., 1978, The Variscan granites of southwest England: A progress report: Proceedings of the Ussher Society, v. 4, p. 158–171.

Hestir, K., Chiles, J-P., Long, J., and Billaux, D., 1987, Three-dimensional modeling of fractures in rock: From data to a regionalized parent-daughter model, *in* Evans, D. D., and Nicholson, T. J., eds., Flow and transport through unsaturated fractured rock: Washington, D.C., American Geophysical Union, p. 133–140.

Huang, Q., and Angelier, J., 1989, Fracture spacing and its relation to bed thickness: Geological Magazine, v. 126, p. 355–362.

Hudson, J. A., and Priest, S. D., 1979, Discontinuities and rock mass geometry: International Journal of Rock Mechanics and Mining Sciences and Geomechanics Abstracts, v. 16, p. 339–362.

Jammallo, J. M., 1984, Use of magnetics to enhance identification of bedrock fracture trace zones for well locations, *in* Proceedings of the National Water Well Association Annual Eastern Regional Groundwater Conference: Worthington, Ohio, National Water Well Association, p. 273–287.

LaPointe, P. R., 1980, Analysis of the spatial variation in rock mass properties through geostatistics, *in* The state of the art in rock mechanics, Proceedings of the 21st U.S. Symposium, Rolla, Missouri: Rolla, Missouri, University of Missouri, p. 570–580.

LaPointe, P. R., 1988, A method to characterize fracture density and connectivity through fractal geometry: International Journal of Rock Mechanics and Mining Science and Geomechanics Abstracts, v. 25, p. 421–429.

LaPointe, P. R., and Hudson, J. A., 1985, Characterization and interpretation of rock mass joint patterns: Boulder, Colorado, Geological Society of America Special Paper 199, 37 p.

Lattman, L. H., and Parizek, R. R., 1964, Relationships between fracture traces and the occurrence of ground water in carbonate rocks: Journal of Hydrology, v. 2, p. 73–91.

Long, J. C. S., Billaux, D., Hestir, K., and Chiles, J-P., 1987, Some geostatistical tools for incorporating spatial structure in fracture network modeling, *in* Proceedings of the 6th International Congress on Rock Mechanics, Montreal, Canada: Boston, A. A. Balkema, p. 171–176.

Miller, S. M., and Borgman, L. E., 1985, Spectral-type simulation of spatially correlated fracture set properties: Mathematical Geology, v. 17, p. 41–52.

Mohammad, M. R., 1987, Jointing and airphoto lineations in Jurassic limestone formations of Al-Adirab area, Tuwayq Mountain, adjacent to Ar-Riyadh, Saudi Arabia, *in* Gardiner, V., ed., International Geomorphology 1986, Part II: Chichester, England, John Wiley and Sons, p. 359–365.

Moore, D. L., and Stewart, M. T., 1980, Geophysical signatures to fracture traces in west-central Florida: Geological Society of America Abstracts with Programs, v. 12, no. 4, p. 202.

Ormerod, G. W., 1869, On some of the results arising from the bedding, joints, and spheroidal structure of the granite on the eastern side of Dartmoor, Devonshire: Quarterly Journal of the Geological Society of London, v. 25, p. 273–280.

Reid, C., Barrow, G., Sherlock, R. L., MacAlister, D. A., Dewey, H., and Bromehead, C. N., 1912, The geology of Dartmoor: London, His Majesty's Stationery Office, Memoirs of the Geological Survey, Sheet 338.

Rinker, J. N., 1974, An application of air photo analysis to a cave location study, *in* Proceedings of the American Society of Photogrammetry 40th Annual Meeting, St. Louis, Missouri: Washington, D.C., American Society for Photogrammetry, p. 281–289.

Rinker, J. N., and Corl, P. A., 1984, Air photo analysis, photo interpretation logic, and feature extraction: Fort Belvoir, Virginia, U.S. Army Engineer Topographic Laboratories, ETL-0329, 375 p.

Robertson, A. MacG., 1970, The interpretation of geological factors for use in slope theory, *in* Rensburg, P. W. J., ed., Planning open pit mines, Proceedings of the Symposium on the Theoretical Background to the Planning of Open Pit Mines with Special Reference to Slope Stability: Johannesburg, Republic of South Africa, The South African Institute of Mining and Metallurgy, p. 55–71.

Segall, P., and Pollard, D. D., 1983, From joints and faults to photo-lineaments, *in* Ramberg, I., and Gabrielson, R. H., eds., Proceedings of the 4th International Conference on Basement Tectonics, Norway: Denver, Basement Tectonics Committee Inc., p. 11–20.

Snee, L. W., 1978, Petrography, K-Ar ages, and field relations of the igneous rocks of part of the Pioneer batholith, southwestern Montana [MS thesis]: Columbus, Ohio, Ohio State University, 110 p.

Snow, R. S., 1992, The Cantor dust model for discontinuity in geomorphic processes: Geomorphology, v. 5, p. 185–194.

Soonwala, N. M., and Dence, M. R., 1981, Geophysics in the Canadian nuclear waste program—A case history, *in* Proceedings of the Society of Exploration Geophysicists Annual International Meeting: Tulsa, Oklahoma, The Society of Exploration Geophysicists, p. 83–98.

Terzaghi, R. D., 1965, Sources of errors in joint surveys: Geotechnique, v. 15, p. 287–304.

Thorp, M. B., 1967a, Joint patterns and the evolution of landforms in the Jarawa granite massif, northern Nigeria, *in* Steel, R. W., and Lawton, R., eds., Liverpool Essays in Geography: London, Longmans, p. 65–83.

Thorp, M. B., 1967b, Closed basins in Younger Granite Massifs, northern Nigeria: Zeitschrift für Geomorphologie, v. 11, p. 459–480.

Thorpe, R., 1979, Characterization of discontinuities in the Stripa granite time-scale heater experiment: Berkeley, California, Lawrence Berkeley Laboratory, LBL-7083, 107 p.

Velde, B., Dubois, J., Touchard, G., and Badri, A., 1990, Fractal analysis of fractures in rock: The Cantor's dust method: Tectonophysics, v. 179, p. 345–352.

Velde, B., Dubois, J., Moore, D., and Touchard, G., 1991, Fractal patterns of fractures in granites: Earth and Planetary Science Letters, v. 104, p. 25–35.

Whittle, R., and McCartney, R. A., 1989, The granites of Southwest England, Section 1, lithology, mineralogy and structure, *in* Parker, R. H., ed., Hot dry rock geothermal energy, Phase 2B Final Report of the Camborne School of Mines Project, Vol. I: Oxford, Pergamon Press, p. 43–117.

Worth, R. H., 1930, Address of the President: The physical geography of Dartmoor: Transactions of the Devonshire Association for the Advancement of Science, Literature and Art, v. 62, p. 49–115.

Zen, E-an, 1988, Bedrock geology of the Vipond Park 15-minute, Stine Mountain 7½-minute, and Maurice Mountain 7½-minute quadrangles, Pioneer Mountains, Beaverhead County, Montana: U.S. Geological Survey Bulletin 1625, 49 p.

Manuscript Accepted by the Society October 29, 1997

Printed in U.S.A.

Geological Society of America
Reviews in Engineering Geology, Volume XIII
1998

Playas and military operations

James T. Neal*
Sandia National Laboratories, Albuquerque, New Mexico 87185-0706

ABSTRACT

Playas are among the most common of desert landforms, with more than 50,000 worldwide and with great variation in size and physical characteristics. They range from a hectare or less to hundreds of square kilometers, and they may support aircraft landings or be totally impassable to land vehicles. Because they possess distinctive physical attributes and are common elements of arid and semiarid landscapes, which occupy nearly 30% of the Earth's land surface, they have had significant effects on military history and operations for centuries. This significance is not apt to diminish.

The single most important determinant of playa-surface conditions is the combination of surface and ground-water hydrology. Playas with ground water that is sufficiently deep to preclude capillary rise favor hard surface crusts. Prolonged capillary or direct ground-water discharge in arid environments readily builds up evaporite mineral assemblages including halite, gypsum, and calcite. Changing hydrologic conditions resulting either naturally or from human activity, for example, ground-water extraction, can result in alteration of relief and surface hardness of playas over both short- and longer-time spans.

Military use by ground forces includes both expeditious and evasive maneuvering of ground vehicles. Air and spaceborne vehicles, including fixed and rotary wing craft and spacecraft, have used numerous playas in their natural state as landing platforms, and many other playas are readily available in remote, unpopulated areas of the globe. Future uses are apt to take advantage of the unique attributes of this arid-region feature—the flattest of all landforms.

INTRODUCTION

At least 50,000 playas occur in the world's arid regions and most extend over 3 km^2 or less. About a thousand exceed 50 km^2, and a few are immense, occupying hundreds of square kilometers. *Playa,* a Spanish word meaning shore or beach, is the term applied in English-speaking countries to the flat and generally barren portion of arid basins of internal surface drainage; when flooded they are termed *playa lakes* (Neal, 1974). Playas around the world are known by a variety of terms, not all of which are equivalent (Table 1). Many playas occupy the sites of ancient lakes that were contemporary or nearly so with the glacial stages of the Pleistocene (Chico, 1968; Morrison, 1968; Snyder et al., 1964). Playas might then be called *dried lakes* (literally), which may explain the colloquial *dry lake* (an oxymoron). Playas also develop in closed depressions without lake histories; the latter tend to be smaller in area and have thinner sediment thickness.

Playas unquestionably are the flattest of all landforms, often sloping a few centimeters per kilometer and having local relief of less than 15 cm. The flatness is attributed to periodic flooding and sheetflow, most typically in arid regions with annual evaporation exceeding precipitation by at least 10:1. Some polar regions are characterized by these conditions, and landforms nearly identical to arid-region playas occur there where clayey sediments are deposited (Krinsley, this volume; Fristrup, 1952, 1953). Human activities in arid lands, and especially military activities, are aided by a geologic and hydrologic understanding of playas and dried lakes. Playas have specific applications in

*Present address: 1911 Crestview Drive, Prescott, Arizona 86301.

Neal, J. T., 1998, Playas and military operations, *in* Underwood, J. R., Jr., and Guth, P. L., eds., Military Geology in War and Peace: Boulder, Colorado, Geological Society of America Reviews in Engineering Geology, v. XIII.

military operations, most notably involving aircraft. The range of playa variations is presented here, together with examples of the use of playas for military and civil purposes.

NATURE OF PLAYA SURFACES

Numerous attempts have been made to classify playas (Rosen, 1994), but from a military viewpoint, the trafficability of the surface often may be all that matters, not its terminology; the discussion that follows recognizes this. The descriptions here follow in general those of Mabbutt (1977).

Hard, dry crusts (Figs. 1–3) develop when admixtures of clay, silt, and sand are dried. Microrelief is very low, except for small mud cracks; one can easily drive an automobile over these surfaces at 100 km/hr, leaving little indentation. Aircraft of almost any size and weight can land and take off from many of these surfaces, but repetitive cycles gradually degrade their use. With hundreds of these surfaces present in the various deserts of the world, a variety of opportunities for their military use becomes evident. When wet these surfaces become very slippery and may be impassable if the moisture penetrates more than about 10 cm. However, some hard clay crusts are sufficiently compact that moisture penetration is extremely shallow.

Rogers Lake (playa), a hard, dry crust at Edwards Air Force Base, California, was the recovery site for early Space Shuttle missions, although confidence and improved navigation has made regular landings on concrete runways routine. During the third test mission the scheduled landing at Edwards was altered when the lakebed flooded. The alternate hard gypsum lakebed at White Sands, New Mexico, the only gypsum surface of its kind in the United States, was used for the recovery on March 30, 1982.

Giant desiccation fissures, up to 1 m wide and 10 m deep, occur preferentially on hard, dry playa crusts, especially where long-term drought has occurred, or where humans have lowered ground-water levels over protracted intervals, as has happened at Edwards Air Force Base. Prior to the establishment of the air base, three-fourths of all wells in Antelope Valley flowed naturally, but now none flow. Intensive drawdown near the southern end of Rogers Lake is sufficient to have induced subsidence of 1 m, and has effectively altered the drainage patterns on the playa (Londquist et al., 1993). This may affect future lakebed operation, as drainage and standing water patterns are significantly altered. The giant fissures frequently are manifested *after* flooding, because the fissures have existed at depth in contraction cracks and appear only when the weight of the water causes the moistened surface crust to collapse into the voids (Neal et al., 1968). Some meter-wide fissures have presented a serious impediment to aircraft landings and to land vehicular traffic. During the X-15 test program, many maintenance personnel were employed to fill and patch the fissures.

Soft, friable surfaces (Fig. 4) accumulate salts through intense surface evaporation from capillary rise of shallow ground water. This may not take long in arid environments, possibly as little as a single season. Microrelief ranges from 5 to 15 cm, often creating a "puffy" appearance that resembles a plowed and disked field. These surfaces can change markedly upon flooding because the surficial salt is released by dissolution, leaving a nearly salt-free surface layer of sediments (Neal and Motts, 1967). The solute in the surface water can then be moved to another location by sheetwash, or the dried residual layer of salts at the surface may be removed by wind, leaving no trace of its former dispersed condition. These surfaces are much less trafficable in their puffy state and may require all-wheel drive vehicles.

TABLE 1. PLAYA TERMINOLOGY*

Locale	General Terms	Clay-silt Playas	Saline Playas
United States	Playa, dry lake, alkali flat	Dry playa, clay playa	Salt flat, salt marsh, salina
Mexico	Laguna, salina	Laguna	Salina
Chile	–	–	Salina (moderate salt), salar (much salt)
Australia	Playa, lake	Clay pan	Salt pan, salina
Russia	Pliazh	Takir	Tsidam
Mongolia	Gobi, nor	Takyr	Tsaka, nor
Iran	Daryacheh	Daqq	Kavir
South Africa	Pan, vloer, mbuga	Clay pan, kalpfannen (lime)	Saltpan, kalahari
North Africa	Sebkha	Garaet, qarat	Sebkha, chott
Arabia	–	Khabra	Mamlahah (salt flat), sabkhah (coastal salt flat)
Jordan	Ghor	Qa	–
Iraq	Hawr	Faydat	Sabkhat
India	Rei	–	–
Pakistan	Hamun	–	–

*After Neal, 1969.

Salt crusts (Figs. 5 and 6) are nearly always associated with near-surface ground water. An exception is the saline crust on the Dasht-I-Kavir in central Iran, which represents an eroded salt canopy of diapiric origin (Jackson et al., 1990). In many instances local kavir-salt surfaces are indistinguishable superficially from many salt crusted playas in Nevada and California, with crusts ranging from 5 to 15 cm and more in thickness (Krinsley, 1970, 1976). Perhaps the best known salt crust is the Bonneville Salt Flats, Utah, where land vehicle speed trials have exceeded 960 km/hr. At the other extreme, the rugged salt crust at the "Devils Golf Course" in Death Valley, California, is virtually impassable to conventional vehicles—its jagged salt spires with relief of 30–50 cm would quickly shred rubber tires. The Bonnevillle crust floods regularly, thus dissolving and smoothing the surface, whereas the Devils Golf Course does not. However, an adjacent area in Death Valley known as the Devils Speedway does flood periodically and is thus leveled more like the Bonneville condition. Another lesser known salt crust at Searles Lake, California, has yielded more than $2 billion in chemical products from its subsurface brines, making it one of the very richest mines anywhere.

Salar (salt) crusts in northern Chile occur at much higher elevations than in North America or Africa and under extreme aridity. Stoertz and Ericksen (1974) identified nine distinctive types of salt crusts there, many of which have no equivalent elsewhere, perhaps owing to the extreme aridity. Some of these crusts are so rugged that they are impassable to ordinary vehicles. However, at Salar de Maracunga and Salar de Pedernales, Stoertz and Ericksen reported smooth crusts that they attributed to seasonal flooding. At Salar de Uyuni in Bolivia the salt crust has sufficient smoothness during the dry season to allow high-speed truck traffic in almost any direction.

Lake Bonneville occupied an area in western Utah and eastern Nevada as large as Lake Michigan until about 17,000 B.P., after which it declined rapidly; Great Salt Lake is now less than 10% of the area of its predecessor (Snyder et al., 1964). Part of this ancient lakebed now contains hard evaporite crusts, for example, Bonneville Salt Flats, whereas other arms of the lake contain dry clay crusts (Wah Wah and Tule Valleys), and still other places (Sevier playa) are soft and not trafficable to most land vehicles. These differences result from variation in ground-water depth and to a lesser extent the periodic flow of surface water.

Figure 1. The hard clay crust in Delamar Valley, Nevada, with 5 km of unobstructed natural runway, was used as an alternate landing site during the X-15 test program in the event that Edwards Air Force Base, California, could not be reached. This typical intermontane basin was formerly occupied by Pleistocene Lake Delamar (Snyder et al., 1964). More than 100 major playas occupy similar basins in the arid West and Southwest, but many are not suitable for landing aircraft.

Figure 2. Hard clay-silt crust at Ibrahimabad Kavir, Iran. These are the smoothest of all playa surfaces and possess superior trafficability.

Figure 3. Emergency landing of B-58 aircraft at Rogers Lake (playa), Edwards Air Force Base, California. Both lives and dollars have been saved repeatedly over the years because of the availability of this landing platform.

Smith (1969), in his study of desert basins on North Africa, summarized past interpretations of the origin of the Qattara depression in Egypt (Fig. 7), maintaining eolian erosion and deflation were dominant processes of excavation, with downwind dune belts representing the material removed, and groundwater levels forming the base level of eolian erosion. Smith saw no evidence in North Africa of Pleistocene lakes comparable to those of the western United States and, in general, saw more differences than similarities. He saw only superficial similarities between the salt crust of the Qattara and the Death Valley salt crust, believing that the only western hemisphere analog might be the much smaller Sechura basin in northern Peru.

Figure 4. Soft, friable surface at Harper Lake, California, where off-runway tests for the C-5A transport aircraft were conducted. This surface was subsequently flooded and converted to a hard crust after saline constituents were dissolved and washed and blown away. Capillary evaporation and efflorescence over time are apt to revert the hard crust to the condition shown here.

HYDROLOGIC ASPECTS

Playas could easily give an impression of hydrologic inactivity, but to the contrary, both surface and ground water are *primary* influences on their development (Neal, 1972; Rosen, 1994). Surface flooding is responsible for perpetuating the extreme flatness and for filling cracks, holes, and erosional scars, even though it may occur only every few years (Neal, 1972).

The depth and circulation of ground water largely determines if capillary discharge will occur through the playa surface. Ground-water depths exceeding about 6–10 m often discharge little or no ground water, but properties of individual aquifers range widely. Hard, dry playa crusts with minimal evaporite concentrations are favored when ground water is deep or where interbasin circulation restricts mineral accumulation by continuing flow out of the basin (Motts, 1965). The composition of the saline constituents often reflects the nature of the host rocks in the source area.

Hydrologic understanding is key to predicting fundamental characteristics of playa surfaces, especially trafficability aspects affecting military operations. Air and satellite photographs in the hands of trained observers are useful in establishing current trafficability indicators, e.g., dryness. Giant desiccation cracks,

Figure 5. The hard salt crust at Sirjan Kavir, Iran, is similar to the Bonneville Salt Flats but less smooth. Salt ridges form from thermal expansion of salt and from the capillary rise of brine; they may be 60 cm high in places and are a clear hazard to traffic of any kind. Krinsley (1970) noted that this playa had multiple surfaces, perhaps reflecting changing ground-water conditions.

Figure 6. Salt crust in Dixie Valley, Nevada, is deceptively similar to some hard crusts but is only 1 cm thick and underlain by soft mud.

which may be hazardous to wheeled vehicles, are caused by specific conditions of sediment dewatering and are usually easy to observe on medium-scale airphotos.

MILITARY USE

Alexander the Great's army, on its return from the Indus Plains, is reported to have watered its animals at Siranda Lake (a playa) in west Pakistan. Few details are known of this incident, and of other military incidents involving playas, until 22 centuries later when General Winfield Scott led the invasion of the capital of Mexico.

Scott's invasion of Mexico City in August and September 1847 is one of the early examples of military encounters in a drying lake bed, that of former Lake Texcoco. The Mexico City basin at 2,400 m elevation is large—51 × 74 km—and semiarid. Although he never had more than 11,000 men of his own, Scott outmaneuvered General Santa Anna's Mexican army of 30,000 through the network of dikes and causeways that crisscrossed the clayey quagmires. Santa Anna had depended on capitalizing on the wetlands and defending what he could, hoping Scott's troops would mire and suffer heavy losses. Scott's engineers and junior officers included Robert E. Lee, P. G. T. Beauregard, Ulysses S. Grant, George B. McClellan, Thomas J. Jackson, and George G. Meade, all of whom would later distinguish themselves in the U.S. Civil War. Following reconnaissance of possible invasion routes, Scott's officers carefully evaluated terrain conditions and concluded the southern route around Lakes Chalco and Xochimilco would reduce losses. Scott was able to avoid the most treacherous areas and outmaneuvered Santa Anna through the myriad canals and causeways. Scott's eventual victory also resulted partly from General Santa Anna's failure to understand the combat limitations of the Mexican Army's defenses (Bauer, 1974).

A detailed evaluation of specific lakebed engagements in this invasion has not been made, but the outcome was decisive and General Scott received exceptional notoriety for his victory and for the treaty signed at Guadalupe Hidalgo that ended the war (Bauer, 1974). Bauer commented that, "nothing like the Mexico City campaign exists in American military history for sheer audacity of concept except for MacArthur's Inchon-Seoul campaign of 1950." The site of the military engagement

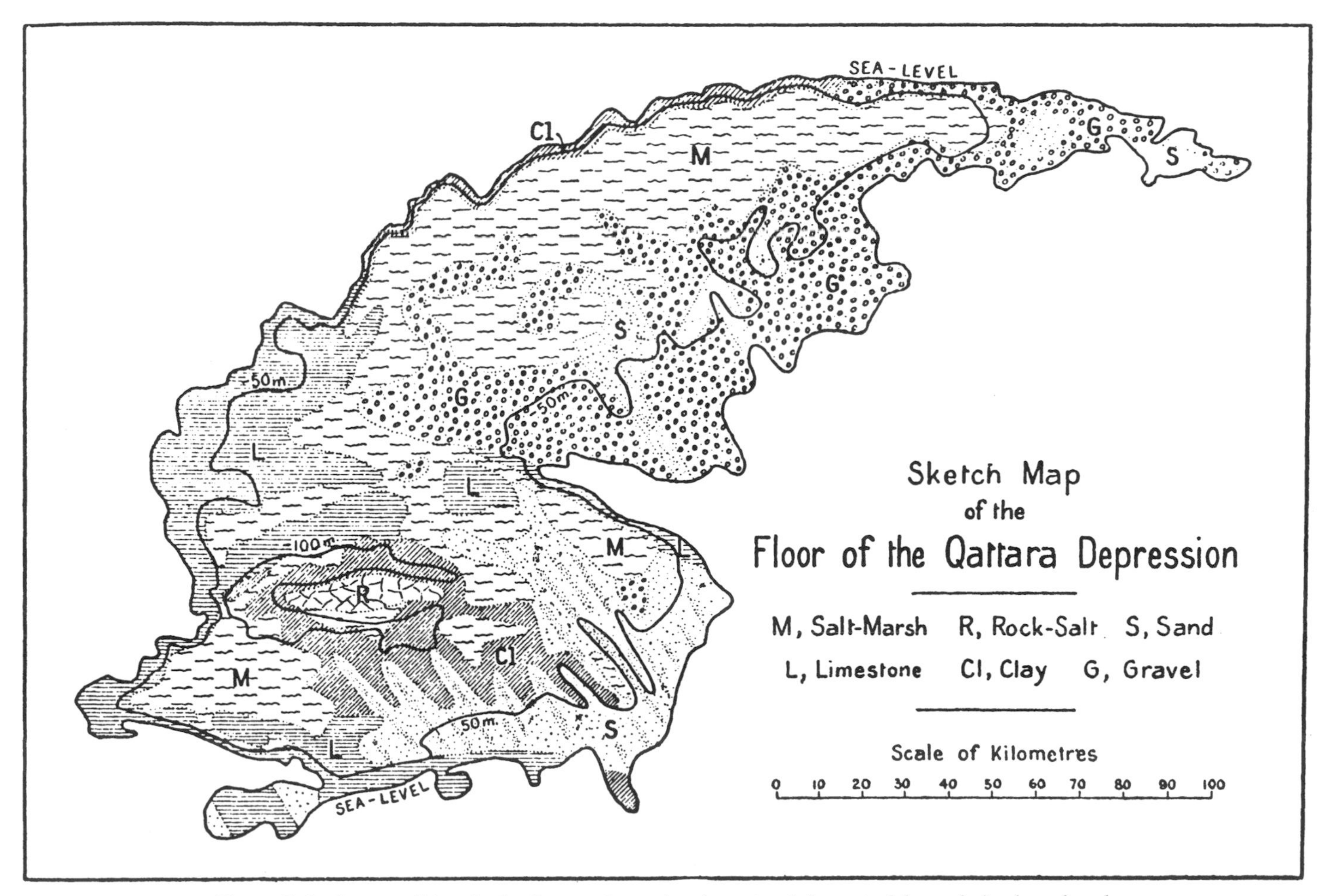

Figure 7. Surface conditions in the Qattara depression, largest and deepest of the undrained, enclosed basins. It is located west of the Nile Valley at latitude 30°N. (After Ball, 1933, fig. 1.)

has since been drained and dewatered, and trafficability conditions are no longer the same—witness the crash of a DC-8 jetliner that occurred before dawn on Christmas morning in 1965. The holiday-bound aircraft crashed on the dried lakebed of the former Lake Texcoco, about six miles from the Mexico City airport. The touchdown point was crisscrossed with earthen farm roads, the aircraft split in two on impact, but no one was killed. Authorities called it a "Christmas miracle," but the explanation appears to be that the landing occurred on the flat and dried lakebed of Lake Texcoco. Just 119 years earlier, Santa Anna's defeat was dominated by different conditions on the same lakebed.

Mexico City in the 20th century has experienced major subsidence of substantial portions of the city, resulting from dewatering and hydrocompaction of the old lake sediments following intensive ground-water withdrawal. The same lakebed sediments amplified seismic waves during the 1985 Michoacán earthquake, whose epicenter was about 400 km west of Mexico City, resulting in the collapse of 300 buildings and 20,000 fatalities within the capital (Rial et al., 1992).

Military training and testing in and around playa basins is a major activity in the arid U.S. Southwest; no less than 10 Department of Defense installations are involved. Some missions date back 70 years and have continually changed with advancing technology. During World War II, General George Patton's armored units trained at Palen Dry Lake in the Mojave Desert, and the training undoubtedly helped during the subsequent Tunisian campaign.

The former commander of the Flight Test Center at Edwards Air Force Base, California, took pride in showing that $4.1 billion had been saved on the Edwards lakebed runways between 1946 and 1966 in emergency landings alone, and regular use of them by X-15, SR-71, XB-70, and Space Shuttle craft has allowed these programs to proceed safely (Manson, 1966). The added savings since 1966 have undoubtedly made this number substantially larger, as the cost of just one B-70 bomber is nearly a billion dollars.

Those of just two generations ago can recall the formidable natural barrier posed by the salt marshes in the Qattara depression, 60 km south of El Alamein in Egypt. Armored units on both sides in World War II were crucially aware of the trafficability limitations imposed by terrain conditions, especially those around playas. Many have speculated on the role that Alamein had in the eventual outcome of the war. Winston Churchill (1950) remarked, "It may almost be said before Alamein we never had a victory; after Alamein we never had a defeat."

The Qattara depression in the western desert of Egypt was prominent in the several battles near El Alamein in World War II, but only indirectly as it was considered impassable to most military vehicles at the time. It served as an effective barrier on the south side of the 60 km corridor between there and the Mediterranean. The armored movements both by Axis and Allied powers were influenced by several other depressions as well—Alinda, Munassib, Muhafid, and Ragil—and General Montgomery took advantage of the terrain, slowing Rommel's Afrika Korps and exposing it to R.A.F. bombing (Lewin, 1968).

The Qattara is the largest of many playa-like depressions in Egypt, extending to 134 m below sea level and occupying more than 18,000 km^2 below sea level. The floors of these depressions, some of which are above sea level, exhibit a diverse assemblage of limestone bedrock, alluvium, sandy areas, dunes, salt crusts, salt marshes, salt lakes, and sometimes cultivated and irrigated ground (Fig. 7). Ball (1933) described the pronounced variations of the Qattara, indicating that about one-third of the bottom is sabkha, a mixture of sand and salt with varied moisture. Some of the sabkha overlies solid ground, which can be traveled over with difficulty in dry seasons, but in much of the area it forms a solid or semisolid crust overlying "salty sludge." Sand, gravel, clay, and limestone cover the rest of the floor.

Ball's map (Fig. 7) shows the salt-marsh is not a single expanse, and it does not cover the deepest part of the depression. Neither does it occur at a constant elevation, even within each continuous tract. The sabkha and salt sludge underlying it result from constant seepage of ground water into the depression, according to Ball. The Qattara has been uninhabited historically, no doubt because of the inhospitable terrain and adverse climate. Mather (1944) told of his difficulty in jeep transit; however, Squyres and Bradley (1964) indicated that oil company personnel crossed the Qattara regularly in all directions. However, military vehicles often travel by convoy; thus if one vehicle becomes immobilized it may sometimes impede all the others.

Rommel's Afrika Korps retreated to Tunisia following defeat at El Alamein and once again encountered severe mobility constraints imposed by the salt marshes of Chott Djerid, a playa superficially resembling some surfaces in the Qattara depression (Coque, 1962; Smith, 1969). The final phase of conflict in North Africa began at the battle of Mareth in March 1943 and proceeded northward to final defeat at Tunis in May. At Mareth, forces on both sides were constrained by the Gulf of Gabes on the east and the nearly impassable salt marshes of Chott Djerid on the west. The funnel between was very constricted and became known as the Gabes Gap. Although the playa was a deterrent to mobility, it probably played only an indirect and minor role in the final outcome (Burne, 1947). Nonetheless, the Qattara depression and Chott Djerid *did* play a role in the desert warfare of North Africa and, ultimately, in the speedy resolution of World War II.

Playa surfaces also have been used for terrorist activities. On September 13, 1970, three civilian jetliners were hijacked from separate European airports by Arab extremists known as the Popular Front for the Liberation of Palestine. Two of the jetliners were diverted to a remote dry clay crust known as Dawson's Field, 40 km northwest of Amman, Jordan, where they eventually were destroyed; the third was flown to Cairo where it too was destroyed. The first two, a Boeing 707 and a Douglas DC-8, had landed on the natural surface without incident; the terrorists knew well what they were doing.

Playas around the world offer similar opportunities for aircraft operations, either on a planned, regular basis or for contingent operations. The latter use must recognize playa vulnerability to changing trafficability conditions resulting from flooding, or intensive, long-term evaporation and efflorescence of evaporite minerals that may degrade the surface hardness (Neal, 1972).

Military interest in playas has increased in the nuclear and rocket age. The world's first atomic detonation, Trinity, occurred along the shoreline of a Pleistocene lake and modern playa near San Antonio, New Mexico (Neal et al., 1983). The bomber crew of the *Enola Gay*, the Hiroshima delivery B-29, trained at Wendover, Utah, adjacent to the Bonneville Salt Flats. Frenchman Flat and Yucca Flat are playas at the Nevada Test Site and were ground zero for several early above-ground atomic tests. Lop Nor, the site of some Chinese nuclear testing, is also a playa. Rocket testing at the White Sands Missile Range in southern New Mexico routinely involves flights over Pleistocene Lakes Otero and Trinity, adjacent playas, and sometimes into targets located on several other lakebeds.

As the end of the 20th century approaches, air-to-ground gunnery ranges are operating at and around at least a dozen playas in California, Nevada, Arizona, Utah, and New Mexico. The presence of the lakebeds at Edwards Air Force Base (then named Muroc) prompted that facility to open in the 1940s; the three major lakebeds still are used daily for a variety of military testing activities.

FUTURE CONSIDERATIONS IN PEACE AND WAR

The Persian Gulf War of 1990 and 1991 served as a reminder that some three-fourths of the world's petroleum resources are concentrated in arid lands, most notably the Middle East. Yergin (1992) reminds us that the thirst for petroleum has been a decisive factor in both motivating and controlling major military conflicts of the 20th century. In addition, conflicts between nations in the oil-rich parts of the world continue, and their irresolution may lead to hostilities by nations beyond their borders. In any future military conflicts in arid lands, playas are apt to be influential—in ways similar to those reviewed in this chapter.

Numerous depressions in North Africa are below sea level and have been studied regarding their power-production potential, wherein sea water would be directed basinward via gravity to drive turbines, then to evaporate in the vast interior basins such as the Qattara. Other locations considered from the same point of view are the Dead Sea; Lake Eyre, Australia; and the Salton Sea, California.

Smaller inland playas are used sporadically in the former Soviet Union and the United States to catch rainwater for animals and agriculture. Means for storing the water for other purposes such as recreation are being applied at many of the small but important 17,000 playas in west Texas. Recharge of the Ogalalla aquifer is an important concern in the High Plains, and some playas may be useful in salvaging surface water for artificial recharge before it is lost to evaporation.

Other uses are legion. The broad flat expanse of former Lake San Agustin, New Mexico, is used for the very large array (VLA) or phased-array radiotelescope; and Goldstone Lake (playa), California, contains a satellite tracking system. Other playas have served as sites for solar-electric panels at several locations. Brine shrimp that hatch in playa lakes after sporadic floodings provide needed protein for numerous East Africans.

New uses for playas appear continually, showing the need to understand their geologic development and sometimes suggesting military implications. Lieutenant George M. Wheeler, U.S. Army, led the geographical expedition 125 years ago to investigate the 100th meridian. His men brought back valuable information on Lake Bonneville and some of the playas in its basin. Their legacy has not been forgotten! History suggests that future military use of playas will carry over into peacetime applications, and vice versa.

CONCLUSIONS

Playas, by virtue of their distinctive trafficability attributes, have played an interesting and, at times, critical role in military history in many different ages, battles, and parts of the world. Because playas are both widespread and numerous, they will very likely see a continued role in military operations. Military schools and military geologists should be well versed in them and know their properties or be able to predict them. Ignorance of playas will give unnecessary advantage to one's adversaries.

ACKNOWLEDGMENTS

I thank Dan Krinsley and Ward Motts for their reviews and helpful suggestions for improving this paper.

REFERENCES CITED

Ball, J., 1933, The Qattara Depression of the Libyan Desert and the possibility of its utilization for power-production: Geographical Journal, v. 82, p. 289–312.

Bauer, K. J., 1974, The Mexican War, 1846–1848: New York, Macmillan Co., 454 p.

Burne, A. H., 1947, The art of war on land: Harrisburg, Pennsylvania, Military Service Publishing Co., 205 p.

Chico, R. J., 1968, Playa, *in* Fairbridge, R. W., ed., Encyclopedia of Geomorphology: New York, Reinhold, p. 865–871.

Churchill, W., 1950, The Second World War; The hinge of fate: Boston, Houghton Mifflin, 1,000 p.

Coque, R., 1962, La Tunisié Présaharienne: Étude Geomorphologique: Paris, Armand Colin, 476 p.

Fristrup, B., 1952, Physical geography of Peary Land I: Meteorological observations for Jørgen Brønlands Fjøord: Meddelelser om Grønland 127, p. 1–143.

Fristrup, B., 1953, High arctic deserts: International Geological Congress, Algiers, 1952, v. 7, p. 91–99.

Jackson, M. P. A., Cornelius, R. R., Craig, C. H., Gansser, A., Stöcklin, J., and Talbot, C. J., 1990, Salt diapirs of the Great Kavir, central Iran: Boulder, Colorado, Geological Society of America Memoir 177, 139 p.

Krinsley, D. B., 1970, A geomorphological and paleoclimatological study of the playas of Iran: Washington, D.C., U.S. Geological Survey Interagency

Report IR-Military-1, 2 v, 486 p.

Krinsley, D. B., 1976, Selection of a road alignment through the Great Kavir in Iran, *in* Williams, R. S., Jr., and Carter, W. D., eds., ERTS-1, A new window on our planet: Washington, D.C., U.S. Government Printing Office, U.S. Geological Survey Professional Paper 929, p. 296–302.

Lewin, R., 1968, Rommel as military commander: London, Batsford, 262 p.

Londquist, C. J., Rewis, D. L., Galloway, D. L., and McCaffrey, W. F., 1993, Hydrogeology and land subsidence, Edwards Air Force Base, Antelope, Valley, California, January 1989–December 1991: Sacramento, California, U.S. Geological Survey, Water Investigations Report 93-4114, 74 p.

Mabbutt, J. A., 1977, Desert landforms: Cambridge, Massachusetts, MIT Press, 340 p.

Manson, H. B., 1966, "General Manson discusses burgeoning role of flight testing at Edwards": Andrews Air Force Base, Maryland, AFSC News Review, July 1966, p. 8–9.

Mather, D. C. M., 1944, A journey through the Qattara Depression: Geographical Journal, v. 103, p. 152–160.

Morrison, R. B., 1968, Pluvial lakes, *in* Fairbridge, R. W., ed., Encyclopedia of geomorphology: New York, Reinhold, p. 873–883.

Motts, W. S., 1965, Hydrologic types of playas and closed valleys and some relations of hydrology to playa geology, *in* Neal, J. T., ed., Geology, hydrology, and mineralogy of U.S. playas: Bedford, Massachusetts, Air Force Cambridge Research Laboratories, Environmental Research Papers No. 96, p. 73–104.

Neal, J. T., 1969, Playa variation; *in* McGinnies, W. G., and Goldman, B. J., eds., Arid Lands in Perspective: Tuscon, University of Arizona Press, p. 13–44.

Neal, J. T., 1972, Playa surface conditions as indicators of environment, *in* Reeves, C. C., Jr., ed., Playa lake symposium, International Center for Arid and Semi-arid Lands Studies, Pub. No. 4: Lubbock, Texas, Texas Tech University, p. 107–132.

Neal, J. T., 1974, Playas, pans, and saline flats: Encyclopaedia Britannica, v. 14, p. 552–558.

Neal, J. T., and Motts, W. S., 1967, Recent geomorphic changes in playas of western United States: Journal of Geology, v. 75, no. 5, p. 511–526.

Neal, J. T., Langer, A. M., and Kerr, P. F., 1968, Giant desiccation fissures of Great Basin playas: Geological Society of America Bulletin, v. 79, p. 69–90.

Neal, J. T., Smith, R. E., and Jones, B. F., 1983, Pleistocene Lake Trinity, An evaporite basin in the Northern Jornado del Muerto, New Mexico: Socorro, New Mexico, New Mexico Geological Society Guidebook, 34th Field Conference, September 1983, p. 285–290.

Rial, J. A., Saltzman, N. G., and Ling, H., 1992, Earthquake-induced resonance in sedimentary basins: American Scientist, v. 80, p. 566–578.

Rosen, M. R., 1994, The importance of groundwater in playas: A review of playa classifications and the sedimentology and hydrology of playas, *in* Rosen, M. R., ed., Paleoclimate and basin evolution of playas: Boulder, Colorado, Geological Society of America Special Paper 289, p. 1–18.

Smith, H. T. U., 1969, Photo-interpretation studies of desert basins in North Africa: Bedford, Massachusetts, Air Force Cambridge Laboratories Report AFCRL-68-0590, 80 p.

Snyder, C. T., Hardman, G., and Zdenek, F. F., 1964, Pleistocene Lakes in the Great Basin: Washington, D.C., U.S. Geological Survey, Miscellaneous Geologic Investigations Map I-416, scale 1:1,000,000.

Squyres, C. H., and Bradley, W., 1964, Notes on the western desert of Egypt, *in* Reilly, F. A., Guidebook to the geology and archeology of Egypt: Tripoli, Petroleum Exploration Society of Egypt, p. 99–105.

Stoertz, G. E., and Ericksen, G. E., 1974, Geology of salars in northern Chile: Washington, D.C., U.S. Geological Survey Professional Paper 811, 65 p.

Yergin, D., 1992, The prize: The epic quest for oil, money, and power: New York, Simon and Schuster, 885 p.

Manuscript Accepted by the Society October 29, 1997

Printed in U.S.A.

Geological Society of America
Reviews in Engineering Geology, Volume XIII
1998

Role of geology in assessing vulnerability of underground fortifications to conventional weapons attack

Thomas E. Eastler
Natural Sciences Department, University of Maine, Farmington, Maine 04938
Donald J. Percious
U.S. Geological Survey, Reston, Virginia 22092
Paul R. Fisher
817 Berwyn Drive, Wilmington, North Carolina 28409

ABSTRACT

The military use of subsurface geologic environments dates back at least 5,000 years to Mesopotamia and Egypt, and continues to be a critical element in planning for both tactical and strategic military activities worldwide. In the context of present-day concerns of "proliferation," the concept of geologic barriers and how best to defeat them has taken on new meaning. Characterization of the geology and the engineering properties of materials surrounding and constituting a deeply buried bedrock underground military facility (UGF) is of great military interest. The degree of success of employing conventional munitions against such UGFs will be limited by our ability to understand the matter/energy interactions between penetrating conventional warheads and rock environments. Geotechnical information that can be used strategically to evaluate the vulnerability of UGFs is herein defined as "strategic geologic intelligence" and includes lithologic characterization; intact mechanical, weight/volume, penetrability; and interpreted in situ engineering properties of geologic units proximal to UGFs. Geologic vulnerability of UGFs can be considered primarily a function of three variables: depth, rock-mass strength, and surface-layer penetrability.

To the degree that any bedrock UGF is vulnerable to conventional weapons attack, the availability of appropriate site characterization data significantly increases one's ability to choose optimal weapons and tactics to defeat UGFs. Thus the role of "strategic geologic intelligence" in future war planning cannot be overstated.

Those expert at preparing defenses consider it fundamental to rely on the strengths of such obstacles as mountains, rivers, and foothills. They make it impossible for the enemy to know where to attack. They secretly conceal themselves as under the nine-layered ground.

———Tu Yu, 735–812

INTRODUCTION

Recent recognition of the degree of proliferation of hardened military underground facilities (UGFs) throughout the world, together with reduced inventories of nuclear weapons, has sparked renewed interest in both the weapons-research and military-targeting communities in assessing the vulnerability of these UGFs to conventional weapons attack. Many nations have been and are currently engaged in massive efforts to harden their military facilities. Any country that expends great effort and expense to harden key military and industrial assets by means of placing them underground also presents a threat by being able to conceal and protect vast quantities of military hardware and supplies up to and including weapons of mass destruction. These actions can transform any country into an

Eastler, T. E., Percious, D. J., and Fisher, P. R., 1998, Role of geology in assessing vulnerability of underground fortifications to conventional weapons attack, *in* Underwood, J. R., Jr., and Guth, P. L., eds., Military Geology in War and Peace: Boulder, Colorado, Geological Society of America Reviews in Engineering Geology, v. XIII.

underground fortress that can serve as a secure base for armed aggression.

The increased use of underground space for military activity has prompted a concomitant increase of research designed to broaden knowledge of UGF vulnerability to conventional weapons attack. Vulnerabilities of UGFs to conventional weapons attack will most certainly be much lower and much more site-specific when compared with vulnerabilities from a nuclear weapons attack. The non-nuclear option does not impart the "brute-force" effect on targets that the nuclear option does and therefore must be more precise in its application.

A great deal is known about the general physical and functional characteristics of the underground environment thanks to a long and arduous history of underground mining endeavors worldwide. Notable publications thereof date back at least as far as Agricola's treatise *De Re Metallica* in 1556 (Hoover and Hoover, 1950). Egyptian papyrus accounts of mining activities date back to the 12th Dynasty, 1900 B.C. (Beall, 1973). Unfortunately, not enough is known about specific physical and functional characteristics of modern-day UGFs to state with great assurity the location of specific vulnerabilities or the single-mode point of failure, if any, without direct inspection of the facility. Knowledge of the vulnerability of such UGF structures to conventional weapons attack is therefore incomplete and drives a continuing research effort toward a greater understanding of the strengths and weaknesses of UGFs in general, and bedrock-contained UGFs in particular.

This chapter will discuss: (1) historical and present-day efforts (fortifications) to protect vital military assets, both strategic and tactical, by use of underground space, (2) efforts (siegecraft) to deny the use of underground space, and (3) the role of military geology in achieving success in these efforts.

HISTORICAL PERSPECTIVE

Those who cannot remember the past are doomed to repeat it.
———George Santayana in *The Life of Reason* (1905)

The use of natural underground caves for protection against the elements and the enemy and for the fabrication of the implements of warfare dates back 400,000 years to the first toolmaker, Pithecanthropus (Peking man) at Chou K'ou Tien near Beijing (Piggot, 1961). Tactical underground chambers and passageways, constructed by humans and used primarily to surprise unwanted intruders and for hiding or escaping or both, were introduced in Mesopotamia and Egypt between 3500 and 3000 B.C., about the same time as the beginning of organized warfare, systematic writing, and the use of worked metals for the manufacture of the implements of war (Dupuy and Dupuy, 1986).

By 600 B.C., the art of war had become highly developed and relatively comprehensive, and continuous records of war, weapons, fortifications, and tactics became available. As the art and science of fortifications developed, so did the art and science of siegecraft (offensive tactics). Between 400 and 1 B.C. the catapult, ballista, and battering ram were developed, and underground tunneling techniques were used by invading forces to gain advantage over defending forces (Dupuy and Dupuy, 1986). Because no techniques existed to detect tunnel-digging activities, the defender was at a great disadvantage. If discovered, the only defense against siege tunnels was to divert manpower from other activities to countersink interdicting tunnels. All of this activity was quite remarkable considering that explosives had not yet been developed and all techniques were manual (Dupuy and Dupuy, 1986). Clandestine tunnel construction in the Korean Demilitarized Zone (DMZ) to achieve the same tactical goals, first discovered in the 1970's and possibly continuing even now, demonstrates how the dimensions of underground warfare have remained timeless in their applicability.

Eventually, powerful castles were built on inaccessible rock terrain, thus negating the use of tunneling against such castles until explosives were developed. Siegecraft techniques lagged far behind the art of fortification, and by A.D. 1300 above-ground defense fortifications became nearly impossible to defeat. Shock and missile weapons, which were no match for the exceedingly "hard" stone fortifications, were limited to manual employment or propulsion systems that were based on tension, torsion, or counterpoise. In fact, for some time fire was the most effective means of winning a siege battle as was amply illustrated by Dupuy and Dupuy (1986) in their discussion of the use of "Greek Fire" by the Byzantines during the first Moslem siege of Constantinople in A.D. 717. The use of fire may have important implications to present-day tactics concerned with neutralizing or capturing the underground equivalent to these Byzantine-like impregnable fortresses, the UGFs.

With the advent of explosives—which were used (1) as early as A.D. 1161 in China, (2) in the 13th century as rockets in Chinese wars against the Mongols, and (3) in Europe by the mid-14th century as gunpowder—the impregnability of stationary, above-ground fortifications disappeared and a new era of offensive and defensive tactics dawned. Not only did the growing strength of siege artillery render all existing above-ground fortifications exceedingly vulnerable, but the use of explosives in mining technique, employed after the introduction of gunpowder, also assisted in physically undermining the once impregnable fortifications. With the advent of increasingly accurate and effective artillery, first employed in fixed locations in the mid-14th century and later converted to mobile platforms by the early 15th century, a new type of fortification evolved that had thick, low walls and offered increased protection for defending against such artillery. Gradually sloping terraces, or glacis, were used to deflect incoming munitions—as do the glacis plates of present-day tanks or the deflector slabs over cut-and-cover UGFs—and personnel entrenchments were incorporated into fortification design (Mallory and Ottar, 1973).

As fortress construction techniques and site-selection improved, tunneling techniques became inadequate because of the long tunnel distances required and because of the lack of adequate ventilation for the tunnelers. The art of siegecraft was once again surpassed by that of fortification, and the repetitive cycle of offense-defense had come full circle. Entrenchment became the key tactic, and both offensive and defensive forces relied heavily on perfecting trench warfare.

As a result of improvements in artillery at the beginning of the 19th century, unprotected trenches and fortress-gun positions became easy prey to vertically falling mortar shells. Subsequently, casements were added to both field and permanent fortifications to give overhead protection from vertically falling mortar shells. With the French development of the explosive shell between 1885 and 1890, masonry construction gave way to the use of concrete and metal cupolas to protect fortress artillery (Mallory and Ottar, 1973). Siegecraft, a form of offensive artillery technology, once again was driving the evolution of the art of fortification from casements to cupolas and from mortar to concrete to reinforced concrete to iron to armor plate. By the early 1900s, detached concrete fortresses with revolving cupolas, designed for the relatively small armies and artillery power of the 1800s, proved to be no match for the heavy artillery produced by the industrial revolution. At Antwerp in October 1914, unreinforced concrete forming the heart of the Belgian fortifications could not withstand direct hits from 210-mm black-powder shells, let alone those from the German 420-mm "Big Bertha" howitzers (Benoit, 1973). The 1.5-m-thick concrete was laid on 40 cm of sand and 2 m of stone and was poured in two layers rather than cast as a single piece. Because these Belgian fortresses were designed before the advent of explosive shells, they were not resistant to powerful munitions.

The French fortresses on the France-German border, on the other hand, were constructed of reinforced concrete 2.6 m thick. These were built in the form of a sandwich of concrete, overlying shock-absorbing sand and covered by 5.5 m of earth (Hautecler, 1973). Moreover, unlike the Belgians, the French had tested 200-mm shells on some of their own fortifications and had strengthened their own cupolas as a result (Macksey, 1970). Consequently, the French fortifications were able to withstand repetitive direct hits from the German 420-mm howitzers with virtually no damage. The success of the concrete structures at Verdun would lead to the construction of the ill-fated Maginot line, perhaps one of the best-known historical examples of a hybrid UGF type of construction.

Trench warfare and heavily armored, detached fortresses would be, apart from some amphibious naval operations, the major theme of land warfare until armored mobility (tanks) and three-dimensionality (dirigibles and airplanes) were introduced in a tactical setting early in the 20th century. Once again, aboveground portions of fortifications became vulnerable, particularly to bomb warfare, and once again the art of fortification would have to advance to match the progress of siegecraft. Table 1 lists a generalized chronology in the oscillatory relationship between fortification and siegecraft.

The Maginot line deserves special mention. A lack of fortification did not precipitate its demise, but rather a tactical error allowed the Germans to circumvent the massive defense by breaking through the unprotected Ardennes Forest. As the Germans showed both at the Maginot line and the Belgian Fort Eben Emael, heavily fortified, buried facilities could simply be bypassed or, as at Fort Eben Emael, could be defeated by lowering incendiary devices down vulnerable pathways, gun-turrets, ventilation shafts, or other similar openings (Mallory and Ottar, 1973). Clearly, this represents a tactic similar to the original Byzantine "Greek Fire," suggesting that when fire has been overtly or covertly deployed, it has been an excellent way to defeat UGFs or quasi-UGFs; fire may well be used effectively against UGFs in the future.

TABLE 1. GENERALIZED CHRONOLOGY OF MAJOR EVENTS IN THE OSCILLATORY RELATION BETWEEN SIEGECRAFT AND FORTIFICATION

400,000 B.C.	Pithecanthropus (Peking Man)
3,500 B.C.	Mesopotamia and Egypt
400 B.C.	Ballista, battering ram, underground tunnels
A.D. 717	Moslem siege of Constantinople—Greek Fire
A.D. 1161	Advent of explosives in China
A.D. 1300	Massive above-ground fortifications
14th century	Use of artillery
19th century	Overhead casements, metal cupolas
20th century	Big Bertha, tactical nuclear weapons, UGF

The aerial bombing of London in 1915 proved that traditional geographical boundaries of war could be transcended. As a result, various types of air-raid precautions were factored into the plans for defense by the British, French, Germans, and all other nations preparing for war. Sir John Anderson, chairman of the United Kingdom Special Air Raid Precautions Subcommittee, concluded in his first annual report, 1924, "that in the next war, it may well be that nation whose people can endure aerial bombardment the longer, and with greater stoicism, that will ultimately prove victorious" (O'Brien, 1955). The standoffs in Korea, in Vietnam, and especially in London during World War II serve as excellent examples of how prophetic Sir John Anderson's comments were.

As bombing techniques were developed and improved throughout the early 20th century, bomb-deflecting shelters and "bombproof" concrete gun emplacements were designed by French and British using the "burster slab" and "airspace" theory (Fig. 1).

What was then thought to be an eccentric idea, the pyramid of concrete balls (Fig. 2), was developed during the 1930s. In the 1980s, techniques to defeat incoming munitions used the burster-slab technology and a new concept quite similar to the concrete ball idea, the rock-rubble overlay. Although the concrete balls of the 1930s were designed to absorb the kinetic energy of a direct hit by momentum transfer and dispersal, one or a few direct hits would have scattered the protective layer and left the shelter exposed to subsequent impact. The rock-rubble concept of the 1980s was similar to the concrete ball approach to protection, but differed in several important ways. Large monolithic rock fragments, whose size was based on the expected caliber of incoming rounds, were embedded loosely in a soil matrix above a cut-and-fill UGF to divert the path of an incoming round and to detonate

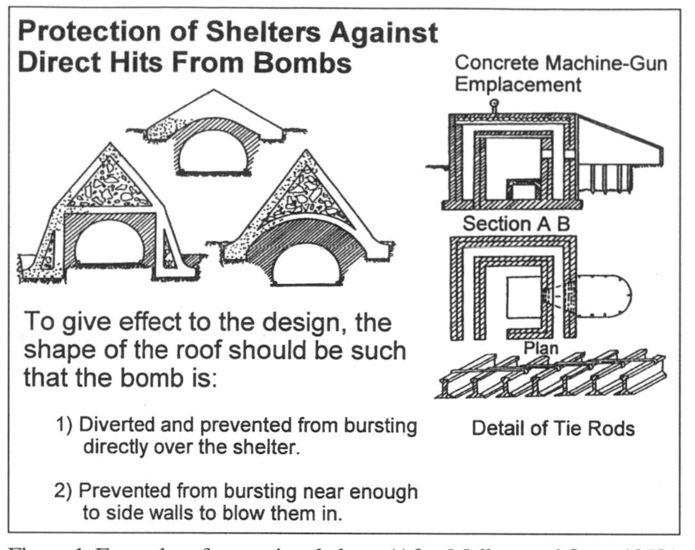

Figure 1. Examples of protective shelters. (After Mallory and Ottar, 1973.)

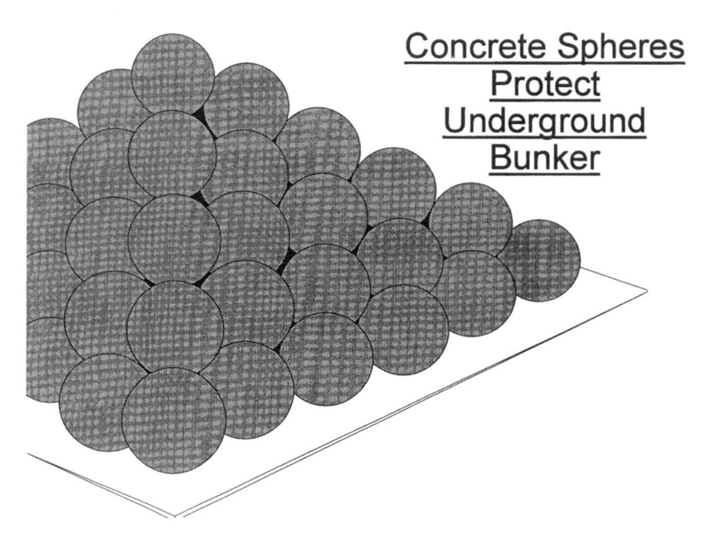

Figure 2. Pyramid of concrete balls placed on top of underground structure to disperse energy from incoming rounds, thus rendering them ineffective in penetrating bunker below. (Modified from Benoit, 1973.)

the round prematurely or cause a weapon-case failure. The weapon thus would be defeated while the rubble mass remained relatively intact owing to the confining action of the soil overlying and surrounding the rock fragments. The same rock-rubble–sand-soil–shelter wall "sandwich" was designed to protect above-ground shelters from horizontal lines of fire (Fig. 3).

As protection against aerial bombardment, the mid-20th century saw the development of dispersed shelters whose varying degrees of effectiveness depended on the type of construction of each shelter. In Britain during World War II, the preferred shelters were those deepest in the ground, the safest being the "tube" or subway stations. Even though the German wartime strategy up to 1943 was all-out attack, the Germans had perhaps the most ambitious bombproof protection program in the world. Adolf Hitler's private bunker, for instance, was capped by nearly 5 m of concrete and 2 m of earth (O'Brien, 1955). But the German underground military excavations were far more ambitious even than Hitler's bunker. By the end of the war more than 108 underground aircraft factories were planned or under construction, of which 78 had reached some stage of production. The German plan of November 1944 called for 4,475,000 m^2 of underground floor space for the aircraft industry alone that, because of closer spacing, was equivalent to nearly 7,400,000 m^2 above ground. The final plan included nearly 10,000,000 m^2 of underground floor space shared by 250 different underground factories.

The Allies had their own underground facilities, but they were trivial in comparison with underground space occupied by Germany's UGFs and with the numerous massive underground caverns used by the Japanese in the Pacific. Some of the free-world military underground fortifications were constructed during World War II. Numerous fortifications, coastal and otherwise, were embedded in solid bedrock tens or even hundreds of meters thick.

As the 20th century wore on, rocket and missile warfare was perfected, and a few of the huge guns of the Allied and Axis forces were reportedly capable of penetrating great thicknesses of concrete and soil. August Coendor's anticoncrete/high-explosive German Rochling round, for example, was credited with penetrating massive thicknesses of soil, burster slabs, and steel-reinforced concrete. The close of World War II brought to an end the impetus to continue intense research into the military applications of geology in developing underground bombproof structures to protect against conventional munitions.

The development of the nuclear age in warfare further curtailed conventional weapons effects research and introduced a new era of nuclear weapons effects research. Some limited testing continued on the penetration of conventional weapons, but most research was devoted to nuclear weapons. The development and testing of these powerful weapons was responsible for the initiation of rigorous research into rock and air blast phenomenology, rock penetration, and a host of other nuclear weapons effects studies. Near-surface, hardened missile silos and command posts built at the core of deep, hard-bedrock mountains became new models of fortifications. But as before, siegecraft also advanced and higher yield, more accurate nuclear weapons were developed, thus negating the fortification effect of even the hardest bedrock silos.

Concurrently, the cold-war conflicts of the 1950s and 1960s demonstrated that the nuclear theater was indeed reserved for all-out war or deterrence of war and did not apply to real-world limited military conflicts. Thus, even with the capability to virtually destroy whole mountains with sufficiently large nuclear weapons, adversaries in reality were limited to applying relatively small and ineffective high-explosive yields against increasingly difficult-to-find underground targets. The political limitations on modern warfare placed restrictions on siegecraft techniques and resulted in significant disadvantages in facing the challenge of modern-day fortification techniques.

Few fortifications are invulnerable to direct nuclear attack, but many fortifications are indeed capable of withstanding continued bombardment even from the most advanced, precision-guided, high-explosive weapons. In the nuclear era, siegecraft has had the upper hand, and great effort has been made to invent new basing modes (i.e., ways to hide offensive missiles, so the enemy will not know where to aim its potent nuclear weapons). One example is the race-track basing mode originally proposed for the MX system. In non-nuclear, conventional warfare, the kind of warfare proposed for the approaching 21st century, it appears that fortification once again has the upper hand. Ways now must be devised to allow siegecraft techniques once again to achieve strategic and tactical balance with deeply buried bedrock UGFs, while at the same time assuring that fortification techniques are adequate. This is a formidable task and one that requires utmost attention as long as nuclear warfare is politically and militarily inappropriate.

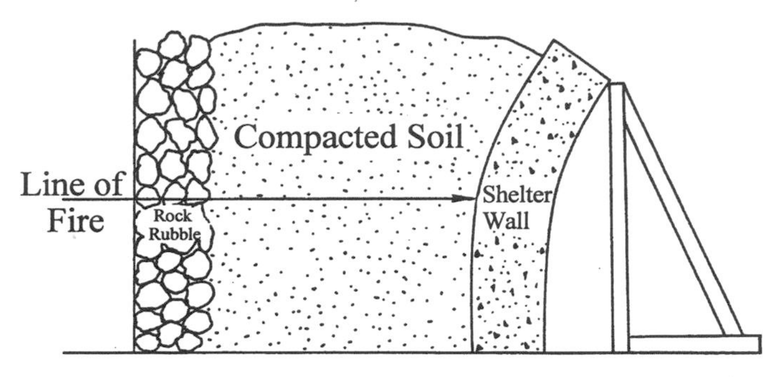

Figure 3. An above-ground bunkered structure designed for protection against penetrating weapons.

GEOLOGY AND UNDERGROUND FACILITIES

Regardless of whether an armed conflict is conducted by conventional or nuclear weapons, the geotechnical characteristics of UGFs are critically important. However, the role that geology—especially material properties—plays will be somewhat different for each weapon strategy. For conventional weapons, the geology of the first few meters and the associated engineering properties are critical, whereas for a nuclear weapons strategy, the general geology and engineering properties down to considerable depths are important. The success of each type of weapon deployment will depend on timely and accurate geotechnical assessment of the rock environment enclosing a UGF.

Historically, UGFs have been built in a wide range of geologic environments, and many have been constructed in hard to very hard rock, making them difficult targets to destroy. As noted above, facilities buried more than several meters generally are invulnerable to conventional, nonpenetrating munitions; targets buried deeper in hard rock may be invulnerable to earth-penetrating weapons (EPWs) fitted either with conventional or with nuclear warheads (Yengst and Deel, 1993). Both military and security advantages result from hardening critical military facilities of all types (command and control, weapons production and storage, and so on) by constructing them underground; the difficulties posed by this class of hardened targets to military analysts are manifest. The proliferation of underground military facilities has resulted in an imposing array of targets buried under diverse geologic conditions having a variety of engineering properties. The success of modern-day siegecraft will depend greatly on timely geotechnical characterization of UGFs.

Geology and strategic intelligence

The abatement of nuclear weapons as an option for warfare has been matched by a proliferation of the use of rock environments to protect underground military facilities (Taubes, 1995). The structural impregnability of deeply buried UGFs that takes advantage of topography and geologic structure requires innovative ways to put these facilities at some risk. Rather than thinking in terms of actually structurally destroying the facility, the concept of "functional kill" or "neutralization" emphasizes the importance of geology in assessing the vulnerability of UGFs to conventional weapons.

A new kind of intelligence, termed "strategic geologic intelligence" in this chapter, has emerged in contrast to the more traditional combat or terrain intelligence. In the past, geologic information has been mainly in the area of combat intelligence where geology was relied upon to evaluate the terrain for troop movements and to determine the availability of construction materials and water supplies (Patrick and Hatheway, 1989). However, the advent of nuclear weapons, the proliferation of UGFs, and the current emphasis on conventional weapons have given rise to the need for information that can be used strategically to evaluate the vulnerability of UGFs that are sited so as to take advantage of topography, rock burial, and geologic structure to increase their survivability. These UGFs range from command and control facilities, to underground aircraft and naval facilities, to those designed to develop and manufacture weapons of mass destruction (WMD).

The so-called "Deutch Report," released by the Office of the Deputy Secretary of Defense (1994), has put into sharp focus the role of geology in the assessment of the survivability of a UGF. This report identified 16 key technological areas involved in countering threats from nations or terrorist groups armed with WMD. Two of the 16 technological areas directly involve military geology: (1) underground-structures detection and characterization; and (2) hard underground-target defeat, including the use of advanced non-nuclear weapons, lethal or nonlethal, capable of holding targets at risk with low collateral effects. Other recommendations in the report, for example, the enhancement of collection and analysis of intelligence and the support of a verifiable comprehensive test ban treaty, also will require geotechnical information for fulfillment, but are beyond the scope of this chapter.

The ability to hold UGFs at some level of risk depends greatly on the availability of site-specific geotechnical information. Knowledge of the site geology and the engineering proper-

TABLE 2. IMPACT OF GEOTECHNICAL SITE-CHARACTERIZATION PARAMETERS ON CONSTRUCTABILITY AND SURVIVABILITY OF UNDERGROUND FACILITIES

Geotechnical Site Characterization Parameters	Geoconstructability/Cost/Schedule									
		Weapons Effects/Hardness/Survivability/Vulnerability								
			Weapons Effects Attenuation							
						Penetrability				
	Excavation Rate	Excavation Method	Opening Size	Opening Depth	Ground Support	Water Control	Endurance Heat Rejection	Shock Attenuation	Penetrator Resistance	Penetrator Lateral Loads
Rock/soil description	x	x	x	x	x	x	x	x	x	x
Formation/layer thickness		x	x		x		x	x	x	
Fracture spacing/orientation	x	x	x	x	x	x			x	x
Water table/saturation	x	x	x	x	x	x	x		x	
Porosity/density	x	x		x	x			x	x	
Unconfined comp. strength	x	x			x				x	
Shear strength	x	x			x				x	
Rock mass quality	x	x	x	x	x				x	
Rock mass strength		x	x	x	x				x	
Constrained modulus								x		
Loading velocity								x		
Seismic velocity		x			x					
Attenuation factor								x		
"S" number									x	

ties of the rocks enclosing UGFs becomes critical in the determination of their degree of vulnerability and the best weapons with which to attack them effectively. The availability of accurate probabilistic geotechnical information is also an essential factor in the research and development of conventional weapons capable of inflicting damage on this class of military facilities; these topics are not treated in this chapter.

Role of geology in assessing vulnerability of UGFs to conventional weapons attack

Several aspects of the rock environment influence the development of UGFs, affect their remote intelligence assessment, and relate to their vulnerability in a military sense. The geology of a given site is a key factor in all three of these considerations. Geologic data are needed to assess the feasibility of construction of an underground facility, its depth, probable configuration, and vulnerability to attack. The influence of geotechnical data on the design, construction, and vulnerability of UGFs is generally manifested in the following: (1) location and layout of the facility, (2) depth, and (3) method of construction. A more complete listing can be found in Table 2.

These factors are interdependent, that is, the depth of the facility has some control on its configuration, particularly when very deep, and also influences the method of construction. The vulnerability, and hence the survivability, of UGFs to attack also are directly related to the facility depth and the strength of the enclosing rock environment. These factors, in turn, depend on the geology of the site, which can vary widely from one location to another; vulnerability and survivability also depend on the range of rock environments available within the borders of a given country.

Location and layout

The location and layout of a facility are linked to the availability of rock suitable for UGF construction, and the kind of rock selected to house the UGF determines the method of construction. For example, limestone and igneous rocks such as granite are ideally suited for construction and are not uncommon, enjoying fairly widespread distribution. These rocks normally range in uniaxial unconfined compressive strength from soft (35 Mpa) to hard (150 Mpa), and during long periods of time no additional support is required to stabilize the structure under construction.

Facility depth

The depth of a UGF is the primary factor affecting its vulnerability to attack. Usually, intelligence on the depth and underground configuration of a facility either is absent, of insufficient detail, or in contradiction with other evidence. In these instances, an analysis of the subsurface geology can place constraints on the depth and possible extent of the underground workings. This is particularly true if information on construction methods, derived from remote sensing, is available. Figure 4 is a simple sketch illustrating one of many possible configurations of a UGF and shows the importance of topography and knowing the "layout" in determining depth and the potential points of attack.

The geological influences on the depth of construction are the local topography, the soil/rock column, and the surface and subsurface hydrology of the site. For example, a tunnel-type excavation takes advantage of the topography to increase the depth of burial and hence the hardness of the facility. Location of a UGF above a deep water table can simplify construction and help in shock-wave attenuation because of the unsaturated media around the facility. A knowledge of regional and local geology can be used to establish and verify the depth, and hence the likely rock environment that is enclosing the UGF.

Method of construction

Various construction techniques are used for UGFs depending mainly on rock strength and the condition of the rock mass. For example, unweathered, sparsely jointed granite will require drill-and-blast techniques or the use of tunnel-boring machines, whereas a softer rock like some limestone usually is easier to excavate using smaller boring and grinding machines such as road headers. On the other hand, highly weathered and fractured granite, depending on the degree of weathering and extent of fractures, can be excavated by hand. Figure 5 depicts the variety of excavation methods as a function of rock strength.

Unique construction achievements include:

Unsupported cavities in salt excavated by humans, with volumes in excess of 17 million cubic meters, have been reported. The two largest known cavities are in Texas—each is large enough to hold the World Trade Center (400 m high) one and a half times (Leith, 1993).

The largest unsupported span in hard rock ever constructed, 61 m, is that of the ice-hockey stadium constructed in Norway for the 1994 Winter Olympics (Barton et al., 1992).

Underground coal and copper mines that are the size of large cities have been constructed in Russia and South Africa (Leith, 1993).

Vulnerability and survivability

The geologic vulnerability of an underground facility can be considered primarily a function of three variables: depth, rock-mass strength, and surface-layer penetrability. The relationship between penetrability, rock-strength, and facility depth is illustrated qualitatively in Figure 6. The upper figure illustrates the interdependence of penetrability, rock strength, and depth of burial. The lower figure is a projection of the upper figure onto the impenetrability and rock-strength plane, and shows how these factors affect the geologic vulnerability of a UGF at any given depth.

Penetrability depends on the compressive strength of the rock, the extent and depth of the weathered rock, and the quality of the rock (i.e., the number of joints, cracks, fissures, bedding planes, and so on). Depending on penetrator configuration (orientation at time of impact) and such flight parameters as velocity and impact angle, the presence of fractures can strongly influence the fate of the projectile. Figure 7 shows the path of a penetrator fired from a stationary platform. The initial impact angle was decreased by the rock joints, clearly illustrating their effect on the ultimate path of the penetrator.

Detailed geologic information and engineering properties for the rock environment above the UGF provide critical input to penetrability calculations; consequently, an estimation of the rock or soil properties of the surface layers is required in geotechnical evaluations. Equations have been developed by Sandia National Laboratories from empirical data for the penetrability of concrete, frozen soil, ice, and rock materials (Young, 1988, 1992).

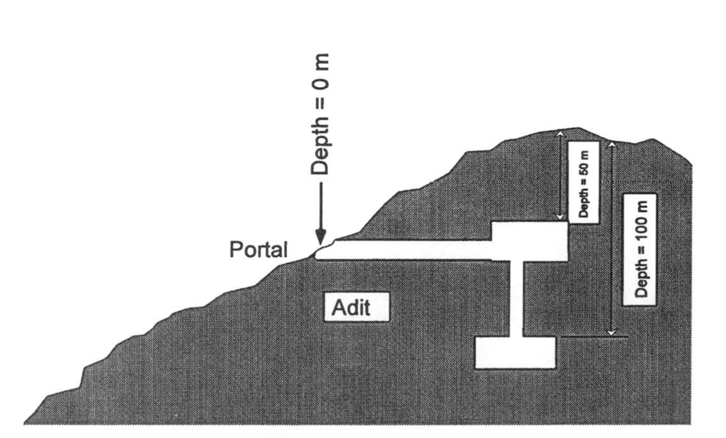

Figure 4. Hypothetical underground facility of varying complexity excavated into the side of a mountain, illustrating the effect of topography on depth of burial.

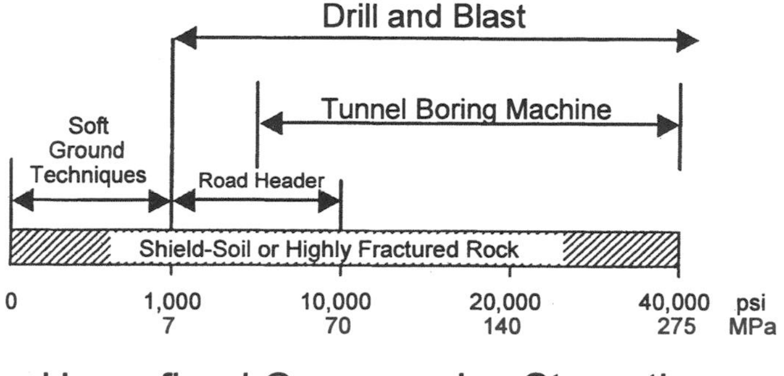

Figure 5. Tunnel-excavation methods as a function of rock strength. The shaded bar signifies the entire spectrum of rock conditions ranging from shield soil to highly fractured rock that is saturated requiring shield methods of excavation. The compressive-strength values do not apply to shield soil and highly fractured rocks in the shaded area.

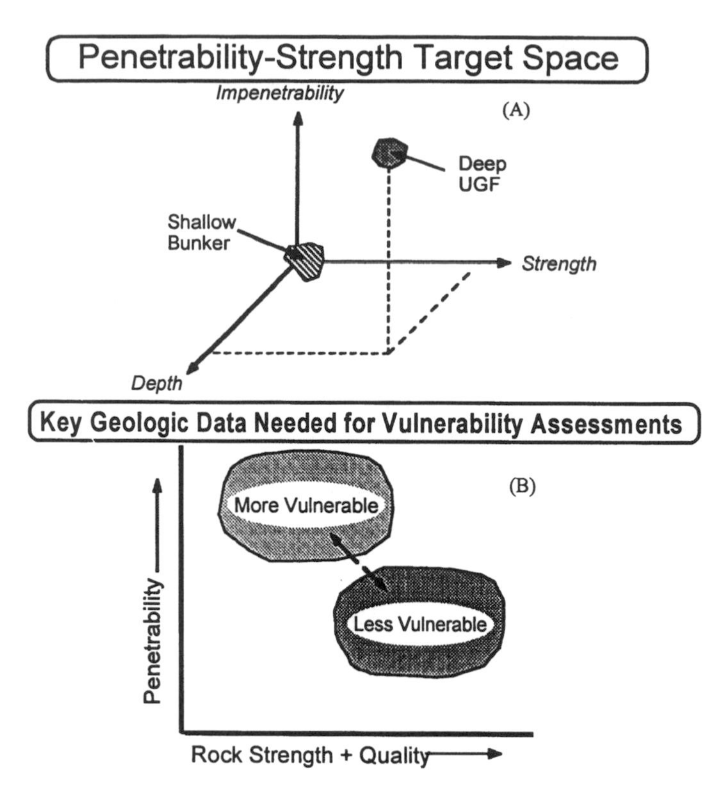

Figure 6. Notional diagram illustrating the penetrability-strength target space of an earth-penetrating weapon. A, Relative position of bunker-type facilities and a tunnel-type UGF. Because penetrability decreases with increasing rock strength, impenetrability is used to label the ordinate. B, Relationship of vulnerability to penetrability and rock strength.

The depth of penetration of an EPW having a certain configuration can be predicted by using established computer models. Using common values for rock strengths and constraining Q or rock-quality values based on the density of fractures and intensity of weathering, the relationship between UGF target strength and penetrability is shown in Figure 8. The zone between the dotted lines defines the "penetrator-target space." All United States war-fighting experience with typical bunker-type targets would occupy the upper left region (penetrable targets of Fig. 8), whereas hard targets in fresh rock—for example, granite—might occupy the lower right region of the zone between the dotted lines. The region of typical weathered granite is located for comparison. The position within the envelope is dictated by the condition of the rock. Penetrability increases upward and to the left along the envelope with increased fracture density and weathering intensity.

The depth to which a projectile penetrates into the earth depends on the weapon configuration, its velocity, impact conditions, and the S-number (Young, 1988). The S-number, in turn, depends mainly on rock strength and rock quality (i.e., whether the rock is fractured and weathered [higher S-number], or whether the rock is massive, nonfractured, and unweathered [lower S-number]). In general, the softer the rock the deeper the penetration and the better the energy coupling, which renders the UGF more vulnerable.

ENGINEERING PROPERTIES FOR ANALYSIS OF STRATEGIC UNDERGROUND FACILITIES

Research and development of conventional weapons and techniques to neutralize UGFs (e.g., to attack tunnel portals, induce landslides, breech tunnels, and other measures) require recognition and understanding of the physics and phenomenology of interactions that occur when the weapon meets the ground. Modeling weapon effects requires the input of a specific group of engineering properties such as those described below. The greatest certainty in the applicability of the geotechnical characterization of a UGF is obtained when the data are site specific; however, when local access is denied, reliance must be placed on regional analysis, foreign geological literature, remote sensing, and engineering geological experience.

During the past decade, the Defense Nuclear Agency (DNA), the U.S. Geological Survey, the U.S. Army Engineer Waterways Experiment Station (WES), and their contractors have developed a methodology for compiling the geological information and geotechnical parameters needed to assess the survivability and vulnerability of UGFs. Because of the nuclear threat during the cold-war period, the development of this methodology was oriented toward the use of nuclear weapons. With the more recent deemphasis of nuclear weapons, the types of geological information and engineering properties data needed to determine the survivability and vulnerability of UGFs under conventional attack are being reassessed, as is the methodology of attack on UGFs.

Site-characterization reports for individual or sets of UGFs include narrative descriptions of the areal geomorphology, stratigraphy and lithology, geologic structure, ground-water conditions, and selected engineering properties for the UGF rock environment. Table 2 shows the impact of selected engineering site-characterization properties on facility construction and survivability/vulnerability. Factors that were considered in the development of this table included: (1) the role of state of stress, rock-mass strength, deformability, anisotropy, and heterogeneity, and the effect of ground-water conditions on the depth, method of construction, and configuration of a facility; (2) the role of rock-mass density, porosity, saturation, rock stiffness, and rock anisotropy and heterogeneity on shock-wave propagation and attenuation; (3) the role of depth of burial, facility opening and configuration, rock-mass strength, deformability, and detailed geologic structure on facility survivability or vulnerability; and (4) the role of rock strength and detailed geologic structure on weapon penetrability.

Consideration of the factors above resulted in the selection of those engineering properties and parameters shown in Table 3 for use in site characterization, usually presented in tabular form to accompany the narrative descriptions. Information on engineering properties may be available either as point values or as ranges of values for a specific geologic unit. Because of the necessity to satisfy the mathematical relationships between certain properties, particularly weight/volume relationships, the values in tables of engineering properties for individual sites are not

Figure 7. Vertical cross-section through the path of a penetrator. Note the flattening of the lowermost part of the penetration path, illustrating the effect of joints in the rock. (From Eastler, 1985.)

necessarily the central value of a given range for a property. A typical tabulation of geotechnical properties for UGFs follows the order shown in Table 3. The major categories in Table 3 are discussed briefly below.

Geology

The description of the geologic environment is furnished in text format and only summarized in the accompanying data set of engineering properties parameters. One geologic parameter that merits discussion is joint spacing. This parameter reflects the spacing between major joints, including horizontal bedding plane joints or fractures. Joint spacing, intended to indicate the degree of heterogeneity of the rock mass, has an overall effect on the mass strength, deformability, permeability, and seismic-wave propagation. It is desirable to obtain the distribution of fracture spacing for the rock mass; where possible the average fracture spacing is presented. With the kind of information usually available for a remote site, it is difficult to estimate with confidence the average fracture spacing for rock masses. As a result, in most instances the fracture spacing is estimated to be greater or less than 0.5 m, a dimension that is one-tenth to one-twelfth of the estimated tunnel size for many deep underground facilities. This tunnel diameter-to-fracture ratio is used in one of the well-known procedures for calculating stresses and strains around a cylindrical tunnel subjected to external loading (Hendron and Aiyer, 1972).

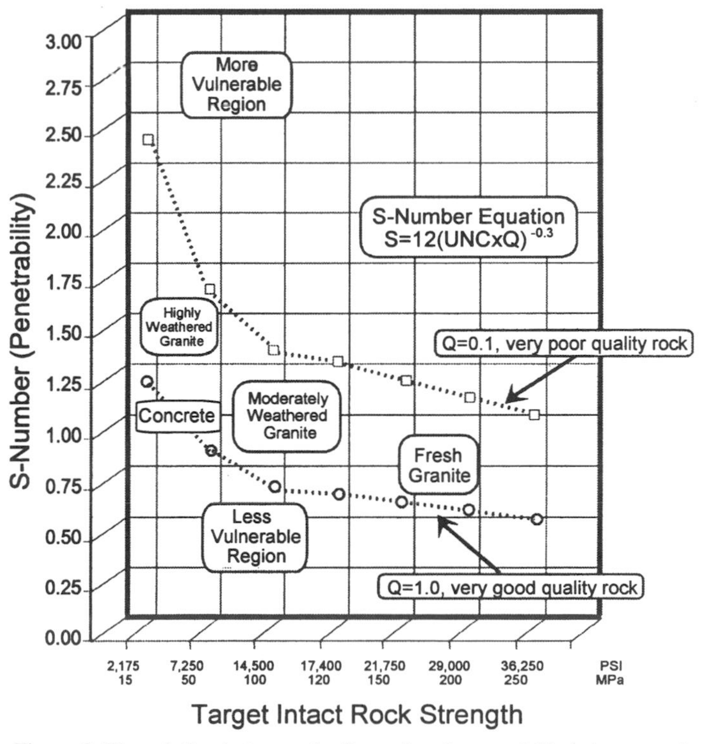

Figure 8. The relation between the S-number (penetrability), intact-rock strength, and UGF vulnerability. Note the influence of rock fracturing and weathering on the vulnerability of a UGF within the envelope. The position of concrete is a reference point.

TABLE 3. SITE-CHARACTERIZATION ENGINEERING-PROPERTIES PARAMETERS

Geology	Intact Mechanical Properties
Formation	Unconfined compressive strength
Description	Friction angle
Thickness	Cohesion
Depth	Sonic velocity
Joint spacing, > or < 0.5 m	Poisson's ratio
Water-table depth	Young's modulus
Weight/volume properties	**Interpreted in situ properties**
Sample porosity	Rock-mass quality
In situ porosity	Rock-mass strength
Grain specific gravity	Modulus of deformation
Sample dry density	Initial constrained modulus
In situ bulk density	Initial loading velocity
In situ water content	Conducitvity
In situ saturation	Peak velocity attenuation exponent
Penetrability properties	
"S" number	
Number variation	
Lateral load	

Weight/volume properties

The weight/volume properties are important to the understanding of strength, deformability, wave propagation, hardness, and constructability. Sample porosity is one of the more commonly available properties, but in situ porosity values are rarely reported for rock masses or soils. As a consequence, judgment is required for increasing the sample porosity to indicate in situ porosity. Grain specific gravity is one of the most frequently available parameters and is considered to be the most accurate. Its accuracy stems from the fact that it is a measurable property intrinsic to the composition of a particular rock type. Whereas bulk, or wet, density of a sample is a frequently available property, in situ bulk density is not. As a consequence, in situ bulk density is almost invariably calculated using in situ porosity, grain specific gravity, and water content. Because of the sensitivity of deformation to small amounts of air-filled voids, approximately 1%, the degree of saturation is important to ground-shock propagation and attenuation. Below the water table, saturation is considered to be 100% with the possible exception of occluded pores in carbonate rocks, which may contain either gas or liquid or both. Where air fills occluded pores in carbonates below the water table, the percentage of air-filled voids will be less than 1% (Weber and Bakker, 1981).

Intact mechanical properties

Measurable mechanical properties of intact rock, such as unconfined compressive strength, the angle of frictional resistance along discontinuities, modulus of elasticity (Young's modulus), and Poisson's ratio are required for calculations of tunnel hardness and stability (Hendron and Aiyer, 1972). When used in such calculations, they are adjusted to account for rock-mass conditions. Unconfined compressive strength, a commonly known property usually used in tunnel-hardness calculations, is an important index parameter for rock strength and deformability. Like porosity, unconfined compressive strength for a specific rock type may vary widely, a variation to be expected because the value depends on the sample condition as well as the intrinsic strength of the rock. It is clear that unconfined compressive strength (material strength) of a laboratory specimen bears little resemblance to the actual in situ bulk strength owing to the likely presence of anisotropy on a larger scale than in laboratory samples. The measured value should be viewed as an upper limit to the real in situ values that are encountered by deforming forces.

For shear strength, a linear Mohr-Coulomb failure envelope with a straight-line relationship between friction angle, cohesion, and unconfined compressive strength is assumed, recognizing that Mohr-Coulomb failure envelopes are curved over any significant normal-stress range. However, the linear concept was commonly accepted for relatively low normal stresses at the time that many of the analytical methods still in use were first developed. Linear envelopes are frequently used that are tangent to the curved envelope for the normal stresses being considered (Stagg and Zienkiewicz, 1968).

Young's modulus is a commonly available parameter and may be used to calculate in situ deformation moduli. Laboratory sonic velocity is used as an aid in estimating field seismic velocities where they are not available. Poisson's ratio, used in the determination of stress and strain around a tunnel subjected to external loading and in calculations of constrained modulus, is infrequently available and must be estimated.

Interpreted in situ properties

When the results from geotechnical explorations are not available, as is common, rock-mass quality is estimated based on the description of the site geologic environment and comparisons with similar geologic environments where rock-mass conditions are known. Rock-mass quality is estimated in the qualitative terms: Very Good Undisturbed (VGU), Very Good Disturbed (VGD), Good Undisturbed (GU), Good Disturbed (GD), Fair Undisturbed (FU), Fair Disturbed (FD), Poor Undisturbed (PU), and Poor Disturbed (PD). These terms are correlatable with popular quantitative Rock-Mass Rating (RMR) and Tunneling Quality Index (Q) systems (Hoek and Brown, 1980). The terms Disturbed and Undisturbed are specifically intended to be used with an empirical rock-mass strength described below. A relationship has been developed between the rock-mass quality descriptors, Rock Quality Designator (RQD), and Seismic Velocity Ratio (Vfield/Vlab), as is shown in Table 4. The square of the velocity ratio is proportional to the ratio of the in situ deformation modulus to the elastic modulus of the intact rock.

Rock-mass quality descriptors allow the assignment of constants for (1) use in an empirical rock-mass strength criterion and

TABLE 4. CORRELATION BETWEEN ROCK-MASS QUALITY, RQD, AND VELOCITY RATIO

Rock-Mass Quality	RQD	$\left(\frac{V_{FIELD}}{V_{LAB}}\right)^2$	$\frac{V_{FIELD}}{V_{LAB}}$
Intact	90	0.9	0.95
Good	75	0.75	0.86
Fair	50	0.5	0.71
Poor	25	0.25	0.5
Very poor	0		

(2) the designation of values of modulus of deformation and initial constrained modulus (Stagg and Zienkiewicz, 1968; Chitty and Blouin, 1990; Deere and Miller, 1966).

Rock-mass strength is important to tunnel hardening and constructability analyses, and because small samples of intact rock are easy to collect and test, considerable information exists on intact-rock behavior. Full-scale tests on fractured-rock masses are extremely difficult because of the problems of preparing and loading samples and because they are very expensive owing to the scale of the operation. As a consequence, test data for large-scale rock-mass behavior will vary between the strength of the intact rock and the strength of the weakest discontinuity in the rock mass. Figure 9 illustrates the variation in strength between an intact granitic rock and the strength along joints in the same rock.

The rock-mass strength used in geologic environment characterization of an underground facility has followed the empirical rock-mass-strength criterion developed by Hoek and Brown in 1980 and updated in 1988 (Hoek and Brown, 1980, 1988). The criterion describes a curved failure envelope and is defined by: (1) the major principal stress; (2) the minor principal stress; (3) the unconfined compressive strength; and (4) two empirically derived dimensionless constants, "*m*" and "*s*," which are analogous to the angle of friction and cohesion, respectively (Hoek et al., 1992). Figures 10 and 11 present the equations associated with this nonlinear failure criterion and provide illustrations of the failure envelopes developed with it. As shown on Figure 11, the failure criterion cannot be applied at normal stresses higher than the brittle-ductile transition. This failure criterion was modified by Hoek et al. (1992), and that modification is still in the process of being integrated into UGF site-characterization efforts.

The modulus of deformation is important to calculations of constructability and hardness of underground openings. The constrained modulus and initial loading velocity are used to predict explosion-produced ground shock. In an elastic material, the constrained modulus relates to the elastic or Young's modulus in the following manner:

$$M = E(1 - v)/(1 + v)(1 - 2v)$$

where: M = Constrained modulus
E = Modulus of elasticity
v = Poisson's ratio of the intact rock or soil skeleton

Samples frequently are not available from the soils and rocks at the sites of many existing deep UGFs. As a consequence, methods were developed for predicting the moduli of deformation, initial constrained moduli, and loading velocities from available intact-rock parameters and assigned rock-quality descriptors (Zelasko, 1990).

For the modulus of deformation, the assigned modulus of elasticity is adjusted by equating the assigned rock-mass quality to an empirically determined ratio of the field modulus of deformation to the laboratory-determined elastic modulus (Stagg and Zienkiewicz, 1968). This methodology is considered applicable to the loosened and drained rock around a tunnel. Figure 12 presents the relationship between rock quality and the ratio of field modulus to seismic modulus; the authors of this figure make no distinction between laboratory modulus and seismic modulus. Using these relationships, Young's modulus multipliers of 0.15 for poor rock, 0.2 for fair rock, and 0.5 for good rock were used to obtain the deformation modulus.

An empirical relationship has been developed by Zelasko (1990) for relating the constrained modulus of drained soils to their porosity. This relationship is quite simple and embodies a considerable practical convenience: Note that the soil constrained modulus (*Ms*) approximately doubles for every 0.05 decrement in porosity (*n*) (i.e., at $n = 0.45$, $Ms = 100$ MPa; at $n = 0.40$, $Ms = 200$ MPa; at $n = 0.35$, $Ms = 400$ MPa; to the final value of $Ms = 51{,}200$ MPa at zero porosity).

Chitty and Blouin (1990) developed a multistep process to estimate the constrained modulus of an in situ rock mass. First, the properties of the intact rock are estimated. Second, an approximate procedure based on rock-quality descriptors is used to esti-

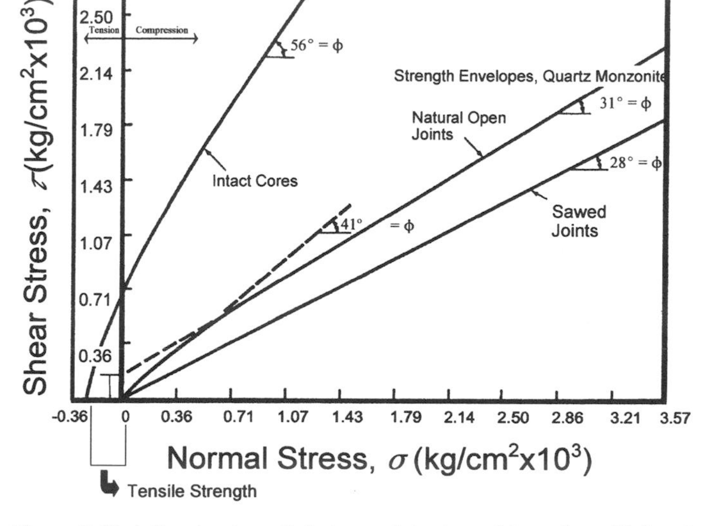

Figure 9. Variation in strength between intact granitic rock and jointed granitic rock. ϕ = angle of internal friction. Note that, as expected, jointed specimens have no tensile strength and lowered shear strength. Unequal increments between numbers along the ordinate and abscissa result from conversion from English to metric units. (After Stagg and Zienkiewicz, 1968.)

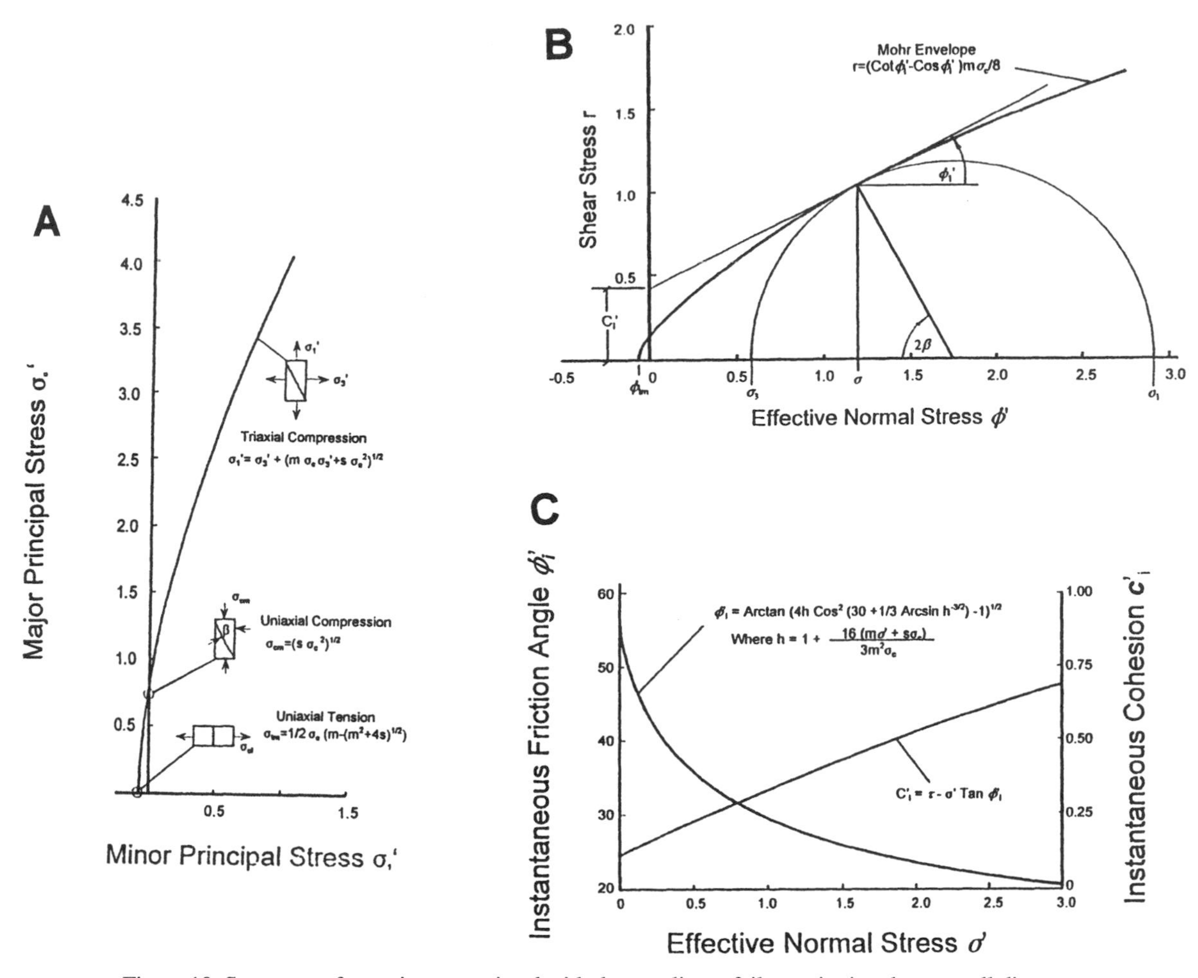

Figure 10. Summary of equations associated with the non-linear failure criterion; because all diagrams are schematic, units of stress and cohesion are not indicated. A, Major principal stress/minor principal stress. B, Shear stress/effective normal stress. C, Instantaneous friction angle/effective normal stress. (After Hoek and Brown, 1980.)

mate the effects of jointing on rock mass properties. Finally, the influence of saturation is considered, where applicable, through use of a two-phase effective-stress model. This multistep process is summarized below.

Two methods for estimating the constrained modulus for dry intact rocks were developed. The first is applicable only to carbonate rocks and is based on extensive testing on a variety of low- to high-porosity carbonate materials. The carbonate model assigns values of constrained modulus based on the rock porosity and has been named the "porosity" model. An extensive testing database has not yet been developed for other rock types. Because many rock types have no significant porosity, a porosity-based model may not always be appropriate. As a consequence, a second, less sophisticated model applicable to a wide variety of rocks was developed based on published results of testing on rocks in the United States. This model relies on the fact that the ratio of elastic modulus to unconfined strength varies over a narrow range and is even more consistent within a given rock type (Deere and Miller, 1966). This approach has been named the "ratio" model.

To use the calculated intact constrained-modulus values to derive values for the rock mass, a set of constrained-modulus reduction factors was developed based on the rock-quality descriptors described above. The constrained-modulus reduction factors were selected from the upper portion of the data envelope shown in Figure 12. The rationale for using the upper portion was that most of the data points shown were from near-surface, conventional civil engineering projects and were obtained, in general, at stress levels significantly lower than those pertinent to ground-shock propagation. Moreover, it was reasoned that the strategic structures of interest are generally sited deeper where geostatic stresses are greater and the joints are less open and less weathered. Table 5 shows the values for the constrained-modulus reduction factor and, for comparison purposes, the values for the deformation-modulus reduction factor.

For saturated and jointed rocks, a method of calculation was developed that accounted for the compressibility of the water in the intact rock and joints.

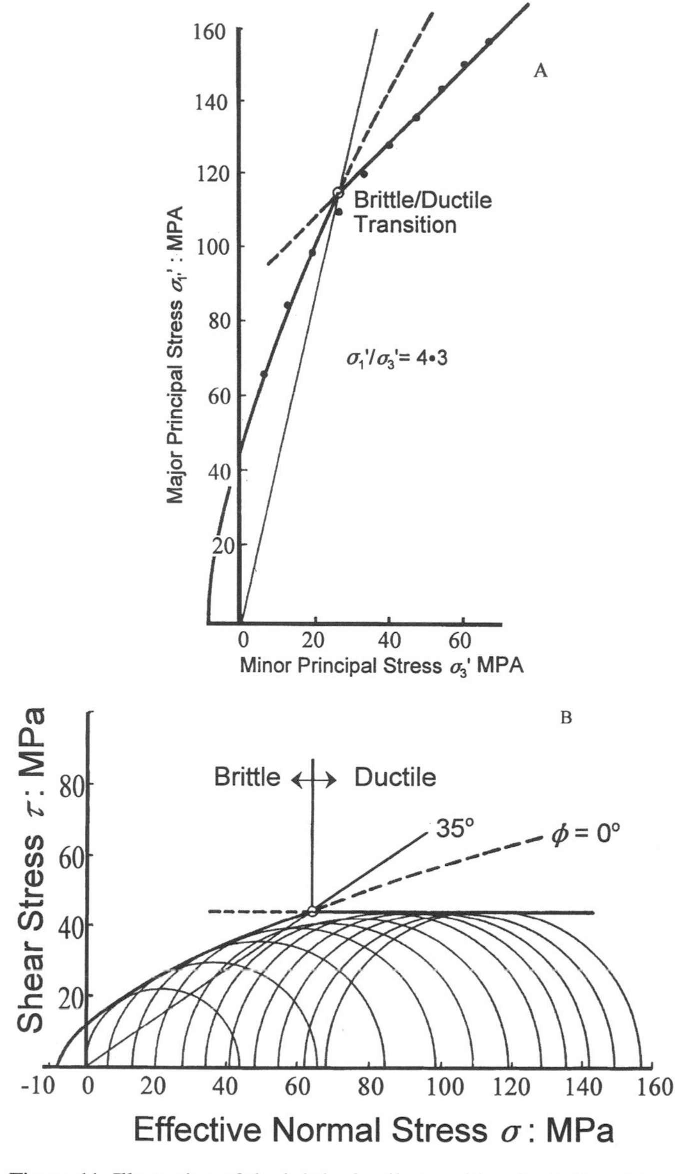

Figure 11. Illustration of the brittle-ductile transition for Indiana Limestone on tests performed by Schwarz. (After Hoek, 1983.)

The loading wave speed is calculated from the initial constrained modulus by use of the standard relationship:

$$Co = [\,Mo/p\,]^{1/2}$$

where: *Mo* = Constrained modulus
p = Bulk density
co = Wave speed

Seismic velocity is important to calculations of ground shock and attenuation. In situ seismic-velocity measurements are common in existing and potential oil- and gas-producing regions, and where available, these measurements characteristically are cited as ranges of values. Under these circumstances, the values presented in an engineering properties table would be selected

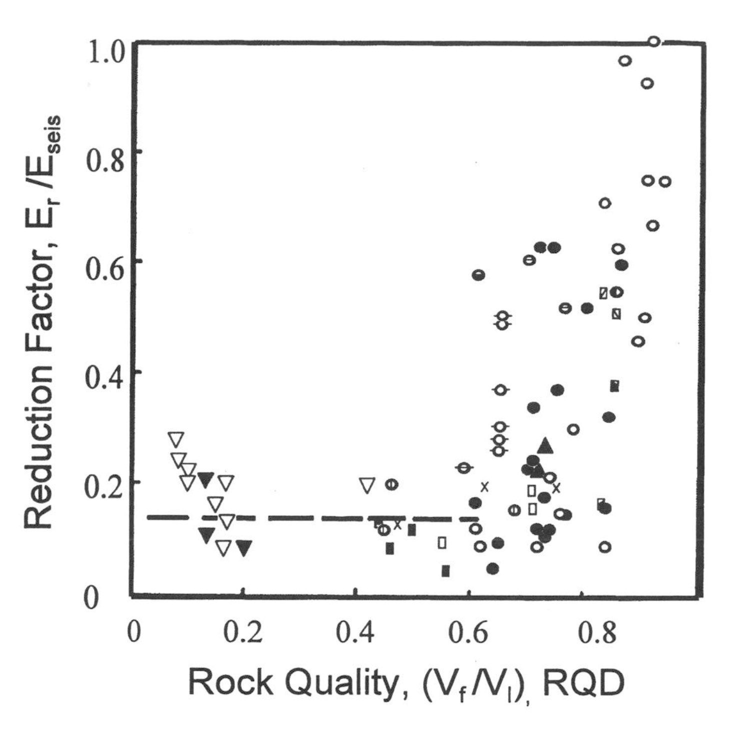

Figure 12. Relationships between rock quality (RQD-Velocity ratio squared) and Reduction Factor, that is, ratio of laboratory modulus to field modulus. Varied symbols in diagram represent different locations of samples as indicated in original text. (Modified from Stagg and Zienkiewicz, 1968.)

from within the range of velocities cited. Where in situ seismic-velocity data are unavailable, the parameters are interpreted from the intact-rock data.

The peak (particle) velocity-attenuation exponent is an important parameter in ground-shock calculations for the determination of the peak ground motions and stresses resulting from a nuclear explosion. Near the explosion source, particle velocity is less influenced by material-property variations than any other measure of the explosion-produced flow field. The attenuation exponent (*n*) would equal 2 for incompressible flow. However, natural materials are not incompressible, especially if they are unsaturated. In saturated materials, the exponent varies between 1.5 and 2, depending on the strength of the material. The criterion for assigning *n* in saturated-rock layers is based on the unconfined compressive strength (*Qu*) and is as follows:

$Qu < 17$ Mpa,	$n = 1.6$
$17 < Qu < 110$MPa,	$n = 1.8$
110 MPa $< Qu$,	$n = 2.0$

For unsaturated surficial materials, *n* was assigned according to the criteria in Table 6.

The electrical conductivity is provided to permit calculation of earth shielding from electromagnetic pulse (EMP). The conductivity value given is applicable at frequencies below 100 hz. Material conductivities have been calculated using given values, where available, for pore-water salinity, formation porosity, and

TABLE 5. REDUCTION FACTORS FOR DEFORMATION AND CONSTRAINED MODULI

Rock Quality	Deformation Modulus Reduction Factor	Constrained Modulus Reduction Factor
Intact	1.00	1.00
Good	0.50	0.75
Fair	0.20	0.40
Poor	0.15	0.20

temperature using Archie's formulae for those nonclayey rocks to which the formulae apply. For rocks with high clay content or where information on pore-fluid salinity was not available, conductivity values have been estimated from measured values made on rocks having similar lithology and porosity.

Detailed geologic information and physical properties for the rock environment above the UGF are critical inputs to calculations of the penetration of projectiles, and an estimation of the rock or soil properties of the surface layers is therefore required in geotechnical evaluations. Equations have been developed from empirical data for the penetrability of concrete, frozen soil, ice, and rock materials by Sandia National Laboratories (Young, 1988, 1992).

The S-number is an empirically derived parameter that represents the material properties in the calculations of weapon penetration; the greater the S-number, the greater the penetration. For sandy soils (<25% clay or silt), S-numbers range from 4 to 7. For silts and clays, S-numbers of 4 to 15 are typical. S-numbers are calculated for rock materials using unconfined compressive strength and rock-quality indices (Young, 1992) as follows:

$$S = 12((UNC)Q)^{-0.3}$$

where: UNC = Rock unconfined compressive strength, psi
Q = Rock quality, as related to weathering, fracturing, and so on. If the rock fracture spacing is <0.5 m, Q may range from 0.3 to 0.5. If the fracture spacing is >0.5 m, Q may range from 0.6 to 0.8. In weathered rock, Q may range from 0.2 to 0.4. The presence of alternating layers tends to lower Q; in rubble, Q may range from 0.1 to 0.3.

CONCLUSIONS

Sophisticated calculations of the vulnerability of UGF to attack by current or future weapons systems require detailed and precise geotechnical data as input. The development of more effective penetrating weapons that take advantage of predicted vulnerabilities will increase the probability of successful functional defeat of UGFs. The degree to which they can be defeated effectively resides both in the knowledge of the key vulnerabilities of such structures and in the understanding of how best to exploit those; geotechnical characterization plays the pivotal role in assessing the vulnerabilities.

As the option of using nuclear weapons becomes increasingly remote, attention may turn toward the development of new "unconventional" non-nuclear weapons for use on UGFs. Some of these weapons may take advantage of the particular physical environment of a given site, for example, the magnetic properties of some rocks, or the construction material of the facility itself. These factors, in addition to the physical features of the rock environment, suggest that more detailed information may be needed, such as the magnetic, thermal, and electrical properties of the rocks. Characterization of UGF environments, therefore, is an evolving task, one that will provide stiff challenges as new ways of defeating such structures evolve.

Finally, the successful destruction of some UGFs, particularly those used for storage and manufacture of nuclear, biological, and chemical weapons, may result in limited to widespread damage to the immediate environment. Advanced assessment of these concerns might require additional information, such as the nature of the local ground-water system, but could help preemptively to mitigate some of the expected collateral effects.

ACKNOWLEDGMENTS

Much of this chapter has resulted from field and laboratory research that has been funded in whole or in part by numerous government agencies and national laboratories during the past 20 years. Individuals too numerous to mention have assisted us in formulating our ideas concerning strategic geologic intelligence. The authors are indebted to Matthew T. Roche of the University of Maine at Farmington for graphical assistance.

TABLE 6. VELOCITY-ATTENUATION EXPONENT FOR UNSATURATED SOILS

Soil Category	Types and Examples	Typical In Situ Density (ρ_o) (kg/m^3)	Typical Seismic Wavespeed (c_o) (m/s)	Typical Initial-Loading Wavespeed (c_o) (m/s)	Velocity-Attenuation Exponent (n)
High density	Alluvium, dense sand	1,900	500 (1,000 cemented)	400	2.1
Medium density	Sand, loam, alluvium	1,700	325	200	2.3
Low density	Loose sand, loam	1,500	200	150	2.5

REFERENCES CITED

Barton, N., Grimstad, E., Aas G., Opsahl, O. A., Bakken, A., Pedersen, L., and Johansen, E. D., 1992, Norwegian method of tunneling: World tunneling and subsurface excavation, v. 5, p. 231–238.

Beall, J. V., 1973, Mining's place and contribution, *in* SME Mining Engineering Handbook, Volume 1: New York, Society of Mining Engineers, p. 1-2–1-13.

Benoit, General, 1973, La fortification permanente pendant la guerre, *in* Mallory, K., and Ottar, A., The architecture of war: New York, Random House, p. 16–17.

Chitty, D. E., and Blouin, S. E., 1990, A methodology for estimating the compressibility of dry and saturated rock and in situ rock masses based on readily obtainable index properties: South Royalton, Vermont, Applied Research Associates, Inc., 28 p.

Deere, D. U., and Miller, R. P., 1966, Engineering classification and index properties for intact rock: University of Illinois, Report No. AFL-TR-65-116: Kirtland Air Force Base, Air Force Weapons Laboratory, New Mexico, 300 p.

Dupuy, R. E., and Dupuy, T. N., 1986, The encyclopedia of military history: New York, Harper and Row, 1524 p.

Eastler, T., 1985, Post-penetration geologic evaluation of Wilson Canyon site, Naval Weapons Center, China Lake, California: Unpublished research note: Albuquerque, New Mexico, Sandia National Laboratory, 8 p.

Hautecler, G., 1973, La rapport du General Leman sur la defense de Liege en Août 1914, *in* Mallory, K., and Ottar, A., The architecture of war: New York, Random House, 307 p.

Hendron, A. J., Jr., and Aiyer, A. K., 1972, Stresses and strains around a cylindrical tunnel in an elasto-plastic material with dilatancy: Report No. TR-10: Omaha, Nebraska, Omaha District, U.S. Army Corps of Engineers, 83 p.

Hoek, E., 1983, Strength of jointed rock masses: Geotechnique, v. 33, p. 187–223.

Hoek, E., and Brown, E. T., 1980, Underground excavations in rock: London, The Institution of Mining and Metallurgy, 527 p.

Hoek, E., and Brown, E. T., 1988, The Hoek-Brown failure criteria, a 1988 update: Proceedings of the 15th Canadian Rock Mechanics Symposium: Toronto, Canada, University of Toronto, p. 31–38.

Hoek, E., Wood, D., and Shaw, S., 1992, A modified Hoek-Brown failure criterion for jointed rock masses, *in* Proceedings of the International Society of Rock Mechanics Symposium on Rock Characterization: Chester, United Kingdom, p. 209–214.

Hoover, H. C., and Hoover, L. H., 1950, Georgius Agricola De Re Metallica: New York, Dover Publications, 638 p.

Leith, W., 1993, Underground construction achievements and decoupling opportunities worldwide, *in* Proceedings, 15th Annual Seismic Research Symposium: Hanscom Air Force Base, Maine, Air Force Office of Seismic Research, 436 p.

Macksey, K., 1970, Fort Douamont: History of the First World War, v. 3, p. 1266.

Mallory, K., and Ottar, A., 1973, The architecture of war: New York, Random House, 307 p.

O'Brien, T., 1955, History of the Second World War civil defense: London, Her Majesty's Stationery Office, p. 19.

Office of Deputy Secretary of Defense, 1994, Report on non-proliferation and counter-proliferation activities and programs: Washington, D.C., U.S. Government Printing Office, 6 p.

Patrick, D. M., and Hatheway, A., 1989, Engineering geology and military operations: An overview with examples of current missions: Association of Engineering Geologists Bulletin, v. XXVI, p. 265–276.

Piggot, S., 1961, The dawn of civilization: London, Thames and Hudson Limited, 403 p.

Santayana, G., 1905, The life of reason, *in* Reason in Common Sense, Volume I: New York, Charles Scribner's Sons, 291 p.

Stagg, K. G., and Zienkiewicz, O. C., editors, 1968, Rock mechanics in engineering practice: New York, John Wiley & Sons, 442 p.

Taubes, G., 1995, The defense initiative of the 1990s: Science, v. 267, February 24, p. 1096–1100.

Tu Yu (735–812), 1983, The art of war, *in* Sun Tzu: New York, Delacarte Press, 197 p.

Weber, K. J., and Bakker, M., 1981, Fracture and vuggy porosity: Paper SPE 10332, presented at the 56th Annual Fall Technical conference of the Society of Petroleum Engineers of the Association of Mining and Metallurgical Engineers, San Antonio, Texas: Richardson, Texas, 11 p.

Yengst, W. C., and Deel, C. C., II, 1993, Hard targets that could not be destroyed by conventional weapons: San Diego, California, Science Applications, Inc., Draft Final Report to Defense Nuclear Agency, Contract No. DNA001-86-C-0024, 226 p.

Young, C. W., 1988, Equations for predicting earth penetration by projectiles: An update: Albuquerque, New Mexico, Sandia National Laboratories, Sandia Report, SAND88-0013, 19 p.

Young, C. W., 1992, Penetration equations update—June 1992: Albuquerque, New Mexico, Sandia National Laboratories, 6 p.

Zelasko, J. S., 1990, Drained constrained modulus-porosity relationship for soils: Vicksburg, Mississippi, U.S. Army Corps of Engineers Waterways Experiment Station/CEWES-SD, unpublished letter report, 4 p.

MANUSCRIPT ACCEPTED BY THE SOCIETY OCTOBER 29, 1997

Printed in U.S.A.

Geological Society of America
Reviews in Engineering Geology, Volume XIII
1998

Location of sites for airstrips in North Greenland

Daniel B. Krinsley
2475 Virginia Avenue, N.W., Washington, D.C. 20037

ABSTRACT

Engineering geological investigations of North Greenland for the purpose of locating sites suitable for airstrips were conducted by earth scientists of the U.S. Geological Survey, Air Force Cambridge Research Laboratories, and the U.S. Army Corps of Engineers in 1956–1960. Potential sites that were approximately 5,000 ft (1,524 m) long with clear approaches, delineated through photogeologic studies and aerial reconnaissance, received on-site examination of their relief, frost features, drainage, depth to permafrost, soil composition, strength, and other engineering properties. The resulting favorable sites that required only a minimal amount of surface modification were selected for the test landings of heavy aircraft such as the C-124 and C-130. Successful landings were made at Bronlunds Fjord in 1957, at Polaris Promontory in 1959, and at Centrum Lake in 1960.

Soils are strong enough to support heavy aircraft at these tested sites except during the spring thaw in June and July. Permafrost conditions at the North Greenland sites do not impose severe restrictions on minor grading and light construction if the surficial materials on and immediately adjacent to the airstrips are not extensively removed.

INTRODUCTION

The site of the future Thule Air Force Base at North Star Bay on Greenland's northwest coast (Fig. 1) was used initially as a small weather station by the U.S. Army during World War II. In 1946, the army built a landing strip for the navy as an operational base for a photographic mission. As a result of the Korean War and the cold war, the defense of the arctic approaches to North America from the immediately adjacent Soviet mainland became increasingly important. On April 27, 1951, an agreement was signed between the United States and the Kingdom of Denmark concerning the common defense of Greenland. The agreement gave the United States the use of a greatly expanded area around Thule (U.S. Air Force, 1972, p. 1). Major construction from 1951 through 1953 resulted in a 10,000-ft- (3,048-m-) long and 150-ft- (46-m-) wide asphalt-surfaced runway with operational apron, hangers, taxiways, and considerable supporting base facilities.

As the air operations increased in North Greenland along with demands for more complete weather information, it became apparent that there was an urgent need for a weather station and an emergency landing strip along the northeast Greenland coast. The Nord area (Fig. 1) was initially studied using aerial photographs by the U.S. Geological Survey in cooperation with the Arctic Section of the U.S. Weather Bureau (Davies, 1961, p. 8). The Nord site was selected in April 1952 by a joint Danish–U.S. Air Force engineering field party, and construction began during the summer of 1953. The 8,000-ft- (2,438-m-) long gravel runway at Nord became operational in 1954 (U.S. Air Force, 1954, p. 21).

In 1955, the Air Force Cambridge Research Center contracted with the Military Geology Branch of the U.S. Geological Survey for a project to locate and investigate areas suitable for airstrips involving little or no construction effort (Davies, 1961, p. 8). It is estimated that about 70,000 aerial photographs from American trimetrogon and Danish oblique photography were examined by the U.S. Geological Survey geologists. In addition, more than 2,000 colored transparencies taken on the ground or from the air provided the photo interpreters with important information (Davies, 1961, p. 9). These photographs and slides were studied together with all of the available scientific and expeditionary reports and maps to produce photogeologic maps of

Krinsley, D. B., 1998, Location of sites for airstrips in North Greenland, *in* Underwood, J. R., Jr., and Guth, P. L., eds., Military Geology in War and Peace: Boulder, Colorado, Geological Society of America Reviews in Engineering Geology, v. XIII.

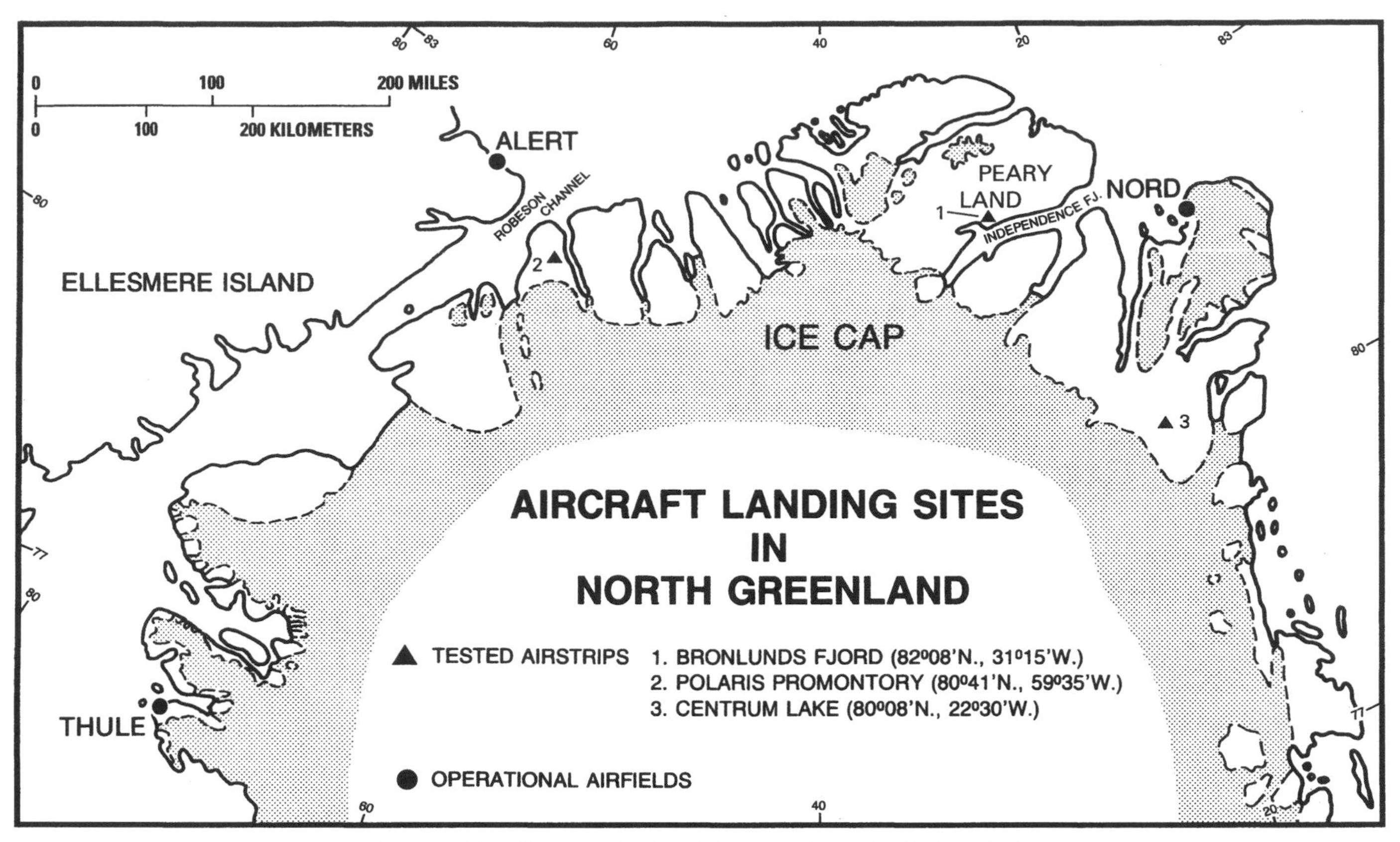

Figure 1. Location map of aircraft landing sites in North Greenland.

16 sites in North Greenland with lengths of 2,000 to 12,750 ft (610–3,886 m) with good approaches and surfaces that appeared to require little or no modification. No small consideration was the accessibility of the site for the emplacement of a small field party to investigate and to test the site. Proximity to water was essential for the approach by ship, amphibious plane, or ski-equipped plane on a frozen surface. The limited availability of suitable aircraft and funding precluded the testing of more than one site per season.

Preliminary field examinations were made at Bronlunds Fjord (Fig. 1) and at Centrum Lake (Fig. 1) in August 1956, and low-altitude aerial reconnaissance was made over the Polaris Promontory site (Fig. 1) on August 31, 1956. The Bronlunds Fjord site, on an inorganic clay, was studied in July–August 1957 and ground-tested by a C-124 on August 16, 1957. The Polaris Promontory site, on a sandy gravel, was studied in August 1958 and in July–August 1959. A C-130 successfully landed at the Polaris Promontory airstrip on August 15, 1959. The site at Centrum Lake, on sand, was studied in May–July 1960. A C-119 landed at the airstrip on July 27, 1960, and a C-130 landed on July 31 and August 1, 1960.

The three tested sites in different soil types confirmed the practicality of locating natural airstrips through photogeologic examination and subsequent field testing employing only a few earth scientists. The field work associated with mapping the surficial deposits and analyzing the soil and testing its strength was shared by U.S. Geological Survey geologists, earth scientists of the Air Force Cambridge Research Laboratories (AFCRL; previously, the Air Force Cambridge Research Center, AFCRC), and earth scientists of the U.S. Army Corps of Engineers.

BRONLUNDS FJORD[1]

Logistics

A four-man reconnaissance party consisting of W. E. Davies and G. E. Stoertz, U.S. Geological Survey; S. M. Needleman, AFCRL; and Lt. Col. R. H. Wilson, Air Force Arctic Advisor, landed on Bronlunds Fjord aboard an SA-16 amphibious aircraft on August 13, 1956, and departed the next day after they determined that the site was a very favorable candidate for detailed soil analyses. Needleman and Stoertz returned to the site with two other geologists on July 29, 1957, via an SA-16 and worked there until they departed aboard the test C-124 on August 16, 1957. The pilot of the C-124 flew into the site aboard the SA-16 several days before his landing on August 16 to inspect the site personally (Stoertz, personal communication, July 6, 1994).

Davies and Krinsley (1961) landed at the site aboard two

[1]This section summarizes the work of Stoertz and Needleman, 1957.

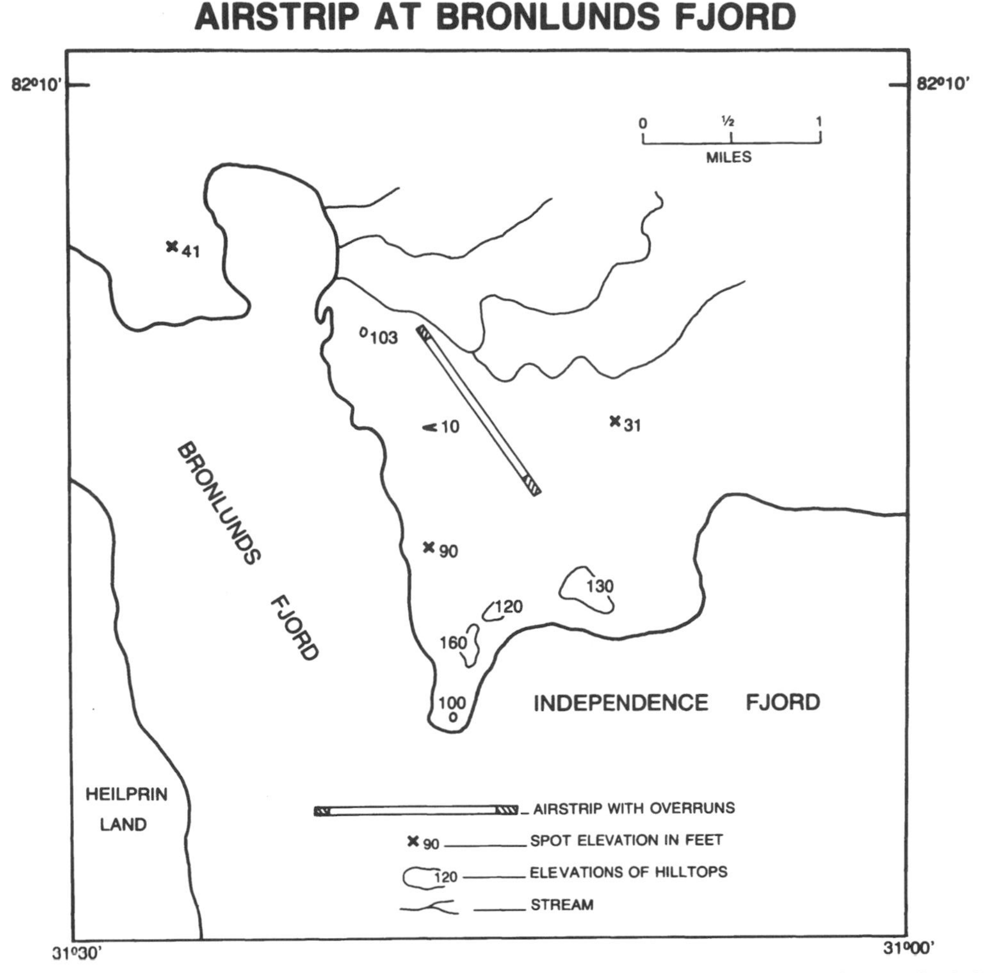

Figure 2. Map of the airstrip at Bronlunds Fjord (after Stoertz and Needleman, 1957, fig. 1). Copied from original map with all measurements in FPS units.

H-34 helicopters on July 7, 1960, and they examined the soil conditions during the next two days. They departed aboard the helicopters on July 10, 1960.

Natural features

The airstrip at lat 82°08′N, long 31°15′W in northeast Greenland (Fig. 1) lies along the air route between Thule and Nord, approximately 607 mi (977 km) from Thule and 131 mi (211 km) from Nord. It is also on the route between Nord and Alert. Seaplanes can use Bronlunds Fjord during most of July and August, and they can taxi near the east shore to within 0.5 mi (0.8 km) of the site (Fig. 2).

The area of the airstrip, less than 10 ft (3 m) above sea level, is part of a coastal lowland from 3 to 5 mi (5 to 8 km) wide at the northeastern junction of the two fjords. The lowland rises northward in a series of low hills and terraces to an altitude of 400 ft (122 m) where it abuts a 1,000-ft (305-m) bedrock escarpment.

The airstrip itself (Fig. 3) occupies a very flat, dry lake plain or marine embayment 0.5 mi (0.8 km) wide and 3 mi (5 km) long

Figure 3. The runway at Bronlunds Fjord, on inorganic clay, is 5,000 ft (1,524 m) long and 200 ft (61 m) wide with flags marking the sides. Emergency overruns at each end increase the total length to 6,200 ft (1,890 m). The view is southeast along the centerline from the extreme northwest end. The dark soil in the foreground is slightly softer than the soils in most of the runway and is part of the overrun area (photograph by Stoertz, August 1957).

and parallel to Bronlunds Fjord. Around the edges of this plain, composed of hard firm clay, undisturbed beds of marine clay are widely exposed.

The landing surface has a slope of less than 0.2% toward the northwest and is very smooth. Hard-packed grass clumps 2 to 4 in (5 to 10 cm) high in the southeastern 500 ft (152 m) of the airstrip were removed; and several gullies as much as 2 in (5 cm) deep and 2 ft (0.6 m) wide, crossing the airstrip diagonally from south to north 1,000 ft (305 m) from the southeast end, were filled and tamped.

The airstrip, oriented N35°W (Fig. 2) is 200 ft (61 m) wide and 5,000 ft (1,524 m) long with a 500-ft (152-m) overrun at the northwest end and a 700-ft (213-m) overrun at the southeast end.

The approach from the northwest, via Bronlunds Fjord, can easily clear a 100-ft (30.5-m) hill 1,500 ft (457 m) west of the airstrip (Fig. 2). A 2,000-ft (610-m) escarpment 4 mi (6.5 km) to the northwest also can be avoided. On the approach over Independence Fjord from the southeast, a 120-ft (37-m) hill is 3,300 ft (1,005 m) southeast of the southeast end of the runway, but it does not obstruct a glide angle of 2% from the southeast.

Climate

As a result of the Danish Peary Land Expedition of August 2, 1948, to August 12, 1950, climatic data were recorded at a station on the south shore of Bronlunds Fjord 6 mi (10 km) west of the airstrip at an elevation of approximately 30 ft (9 m). The data from this two-year record (Fristrup, 1952) were studied by Stoertz, and his analysis is summarized below and in Table 1.

From September until May, the mean monthly temperatures are below freezing. By late September, the ground is frozen sufficiently to support aircraft along the entire 6,200 ft (1,890 m) of runway (Fig. 2). The thaw starts in June and the frost-free period extends from mid-June to late August.

Almost all of the 2.3 in (5.8 cm) of annual precipitation occurs in the summer and fall, and that is recorded on only 20% of the 365 days. The sparse, dry snow is blown free of the exposed airstrip, and deep snow during the winter is unlikely. Low precipitation and a high evaporation rate result in considerable aridity. Probably one-third of the winter snow disappears as a result of evaporation and sublimation before the summer thaw.

Although high winds occur in every month, gales (winds in excess of 31 mph [50 km/hr]) are especially prevalent on an average of 50% of all days from February to July. The highest winds from the west, out of Bronlunds Fjord, are recorded from December to March with a maximum sustained wind of 65 mph (105 km/hr) with gusts up to 74 mph (119 km/hr). The intensity of the wind is reflected in the surface rocks that are polished or etched on their western sides. Wind-blown sand and dust from

TABLE 1. COMPARISON OF CLIMATIC DATA FROM THULE, ALERT, BRONLUNDS FJORD, AND NORD

	Thule*	Alert*	Bronlunds Fjord†	Nord*
Annual absolute maximum temperature (°F/°C)	68 (20)	67 (19)	64 (18)	61 (16)
Annual mean daily maximum temperature (°F/°C)	17 (-8)	4 (-16)	8 (-13)	5 (-15)
Annual mean temperature (°F/°C)	12 (-11)	0.0 (-18)	4 (-16)	2 (-17)
Annual mean daily minimum temperature (°F/°C)	6 (-14)	-4 (-20)	1.0 (-17)	-1.0 (-18)
Annual absolute minimum temperature (°F/°C)	-47 (-44)	-53 (-47)	-45 (-43)	-53 (-47)
Annual mean precipitation (in/cm)	4.5 (11.4)	6.1 (15.5)	2.3 (5.8)	15.1 (38.4)
Annual mean number of days with precipitation	179	196	68	244
Annual mean days with snowfall (≥ than a trace)	156	165	56	231
Annual mean relative humidity (%)	62	73	72	82
Annual maximum wind speed and direction (mph/kmph)	E65 (105)	N74 (119)	W65 (105)	SSW96 (155)
Annual mean number of days with gales	29	29	105	n.d.§
Annual mean number of clear days (cloud cover ≤3/10)	130	143	130	132
Annual mean number of cloudy days (cloud cover ≤8/10)	179	189	120	190

*National Climatic Data Center, 1994.
†Stoertz and Needleman, 1957, Table 1, after Fristrup, 1952.
§No data.

adjacent areas may create a hazard to aircraft although the airstrip itself is clear. A high mean barometric pressure of 1,012 to 1,015 mb or higher prevails at sea level in Bronlunds Fjord.

Only one-third of the days at Bronlunds Fjord are cloudy, but one-half of all days are cloudy at Thule and at Nord. All reports suggest that Bronlunds Fjord enjoys better weather than Nord, making it an attractive alternate airstrip.

During the summer thaw from mid-June to mid-July, streams draining the valleys and gullies to the northeast and southwest occasionally flood the airstrip (observed by Davies and Krinsley, 1961, see below). However, the soil is sufficiently dry to a depth of 6 to 8 in (15–20 cm) with enough shearing strength for landings by the first week in August.

The clay surface may become slippery after rains in August and September. Although this may not appreciably affect the shearing strength of the soil, it may adversely impact aircraft braking, steering, and control of the aircraft on takeoffs and landings. However, after 24 to 36 hours following a moderate rainfall, the site should be suitable for landings. The clay soil forms a hard, smooth surface with an exceptionally high shearing strength from early August until early June of the following year when the airstrip is dry or frozen.

As a result of its high latitude (82°N), Bronlunds Fjord has continuous sunlight from April 7 to September 5. From October 14 to February 28 the sun is below the horizon, but no entire day in the year is without some astronomical twilight.

Construction materials and water supply

Outwash deposits of sand and gravel lie about 1.0 mi (1.6 km) northeast of the airstrip. These also may be obtained from the beach or from glacial moraine 0.5 mi (0.8 km) to the southwest. Two miles (3.2 km) to the east, near Independence Fjord, dolomite is exposed. Potable water near the surface of Bronlunds Fjord is available year-round, but it may occasionally become brackish. A large fresh water lake lies 2 mi (3.2 km) north of the airstrip. Its depth is unknown and a surface cover of at least 6 ft (1.8 m) of ice can be expected during the winter. From early June to mid-August, small amounts of water can be obtained from the stream that flows along the northeast border of the area.

Soil investigations

Starting from a point along the southwest border of the 5,000-ft (1,524-m) airstrip and 500 ft (152 m) from the southeast end, six pits (Fig. 4) 900 ft (274 m) apart on alternating sides of the runway were dug to permafrost. The depth to permafrost increased from 20 in (51 cm) at the southeast end to 30 in (76 cm) at the northwest end in pit number six. At each pit, samples were collected and measurements made for the soil profile, size distribution, temperature, moisture content, plasticity, compaction characteristics, bulk density, and soil strength. The tests for soil strength were conducted with two soil penetrometers (Fig. 4), which were used at all three sites and are described in Appendix 1. The index values obtained from these penetrometers were then converted to "equivalent" CBR values (see Appendix 1).

Figure 4. Test pit dug along the runway to permafrost at a depth of 24 in (61 cm). The airfield penetrometer is upright on the left, and the cone penetrometer is on the right (photograph by Stoertz, August 1957).

A generalized soil profile of the airstrip prepared from the considerable data obtained from the pits is shown in Table 2. Most of the airstrip (94%) was classified as inorganic clay (CL, Unified Soil Classification System, Waterways Experiment Station, 1953), and there was a close relationship between its moisture content and its shearing strength. At a moisture content of approximately 17%, its equivalent CBR value was 1, but when the moisture content was reduced to 8%, the equivalent CBR value increased to almost 8.

With an average moisture content of 6%, the average remolding index for nine generally dry surface samples of clay (CL) was 1.8. These data suggest that, when the airstrip is dry, aircraft traffic could sufficiently improve the clay surface to nearly double its strength. Since 94% of the airstrip had an average equivalent CBR at the 3 in (7.6 cm) depth of greater than 20 by mid-August (Table 2), most of it is strong enough to receive some of the heaviest aircraft at that time.

At the southeastern end of the airstrip, a remolding index of 0.5 was obtained from the surface of a soft moist area in silty clay (CL-ML). When wet, these areas cannot be strengthened by compaction and probably would be weakened by traffic. Fortunately, these wet areas are located at the ends of the runway and occupy only 6% of the entire runway surface (Table 2).

Surface conditions and soil strength along the airstrip

The surface conditions and soil strength along the airstrip in mid-August 1957 and on July 8, 1960 (see Subsequent observations) are compared in Figure 5. The soft areas at opposite ends of the runway (Fig. 5, 1A) were avoided during test landings and takeoffs by placing touchdown markers 150 ft (46 m) from the southeast end and 900 ft (274 m) from the northwest end. A small area had its marginal strength raised to the strength of the rest of

TABLE 2. GENERALIZED SOIL PROFILE OF THE AIRSTRIP AT BRONLUNDS FJORD*

Depth	Most of the Airstrip (94%)	North and South Ends of the Airstrip (6%)
0–24 in. (0–61 cm)	Surficially mud-cracked inorganic clay normally containing 8% sand (CL)	Silty clay, wetter than adjacent clay areas, and frequently containing considerable sand (CL-ML)
	Average equivalent CBR at the 3 in. (7.6 cm) depth in excess of 20† in mid-August 1957	Average equivalent CBR at the 3 in. (7.6 cm) depth was 2.5 in mid-August 1957
24 in. (61 cm)	Permafrost	

*After Stoertz and Needleman, 1957, Table 2.
†This value is much higher than the minimum CBR of 6 within a depth of about 3 in. (7.6 cm) required for two landings and two take-offs (two cycles = one coverage) of the C-124 test plane that landed at Bronlunds Fjord on August 16, 1957 (Stoertz and Needleman, 1957, p. 25).

the runway by hand tamping. As a result of these measures, 4,700 ft (1,433 m) of the runway were considered to have an equivalent CBR greater than 20 at a depth of 3 in (7.6 cm) by August 15, 1957 (Fig. 5, 1B; Table 2).

Landing tests

A C-124 heavy cargo plane with a gross weight of 145,000 lb (65,772 kg), weight per tire of 33,750 lb (15,309 kg), and tire pressure of 58 psi (4.1 kg/cm^2) made two landings and two takeoffs (one "coverage") on August 16, 1957, (Fig. 6) on the well-marked airstrip. The approaches were from the northwest with wind speed of 5 knots toward the southeast. The first touchdown was 1,000 ft (305 m) from the northwest end, and a ground-run of 1,900 ft (579 m) was made. The second touchdown was 900 ft (274 m) from the end of the runway with a ground-run of 1,800 ft (549 m). After the plane had parked one hour on the runway shoulder, the clay at the wheels was compressed to a depth of 0.5 in (1.3 cm). With the exception of the 2-in-(5-cm-) deep rut at the touchdown points, the tires compacted the clay an amount that ranged from 0.25 to 0.50 in (0.6 to 1.3 cm). The pilots reported that the surface was exceptionally smooth and the approaches good.

Subsequent observations

Davies and Krinsley (1961, p. 37–42) visited the site from July 7 to July 10, 1960. On July 8, 1960, most of the northern half of the runway (Fig. 5, 2A, 2B) was dry and strong with equivalent CBR values up to 25. The southern half had several wet areas where streams from the south crossed the airstrip in the shallow ruts previously made by the C-124. Soil strength in the wet areas was very low (CBR 0 to 6). Davies and Krinsley considered the airstrip unsafe at that time and recommended the cutting of shallow ditches parallel to the runway to divert drainage from the hills to the south.

Airstrip availability

The clay surface of the airstrip is generally moist and soft with little or no shearing strength from early June to early August. During the following 10 months, the smooth, hard surface has high shearing strength and is safe for the operations of heavy aircraft. Deep snow does not accumulate at the site. Consequently, it is an emergency option for wheeled aircraft throughout the winter.

POLARIS PROMONTORY[2]

Logistics

A nine-man field party whose principals included W. E. Davies, U.S. Geological Survey, and D. W. Klick and S. M. Needleman, AFCRL, landed at Polaris Promontory from the icebreaker USS *Atka*, on August 8, 1958, and worked at the site and in adjacent areas until August 17, 1958, when they departed aboard the icebreaker. Icebreakers can use the adjacent Robeson Channel from August through September in most years. Landing by amphibious aircraft adjacent to the coast is risky because of heavy pack ice in the vicinity at all times during the summer. The presence of pack ice also requires the exercise of caution when using landing craft to disembark from icebreakers during the summer.

Helicopters from the icebreaker were used to move personnel and most of the equipment to the site. Vehicles were transported to shore via landing craft and then driven overland to the site.

A two-man advance party including D. W. Klick and D. A. Craven, Arctic Institute of North America, flew into the site on July 6, 1959, aboard a Piper Super Cub aircraft with balloon tires to record soil conditions during the snow-melt period. They were followed by the main party of five who disembarked from the icebreaker and whose principals included C. E. Molineux and S. M. Needleman, AFCRL. The party worked at the site until they departed aboard the C-130 test aircraft on August 15, 1959.

Natural features

The airstrip at lat 80°41′N, long 59°35′W in northwest Greenland (Fig. 1) is 375 mi (603 km) from Thule and 70 mi (113 km) from Alert. It lies close to the air route between Alert and Nord.

The airstrip site (Fig. 7) occupies a north-south–trending, 10-mi- (16-km-) long terrace of the Graystone River that flows

[2]This section summarizes the work of Needleman et al., 1961.

SURFACE CONDITIONS AND SOIL STRENGTH ALONG THE AIRSTRIP

BRONLUNDS FJORD

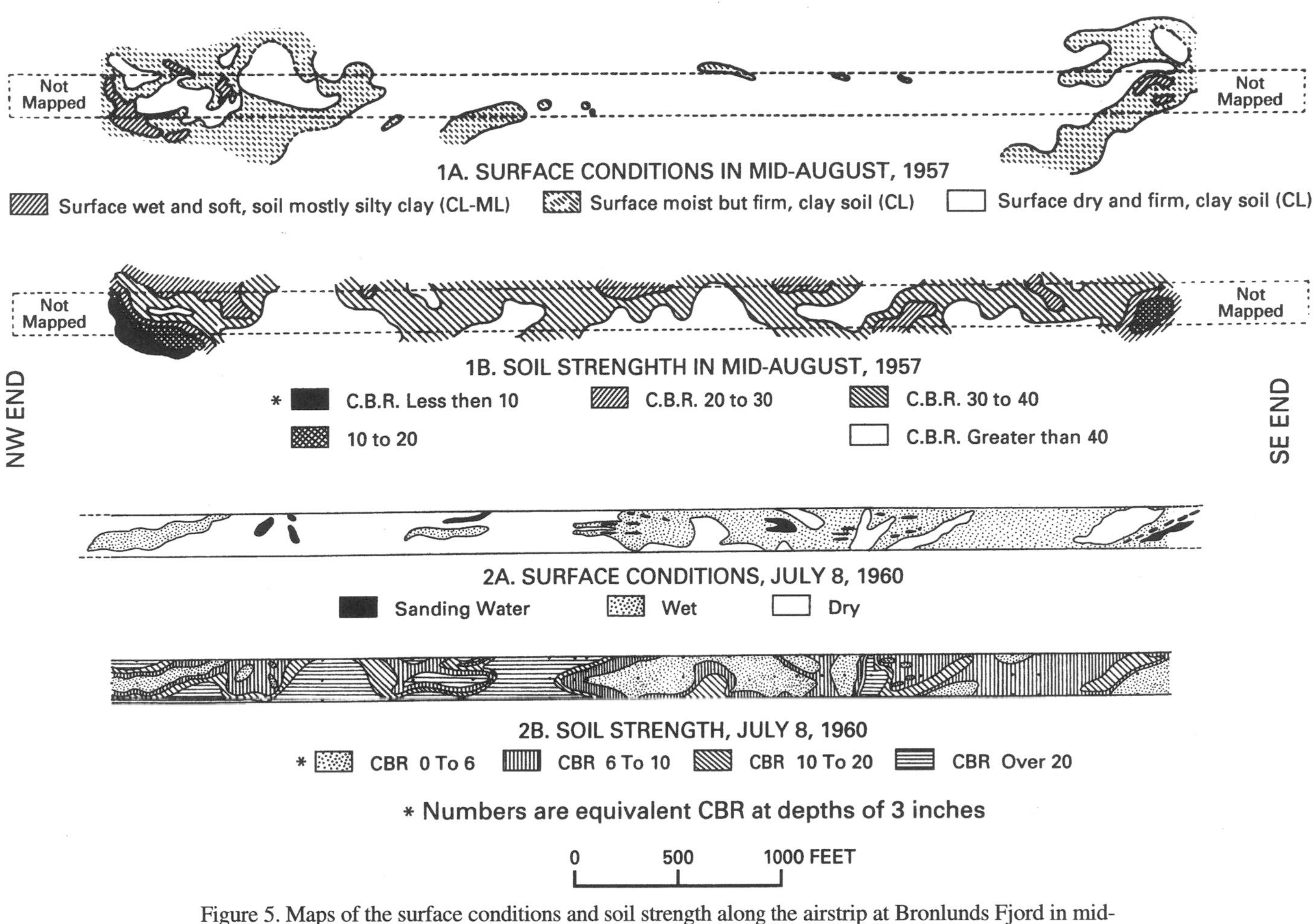

Figure 5. Maps of the surface conditions and soil strength along the airstrip at Bronlunds Fjord in mid-August 1957 and on July 8, 1960 (after Stoertz and Needleman, 1957, fig. 2, and Davies and Krinsley, 1961, fig. 32). Copied from original maps with all measurements in FPS units.

along its east bank. The surface slopes uniformly downward to the north at the rate of 50 ft/mi (9.5 m/km). The terrace is bordered on the north and west by low morainal hills up to 150 ft (46 m) high that are cored by marine clay, and the terrace is mantled by 5 to 20 ft (1.5 to 6 m) of pebbles and cobbles in a sand matrix. A belt of morainal hills up to 700 ft (213 m) high lies east of the Graystone River and extends to Newman Bay, 6 mi (10 km) east of the airstrip.

The landing-area surface (Fig. 8) contains polygons whose depressed boundaries are 1 to 2 ft (0.3 to 0.6 m) wide with gentle slopes and depths mostly less than 2 in (5 cm), with a maximum depression of 6 in (15 cm). Relict drainage channels 50 to 1,000 ft (15 to 305 m) long (Fig. 7) cross the site and are 5 to 30 ft (1.5–9 m) wide. Most are less than 2 in (5 cm) deep, but some are as much as 6 in (15 cm) deep. Lemming mounds up to 10 ft (3 m) in diameter and 6 in (15 cm) high were generally covered with grassy plants. All of the above features that had more than 2 in (5 cm) relief were removed by scraping and filling. Remaining microrelief features were not considered critical to aircraft operations. All drainage flows into the Graystone River; however, precipitation is so low that most rainfall soaks into the ground and little runoff reaches the river.

The airstrip, oriented N27°E (Fig. 7), is 200 ft (61 m) wide and 5,000 ft (1,524 m) long with a 300-ft (91-m) overrun on the north end and a 200-ft (61-m) overrun on the south end. The approach from the northeast via Newman Bay passes a low hill

Figure 6. C-124 taxiing to the campsite along the runway on August 16, 1957. After parking for an hour, the weight of the aircraft, 145,000 lb (65,772 kg), had produced a rut that was about 0.5 in (1.3 cm) deep. The field party departed for Thule aboard the plane (photograph by Stoertz, August 16, 1957).

50 ft (15 m) above and 1 mi (1.6 km) from the north end of the runway. A scarp with a 1,200-ft (366-m) elevation lies one mile (1.6 km) to the west and parallel to this approach. Approach from the southwest is clear of all obstruction.

Climate

Except for weather observations made at the site, three times daily from July 7 to August 15, 1959, no continuous weather records from the immediate vicinity exist. However, daily weather reports received by radio from Alert, 70 mi (113 km) away, compared closely with the conditions at the site. These data indicate that the climate at Polaris Promontory is very similar to that at Alert (Table 1) with some exceptions.

From September until May, the mean monthly temperatures at Alert are below freezing. By late September, the ground is frozen sufficiently to support aircraft at the Polaris Promontory site over the entire runway and overruns. The thaw starts in June and the frost-free period generally extends from mid-June to mid-August.

More than 70% of the 6.1 in (15.5 cm) of annual precipitation at Alert falls from June to October. More than 90% of the precipitation is snow (65 in [165 cm]) and more than half of this falls during August, September, and October. Snowfall at the Polaris Promontory site is very light and is frequently blown clear. Maximum accumulation at any one time probably does not exceed 24 in (61 cm). A low annual precipitation rate and a high evaporation and sublimation rate at Polaris Promontory contribute to the aridity of the area.

Long periods of calm winter weather occur at Alert; the highest wind velocities are from June to October. Gales have been recorded every month except January; the strongest winds, which come from the southwest, commonly exceed 60 mph (96 km/hr). However, winds up to 30 mph (48 km/hr) from the southwest were recorded at the Polaris Promontory site during 1959.

Polaris Promontory has fewer cloudy days than either Thule or Alert. Observations and radio reports during the summers of 1958 and 1959 confirmed that the weather at Polaris Promontory was generally clear while aircraft were grounded at Alert and Thule. At Alert, the mean monthly pressure ranges from 1,010 mb in July to 1,028 mb in March, with a very high mean annual pressure of 1,017 mb. As at Bronlunds Fjord, continuous sunlight extends from April to September. The sun is below the horizon from October to February, but there is no day without some astronomical twilight.

Construction materials and water supply

Gravel fill is readily available adjacent to the airstrip to a depth of 2 ft (0.6 m) to permafrost in summer. Large quantities of clean limestone-pebble and cobble gravel can be obtained from the Graystone River, 250 to 600 ft (76 to 183 m) from the airstrip (Fig. 7). Platy limestone can be extracted from the hill 2 mi (3.2 km) northwest of the airstrip and fine-grained binder from the hill 1 mi (1.6 km) to the north. Only pockets of sand are available from the bed of the Graystone River.

From May through early October, clear potable water in large quantities is available from the Graystone River. During the rest of the year, the water supply is limited to snowbanks or to two small lakes, on the east shore of the Graystone opposite the airstrip. These lakes, which are probably 20 ft (6 m) deep, are covered with 6 to 8 ft (1.8 to 2.4 m) of ice during the winter.

Soil investigations

Five test pits in areas of different surface conditions were dug to permafrost on either side of the runway in August 1958. The soil samples from the pits received the same scrutiny as those from the Bronlunds Fjord site (see above), and 2,800 shearing strength readings were taken at 93 locations with both a cone penetrometer and an airfield penetrometer (Appendix 1). The test pits were extended during the summer of 1959 and an additional 3,300 airfield penetrometer readings were recorded. These studies were conducted just after the meltwater had drained from the surface and the soil was in its wettest condition and again later when the soil had dried.

A generalized soil profile of the airstrip was prepared from the extensive data collected from the pits and is shown in Table 3. Ninety percent of the airstrip is surficially covered to a depth of 6 in (15 cm) by a poorly graded gravel and sand mixture with a silt binder (GP-GM). Below that, the soil is a poorly graded sandy gravel (GP) to permafrost at a depth of 36 in (91 cm), and this coarser, stronger soil underlies the entire runway. Ten percent of the airstrip contains lemming mounds and drainage channels that consist of very fine sandy silt (ML) to a depth of approxi-

AIRSTRIP AT POLARIS PROMONTORY

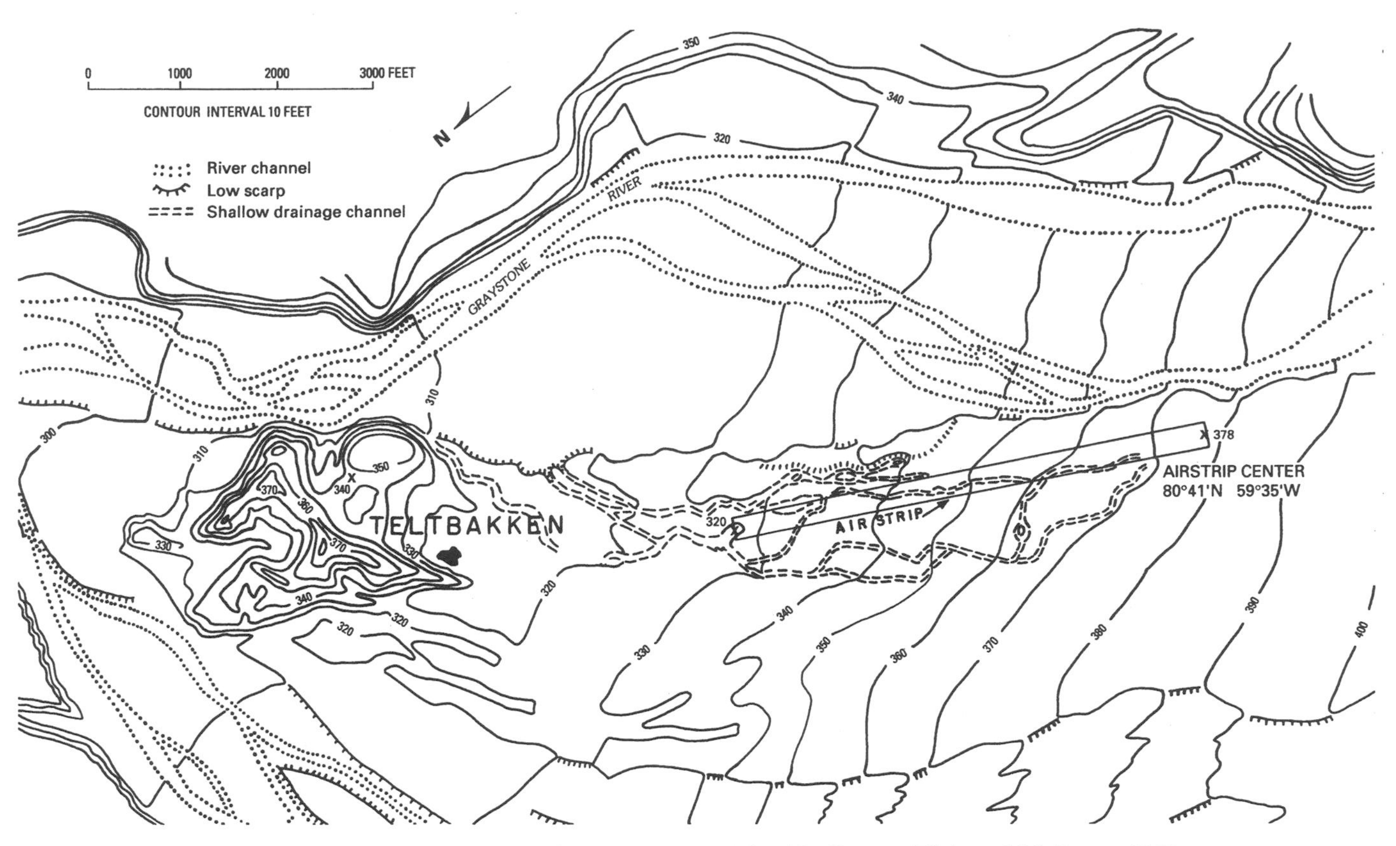

Figure 7. Map of the airstrip at Polaris Promontory (after Needleman, Klick, and Molineux, 1961, fig. 23). Copied from original map with all measurements in FPS units.

mately 15 in (38 cm). Below that depth to permafrost lies the generally continuous poorly graded sandy gravel (GP) of great shearing strength.

On July 9, 1959, the soils in more than 90% of the area had moisture contents approaching the plastic limit at depths from 3 to 6 in (8–15 cm), and they had very little shearing strength. Soils in the drainage channels had moisture contents exceeding the liquid limit at depths from 6 to 9 in (15–23 cm) and had no shearing strength at all.

The soils had dried by July 24, 1959, to about the same extent they were in August 1958 (Table 3) as reflected in the recovery of their shearing strengths. The greatest recovery occurred in the upper 3 in (8 cm) for most of the runway soils, and in the upper 6 in (15 cm) for the soils in the drainage channels.

Soil strength along the airstrip

The map of the soil strength conditions along the airstrip at Polaris Promontory in early August 1958 is shown in Figure 9. It is important to note that, even within the map unit of equivalent CBR 0 to 10, the average equivalent CBR at the 3 in (8 cm) depth

Figure 8. Poorly graded gravel and sand soil at the surface of the Polaris Promontory airstrip. The view is northeast along the runway (photograph by Davies, August 13, 1958).

TABLE 3. GENERALIZED SOIL PROFILE OF THE AIRSTRIP AT POLARIS PROMONTORY*

Depth	Most of the Airstrip (90%)	Depth	Lemming Areas and Drainage Channels (10%)
0–6 in. (0–15 cm)	Poorly graded gravel and sand mixture with a silt binder (46% gravel, 42% sand, and 12% fines as an average; GP-GM)	0–15 in. (0–38 cm)	Very fine sandy silt (ML)
	Average equivalent CBR at the 3 in. (7.6 cm) depth more than 10† in early August 1958		Average equivalent CBR at the 3 in. (7.6 cm) depth 8.5–10 in early August 1958
	Average equivalent CBR, at the 6 in (15 cm) depth more than 25 in early August 1958		
6–36 in. (15–91.5 cm)	Poorly graded sandy gravel with very few fines (69% gravel, 29% sand, and 2% fines as an average; GP)	15–36 in. (38–91.5 cm)	Poorly graded sandy gravel with very few fines (69% gravel, 29% sand, and 2% fines as an average; GP)
36 in. (91.5 cm)		Permafrost	

*After Needleman et al., 1961, Table 6.
†This value is more than the minimum CBR of 5 within an average depth of 6 to 12 in. (15.2–30.5 cm) required for four landings and four take-offs (two coverages) of the C-130 test plane (Waterways Experiment Station, 1960, Table 4).

was 8.5 to 10 in early August 1958 (Table 3). At the 6 in (15 cm) depth, the average equivalent CBR was greater than 25 at that time, and the runway was strong enough to receive some of the heaviest aircraft.

Landing tests

A C-130 heavy cargo plane with a gross weight of 90,000 lb (40,824 kg) and main tire pressure of 58 psi (4.1 kg/cm^2) made three landings and three takeoffs on August 15, 1959, at the well-marked airstrip (Fig. 10). Approaches as well as takeoffs were made from both directions, and all of the runway was tested by landing and by taxiing.

The average rut depth along the entire runway ranged from 1 to 2 in (2.5–5 cm), with the deepest ruts of 4 to 6 in (10–15 cm) occurring in the drainage channel areas. The pilot was completely satisfied with airstrip approaches, markings, soil surface conditions, and, most important, its strength.

Airstrip availability

The thaw starts in June, and from late June to early July (two to three weeks), the site is saturated and is definitely not safe for all but the lightest aircraft with low-pressure tires. The soil recovers its strength by late July and then heavy, wheeled aircraft can use the airstrip until the next June unless accumulations of snow, most likely in September, necessitate the use of skis. The period in mid-July when the soils have not fully recovered their strength is more difficult to assess. At that time the airstrip is an emergency landing site, although its still-soft surface may render any subsequent takeoff problematic.

CENTRUM LAKE[3]

Logistics

A four-man reconnaissance party consisting of W. E. Davies and G. E. Stoertz, U.S. Geological Survey; S. M. Needleman, AFCRL; and Lt. Col. R. H. Wilson, Air Force Arctic Advisor, landed on Centrum Lake aboard an SA-16 amphibious aircraft on August 15, 1956, and departed on August 18, 1956, after deciding that the site was a favorable candidate for detailed soil analyses.

From May 5 to May 8, 1960, a ski-equipped C-130 made four ski-landings on the frozen surface of Centrum Lake, transporting the 10-man field party and 74,000 lb (33,566 kg) of supplies including four vehicles. The principals included S. M. Needleman, AFCRL; C. A. Blackmon, U.S. Army Corps of Engineers; and D. B. Krinsley, U.S. Geological Survey. On July 2, 1960, Krinsley departed the Centrum Camp with W. E. Davies for a 10-day helicopter reconnaissance of the northeast coast of Greenland for potential airstrips. The main party departed on August 1, 1960, aboard the wheeled C-130 after the test landing on the site near Centrum Lake.

[3]This section summarizes the work compiled in Needleman, 1962.

SOIL STRENGTH ALONG THE AIRSTRIP AT POLARIS PROMONTORY

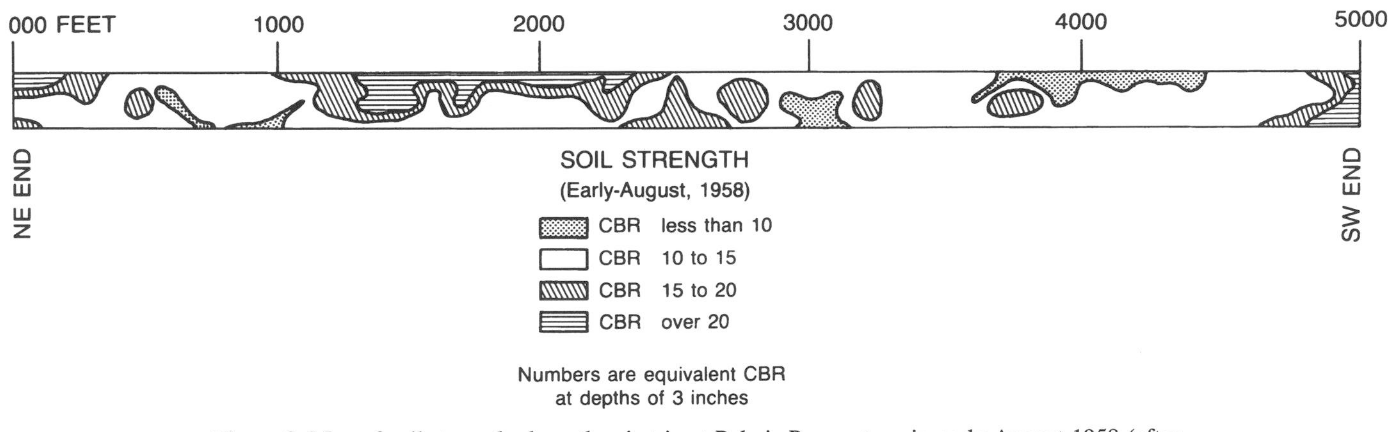

Figure 9. Map of soil strength along the airstrip at Polaris Promontory in early August 1958 (after Needleman, Klick and Molineux, 1961, fig. 9). Copied from original map with all measurements in FPS units.

Natural features

The airstrip at lat 80°08′N, long 22°30′W in northeast Greenland (Fig. 1) is 675 mi (1,086 km) from Thule and 140 mi (225 km) south of Nord. It is accessible by tracked vehicles from the ice cap 25 mi (40 km) to the southwest, by ski-equipped aircraft on the adjacent snow-covered, frozen lake surface November through May, and by amphibious aircraft on the lake when free of ice from July to September.

The site of the airstrip (Fig. 11) occupies a sand terrace whose surface rises from 12 ft (3.6 m) above the lake at its east end to 27 ft (8.2 m) above the lake at its west end 1.5 mi (2.4 km) away along the base of a plateau.

This terrace was part of a delta built by the deposition of glacially derived materials from the Saefaxi and Graeselv rivers and deposited when Centrum Lake was higher. The maximum relief in the vicinity of the airstrip is 6 ft (1.8 m), with a change in grade of less than 1%.

The northern part of the terrace is dissected by four northeast-southwest gullies that are 4–8 ft (1.2–2.4 m) deep, up to 100 ft (30 m) wide, and 500–1,000 ft (152–305 m) long. They are the principal drainageways for the snow meltwater. The surface of the terrace is free of active frost polygons or other evidence of frost action. Minor gullies were filled with soil removed from nearby high spots.

The airstrip, oriented N90°W (Fig. 11), is 200 ft (61 m) wide and 4,600 ft (1,402 m) long. It could be lengthened to 6,000 ft (1,829 m) by filling in the gully at its western end (Fig. 11), but this would require at least 3,000 yd^3 (2,294 m^3) of soil. An unobstructed approach from the east over the lake to the east end of the runway allows a glide path of 14 mi (23 km) with a glide angle of less than 1%. A cliff 1,800 ft (549 m) high adjacent to the west end of the runway precludes an approach from that direction.

Climate

Except for weather observations made at the camp at three-hour intervals from May 16 to July 26, 1960, there are no continual annual weather records from the immediate vicinity. Comparison with conditions at Thule and at Nord (Table 1) by radio reports during this period suggests that Centrum Lake has a more "continental" climate that is favorable for air operations. Temperature is higher, wind velocity is lower, and cloud cover is comparable with the other two stations.

In comparison with Nord, 140 mi (225 km) north, there was slightly more precipitation, less humidity and almost half the maximum wind speed (28 versus 52 mph [45 versus 84 km/hr]). Nord experienced 11 gales during this period (May 16 to July 26, 1960) while none were recorded at Centrum Lake. Although no records exist for the winter months at Centrum Lake, observations from overflights and by explorers suggest that there is less annual precipitation than at Nord.

The frozen surface of the lake is considered safe for ski-equipped aircraft from November through May. The ice was 6 ft (1.8 m) thick on the lake near where the ski-equipped C-130 landed on May 5, 1960 (Krinsley, 1962, fig. 45). Approximately 1 ft (0.3 m) of snow lay above a surface of ice that was as polished as glass. With its latitude only slightly less than those at

Figure 10. C-130 on the Polaris Promontory runway in 2 in (5 cm) of snow after its first test landing on August 15, 1959 (photograph by Klick, August 15, 1959).

Bronlunds Fjord or Polaris Promontory sites, Centrum Lake has continuous sunlight from April to September. The sun is below the horizon from October to February, but each day has some astronomical twilight.

Construction materials and water supply

Large quantities of high-quality limestone can be obtained in the hills 0.75 mi (1.2 km) from the west end of the airstrip. Sand in all sizes is available in large quantities near the airstrip. Extensive deposits of binder soil are located at the margin of the lake 0.25 mi (0.4 km) from the site.

Enormous quantities of potable water are available from Centrum Lake throughout the year. During the summer, large quantities of water are available from the Saefaxi and Graeselv Rivers, but they also carry a large amount of suspended silt from meltwater draining the Greenland Ice Cap 25 mi (40 km) to the west.

Soil investigations

Thirteen test pits were dug to permafrost in areas of different surface conditions on either side of the runway, and the soil samples received comparable field analyses as those from the Bronlunds Fjord and Polaris Promontory sites. Thousands of readings were taken at different depths at several hundred locations with both penetrometers (see Appendix 1) and their index values converted to equivalent CBR values.

A generalized soil profile of the airstrip was prepared from the extensive data collected from the pits and is shown in Table 4. Most of the airstrip contains soils within the polygon boundaries (80% of the airstrip), and these soils are classified as poorly graded sand (SP) to a depth of 12 in (30 cm). Below that and extending to permafrost at a depth of 36 in (91 cm) is a poorly graded medium-to-coarse sand (SP). Approximately 20% of the airstrip contains interpolygonal areas and drainageways consisting of well-graded fine sand (SW) to a depth of 12 in (30 cm). Below that and extending to permafrost is a well-graded, medium-to-coarse sand (SW).

The soil had little shearing strength when it was saturated by the melting snow and rain in June. Recovery of strength occurred in early July with drying of the soil through surface runoff and evaporation. Within the polygons the strength increased as much as 30%, and average increases were from 30 to 50% in the interpolygonal areas. Within the polygon areas (80% of the area of the airstrip), the average equivalent CBR values were 8 at the 6 in (15 cm) depth (Table 4), 15 at the 12 in (30 cm) depth, and more than 20 at the 20 in (51 cm) depth by mid-July. The average equivalent CBR value at the 6 in (15 cm) depth was increased from 8 to 10 with compaction by a 1,000-lb (454-kg) roller towed by a jeep. The average equivalent CBR values at the 6-in (15-cm) depth was 5.5 within the softer interpolygonal and drainage areas (Table 4).

Soil strength along the airstrip

The map of the soil-strength conditions along the airstrip at Centrum Lake in mid-July 1960 is shown in Figure 12. The poorly graded sand has little strength in the first few inches at the surface, but shearing strength rapidly increases with depth. Consequently, even in the interpolygonal and drainage areas that are more prominent in the map unit with equivalent CBR of 5 to 8, the soil strength at a depth of 6 in (15 cm) had an average equivalent CBR of 5.5. That value was greater than the minimum CBR of 5 within an average depth of 6–12 in (15–30 cm) required for four landings and four takeoffs of the C-130 (Waterways Experiment Station, 1960, Table 4).

Landing tests

An unscheduled landing of a Royal Canadian Air Force C-119 was made on July 27, 1960, to medically evacuate S. M. Needleman. The gross weight of the plane was approximately 60,000 lb (27,216 kg) with tire pressures of 65 psi (4.6 kg/cm^2) in the main gear and 75 psi (5.3 kg/cm^2) in the nose gear. Roll distance in the landing was 2,800 ft (853 m) and takeoff distance was 3,200 ft (975 m). The rut depth over the entire used part of the runway was approximately 3 in (8 cm). Rutting in a soft spot at the west end of the runway increased to a depth of 6 in (15 cm) at the turnaround point.

From July 31 to August 1, 1960, the C-130 test plane (Fig. 13) made two landings and two takeoffs with an average gross weight of approximately 95,000 lb (43,092 kg) and tire pressures of 50 psi (3.5 kg/cm^2) in the main gear and 55 psi (3.9 kg/cm^2) in the nose gear. Approaches were from the east with an east wind of 6–8 mph (10–13 km/hr). Rutting along the entire length of the runway averaged less than 3 in (8 cm), except at the turnaround points in soft spots where the ruts reached a maximum depth of 6 in (15 cm).

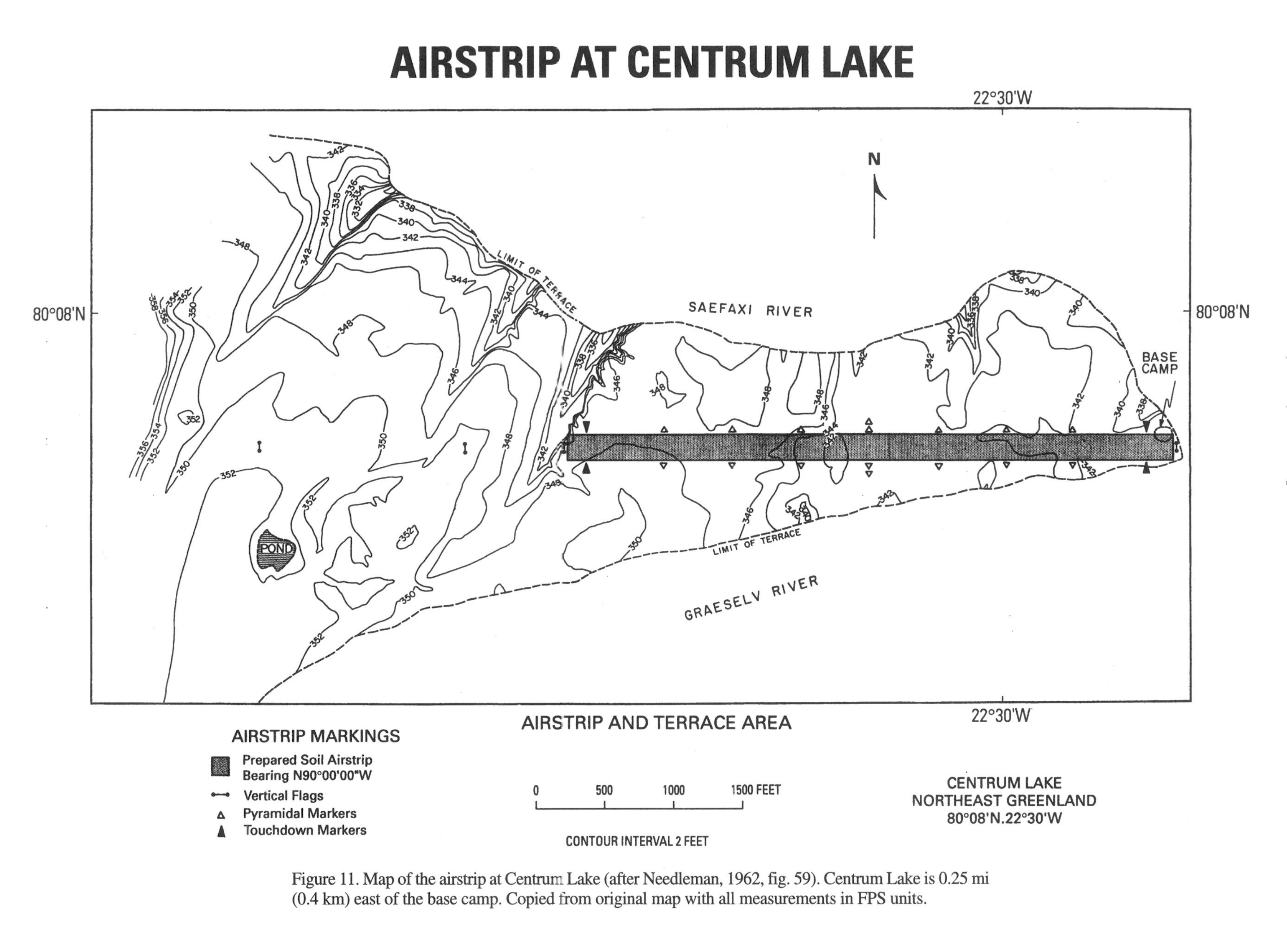

Figure 11. Map of the airstrip at Centrum Lake (after Needleman, 1962, fig. 59). Centrum Lake is 0.25 mi (0.4 km) east of the base camp. Copied from original map with all measurements in FPS units.

TABLE 4. GENERALIZED SOIL PROFILE OF THE AIRSTRIP AT CENTRUM LAKE*

Depth	Most of the Airstrip Areas within Polygons (80%)	Depth	Interpolygonal and Drainage Areas (20%)
0–12 in. (0–30.5 cm)	Poorly graded sand (5% fine gravel, 12% coarse sand, 45% medium sand, 30% fine sand, 8% fines; SP)	0–12 in. (0–30.5 cm)	Well-graded fine sand (2% fine gravel, 10% coarse sand, 30% medium sand, 48% fine sand, 10% fines; SW)
	Average equivalent CBR at the 6 in. (15.2 cm) depth was 8† in mid-July 1960		Average equivalent CBR at the 6 in. (15.2 cm) depth was 5.5 in mid-July 1960
12–36 in. (30.5–91.5 cm)	Poorly graded medium-to-coarse sand (SP)	12–36 in. (30.5–91.5 cm)	Well-graded medium-to-coarse sand (SW)
36 in. (91.5 cm)		Permafrost	

*After Needleman, 1962, Table 8.
†This value is more than the minimum CBR of 5 within an average depth of 6 to 12 in. (15.2–30.5 cm) required for four landings and four take-offs (two coverages) of the C-130 test plane (Waterways Experiment Station, 1960, Table 4).

Airstrip availability

The thaw starts in June, and from mid-June to early July (two to three weeks), the surface is saturated and the soil has very low shearing strength. During that period, the airstrip is not safe except for the lightest aircraft with low-pressure tires. The soil recovers its strength by mid-July and then heavy, wheeled aircraft can operate with safety until accumulation of snow, most likely in late September, might necessitate the use of skis.

The airstrip in its present condition could support more than 30 cycles (landings and takeoffs) during the summer, with only minor repair and stabilization to the surface at the west end where the surface material is displaced by the aircraft's turnaround action. A smooth surface can be maintained by passing over the ruts with a 1,000-lb (454-kg) roller towed behind a jeep. In winter, the frozen strip, with less than 1.0 ft (0.3 m) of snow, could withstand more than 100 cycles by heavy aircraft before any significant rutting would require some repair. Thicker snow would require the use of skis.

SUBSEQUENT USE OF THE THREE AIRSTRIPS

Bronlunds Fjord

In late 1957, the Bronlunds Fjord airstrip was added to the "Atlas of Arctic Ice-free Land Runways" by the AFCRC and distributed to U.S. Air Force Commands. At that time, the airstrip also was established by the U.S. Air Force Strategic Air Command (SAC) as an alternate landing site to Nord (Stoertz and Needleman, 1957, p. 31).

In September 1968 and September 1969, a British Royal Air Force C-130 landed at the Bronlunds Fjord airstrip initially to test the site and then to use it in support of the Joint Services Expedition to North Peary Land in 1969 (Joint Services Expedition, 1969, p. 4, 20). Subsequently, the Danish Ministry of Greenland moved a research station to the site to take advantage of the easy access by wheeled aircraft for at least 10 months of the year (Stoertz, written communication, July 14, 1994).

Currently, the Bronlunds Fjord airstrip is used regularly by a Royal Danish Air Force C-130 with a landing weight of as much as 130,000 lb (58,968 kg) to supply the research station, the Danish Sledge Patrol Sirius, and expeditions. Twin-Otter aircraft also use the site (Brigadier General Erik Lyngbye, Royal Danish Embassy, Washington, D.C., personal communication, October 13, 1994).

Polaris Promontory

The airstrip was used once in 1972 and again in 1975 by a Royal Danish Air Force C-54. It was used twice in 1976 by a Royal Danish Air Force C-130 with a landing weight of 120,000 lb (54,432 kg). Currently, the airstrip is used regularly by Twin-Otter aircraft to supply the Danish Sledge Patrol Sirius and expeditions (E. Lyngbye, personal communication, 1994).

Centrum Lake

The airstrip is currently used regularly by Twin-Otter aircraft to supply the Danish Sledge Patrol Sirius and expeditions (E. Lyngbye, personal communication, 1994).

SOIL STRENGTH ALONG THE AIRSTRIP AT CENTRUM LAKE

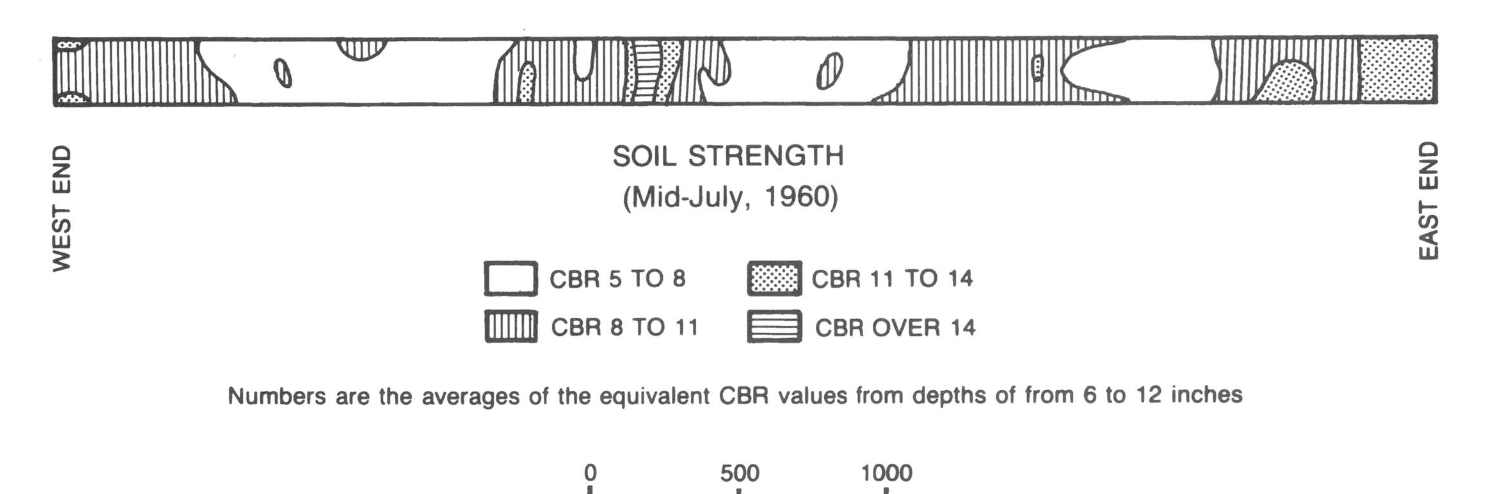

Figure 12. Map of soil strength along the airstrip at Centrum Lake in mid-July 1960 (after Needleman, 1962, fig. 25). Copied from original map with all measurements in FPS units.

CONCLUSIONS

Wheeled heavy aircraft can land on smooth, wind-swept frozen surfaces with less than 12 in (30 cm) of snow. Such conditions prevail at the Bronlunds Fjord airstrip, as well as on the ice of Centrum Lake, during the winter and through the first week in June. Similar conditions prevail during most of this period at the Polaris Promontory airstrip. From early June through early July, all of the tested sites are saturated by melting snows and have low soil strength.

With its well-drained sandy soils, elevated above its adjacent streams, the Centrum Lake airstrip recovers its soil strength by mid-July and can then accept heavy wheeled aircraft until late September when accumulating snow might necessitate the use of skis.

The Polaris Promontory airstrip recovers its soil strength by late July and can then accept heavy wheeled aircraft. At Bronlunds Fjord, the inorganic clay remains wet and soft until the first week in August when it becomes dry, and then the airstrip provides the smoothest and hardest surface of the three tested sites.

Construction materials and water supplies are available at all sites for airstrip improvement or some expansion, and permafrost is not an obstacle to minor grading or light construction if the soils on and adjacent to the airstrips are not extensively removed.

ACKNOWLEDGMENTS

This chapter is dedicated to William E. Davies and Stanley M. Needleman whose vision, pioneering efforts, and leadership qualities were responsible for much of the success of the projects described herein. Mrs. William E. Davies, Donald W. Klick, and George E. Stoertz have given the author access to numerous ancillary publications and colored slides of the airfield sites. Brigadier General Erik Lyngbye, Royal Danish Embassy, Washington, D.C., was instrumental in obtaining climatic and historical information concerning Station Nord and the three sites, and

Figure 13. C-130 at the east end of the Centrum Lake airstrip on August 1, 1960. The average rut depth in the poorly graded sand was 3 in (8 cm) over the entire runway and 6 in (15 cm) at the turnaround point at the west end of the runway (photograph by Needleman, August 1, 1960).

Chief Alan M. Zahnle, Presidential Support Unit, Andrews Air Force Base, provided current airfield and climatic data from Thule, Alert, and Nord.

The preparation of the illustrations was supported by the U.S. Geological Survey and greatly facilitated through the good offices of John Aaron and John Keith, the drafting skills of George Ward, and the computer graphics of David Murphy.

The careful review of this paper by Allen F. Agnew, Donald W. Klick, and George E. Stoertz is very much appreciated.

APPENDIX 1. MEASUREMENT OF SHEAR STRENGTH OF SOIL

One of the most reliable measurements of the strength of soils is the California Bearing Ratio (CBR) test, which evaluates the shearing resistance of a soil under controlled conditions of density and moisture. The CBR test may be conducted on a laboratory-compacted specimen, undisturbed sample, or in place in the field. The bulkiness of the CBR test equipment and the necessity of securing it to a heavy vehicle make it difficult or impractical to use in remote areas such as those described in this chapter.

A practical solution to this problem was the development by the Waterways Experiment Station, U.S. Army Corps of Engineers, of two manually operated, probe-type instruments, the cone penetrometer and the airfield penetrometer.

The cone penetrometer consists of a 30-degree cone with a 1.0-in (2.5-cm) base diameter, two 18-in (46-cm) extension rods with etched 1.0-in (2.5-cm) graduations, providing a 36 in (91 cm) length, a proving ring, a dial gauge, and a handle (Fig. 4). When the cone is forced into the ground, the proving ring is deformed in proportion to the force applied. That force is registered on the dial inside the proving ring as cone index. The range of the cone index on the dial is 0 to 300. The general relation of cone index to CBR is about 30:1. Therefore, the range of the "equivalent" CBR (see below) of the cone penetrometer is 0–10. At the same time that the cone index is read, the depth of the cone below the surface of the soil is recorded from the positions of the rod's etch marks.

The airfield penetrometer consists of a 30-degree cone with a 0.5-in (1.3-cm) base diameter mounted on a staff with 1.0-in (2.5-cm) graduations, on top of which are a spring and handle (Fig. 4). When the cone is forced into the ground, the tension spring is stretched directly in proportion to the force applied to the handle. A device on the barrel just below the handle records the force required to move the cone slowly through the soil. That force is read as the airfield index. The range of the airfield index readings is from 0 to 40. The relation between the airfield index readings and CBR is approximately 1:1. Therefore, the range of the "equivalent" CBR (see below) of the airfield penetrometer is 0 to 40. As with the cone penetrometer, the depth of the cone beneath the soil surface is noted when the airfield index is recorded.

The readings obtained from these instruments, referred to as "cone index" and "airfield index" values have been correlated with values obtained by CBR test equipment in the same soils and field conditions. Numerous graphs have been prepared showing these correlations for many soil types and families of soils. The index readings obtained at the North Greenland test sites were compared to the appropriate graphs (Molineux, 1955, figs. 4, 18) to obtain the "equivalent" CBR values used in this chapter.

Considerable data relating minimum CBR values to various aircraft gross weights, wheel loads, tire pressures, and the number of passes over the same track are available (Bredahl and Kiefer, 1957, p. 101).

REFERENCES CITED

Bredahl, A. R., and Kiefer, E. P., 1957, A classification system for unprepared landing areas, Air Research and Development Command—TR-57-20, U.S. Air Force: Los Angeles, Planning Research Corporation, 144 p.

Davies, W. E., 1961, Arctic terrain research for low cost airfield sites; History and results, *in* Rigsby, G. P., and Bushnell, V. C., eds., Proceedings of the Third Annual Arctic Planning Session, November 1960: Air Force Cambridge Research Laboratories no. 436, Geophysics Research Directorate Notes no. 55, p. 8–14.

Davies, W. E., and Krinsley, D. B., 1961, Evaluation of arctic ice-free land sites, Kronprins Christian Land and Peary Land, North Greenland, 1960: Air Force Cambridge Research Laboratories, Air Force Surveys in Geophysics no. 135, 51 p.

Fristrup, B., 1952, Physical geography of Peary Land, I; Meteorological observations for Jorgen Bronlunds Fjord: Meddelelser om Gronland, v. 127, no. 4, 143 p.

Joint Services Expedition, 1969, Joint services expedition North Peary Land 1969, 46 p., Annexes A–J.

Krinsley, D. B., 1962, Limnology, *in* Needleman, S. M., ed., Arctic earth science investigations, Centrum So, Northeast Greenland, 1960: Air Force Cambridge Research Laboratories, Air Force Surveys in Geophysics, no 138, p. 47–55, 107–119.

Molineux, C. E., 1955, Remote determination of soil trafficability by the aerial penetrometer: Air Force Cambridge Research Center, Air Force Surveys in Geophysics no. 77, 46 p.

National Climatic Data Center, 1994, International station meteorological climate summaries for Thule, Alert, and Nord: Andrews Air Force Base National Weather Service, 6 p.

Needleman, S. M., editor, 1962, Arctic earth science investigations, Centrum So, Northeast Greenland, 1960: Air Force Cambridge Research Laboratories, Air Force Surveys in Geophysics no. 138, 132 p.

Needleman, S. M., Klick, D. W., and Molineux, C. E., 1961, Evaluation on an arctic ice-free land site and results of C-130 aircraft test landings, Polaris Promontory, North Greenland, 1958–1959: Air Force Cambridge Research Laboratories, Air Force Surveys in Geophysics no. 132, 70 p.

Stoertz, G. E., and Needleman, S. M., 1957, Report on Operation Groundhog, North Greenland: Investigations of ice-free sites for aircraft landings in northern and eastern Greenland and results of test landing of C-124 at Bronlunds Fjord, North Greenland: Air Force Cambridge Research Center, Air Research and Development Command, 40 p.

U.S. Air Force, 1954, History, Thule Air Force Base: 6607th Air Base Wing, 1 June 1954–31 December 1954.

U.S. Air Force, 1972, Base civil engineering information brochure: Thule Air Base, Aerospace Defense Command Greenland.

Waterways Experiment Station, 1953, The unified soil classification system: Vicksburg, Mississippi, U.S. Army Corps of Engineers, Technical Memorandum 3-357, 3 volumes, 50 p.

Waterways Experiment Station, 1960, Validation of soil strength criteria for aircraft operations on unprepared landing strips: Vicksburg, Mississippi, U.S. Army Corps of Engineers, Technical Report 3-554, 25 p.

Manuscript Accepted by the Society October 29, 1997

Printed in U.S.A.

Geological Society of America
Reviews in Engineering Geology, Volume XIII
1998

Selected military geology programs in the Arctic, 1950–1970

Louis DeGoes†
17215 Northeast 8th Street, Bellevue, Washington 98008
James T. Neal
1911 Crestview Drive, Prescott, Arizona 86301

ABSTRACT

The military environment of the post–World War II years was dominated by the principal threat of air and missile attacks over polar routes between North America and the former Soviet Union. The extremities of the arctic environment demanded unique adaptations to normal military operations, which provided military and engineering geologists an opportunity to exercise a broad range of technical applications. New knowledge and skills in ice physics and engineering, glaciology, arctic geomorphology, meteorology, oceanography, and permafrost engineering were all relevant, often requiring on-the-job learning. Such adaptability could be needed in the future and would once again challenge military geologists.

INTRODUCTION

Between 1950 and 1970, military readiness was dominated by the cold war between the United States and the former Soviet Union. The formidable environment of the high-arctic regions above 80° N latitude became militarily significant as nuclear engagements between aircraft or submarines, or both, became possible. The United States gave high priority to construction of air bases and electronic early warning systems that would enhance the survival of personnel and aircraft during attack or inclement weather. Aircraft operations from floating shelf "ice islands" and sea ice became routine.

However, climatic extremes of the Arctic were appreciated by few, even though many early explorers had popularized and even glamorized these barren regions. In addition to extremes of cold, aridity equal to or greater than Death Valley, California, created unique landscapes. Ice-free land areas throughout the high Arctic owe their existence to such aridity and to the sublimation of reduced snowfall combined with nearly constant high winds. Such conditions led to the identification of numerous natural landing areas that could be used by heavy aircraft with minimal preparation (Aero Services Corp., 1961). The Strategic Air Command (SAC) adopted a "live-aboard concept" wherein heavy bombers could be dispersed from Thule Air Base to such temporary runways during alerts of hostile forces (or natural disasters). These landing areas, combined with arctic survival kits, could have sustained aircraft and crews alike until "all clear" signals were given.

Geologists, oceanographers, and meteorologists contributed significantly to pressing military needs, often under considerable urgency. The Distant Early Warning (DEW) Line—an electronic barrier of more than 30 radar stations extending from Point Barrow, Alaska, to the east coast of Greenland—was constructed during the mid-1950s at great expense for the purpose of detecting hostile aircraft (Fig. 1). The ever-present permafrost complicated construction and tested geotechnical knowledge of this environment (DeGoes and Needleman, 1960).

In 1955 several "ice teams" tested and prepared natural aircraft operational sites on fresh ice and sea ice along the DEW Line. More than 16,000 metric tons of heavy equipment, supplies, and personnel were subsequently airlifted to these strips by C-124 Globemaster transports. Overland transit or ice landings after the spring thaw would be impossible. The ice thickness criteria for calculating safe aircraft operation were developed by the U.S. Army Snow, Ice, and Permafrost Research Establishment (SIPRE), later to become the Cold Regions Research and Engineering Laboratory (CRREL). The prepositioning of those large tonnages of cargo allowed the construction of the DEW Line facilities to proceed, meeting critical schedules.

†Deceased

DeGoes, L., and Neal, J. T., 1998, Selected military geology programs in the Arctic, 1950–1970, *in* Underwood, J. R., Jr., and Guth, P. L., eds., Military Geology in War and Peace: Boulder, Colorado, Geological Society of America Reviews in Engineering Geology, v. XIII.

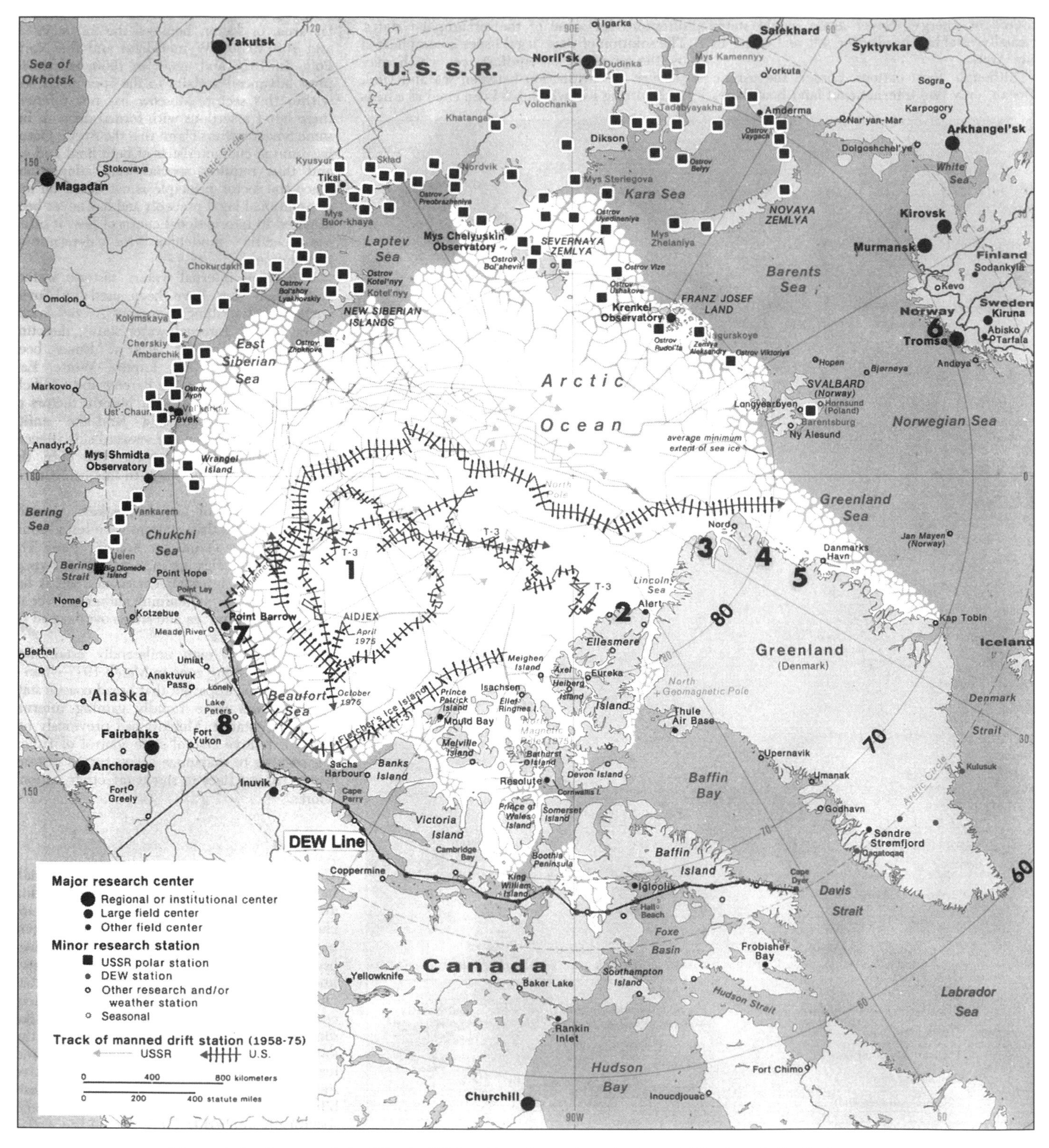

Figure 1. Arctic research facilities in 1970 and locations referred to in text. Numbers 1–8 refer to numbered locations in Table 1. From Polar Regions Atlas (CIA, 1978). Tortuous, counterclockwise rotation of floating polar ice is shown by migration of U.S. and Soviet research stations.

IMPETUS FOR INITIATING COMPREHENSIVE RESEARCH IN THE HIGH ARCTIC

At the Air University, Maxwell Air Force Base, Alabama, long-term interest in climatic extremes existed even before World War II. The Arctic, Desert, and Tropic Information Center had archived extensive data and published flight manuals for global air operations, primarily to be used for evasion and escape of downed aircraft crews. In 1953 Fletcher and DeGoes (1956) began concentrating on their four-volume "Arctic Study Series," as the threat of military engagement in those regimes intensified. Fletcher had discovered Fletcher's Ice Island (T-3) in 1946. He recognized its potential use as an emergency landing platform or staging base and spent three months there in 1952 (Fig. 1). Those efforts expanded into widespread studies of the polar environment, such as those undertaken by the Air Force's Terrestrial Sciences Laboratory beginning in 1957. The primary purpose was to describe, investigate, and evaluate ice, snow, water, and natural land areas of the Arctic that would significantly affect Air Force operations. Moreover, the fundamental understanding of this major climatic extreme of Earth needed to be documented for a variety of engineering and military geology applications, regardless of any specific military purpose. At this point, such information was minimal. Another project with military application—Military Applications of Seismology—had already been under way at the Terrestrial Sciences Laboratory to identify, explore, and develop seismic and acoustic techniques for surveillance and detection, especially of nuclear tests. There was much interaction and overlap of both projects; engineering geology, geophysics, and related disciplines all contributed to their success.

THE ARCTIC TERRAIN PROJECT, TERRESTRIAL SCIENCES LABORATORY

The Terrestrial Sciences Laboratory of the Air Force Cambridge Research Laboratories, Bedford, Massachusetts, was the principal Air Force establishment that addressed geotechnical issues involving the Arctic. A staff of 14 civilians and six military officers having diverse technical backgrounds were augmented by contract support from government agencies, universities, research institutes, and industry. The Arctic Institute of North America, a private research group staffed largely by university personnel under the direction of Walter Wood and John Reed, was notable in providing many scientists and support staff. Scientists and engineers from many disciplines, often working together, brought their expertise to bear on technical problems that were related to the Arctic operational environment. Publication of results in journals was encouraged and several doctoral dissertations resulted.

Technical direction and advice on ice-free land projects was provided by the Military Geology Branch of the U.S. Geological Survey, first under the direction of Frank Whitmore and later directed by Donald Dow. Many projects by a number of other organizations and governments required diligent coordination and participation on advisory panels. The Office of Naval Research, Naval Civil Engineering Research Laboratory, and the Army Cold Regions Research and Engineering Laboratory all had related programs; coordination with them was achieved directly through membership on the Department of Defense Research and Development Board. The Geological Survey of Canada and the Defense Research Board of Canada were valuable partners at Ellesmere Island and elsewhere in the Queen Elizabeth Islands. The Ministry of Greenland (Denmark) provided helpful guidance and data for projects in Greenland.

Each summer during the late 1950s and early 1960s, 60 personnel fielded five to eight projects in the high Arctic. The 1960 field efforts, to locations principally north of 80° (Fig. 1), are listed in Table 1.

PROJECT ICE WAY

Project Ice Way involved the construction, operational evaluation, and scientific study of a sea-ice runway and artificial parking pads, which included evaluation of mechanical properties, salinity, deformation, and deterioration. It was conducted

TABLE 1. 1960 FIELD PROJECTS SPONSORED BY TERRESTRIAL SCIENCES LABORATORY*

Location; Staffing	Subject Matter and Duration
1. **Fletcher's Ice Island (T-3);** Arctic Ocean (12 people)	Oceanography, glaciology, hydrology, meterology, ice physics, submarine geophysics (year round)
2. **Ellesmere Ice Shelf; NWT, Canada** (23 people)	Glaciology, meterology and micro-meteorology, glaciology, geomorphology, seismology, oceanography, and ice engineering (April to September)
3. **Pearyland, North Greenland (Denmark)** (3 people)	Geological reconnaissance, glacial geology, geomorphology, and engineering geology (July)
4. **Centrum Lake Area Northeast Greenland (Denmark)** (10 people)	Meterology, soils, permafrost, glaciology, limnology, geomorphology, and engineering geology (May to August)
5. **Storelv Area, East Greenland (Denmark)** (4 people)	Geological reconnaissance, soil studies, and engineering geology (July)
6. **Finnmark, Northern Norway** (6 people)	Glacial geology, geomorphology, and sedimentation (September and October)
7. **Point Barrow, Alaska** (4 people)	Sea ice physics and ice engineering (March)
8. **Peters Lake, Arctic Slope, Alaska** (4 people)	Limnology, ice and snow physics, meterology and micrometerology, glaciology, and geology (May to September; Dutro, 1985)
9. **Climatic Laboratory, Eglin Air Force Base, Florida** (12 people)	Ice engineering, including processing, additives, spray solidification, ablation control, and sea ice physics (June to September)

*Locations 1–8 are keyed to map locations on Figure 1.

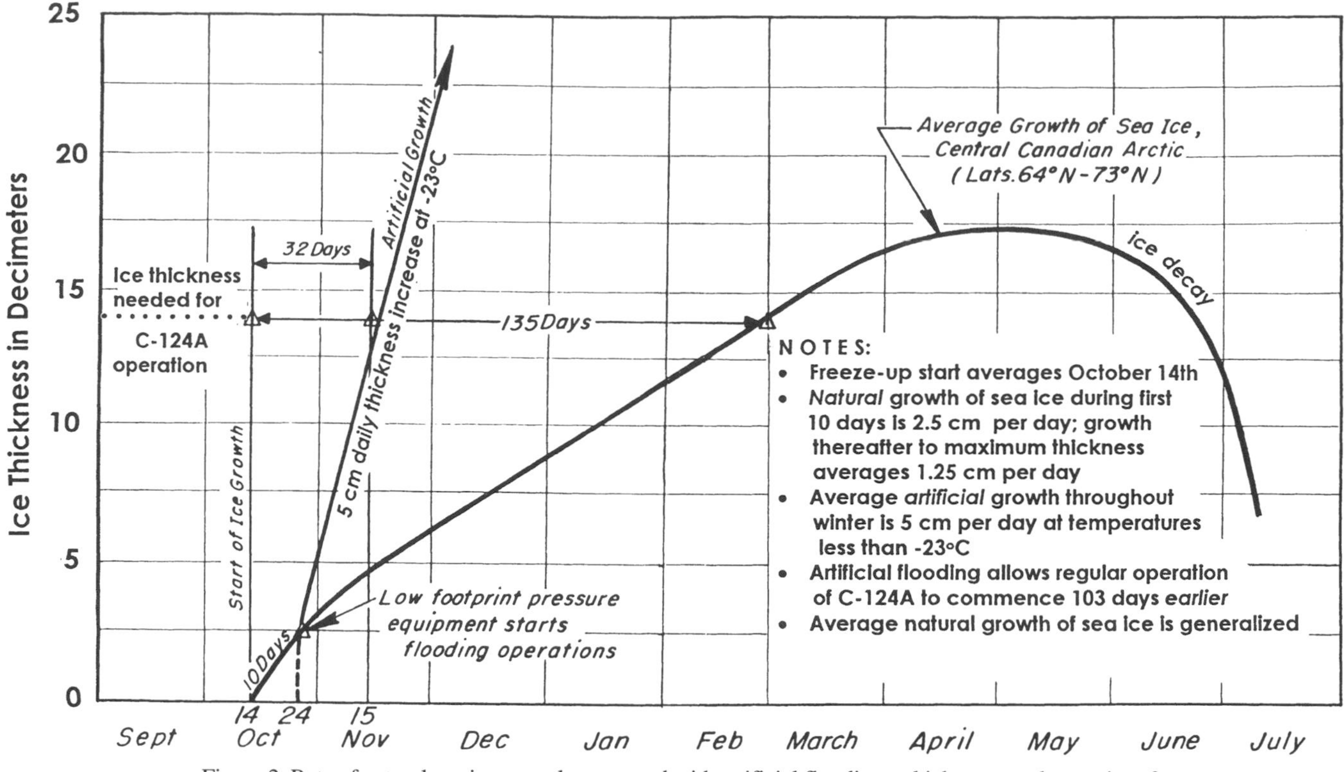

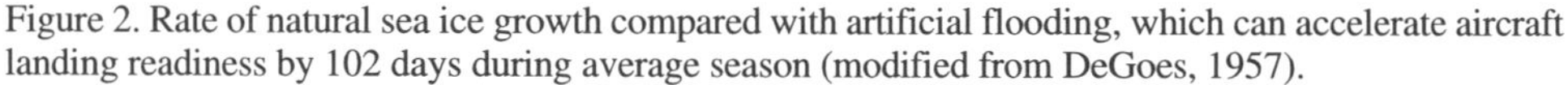
Figure 2. Rate of natural sea ice growth compared with artificial flooding, which can accelerate aircraft landing readiness by 102 days during average season (modified from DeGoes, 1957).

during January–May 1961 on floating sea ice on North Star Bay, near Thule Air Base, Greenland, by the Air Force Terrestrial Sciences Laboratory and the Naval Civil Engineering Laboratory, Port Hueneme, California (Klick, 1961).

Ice, in its many forms, was treated as a readily available and virtually inexhaustible local construction material (DeGoes et al., 1957). Fiberglass mats were added to some of the parking pads to increase the ice strength, similar to reinforcing rods in concrete (Kingery, 1963). Although floating ice runways have a limited life span of six months (November to April) or less at the latitude of Thule, the ice runways and parking apron could have provided temporary refuge for aircraft from Thule Air Base or elsewhere, should their primary installations have been closed for weather reasons or if other exigencies occurred. The advantages of inducing artificial growth are dramatically shown in Figure 2. Because fresh ice is considerably stronger than sea ice (Kingery, 1963), techniques were developed in the Air Force's Climatic Hanger at Eglin Air Force Base, Florida, to remove most of the brine from sea water before freezing, thereby increasing the strength of the ice.

ICE WAY CONSTRUCTION AND AIRCRAFT TEST OPERATIONS

Because heavy loads gradually deform ice sheets, three parking pads were built on top of the natural ice (Kingery, 1962). One pad was flooded with sea water in layers to a total thickness of nearly 3 m. A second pad, nearly 2 m thick, was reinforced with fiberglass mats at the top and bottom of the built-up layer. A third pad was constructed with ice aggregate of various sized ice chips and saturated with sea water; it was about 1.5 m thick and also contained fiberglass reinforcement. At all three pads, sea water was pumped to the surface through holes in the ice and allowed to freeze in place as it flooded the area. In this manner, ice accumulated in 10 cm increments on top of the 1.2-m-thick natural layer, and the resulting cross section of each pad was lens shaped (Fig. 3). The weight of frozen water deformed the natural ice sheet so that the surface of the built-up layers was only inches above the level of the surrounding natural ice.

Aircraft landing, parking, and takeoff tests were conducted from March 22–31, 1961, on the ice runway and included B-52

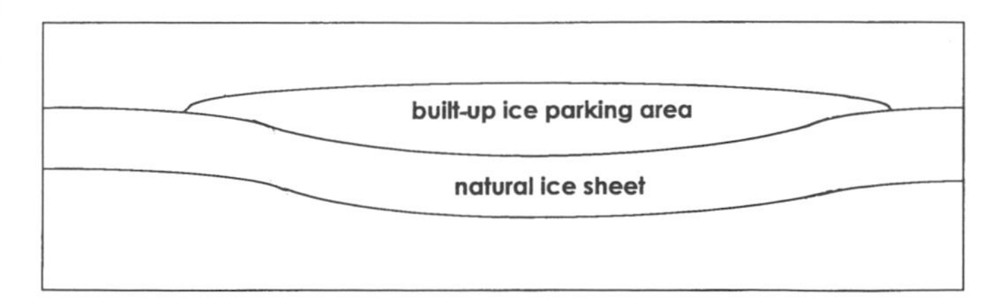

Figure 3. Lens shape of built-up area from artificial flooding and deflection of natural ice under pad. Artificial flooding greatly expedites and enhances weight-bearing capacity of natural ice surfaces.

Figure 4. Aerial view of B-52 parked on Pad #1, with camp facilities just beyond aircraft and Thule Air Force Base in distance. Chevron markings at lower left mark east overrun.

and B-47 bombers (Fig. 4). The 4,270-m runway was one of the longest in the world and could have been extended westward several miles because icebergs were absent and the snow surface was moderately thick. It was constructed on the natural sea ice 1.2–1.4 m thick over water in the bay that was 15–24 m deep. The runway construction involved no artificial buildup or processing, but merely removal of most of the uniformly 15-cm-thick snow cover on the ice surface. Removal of this insulating layer promoted deeper freeze, but minor snow was left on the surface to provide traction for test aircraft and vehicles. Ice cracks occurred but did not adversely affect the tests because ice is remarkably insensitive to cracks unless they are open and wet in the immediate area of loading. The origins of cracks can be complex: Open, wet cracks can result from winds, tides, waves, and excessive loading, whereas superficial cracks may be created by thermal stress, applied loads, or snow-load reduction.

MILITARY GEOLOGY APPLICATIONS, POST-COLD WAR

Despite the reduction in military forces in the post–cold war period, many opportunities lie ahead for environmental and engineering geologists, as well as for those in related disciplines. For example:

Continuing military engagements in past trouble spots such as Haiti, Grenada, Panama, Iraq-Kuwait, Somalia, and Bosnia-Hercegovina are likely. Engineering geology and ground-water requirements should be anticipated.

Humanitarian rescues such as the April 1980 unsuccessful rescue attempt of 53 hostages held by militants in Tehran, Iran, may be necessary. An unprepared landing strip in the desert 400 km southeast of Teheran had been identified, but the mission was terminated after a C-130 and a helicopter collided. Identification of appropriate sites for such operations should be anticipated in numerous areas in the future.

Humanitarian projects may require the use of austere airfields when no others exist. Natural disasters will continue, and food, water, medicine, and emergency assistance will be needed. Intervention against despotic regimes also may require similar support.

Clandestine anti-terrorism projects in regions lacking airfields will require ongoing support.

Environmental remediation of nuclear and other toxic wastes on some military reservations may be needed.

CONCLUSIONS AND IMPLICATIONS FOR FUTURE USE OF GEOLOGISTS

Military technical applications during the period 1950–1970 required flexibility in all areas of science by virtue of global operations in remote and inhospitable terrain, including arctic, desert, and tropical regimes. Military and engineering geology were often employed in arctic operations, but frequently on the fringes of "conventional practice" such as snow and ice physics. Lessons learned included that of the education of geologists—a broad understanding of the many subdisciplines of geoscience combined with the basic sciences is most apt to contribute to the solution of a broad range of military technical problems. Military geology applied to arctic problems will continue to be distinctive and challenging as the 21st century approaches, pointing to ongoing challenges in other environments worldwide that are both physically and politically difficult.

ACKNOWLEDGMENTS

We thank Donald Klick and Daniel Krinsley for their helpful reviews of this paper and for providing some of the illustrations, both of which materially improved the presentation.

REFERENCES CITED

Aero Services Corp., 1961, Terrain analysis of ice-free land sites (north of 65°): Bedford, Massachusetts, Air Force Cambridge Research Laboratories, Contract report to Terrestrial Sciences Laboratory, Contract AF19(604)6182), 203 p.

CIA, 1978, Polar regions atlas: Washington, D.C., U.S. Central Intelligence Agency, 68 p.

DeGoes, L., 1957, Aircraft operations on floating ice sheets: Discussion: Journal of the Air Transport Division, American Society of Civil Engineers, v. 83, AT-1, Paper 1328, p. 11–26.

DeGoes, L., and Needleman, S. M., 1960, Permafrost, *in* Handbook of geophysics, Chapter 12, Sect. 4: New York, The Macmillan Co., p. 1253–1262.

DeGoes, L., Weeks, W. F., and Anderson, D. F., 1957, Aircraft landing tests on sea ice in North Star Bay, Thule, Greenland: Bedford, Massachusetts, Air Force Cambridge Research Center, Geophysics Research Directorate GRD-TM-57-20, 8 p.

Dutro, J. T., Jr., 1985, The G. William Holmes Research Station, Lake Peters, northeastern Alaska, and its impact on northern research, *in* Drake, E. T., and Jordan, W. M., eds., Geologists and ideas: A History of North American Geology, Boulder, Colorado, Geological Society of America, Centennial Special Volume 1, p. 301–311.

Fletcher, J. O., and DeGoes, L., 1956, Arctic study series, 4 volumes, 60 chapters, Montgomery, Alabama, The Air University, 512+ p.

Fletcher, J. O., 1953, Three months on an Arctic ice island: National Geographic Magazine, vol. 103, p. 489–504.

Kingery, W. D., editor, 1962, Summary report—Project Ice Way: Bedford, Massachusetts, Air Force Cambridge Research Laboratories, Air Force Surveys in Geophysics No. 145, Terrestrial Sciences Laboratory, 216 p.

Kingery, W. D., editor, 1963, Ice and snow: Properties, processes, and applications: Cambridge, Massachusetts, The MIT Press, 684 p.

Klick, D. W., 1961, Project Ice Way: Bedford, Massachusetts, Air Force Cambridge Research Laboratories, Terrestrial Sciences Laboratory open-file research report # 1, 15 p.

Manuscript Accepted by the Society October 29, 1997

Printed in U.S.A.

Geological Society of America
Reviews in Engineering Geology, Volume XIII
1998

Hydrogeological assessments of United Nations bases in Bosnia Hercegovina

C. Paul Nathanail
CRBE, Nottingham Trent University, Burton Street, Nottingham NG1 4BU, United Kingdom

ABSTRACT

The need to ensure reliable and secure water supply at United Nations bases in Bosnia Hercegovina led to a hydrogeological and water supply reconnaissance (recce) of each base to determine the feasibility of constructing boreholes within the base perimeter. The recce was conducted under full security measures at a time when the armed conflict in central Bosnia was at its fiercest. Fieldwork was severely limited by the ongoing war. Facilities such as generators, vehicle repair yards, and latrines excluded certain areas within the perimeter wire and restricted where boreholes could be sited.

Despite the disruption to the infrastructure and government caused by several years of conflict, some site-specific information was obtained. All sites were assessed as having a good potential for ground water. Preliminary well designs were determined. These were accepted and well drillers were deployed to Vitez, Gornji Vakuf, and Tomislavgrad. Following the deployment of the drillers to Bosnia, the author was able to use his experience of the ground to advise during the well construction phase.

The author also provided advice on slope stability along parts of the main supply route (MSR) near Prozor and on the location of suitable borrow pits for material to maintain the surface of the MSR near Redoubt Camp.

The participation of the Royal Engineer Specialist Advice Team geologist in Operation Grapple continued the tradition of geological input into recent major British army operations such as those in the Falklands, the Gulf, and northern Iraq.

INTRODUCTION

The need to ensure reliable and secure water supply at five U.N. bases in Bosnia Hercegovina led to a hydrogeological assessment of each base to determine the feasibility of constructing boreholes within the perimeter. Desk studies of available geological information were carried out in the United Kingdom. The findings enabled the regional geological and hydrogeological conditions to be determined. However, mapping was at a small scale and often dated back to World War II. The reliability of some mapping was uncertain as it was suspected that maps produced by Yugoslav geologists for the then-occupying German forces had been deliberately doctored. A field reconnaissance was therefore conducted to assess the hydrogeology and determine the most suitable location for boreholes within the perimeter fencing.

TRAIN OF EVENTS

In June 1993 a Royal Engineer Specialist Advisory Team (RESAT) geologist volunteered for a short service voluntary commission in order to be allowed to take part in a Military Works Force Water Supply Recce under the auspices of Operation Grapple, the British contribution to the U.N. mission in Bosnia Hercegovina. A limited desk study was undertaken within the short time frame available.

The Recce Party examined U.N. Camps at Tuzla, Vitez (School and Garage sites), Gornji Vakuf, and Tomislavgrad (Table 1). A Dutch base close to Vitez was also evaluated for its suitability for a well. The current water supply at each of these localities was either insecure or of dubious quality (Table 1, Fig. 1). The author was able to use the desk study information,

Nathanail, C. P., 1998, Hydrogeological assessments of United Nations bases in Bosnia Hercegovina, *in* Underwood, J. R., Jr., and Guth, P. L., eds., Military Geology in War and Peace: Boulder, Colorado, Geological Society of America Reviews in Engineering Geology, v. XIII.

TABLE 1. UNITED NATIONS BASES EXAMINED DURING THE WATER SUPPLY RECCE

U.N. Base	Geologic Conditions	Water Supply at Time of Recce	Sources of Site Specific Geological Information
Gornji Vakuf	Alluvial sediments	Pumped supply (unreliable)	Walkover survey
Tomislavgrad	Alluvial sediments	Pumped supply (unreliable)	Walkover survey
Tuzla Airfield	Alluvial sediments	Pumped supply (unreliable)	Walkover survey; hydrogeologic maps and cross sections (view only)
Vitez 'School'	Alluvium over limestone	By tanker from river	Walkover survey and large scale topographic mapping
Vitez 'Garage'	Alluvial sediments	By tanker from river	Walkover survey, logs from two borehole within 400 m
Busovace (Netherlands)	Alluvial sediments	Pumped supply (unreliable)	Walkover survey

assessments of the geomorphology and hydrogeology from the surrounding terrain, and, in the case of Tuzla and Vitez, borehole logs obtained from local sources to evaluate the potential for obtaining ground water. All sites were assessed as having a good potential for ground water. Preliminary well design recommendations were made. These were accepted and 521 Specialist Team Royal Engineers (Water Development) (521 STRE) deployed to construct wells, initially at Vitez, Gornji Vakuf, and Tomislavgrad. Following the deployment of 521 STRE to Bosnia, the author was able to use his experience of the ground to advise during the well construction phase.

GEOLOGICAL SETTING

The geology of the former Yugoslavia is a result of the collision between the African and European plates and the associated closure of the Tethys ocean (Anonymous 1970, 1990). The predominantly carbonate ocean floor sediments have been compressed, and a series of northeastward-trending overthrusts developed. The southwest to northeast compression has given rise to the marked northwest to southeast grain in the Yugoslav terrain. Mountains of limestone are separated by steep-sided valleys infilled with Quaternary fluvial and slope-wash deposits. The limestone is heavily karstified and the land surface resembles bomb-cratered terrain. This is the result of the collapse of numerous cave systems. Characteristic red-brown soils, the product of complete weathering of the limestone, infill the circular doline, and much larger polje depressions give the mountain tracks their characteristic pale-brown tint. The region is still seismically active (a large earthquake affected Sarajevo in 1979), and thermal springs are common.

THE RECONNAISSANCE

The reconnaissance took place during the period June 24, 1993, to July 3, 1993. The aim was to assess the need for and feasibility of constructing water supply wells within each of the U.N. bases visited using the factors in Table 2. A limited desk study was undertaken by Territorial Army (TA) geologists within the short time available. Information was obtained from the British Geological Survey and the Royal Engineer Map Library at Tolworth. This included small-scale geologic mapping, satellite imagery, published papers—many in Bosnian, Serbian or Croatian—and extracts from textbooks. A computer literature search revealed only six further references, all in Bosnian, which were obtained by the author immediately prior to departure for Bosnia.

Once in Bosnia the recce was conducted under full security measures at a time when the armed conflict in central Bosnia was at its fiercest. Fieldwork was severely limited by the ongoing war (Nathanail, 1996). Facilities such as generators and latrines excluded certain areas within the perimeter and restricted where boreholes could be sited (Table 3). The disruption to the infrastructure and government caused by several years of conflict has resulted in the loss of much information. Nevertheless, some site-specific information was obtained from the remnants of local authority records in Vitez and the mining research department in Tuzla.

Figure 1. Location of sites in Bosnia Hercegovina mentioned in text. (After http://www.cco.caltech.edu/~bosnia/status/terrain.html)

The author was able to use the desk study information and assessments of the geomorphology and hydrogeology from the surrounding terrain to evaluate the potential for obtaining ground water. For Tuzla, hydrogeologic maps and cross sections were obtained from local sources, and, for Vitez and Tuzla, borehole logs were obtained. All sites were assessed as having a good potential for ground water. The reconnaissance team recommended that boreholes were feasible at all bases.

Tuzla airfield

The Tuzla Airfield lies in the Spreca Valley at an elevation of 250 m above sea level. The area is generally flat lying but rises

TABLE 2. FACTORS TO CONSIDER IN ASSESSING FEASIBILITY OF AND NEED FOR WELLS

Required quantity of water
Adequacy and security of present water supply
Security of well drillers and maintenance personnel
Access to base by drilling rig
Location and condition of existing wells
Environmental constraints
Previous land use and potential for pollution (expecially by hydrocarbons)
Proximity to surface water
Likely soil profile
Nature of bedrock
Likely depth to rockhead
Likely depth to quantity and quality of ground water

gently to the northeast and drops gently to the Rijeka River to the northwest. Unpublished geological information comprising borehole logs, test results, geologic maps, and cross sections was examined at the Mining Research Institute in Tuzla.

The geology in the vicinity of Tuzla airfield comprises Holocene sediments over Miocene and Pliocene clastites, limestones, and coal (Anonymous, 1970). The airfield is underlain by Quaternary alluvial sediments >100 m thick, which was proven in a borehole. A continuous clay layer between 8 m and 15 m thick at the surface is underlain by a 40-m thick upper zone of intercalated clay, silt, sand, and gravel horizons and lenses. A second continuous clay layer, only 1 m thick, separates the upper mixed sediment unit from a lower similar unit. Water for Tuzla is drawn from gravel horizons on the upper mixed unit by a series of boreholes <50 m deep.

A borehole log of one of the existing water supply boreholes indicated that water was being drawn from a sandy gravel (particle size distribution ranging between 1 mm and 8 mm) at 14 m depth to top and 6–8 m thick. Pumping tests at commissioning in 1988 gave-flow rates of between 150 liters and 250 liters per minute.

Vitez

Vitez lies in the northwest- to southeast-trending Lasva valley. The hills to the southwest are composed of Cretaceous limestone underlain by Carboniferous slates and sandstones. The Lasva valley is shown on geologic maps as a fault (Anonymous, 1970). A limestone quarry has been excavated into the slope on the south side of the Lasva valley. Water for Vitez and nearby Zenica is drawn from Krusice, which also has a thermal spa. The geology in the vicinity of Vitez comprises Holocene sediments over upper Miocene clastites, limestones, and coal (Anonymous, 1970).

Two sites were examined in Vitez: Vitez School and Vitez Garage. Vitez School is situated at an elevation of 450 m a.s.l. between the south-flowing Lasva and Bila rivers. The Bila river joins the Lasva 800 m southeast of the School. The site is flat lying with a small hill to the east that contains outcrops of limestone. The geology at the site was interpreted as superficial fluvial sediments over limestone.

Vitez Garage is at an elevation of 425 m a.s.l. on the south bank of the Lasva River in flat-lying terrain. Two borehole logs for wells sunk 200 m north of the site, on the north bank of the river, were made available by local government officials. The logs showed 4.5 m of topsoil, silt, and sand underlain by 11 m of sand and gravel. Rockhead was encountered at 16 m, and 26 m of karstified limestone was penetrated.

Gornji Vakuf

The geology in the vicinity of Gornji Vakuf comprises Quaternary sediments of overlying upper Miocene clastites, limestones, and coal (Anonymous, 1970). The U.N. compound in Gornji Vakuf was at an elevation of 670 m a.s.l. on flat-lying terrain on the north bank of the east-to-west–flowing Kruscica river. Weathered fissured marls and mudstones were exposed in a gentle slope along the north boundary of the compound perimeter. Two 1.5 m deep excavations revealed sandy silty clay with occasional gravel.

Tomislavgrad

The geology in the vicinity of Tomislavgrad, formerly known as Duvno, comprises Quaternary sediments over undifferentiated Oligocene and Miocene (Anonymous, 1970). The base is at an elevation of 870 m a.s.l. at the northwest-southeast–trending basin. A small, unnamed stream flows north to south 100 m beyond the base perimeter. The surrounding hills probably contain limestone.

WELL DRILLING

The British army's regular well drilling team, 521 STRE, was deployed in the autumn of 1993. The team constructed wells at Vitez School, Vitez Garage, Tomislavgrad, and Gornji Vakuf (Wye, 1994). The location of the wells at Vitez School had to be changed in response to changes in the security situation since the time of the recce. One of these wells was found to be severely polluted by diesel fuel spilt from the generators being used to supply power to the base. The geology encountered at Vitez Garage differed greatly from that predicted during the recce. In both cases the author was able to advise 521 STRE as to appropriate courses of action. No drilling was carried out at Tuzla as the British army was no longer deployed there. At the time of this writing (summer 1996), 521 STRE had been back in Bosnia on two other tours of duty constructing wells at additional sites.

GEOLOGICAL INPUT TO MAINTENANCE OF MAIN SUPPLY ROUTE

The main supply route (MSR) was being used to transport humanitarian aid from the port of Split, Croatia, to refugee camps

TABLE 3. CONSTRAINTS ON SITING WELLS WITHIN UNITED NATIONS BASES

Constraint	Reason
Away from generators, POL, latrines, vehicle repair areas	Pollution of ground water is likely
Close to existing water distribution pipes	Minimize infrastructure work required
Disturbance during drilling acceptable	Minimize impact on base personnel
Away from unfriendly forces	Security of drilling personnel and maintenance operatives
Accessible by and working space for proposed drilling rig	Operational and safety aspects of drilling operation
Away from antenna farms	Avoid interference with signals traffic
Within perimeter wire	Security of drilling personnel and maintenance operatives

and enclaves in central Bosnia. To avoid sensitive areas and damaged bridges, the MSR followed mountain tracks created to assist foresters. The loose surface therefore required continual maintenance to permit safe passage of loaded heavy-goods vehicles. The author was asked to provide advice on slope stability along those parts of the MSR near Prozor that were being widened and on the location of suitable borrow pits for maintaining the surface of the MSR near Redoubt Camp. This advice was given during a one-day recce of relevant sections of the MSR with the military plant foreman (MPF).

The assessment of slope stability within cuttings in the limestone through which the road passed revealed that bedding planes dipping steeply out of cut faces were recemented and steeper slope angles than originally intended could therefore be adopted. The benefit to the MPF was the reduced time required to complete his task and therefore to achieve more within the time available.

An assessment of the geomorphology enabled several potential borrow pits to be identified and the likely extent of usable material within current borrow pits to be delineated. This enabled the MPF to phase out the deployment of excavators, scrapers, and earth moving plant and minimize the distances over which aggregate had to be hauled.

CONCLUSIONS

This project represents the first time in recent years that a TA geologist has had to join the ranks of the Regular Army to serve "in theater." Several lessons may be learned from this experience. A geologist on the ground can contribute in several ways, and extra work will usually appear once his or her presence in theater becomes known. Even short periods of time in theater can greatly expand geologists' ability to provide relevant advice. This benefit may be magnified by ensuring that geologists in theater cover and see as much ground as possible within the time available. Geological information can often be obtained, or at least examined, locally. This requires a geologist in theater to rapidly assess and interpret such information.

The participation of the RESAT geologist in Operation Grapple continued the tradition of geological input into recent major operations such as those in the Falklands and the Gulf.

REFERENCES CITED

Anonymous, 1970, Geological map of the SFR Yugoslavia, Belgrade, at Federal Geological Institute, scale 1:500,000, 7 sheets.

Anonymous, 1990, Ground water in eastern and northern Europe, Natural Resources/Water Series No. 24, New York, United Nations Department of Technical Co-operation for Development.

Nathanail, C. P., 1996, Environmental constraints on United Nations operations in Bosnia Hercegovina, *in* Coulson, M., and Baldwin H., eds., Proceedings of the Environment and Defence Symposium: Swansea, University of Wales, September 13–15, 1995, University of Wales, Swansea.

Wye T, 1994, Well drilling in Bosnia: Royal Engineers Journal 108 (2), p. 149–153.

Manuscript Accepted by the Society October 29, 1997

Printed in U.S.A.

Geological Society of America
Reviews in Engineering Geology, Volume XIII
1998

Military geology in support of nation assistance exercises in Central and South America

Robert B. Knowles
U.S. Army Topographic Engineering Center, 7701 Telegraph Road, Alexandria, Virginia 22315-3864

ABSTRACT

Engineer construction training exercises throughout Central and South America have provided U.S. soldiers with construction experience and aided the host nation by developing infrastructure. Roads, airfields, bridges, schools, hospitals, and water wells are constructed in remote areas with maximum use of local materials. Military geologists support construction by locating potential sources of aggregate, evaluating suitability of materials for road metal and fill, designing roadcuts, developing quarries, and evaluating geologic hazards that may affect the construction effort.

In northern Honduras, excavation problems in a karstic limestone of the Mesozoic Yojoa Group delayed completion of a mountain road. Clay-filled solution cavities disrupted the blasting operation, jeopardizing the completion of the project. Geologists recommended rerouting the road through a rippable schist. Although longer, the new road was completed on time.

In southern Honduras the lack of a suitable aggregate for a new airfield required a thorough geologic reconnaissance for new sources. Local ignimbrites were unsuitable due to their low compressive strength. Rhyolite and andesite from nearby volcanic erosional remnants were evaluated for their suitability. A quarry was subsequently developed and provided the aggregate needed to complete the project.

In the Oriente of Ecuador a scarcity of suitable construction materials and unstable slopes created difficult conditions for a road and bridge construction project. Roadcuts into alternating layers of shale and limestone of the Cretaceous Lower Napo Formation created unstable slope conditions. Military geologists recommended new sources of aggregate and advised engineers on slope stability. Local construction methods were adopted to overcome these conditions.

INTRODUCTION

The U.S. Army has conducted numerous engineer training exercises throughout Central and South America in recent years. The objectives are to provide engineer units the opportunity to deploy to an overseas location, conduct training in austere environments, enhance their unit readiness, and develop a positive image in the host country toward the U.S. government and its armed forces. Traditionally, U.S. Army Engineer forces have been structured for combat in Europe, not for large-scale construction in undeveloped theaters. In the last 10 years a significant effort has been made to train engineer units in these more austere environments. These engineer training missions are called nation assistance exercises because they help develop the host nation infrastructure while engineer soldiers practice their wartime missions.

Nation assistance missions include constructing roads, bridges, airfields, schools, hospitals, and water wells, and they require engineer units to establish base camps in remote locations and support themselves logistically. These exercises require the maximum utilization of local construction materials. They train active duty, Reserve, and National Guard engineers and typically

Knowles, R. B., 1998, Military geology in support of nation assistance exercises in Central and South America, *in* Underwood, J. R., Jr., and Guth, P. L., eds., Military Geology in War and Peace: Boulder, Colorado, Geological Society of America Reviews in Engineering Geology, v. XIII.

last for three to six months. Reserve and National Guard units may rotate their entire workforce every two weeks during the duration of an exercise.

Military geologists support these exercises by locating potential sources of construction aggregate, evaluating the suitability of earth materials for road metal and fill, designing roadcuts, developing military quarries, and evaluating geologic hazards that may adversely affect the construction effort. The geologists may be engineer staff officers or troop leaders in the deploying unit, military instructors from the U.S. Army Engineer School, or civilian professionals in the U.S. Army Corps of Engineers. The success of the construction mission often depends on their efforts. Three examples of nation assistance exercises in Central and South America requiring significant, yet typical, geologic input are described in this chapter.

NORTHERN HONDURAS

Beginning in 1985 and continuing over a period of several years, a farm-to-market road was constructed across the Central Highlands of Honduras. The road connected the provincial capital of Yoro with the Río Aguan valley and opened up a large agricultural area and port. In 1988, Task Force 111, West Virginia National Guard, was responsible for building 11.5 km of new road between the hamlet of Puente Grande and the village of Jócon (Fig. 1). In 1987, an engineer task force encountered excavation problems in a karstic limestone, which delayed their portion of the road project. A military geologist from the U.S. Army Engineer School was requested by Task Force 111 to evaluate the geologic conditions that could adversely affect construction.

Existing detailed geologic studies of the area were sparse. Regional studies were available by Weyl (1965), Williams and McBirney (1969), and Mills et al. (1967). A country scale geologic map by Aceituno (1974) was also available. The area lies within the Central Highlands geomorphic province of Honduras. The Central Highlands is composed of long, rugged, irregular mountain ridges with up to 600 m of relief. The underlying rock is composed of folded and faulted Mesozoic sedimentary rocks and older Paleozoic metamorphic rocks. The

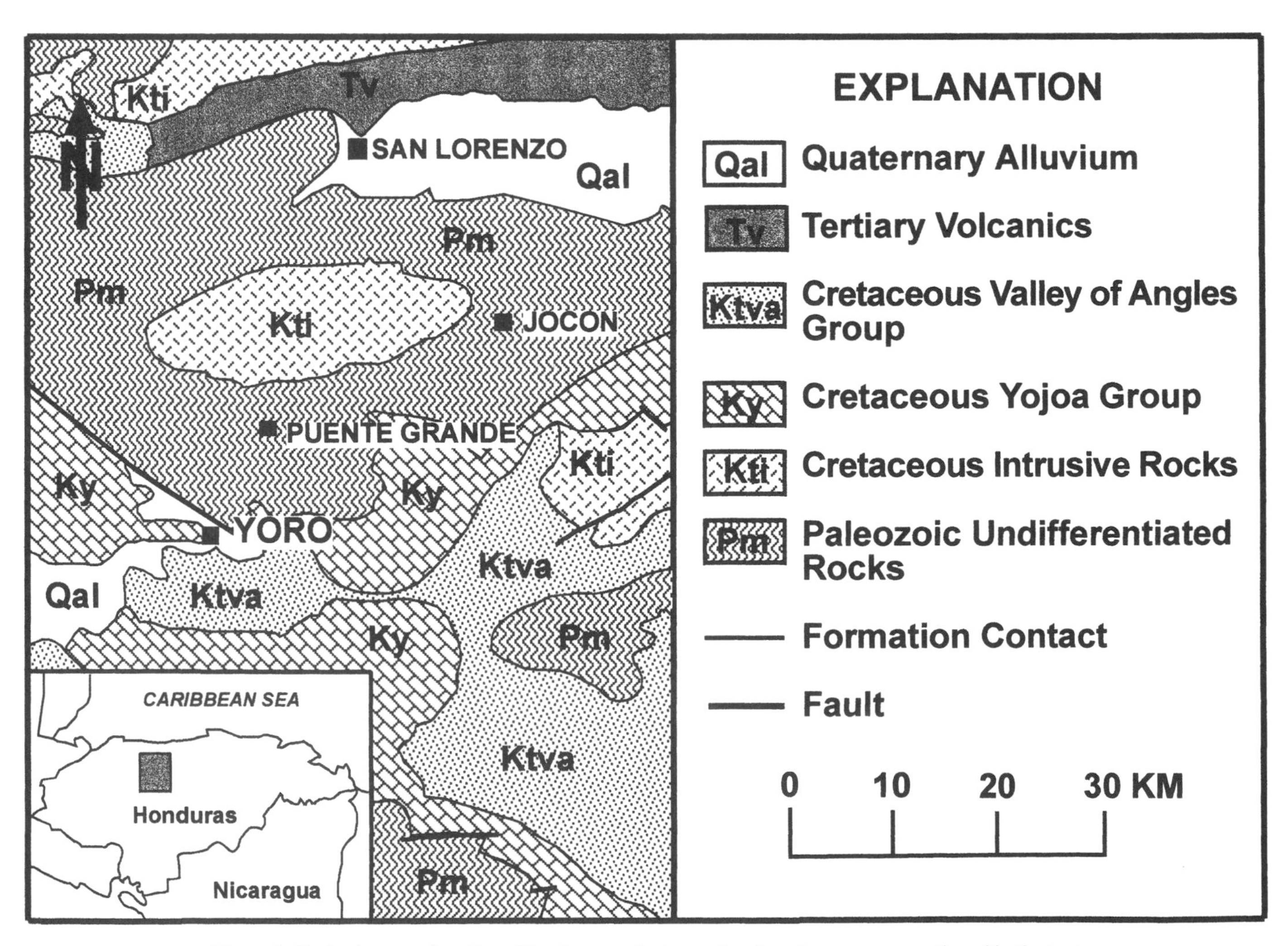

Figure 1. Geologic map of northern Honduras project area showing place names mentioned in the text. Modified from Aceituno (1974).

Highlands are extremely rugged owing to the presence of a resistant limestone.

Aceituno (1974) indicated that the road centerline would cross the Jurassic Yojoa Group and an undifferentiated Paleozoic series. Not available at the time of the project, the 1991 geologic map of Honduras (Kozuch, 1991) explanation describes the Yojoa Group as limestones and shales and names the undifferentiated Paleozoic series shown on the 1974 map, the Cacaguapa Schist. The country scale geologic map available at the time of the project did not provide the detail needed to support road construction. As originally designed by the Honduran Secretaria de Estado en el Despacho de Comunicaciones, Obras Publicas y Transporte (SECOPT), the road followed a mountain ridge line passing north of the village of Jócon. From April 3 to 14, 1987, a geologic reconnaissance was conducted to establish the predominate rock types and their engineering significance along the proposed centerline. The SECOPT proposed centerline and surrounding region was investigated to determine the feasibility of excavation by ripping or blasting. A reconnaissance geologic map was developed from surface float and intermittent outcrops over very difficult terrain. A topographic map (Defense Mapping Agency, 1:50,000 scale, series E752, sheet 2861 IV, Jócon) with 30 m contour intervals was used as the base map.

Two primary rock types were encountered during the reconnaissance: a resistant limestone and a weathered schist. The limestone is gray to buff, massive, dense, and pitted at the surface from solution activity. Some calcite veins and small crystals are present. The limestone formation forms a caprock on the ridge top. It has weathered to numerous rounded outcrops where numerous solution features and sinkholes occur. No bedding is visible at the surface exposures. It was very difficult to ascertain the structural orientation because of weathering.

During traverses along the southern flanks of the mountain ridge, numerous outcrops of an underlying schist were discovered. The schist is phyllitic and weakly jointed. It has some slip planes and includes veins and pods of quartzite. It is exposed in small outcrops on steep mountain slopes below the limestone cap rock. Much of the contact could be approximated from float on the mountain side. Where the contact is exposed, a white chalky clay approximately 1 m thick separates the rock units. The schist was highly weathered and weakly jointed and appeared to be readily rippable by D-8 bulldozers. D-8 bulldozers have ripping teeth for soft rock excavation and were the heavy earthmovers available to the engineer task force.

A geologic reconnaissance map delineating the limestone and exposures of the underlying schist was prepared (Fig. 2). Recommendations from the geologic reconnaissance to the task force engineer were: (1) Avoid large excavation in the limestone areas if possible. (2) Route roadcuts through schist areas for ease of excavation. (3) Use crushed limestone for road and concrete aggregate.

Clay-filled solution cavities in the Yojoa Group limestone had adversely affected the roadcut excavation effort in 1987, delaying the construction effort by a month. Clay seams filling the solution cavities entrapped drill steels or deformed them and detracted from the blasting effort. The clay seams absorbed blast energy, which allowed the rock to move preferentially along the clay seams without the necessary flexure to fragment the rock into manageable sizes. To continue along the projected route, several more kilometers of limestone roadcut had to be excavated. Some of the roadcuts would be 30 m deep. Based on the geologic reconnaissance map, a new road alignment was recommended to avoid the limestone excavation problems by traversing the lower flank of the ridge through the weathered schist. This new alignment almost doubled the length of the road. Although longer, the new road cut across the highly weathered schist allowing maximum excavation progress by ripping with a minimum of blasting. The new road was completed on time during the 1988 construction season.

Although easily excavated, the schist was unsuitable for a road aggregate because of low compressive strength, so a quarry was developed in the limestone caprock to produce road metal. The quarry operation was continually hampered by drilling and blasting difficulties due to the persistent clay seams but was able to produce the 29,000 cubic m^3 of two-inch gravel needed to metal the road.

SOUTHERN HONDURAS

In 1989, Task Force Tiger, 27th Engineer Battalion (Corps Combat) (Airborne) was responsible for designing a new airfield on the Pacific Coastal Plain of southern Honduras near Choluteca (Fig. 3). The new airfield was to be constructed during the next construction season in 1990. The scope of the mission included a 1,280 m flight landing strip with 0.45 m of compacted gravel over a geotextile fabric surfaced with a single surface treatment, and an aircraft parking apron with gravel surface large enough for several C-130 cargo aircraft. Other units nearby were constructing a cargo staging area near the port of San Lorenzo and a raised causeway through mangrove areas near Punta Ratón. Large quantities of construction aggregate were needed for each of the three projects.

At the time of the exercise, only two operating quarries in Honduras produced coarse aggregate. Both quarries are in the Tegucigalpa area, one in basalt and the other in rhyolite. The cost of transporting aggregate from Tegucigalpa was prohibitive. High-quality local materials seemed scarce. Most of the rock types exposed near the project areas are ignimbrites with low compressive strengths that are unsuitable for base course materials. The engineer design team for the exercise requested a military geologist from the U.S. Army Engineer School to evaluate local sources of construction material.

The project area lies on the Honduran Pacific Coastal Plain. Volcanic mountains lie to the north, but only isolated hills rise above the generally flat-lying coastal plain. The 1974 country geologic map indicates that the coastal plain is comprised of Quaternary alluvium and a few large areas of Tertiary volcanics (Fig. 4). Topographic maps show numerous isolated hills that do not appear on the geologic map. These monadnocks occur in the

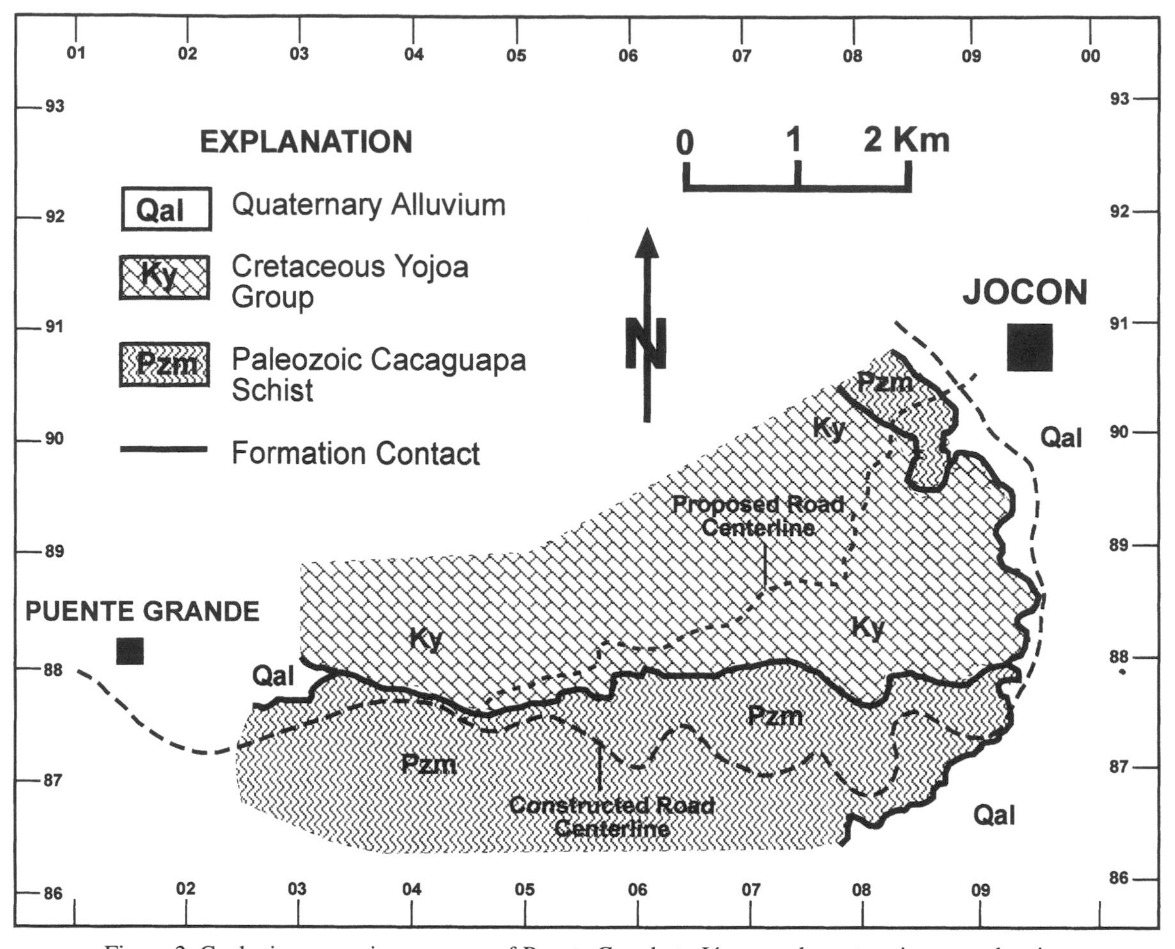

Figure 2. Geologic reconnaissance map of Puente Grande to Jócon road construction area showing proposed centerline and constructed road centerline selected on the basis of more favorable geologic conditions.

vicinity of all three project areas. The absence of detailed geologic mapping in the area necessitated geologic reconnaissance for a suitable aggregate source. From February 14 to 22, 1989, monadnocks in the vicinity of the Choluteca, the port of San Lorenzo, and the Punta Ratón causeway were evaluated for their suitability for quarrying.

Geologic reconnaissance determined that the monadnocks have little soil cover and consist mostly of extrusive igneous rocks. Rock types were determined from float and small rock outcrops. Some of the monadnocks had small abandoned quarry sites from previous local construction projects. Many monadnocks are composed of ignimbrites that were considered undesirable for use as a road metal while others were remnants of more competent lava flows. Monadnocks composed of rhyolite and andesite lavas were evaluated for their suitability as sources for construction aggregate. Rock specimens at each monadnock were collected and classified and field tested in accordance with U.S. Army Engineer School and U.S. Department of the Army (1967, 1971) publications. These tests may seem to be an oversimplification to professional geologists but in many cases represent the practical knowledge of many military engineers in dealing with rock materials. These field tests include:

(1) Toughness (mechanical strength): resistance to crushing or breaking; estimated in the field by attempting to break the rock with a hammer or listening to the sound of a hammer impact. A ringing sound indicates a more competent material than a dull thud.

(2) Hardness: resistance to scratching or abrasion; estimated by scratching the rock with a steel knife blade. Soft materials are readily scratched; hard materials are difficult or impossible to scratch. Mohs' hardness scale is the basis for this test.

(3) Durability: resistance to slaking or disintegration due to weathering; estimated in the field by observing the effects of weathering on natural exposures of rock.

(4) Chemical stability: resistance to reaction with alkali materials in portland cement. Several rock types contain forms of silica that react with alkalies in cement to form a gel that absorbs water and expands to crack or disintegrate the hardened concrete. Estimated in the field by identifying the rock and comparing it to known reactive types. There were no known concrete problems

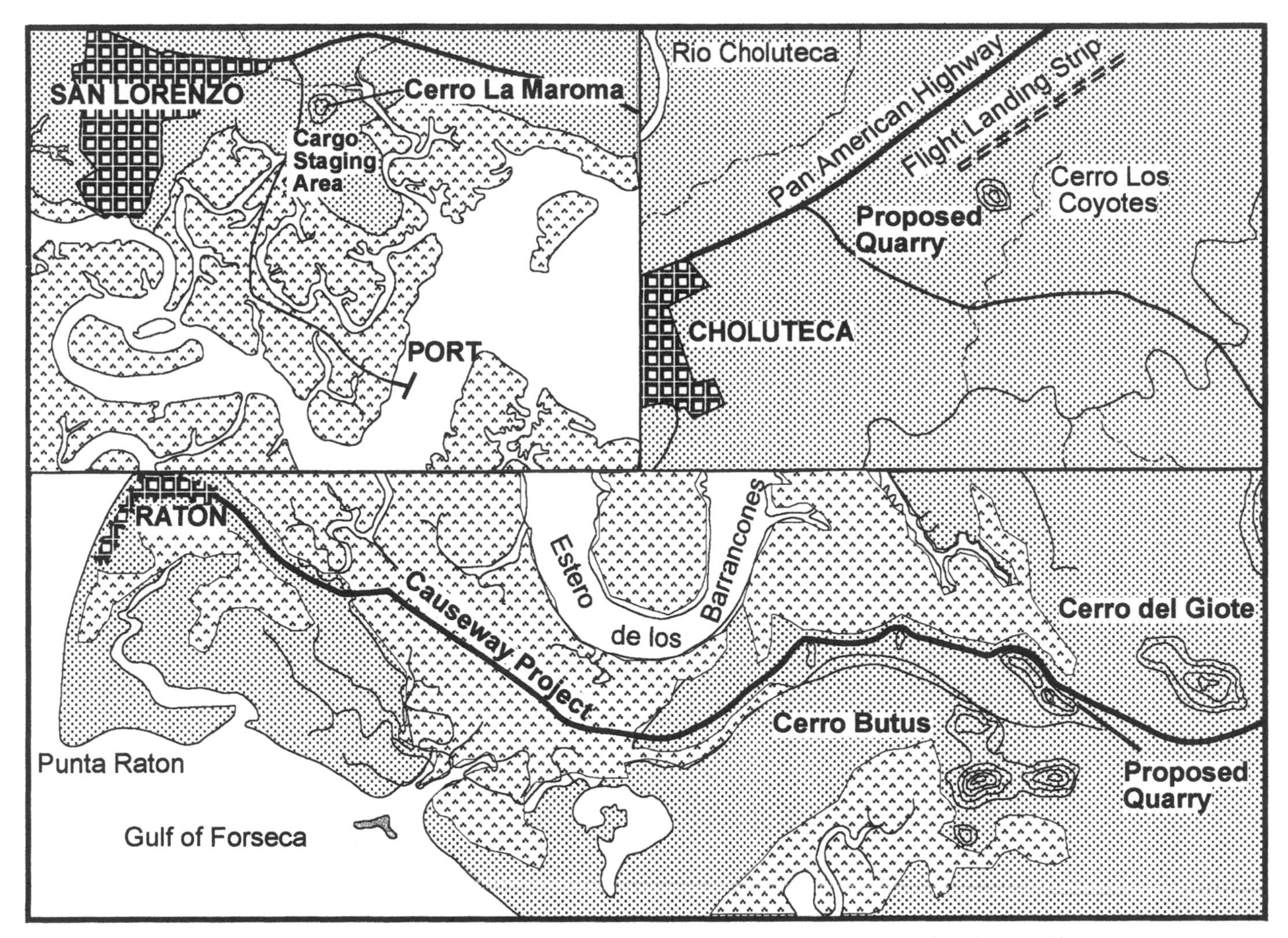

Figure 3. Maps of southern Honduras project areas showing place names mentioned in the text. The waffle-like pattern represents cities, the triangular dots represent mangroves, the screens represent open land, and the white represents water.

due to alkali-aggregate reaction in Honduras, but the silica rock types indicate a potential for problems.

(5) Crushed shape: characterized by breaking into irregular bulky-angular fragments that provide the best aggregates for construction because the particles compact well, interlock to resist displacement, and distribute loads that are nearly equal in strength in all directions.

(6) Surface character: bonding characteristics of the broken rock surface. Excessively smooth, slick, nonabsorbent aggregate surfaces bond poorly with cementing materials and shift readily under loads.

(7) Density: weight per unit volume; estimated in the field by "hefting" a rock sample. Heavy generally indicates both toughness and durability.

Based on these tests three monadnocks were recommended for quarry development, one in each project area. In addition, a quarry reconnaissance for each site was developed describing the nature of the deposit, overburden, existing facilities, potential water supply, ground-water and drainage problems, environmental constraints, and recommendations on site development.

To support airfield construction near Choluteca, the Cerro Los Coyotes monadnock was recommended as a quarry site. On Defense Mapping Agency, 1:50,000 scale topographic map, sheet 2755 I, Ciudad Choluteca, it appears as a prominent hill in the middle of the Choluteca valley located approximately 4 km east of Choluteca. It was conveniently close to the proposed airfield and provided an aggregate suitable for base course. The rock type is a light purple and gray felsite, with fine brown flow banding. It weathers to chalk white at the surface. Aggregate suitability field test results are included in Table 1. Although this rock type is not considered ideal for a construction aggregate, it was by far more suitable than many of the weaker ignimbrites in the surrounding area. The ignimbrites generally fail both the toughness and hardness tests.

To support construction of a cargo staging area near the port of San Lorenzo, a quarry site was recommended at Cerro

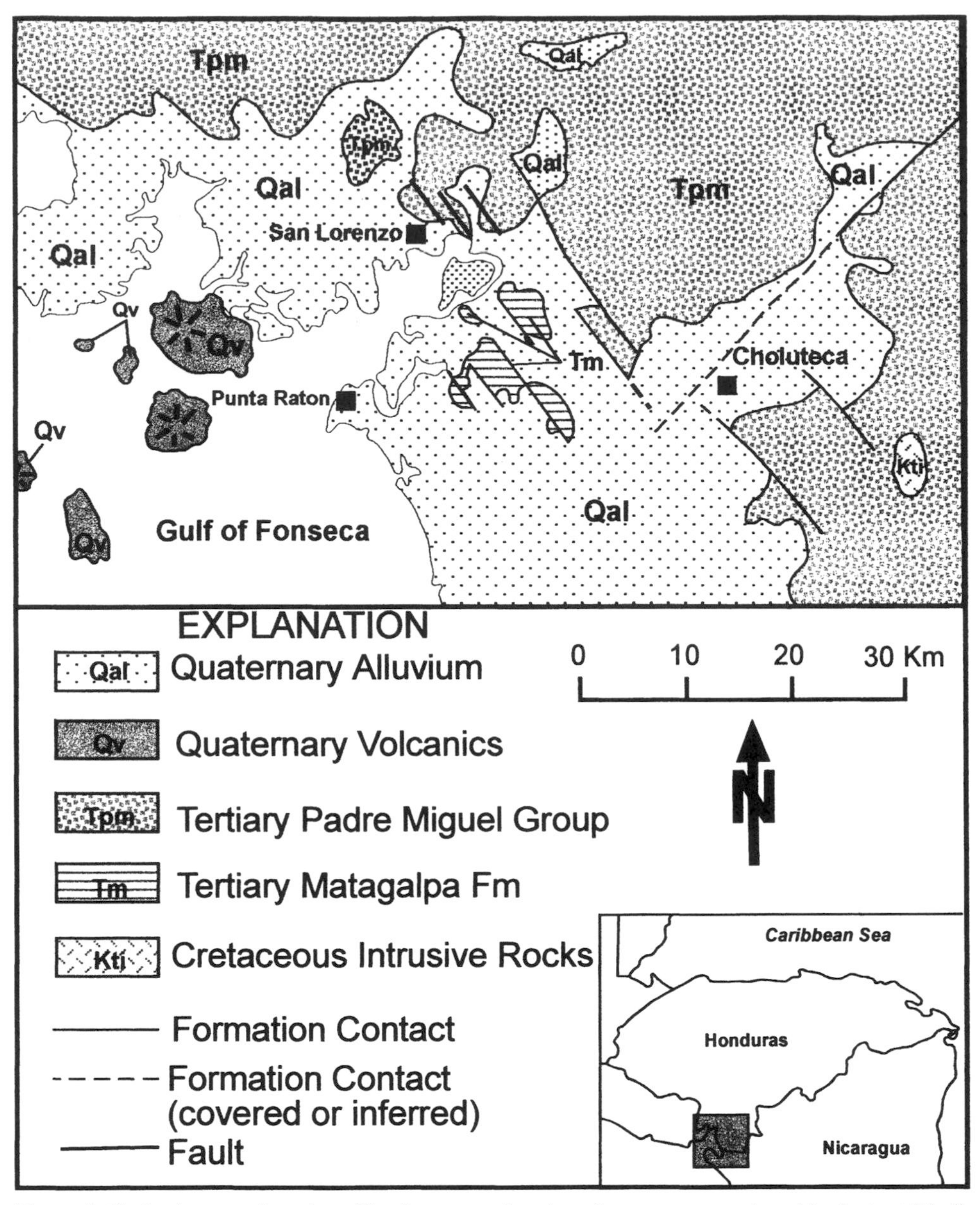

Figure 4. Geologic map of southern Honduras area showing place names mentioned in the text. Modified from Kozuch (1991).

La Maroma, an imposing volcanic erosional remnant composed of andesite. On Defense Mapping Agency, 1:50,000 scale topographic map, sheet 2756 III, San Lorenzo, it appears as a prominent hill located approximately 1 km east of San Lorenzo. The rock was described in previous U.S. Army Corps of Engineers study by Willingham (1983) as a fine-grained, porphyritic, brownish-gray, fresh, hard, dense, layered extrusive-flow rock. Aggregate suitability field tests results are included in Table 1. The andesite overlies an ignimbrite that limited the amount of suitable aggregate that could be quarried, although more than enough was available to support the construction of the cargo marshaling area. The ignimbrite was unsuitable for military construction.

To support construction of a causeway in the Punta Ratón area, a quarry site was recommended at Cerro Butus, a monadnock composed of felsite. On Defense Mapping Agency, 1:50,000 scale topographic map, sheet 2755 IV, Marcovia, it appears as a prominent ridge located approximately 8 km east of Ratón. The felsite is purple, with small white inclusions and prominent light-brown flow bands. Cerro del Giote, another monadnock 1 km east of Cerro Butas, had been utilized in 1987 for a military quarry, although the rock material was deemed unsuitable. The 1987 quarry leveled a small rock knob located immediately to the south of the hill before developing the small quarry existing face in hillside. This small face was developed in the center of the primary drainage channel on the flank of the

TABLE 1. AGGREGATE SUITABILITY FIELD TEST RESULTS FOR SOUTHERN HONDURAS PROJECT AREAS

Engineering Property	Coastal Plain Ignimbrites	Choluteca Felsite	San Lorenzo Andesite	Punta Ratón Felsite
Toughness	Poor	Good	Good–fair	Good
Hardness (Mohs')	3	5.5	5.5	5.5
Durability	Poor	Good	Fair	Good
Density	Light	Fair	Fair	Fair
Chemical Stability	Alkali reaction possible	Alkali reaction possible	Alkali reaction possible	Alkali reaction possible
Surface Characteristics	Good	Good	Good	Good
Crushed Shape	Good	Excellent (bulky-angular)	Excellent (bulky-angular)	Excellent (bulky-angular)

mountain in very weathered ignimbrites. Aggregate suitability field test results for the felsite are included in Table 1.

In the Choluteca region the engineer task force constructing the new airfield was primarily composed of Airborne Engineers. They and most of their equipment were parachuted into the area. Based on time constraints that the airborne engineers were working under, the quarry supporting airfield construction was established well before the main body arrived with the intent of stockpiling aggregate so that it would be available when needed. An army quarry team developed a quarry in Cerro Los Coyotes to produce 33,000 m^3 of aggregate to support a 60-day construction effort.

Military quarries were also developed at Cerro La Maroma to produce 6,200 m^3 of aggregate for the Port of San Lorenzo staging area, and near Cerro Butus to produce 20,600 m^3 of aggregate for the Punta Ratón causeway. From the geologic reconnaissance, suitable aggregate sources were identified and developed to support military construction in a region where few suitable aggregate sources exist, contributing to the success of each of these projects.

ECUADOR

In the spring of 1987, Task Force 1169 planned to upgrade 18 km of coastal road to a two-lane road with a surface treatment on the Pacific coast of Ecuador. Three weeks before the equipment was to sail for Ecuador, however, the mission was changed because large earthquakes severely damaged many areas of the country. The task force mission was changed to build roads in the Napo Province, which was isolated by the earthquakes.

The Napo Province of Ecuador is in the Oriente (eastern jungle) physiographic belt (Fig. 5). The Oriente physiographic province is subdivided into the sub-Andean zone and the Amazon lowlands. The sub-Andean zone consists of the eastern slopes of the Andes mountains and the Amazon lowlands includes the flat valleys of the upper reaches of the Amazon tributaries. The new mission required road building in the sub-Andean zone. The sub-Andean zone is composed of dissected mountainous plateaus

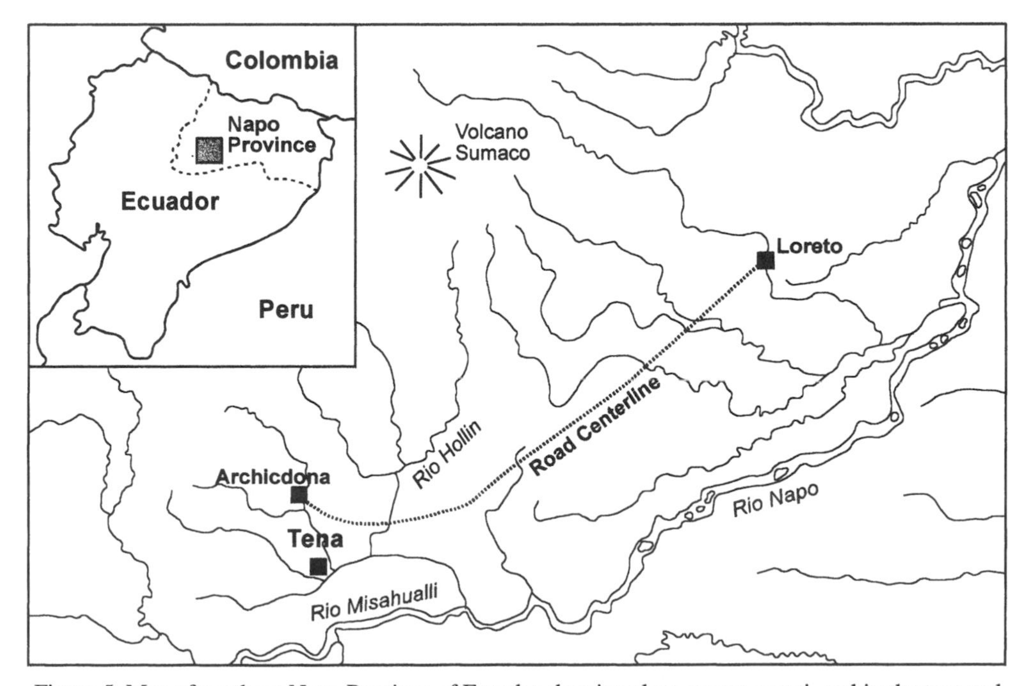

Figure 5. Map of southern Napo Province of Ecuador showing place names mentioned in the text and road and bridge project general alignment. Modified from Instituto Geográfico de Militar (1985).

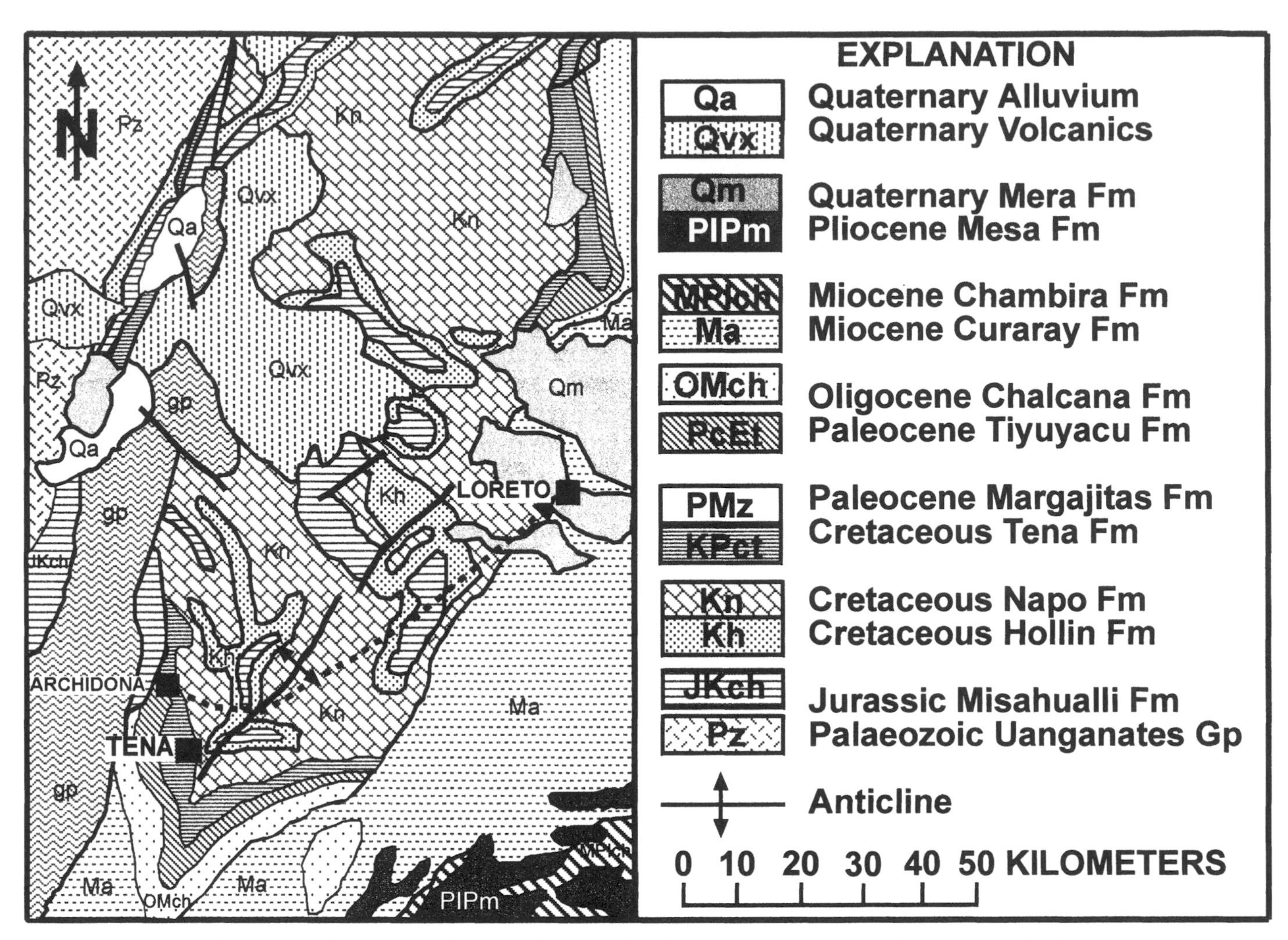

Figure 6. Geologic map of Napo uplift area in Ecuador showing general road alignment crossing the axis of the anticline. Modified from Ministerio de Recursos Naturales y Energéticos (1982).

with steep jungle-covered slopes and fast-moving streams and rivers. The new exercise area receives more than 500 cm of rainfall per year.

The new mission was to upgrade an existing trail approximately 11 km from Archidona to the Río Hollín to a one-lane military standard road, bridge the river, and construct as much of a new one-lane road in the general direction of Loreto as possible. Eleven km of road were upgraded, a 91 m (300 ft) panel bridge was built over the Río Hollín, and 4 km of new road on the far bank were completed under very difficult environmental conditions. The Río Hollín river initially provided aggregate, but was soon nearly exhausted. New sources of aggregate were needed to continue the road construction. The follow-on engineer task force requested a military geologist from the U.S. Army Engineer School to accompany a surveying party to identify aggregate sources for road construction and develop a quarry plan for locations east of the Río Hollín.

Previous studies by Kennerley (1980), Wasson and Sinclair (1927), and the Ministerio de Recirsos Naturales y Energéticos (1982) provided a basic overview of the region. The geologic map showed the projected road crossing the axis of the Napo anticline. The major rock units exposed in the structure are the Cretaceous Napo and the Hollín Formations (Fig. 6). Tschopp (1953) described the structure of the anticline and the stratigraphy in some detail, which was instrumental in determining the availability of higher quality construction materials.

From August 16 to 26, 1987, a geologic reconnaissance was conducted to find a suitable quarry site for new sources of aggregate. Ecuadoran Instituto Geográfico Militar (IGM), 1:50,000 scale, series CT-0111-E3, Tena, and CT-0111-E4, Lushanta, map sheets with 40 m contour intervals were used in conjunction with stereo pair mapping aerial photography. According to Tschopp (1953) the Río Hollín cuts through the Lower Napo Formation east of Archidona, exposing black shales. The shales and limestones of the lower member of the Napo Formation were exposed in roadcuts near the river during the road construction. Tschopp (1953) divided the Napo Formation into three main units:

(1) Upper Napo: gray-green, dark gray, and black shales.

(2) Middle Napo: main limestone, a zone of thick-bedded to massive limestones that maintains a remarkably constant thickness.

(3) Lower Napo: gray-green to dark-gray and black or gray sandy shales with glauconitic sandstone nodules, lenses, and beds, and a few subordinate limestones.

The shales and limestones of the Lower Napo Formation were unsuitable for construction. The limestones were thin and weathered, and the shales slaked upon exposure to the atmosphere. Deliberate exploration on the far side of the Río Hollín with the help of a machete led to the discovery of limestone outcrops with sinkholes at higher elevations. A series of massive limestone benches comprise part of the plateau field tested between the Río Hollín and the Río Copayacu. Field tests showed this limestone to be a high-quality construction aggregate. It matched Tschopp's description of the Middle Napo Formation. A suitable quarry site was selected after further exploration, and a quarry reconnaissance report developed. This source ensured that aggregate would be available to complete the road construction project.

Other problems, particularly unstable slopes, were encountered with the Lower Napo Formation. Saturation from continual rainfall, combined with unfavorable structural orientation of the bedding planes into roadcuts, resulted in numerous slope failures. Rock excavations were being cut almost vertically without taking into consideration the structural orientation of the bedding planes. In general, the bedding dipped parallel to the roadcuts climbing the ridge on the far side of the river. Where the road curved, the bedding planes would dip into the road and fail in a series of small slides and slumps. Construction engineers were advised on aspects of slope stability especially with regard to the structural orientation of dipping sedimentary rocks.

Because of the difficulties encountered in excavating in the Lower Napo Formation, it was necessary to determine what other types of geologic difficulties might be encountered during subsequent road segments. The geologic map indicated the major rock types and formation names, but it was important to relate these geologic mapping units to aggregate suitability and the potential for excavation problems. Reconnaissance to the north across the Napo structure indicated what could be expected during the remaining construction effort. The Hollín Sandstone, which underlies the Napo Formation, occurs closer to the anticline crest along the proposed route. It is exposed in existing roadcuts across the north end of the Napo anticline. The Hollín Formation is generally a clean quartz sandstone and is responsible for numerous tar and oil seepages that occur along the Río Hollín. The basal sandstone in this formation oozes natural asphalt. The Ecuadorans excavate the natural asphalt and use it as a natural paving material. The material is generally rippable. Beneath the Hollín Formation is the Río Misuallí Formation of ignimbrites and basalts. Drilling and blasting would be required in these units. Because of this reconnaissance, follow-on engineer task forces had a good idea of what types of construction aggregate and excavation difficulties they could expect as the construction exercises continue the road toward Loreto.

CONCLUSIONS

Military engineers training in a wide variety of difficult conditions in austere regions of Central and South America owe much of their success to the efforts of military geologists who identify sources of construction aggregate and provide advice on geologic conditions that can ultimately dictate success or failure of the construction exercises. These geologists may come from the deploying unit or be advisors from the U.S. Army Engineer School or Corps of Engineers. They must turn previous studies and generalized geologic maps into practical engineer assessments based on firsthand observations and present their findings to engineer staff officers in a manner that will be understood. Nation assistance exercise success in developing host nation infrastructure, building a better relationship between these countries and the United States, and creating opportunities for a better life for the local inhabitants often depends on such efforts.

ACKNOWLEDGMENTS

The author gratefully acknowledges the assistance provided by the U.S. Army Topographic Engineering Center to the U.S. Army Engineer School, and U.S. Army engineers everywhere. During my assignment at the U.S. Army Engineer School, their support in providing maps, aerial photography, and special studies was instrumental in the success of numerous nation assistance exercises.

REFERENCES CITED

Aceituno, R. E., 1974, Mapa Geologico de Honduras: Ministerio de Recursos Naturales Direccion General de Minas E Hidrocarburos, scale 1:500,000, 1 sheet.

Defense Mapping Agency Hydrographic/Topographic Center, Maps, scale 1:50,000, series E752, sheets 2755 I, 2755 IV, 2756 III, and 2861 IV, Washington, D.C.

Instituto Geográfico de Militar, 1985, Republica del Ecuador, scale 1:1,000,000, 1 sheet.

Kennerley, J. B., 1980, Outline of the Geology of Ecuador, Overseas Geology and Mineral Resources: No. 55: London, Institute of Geological Sciences, 17 p.

Kozuch, M. J., 1991, Mapa Geologico de Honduras 2d ed.: Instituto Geográfico de Nacional, scale 1:500,000, 3 sheets.

Mills, R. A., Hugh, K. E., Feray, D. E., and Swolfs, H. C., 1967, Mesozoic Stratigraphy of Honduras: Bulletin of the American Association of Petroleum Geologists, v. 51, p. 1711–1786.

Ministerio de Recirsos Naturales y Energéticos, 1982, National Geological Map of the Republic of Ecuador: Instituto Geográfico Militar, scale 1:1,000,000, 1 sheet.

Tschopp, H. J., 1953, Oil Exploration in the Oriente of Ecuador 1938–1950: Bulletin of the American Association of Petroleum Geologists, v. 37, p. 2302–2347.

U.S. Department of the Army, 1971, Geology: U.S. Army Technical Manual (TM) 5-545, 125 p.

U.S. Department of the Army, 1967, Pits and Quarries: U.S. Army Technical Manual (TM) 5-322, 142 p.

Wasson, T., and Sinclair, J. H., 1927, Geological Explorations East of the Andes in Ecuador: Bulletin of the American Association of Petroleum Geologists, v. 11, p. 1253–1281.

Weyl, R., 1965, Die Geologie Mittelamerikas: Berlin, Gebruder Borntraeger, 226 p.

Williams, H., and McBirney, A. R., 1969, Volcanic History of Honduras: Los Angeles, University of California Press, 101 p.

Willingham, R., 1983, Petrographic Examination of Honduras Ledge Rock for Mobile District: U.S. Army Corps of Engineers, South Atlantic Division, Lab No. 57/9933, p. 1–3.

Manuscript Accepted by the Society October 29, 1997

Printed in U.S.A.

Geological Society of America
Reviews in Engineering Geology, Volume XIII
1998

Potable water well design for Humanitarian Civic Action well drilling missions

John N. Baehr
U.S. Army Corps of Engineers, Geotechnical and Environmental Branch, P.O. Box 2288, Mobile, Alabama 36628-0001

ABSTRACT

Geologists from the U.S. Army Engineer District in Mobile, Alabama, have supported military water well drilling missions throughout the world. Many of these missions supported design requirements for Humanitarian Civic Action (HCA) wells but did not follow standard military water well construction practice. These design requirements have often been the result of local well construction regulations or the need for well yield that exceeded typical design.

Each branch of our military has water well drilling capability, and most drilling systems are similar in depth and hole size ratings. Standard well completion kits for mobilization have been developed for construction of wells up to 455 m (1,500 ft) deep. Normal training for military well drillers has been limited, and the emphasis is on completion of a tactical, low-yield well where many potable well construction practices are not required.

Humanitarian Civic Action well drilling missions have become an integral part of Nation Assistance exercises. Some HCA wells required special training, modifications to drilling equipment, and special well designs to meet the goals of the exercise. The probability of success had to be high for these missions to be approved. Consequently, civilian geologists were used to support the siting, well design, and procurement of materials. Some of these complex missions required on-site consultation. A specialized team of personnel including geologists, hydrogeologists, and geophysicists, designated the Water Detection Response Team, was assembled during the 1980s by the Corps of Engineers to site well drilling locations for military drilling operations and is used for many HCA missions.

INTRODUCTION

The U.S. Military has become extensively involved with construction of potable water wells for developing nations during various types of exercises. Our military well drilling capability has been maintained to provide a potential source of water for base camps where existing surface water supplies or water wells may not be readily available. Each branch of the military has its own drilling system, generally suited to its mode of transportation. The depth and hole size limitations of the equipment are generally similar, approximately 455 m (1,500 ft) and 300 mm (12 in), respectively. All of the military systems are capable of both mud and air rotary drilling (Multiservice Procedures for Well-Drilling Operations, 1994).

As is true with any skill, continual practice is required to maintain expertise. The well drilling training that is provided to soldiers is brief by most civilian standards and is centered on the construction of what may be classified a "tactical" well. A tactical well must be completed in a short period of time, possibly under adverse conditions. The water produced from the well has no quality criteria, only quantity. The water is always provided with some type of on-site treatment, usually through a reverse osmosis system. For this reason, wells that are contaminated or produce sand or turbid water can successfully be used by the mil-

Baehr, J. N., 1998, Potable water well design for Humanitarian Civic Action well drilling missions, *in* Underwood, J. R., Jr., and Guth, P. L., eds., Military Geology in War and Peace: Boulder, Colorado, Geological Society of America Reviews in Engineering Geology, v. XIII.

itary as a water source. A general comparison of an HCA well and a tactical well is shown in Table 1.

Through the years, a tactical well for the military became either a 150-mm (6-in) or 200-mm (8-in) diameter well using either polyvinyl chloride (PVC) or steel construction components. The well yields required for most base camps are small, generally about 3 lps (50 gpm). A 150- or 200-mm diameter well is more than sufficient. PVC well constructions are often used in shallow wells up to 180 m (600 ft) in depth, and steel constructions for wells up to 455 m (1,500 ft) in depth.

To facilitate the construction of tactical wells, two well completion kits have been developed for the U.S. Army as indicated in Table 2. For wells up to 180 m (600 ft) deep, a kit containing 200-mm (8-in) diameter PVC casings and screens was developed. This kit is lightweight and easily transportable. A second kit contains 150-mm (6-in) diameter stainless steel screens and steel casing. This kit can be used for depths of up to 455 m (1,500 ft). These kits have been designated the "600 Foot Well Completion Kit" and "1,500 Foot Well Completion Kit," respectively.

To maintain proficiency in well drilling, military well drilling detachments are attached to many overseas engineering exercises. As a part of these exercises, HCA wells are constructed to provide reliable sources of potable water for towns or villages in the host nation. These wells also serve as important training exercises for military well drillers. The best training is provided when the well construction is similar to what would be expected in the event of an actual military mobilization.

For the last several years, the well drilling program has grown tremendously. The requests for HCA wells during exercises are now receiving the attention of general officers and ambassadors. Although the program is designed for the training of our troops, the success of each drilling mission has become extremely important. Notable differences in the definition of success became obvious as wells were constructed and turned over to host nations. To be successful, the wells had to produce a high quality water that required little or no treatment. The well could not produce sand or other solids that fouled plumbing and ruined pumps. The most important consideration for a successful well is that the location has the hydrogeologic potential to produce water. In the early 1980s, the poor success rate of HCA wells became a concern and the various military commands looked for assistance.

TABLE 1. COMPARISON OF WELL DESIGN CRITERIA

Parameter	HCA Well	Tactical Well
Yield per person per day	50 to 150 gallons	10 to 50 gallons
Well siting	Must have rights of entry or be public property	Well placement less restricted
Well grouting	Required to be sanitary	Inadequate seal
Sand content	<5 ppm	Not a concern due to short well life
Water quality	Must be good; water generally not treated	Less important; water processed with reverse osmosis unit

TABLE 2. ARMY WELL COMPLETION KITS

Standard Kit	Depth (m)	Diameter (mm)	Material
600 ft well completion kit	180	200	PVC
1,500 ft well completion kit	455	150	Steel

WELL DRILLING ASSISTANCE

Once the village or town in the host nation is nominated for an HCA well, the local populace fully expects a successful well. Some past failures with well construction have created strained relations between the host nation and our military, bringing the program under close examination.

Within the Corps of Engineers, a team designated the Water Detection Response Team (WDRT) exists to assist the U.S. military in the selection of water well drilling locations and well designs during any type of mobilization. This team has several components, but two are of special interest to HCA missions: the Worldwide Water Resources Database and the supporting specialists who assist in final well design and construction. Using these two elements, the success rate of HCA wells has risen sharply. The database provides the information necessary for site selection, potential well depths, and types of material to be penetrated during drilling (U.S. Army Corps of Engineers, Mobile, unpublished open file reports). With this information, potential drilling sites are selected that assure a high rate of success for the military drillers. The wells can also be designed to suit the needs of the drilling detachment with respect to their available equipment and supplies. The materials used in construction of the wells can be estimated fairly accurately, which prevents excessive expenditures for shipping and materials. The following items are considered for each site: (1) hydrogeologic conditions, (2) limitations of drilling equipment, (3) accessibility of equipment, and (4) water requirements per capita.

After site selection, most of the wells do not require extensive services from a civilian well design expert. Ideally, wells are located in areas where the maximum number of wells can be completed during an exercise. This translates to well-defined aquifers in unconsolidated formations with maximum depths of a few hundred feet. There have been many exceptions to this "ideal" site, causing some problems for the program.

SPECIAL WELL DESIGNS

The limitations of military drilling equipment can be an important factor for HCA well designs. During the 1980s, a series of wells were promised to a South American country to supply water to highly populated coastal areas. These areas were lacking potable water supplies. The host nation requested well construction similar to other public wells near the capital city, which was initially promised. These promises were made by persons not familiar with well construction and the military drilling systems. The mission, as tasked, was supposed to produce wells 275 m (900 ft) deep and capable of yielding 93 lps (1,500 gpm) of water. Wells were to be completed at a rate of one well every three days. After much debate and discussion, the unrealistic mission was drastically changed. An army well drilling detachment was deployed to the host country to construct a limited number of wells based on a well design from the Mobile District, Corps of Engineers. The basis of the design was completion of a single-string, naturally developed well with an upper, inside diameter of 300 mm (12 in), reduced to 262 mm (10.5 in) in a thick, confining layer overlying the aquifer. Estimates were to complete a well in 30 days with a design yield of 70 lps (1,100 gpm). This design, with a hole up to 400 mm (16 in) in diameter, was considered the largest capacity well that could be drilled with the available military equipment. In spite of the unconsolidated formations underlying the site, the limited size of the military drilling machine made some tasks difficult that would have otherwise been simple with larger drilling equipment. Because of the low capacity mud pumps on the drilling machine, the uphole velocity of the drilling fluid was only a fraction of the normal minimal rate. This made maintaining the proper characteristics of the drilling fluid important to provide the viscosity necessary to remove the drill cuttings without becoming so thick as to overload the pump.

To complicate the well design, the upper 65 to 100 m of overburden strata was Quaternary coastal plain sediments containing brackish water, and the underlying confined Tertiary aquifer contained water with extremely high dissolved CO_2 and a low pH, making it highly corrosive. The selected well screen was wire-wrapped stainless steel, but the casing selected was fiberglass with threaded joints. The fiberglass provided a material with high collapse strength that was unaffected by the corrosive water. The reduction in the casing diameter corresponded with a reduction in hole diameter. These size reductions, used in combination with a grouting basket, were made in a thick confining layer that allowed for a deep grout seal to isolate the brackish water overlying the aquifer.

Another example was a location where the hydrogeologic potential for a successful well was low and the type of required well construction was unknown. In 1992, a well was promised to a Central American nation without first investigating the possibility of success at the site. The village was in a mountainous region at an elevation more than 2,500 m. A political decision was made to drill at this site, and the ultimate failure of the initial well attempt brought the military well drilling program under unfair criticism. After the initial unsuccessful attempt, the Corps' WDRT was asked to investigate the possibility of siting, designing, and drilling a successful well. The WDRT recommended a well site approximately 1 km from the first attempted site.

Because of the scarcity of available data and many unknowns, the well depth could only be estimated at between 180 and 400 m; the final well design was to be determined after completion of a test hole. As is true of most engineer exercises, all well construction materials had to be purchased before the mobilization. Based on the initial investigation, the use of one of the standard military well completion kits would not be sufficient for an HCA well at this site. Utilizing advanced geophysical investigative methods would have greatly enhanced the ability to design a well for this site. Time and budget constraints precluded the use of geophysical exploration.

The final well design was based on a potential 180-m thickness of unconsolidated Quaternary pyroclastics overlying fractured Tertiary volcanics. It was not known if a sufficient aquifer existed in the unconsolidated material. This was due to the topography of the intermountain valley where the site was located with steep grades and numerous intermittent stream channels dissecting the pyroclastics. Considering these unknowns, the well casing procured was sized to be used in either of two ways: first, as well casing attached directly to well screen in a gravel-packed well design; or alternatively, casing set through the pyroclastics with a Halliburton-type grouting method for a deep well into the volcanic rock using down-hole, air percussion drilling. These optional designs are illustrated in Figure 1. A detailed procedural report was sent to the drilling detachment on recommended well completion, based on the results of the test hole drilled through the unconsolidated materials. The required well yield was 4 lps (60 gpm) to meet the needs of the village. However, with the limited amount of available information, it was not possible to guarantee this yield. When pressed for an opinion without advance geophysics or a test drilling program, a 50% probability of success was estimated. Because of the low probability of success, this mission was canceled at the last minute. I supported this decision because of the unknowns and the potential political fallout of another unsuccessful well. This site has been recommended for a geophysical survey and exploratory test boring in the future.

Another successful aspect of the HCA well drilling program involved the procurement of a complete mud/air rotary well drilling system for the El Salvadorian army in 1989. Using translated well drilling manuals, a drilling team of 10 soldiers was trained for one month to operate the equipment and to design and construct small wells. This system is able to complete wells up to 150 mm (6 in) in diameter and 65 m (200 ft) deep. This detachment drilled HCA wells during the latter years of the civil war in El Salvador in areas where contractors and various relief agencies could not operate due to hostilities. This detachment is still drilling today.

NEW GUIDELINES FOR WELL DRILLING OPERATIONS

The entire operation for well site selection, well design, procurement of construction materials, and actual drilling and construction has been streamlined through the years. One of the past

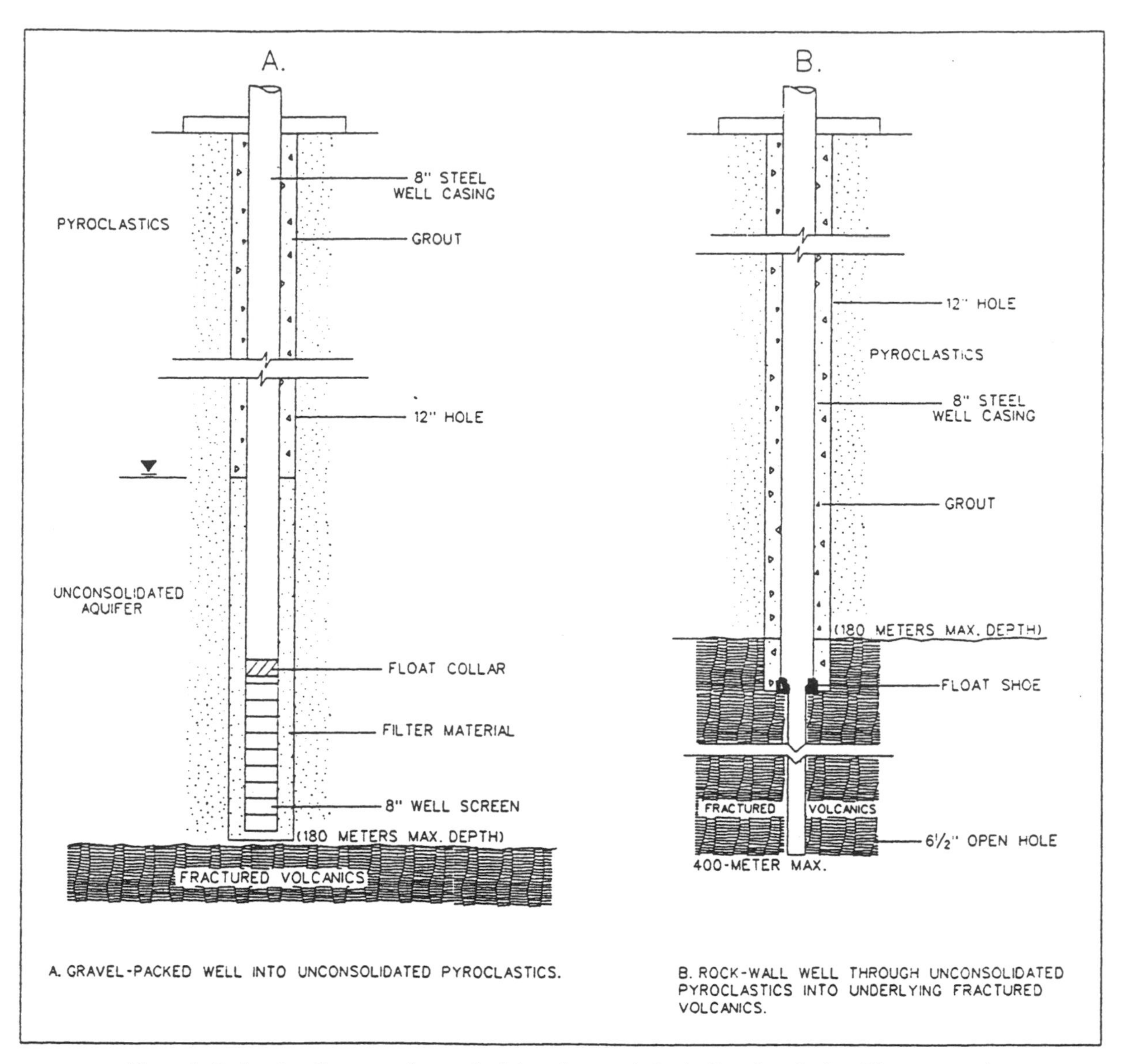

Figure 1. Optional well construction methods based on test hole. A. Gravel-packed well into unconsolidated pyroclastics. B. Rock-wall well through unconsolidated pyroclastics into underlying fractured volcanics.

problems with well drilling has been the rapid turnover of engineer officers associated with this mission. To help overcome the lack of experience in the U.S. Southern Command's exercise personnel, the Mobile District was asked to hold short well drilling conferences in Panama. As procedures were developed within the military system, Mobile District was asked to complete an HCA well drilling operations manual. This manual, Operation Guidelines For Humanitarian Civic Action Water Well Drilling, was completed in 1994 and was distributed to all branches of the military involved in the Southern Command's engineer exercises. This manual is not a typical drilling manual, but a manual for the personnel who will be directing the entire operation. One point the manual makes repeatedly is to "know before you go" when deploying and seek technical assistance to solve problems. Some important planning and operational procedures have been adopted because of the manual (Baehr, 1994).

Because there are usually so many more potential sites where communities need water than there are resources to provide them, selection lists are generally reduced to locations where the chance of success is high. Military planners desire a 90% success rate and have been achieving that goal for the last two years.

The U.S. Southern Command in Panama is including test holes as part of the HCA program. These test holes are not considered water well drilling missions, but rather a means to explore the potential for later well construction. A byproduct of the test holes may be a small diameter well installed in the test hole. This well can be supplied with either a hand pump or a small submersible pump to produce a limited amount of water. The data from the test hole is used to properly design a successful well that will provide maximum benefit to the surrounding community.

The new well drilling operations manual is the standard operating procedure for future missions in the Southern Command.

CONCLUSIONS

The U.S. military's HCA well drilling program has evolved into an effective approach for providing a safe source of drinking water in many countries. Data provided by civilian hydrogeologists prior to the missions ensure proper siting and design of most wells. Specific well design problems can be addressed by specially trained personnel while maintaining useful training for drilling detachments. New procedures that will allow for preliminary test drilling will enhance the efficiency of the program while providing useful data for any future development of ground-water resources.

REFERENCES CITED

Baehr, J. N., 1994, Operational Guidelines for Humanitarian Civic Action Water Well Drilling: Mobile, Alabama, U.S. Army Corps of Engineers, 128 p.

Multiservice Procedures for Well-Drilling Operations, 1994, Headquarters, Department of the Army, FM 5-484, Department of the Navy, NAVFAC P-1065, Department of the Air Force, AFP 85-23: Washington, D.C., U.S. Government Printing Office.

MANUSCRIPT ACCEPTED BY THE SOCIETY OCTOBER 29, 1997

Printed in U.S.A.

Geological Society of America
Reviews in Engineering Geology, Volume XIII
1998

Military geology should be upgraded as the U.S. Army stands down

Allen W. Hatheway and M. Merrill Stevens
Department of Geological and Petroleum Engineering, 129 McNutt Hall, University of Missouri, Rolla, Missouri 65401-0249

ABSTRACT

Military geologists have provided essential but little-known military intelligence and combat engineer support to the U.S. Army since the grand days of Lieutenant Colonel Alfred H. Brooks' assemblage of U.S. Geological Survey personalities on the World War I western front. Regrettably, since 1918, active-duty practice of military geology and topographic engineering has been career-killing, and therefore most commanders do not establish such a technical proficiency.

Germany found, at least by 1914, geologic knowledge to be essential to the advantageous commitment of troops. Works of German military geologists have never been equaled. Von Bulow's *Wehrgeologie* (Berlin, 1938) today is a superior manual of military/engineering geology. Germany's superior use of military geology employed professional geologists, many of whom were leading academics, through its reserve forces structure.

The authors advocate training of Army Reserve and Army National Guard military geologists, employed in a regular paid-drill augmentation to Regular Army combat units, down to maneuver battalion level, serving the Operations (S-3) Sections. These reservists should be treated in the manner of the health-science professionals and promoted as technical specialists rather than as troop leaders. The career ladder should run from second lieutenant to colonel, and the officers should be integrated through the Corps of Engineers.

INTRODUCTION

Students of military history know full well that the equation of combat force is made up of troop numbers and various multipliers relating to the relative advantages accrued to offensive or defensive tactics. Along with numbers of fresh, experienced troops and various types of weapons and firepower, *terrain* constitutes the third overwhelming triumvirate of most-important combat multipliers. Terrain in every sense of the word is *geology*. Master military tacticians throughout history have possessed some formal knowledge of, or innate appreciation for, terrain. Terrain is the "chessboard of the battlefield." The terrain factor is built into every screened tactical combat option presented to the commander by the supporting field staff. These officers are assigned to units commanded by officers up to and including colonel, or by the general staff for units commanded by general officers.

There are relatively few instances today in which tactical and command doctrine among the armies of the first- and second-world nations have actually recognized the value of formal military geologic evaluations and assessment of tactical options.

The very nature of military command and staff structure in most armies precludes the selection and training of technical specialists in the combat arms (such as infantry, artillery, armor, and engineer) and the combat support arms (such as intelligence, signal, and aviation). Every successful officer is considered first on merit of "troop leading" experience and capability. In the U.S. Army, commissioned officers are faced with a demanding series of formal military education, beginning with the Branch Basic school for second lieutenants, followed in order by Advanced Branch training for first lieutenants and captains, Combined Arms and Service Staff School (CAS3) for captains, the Command and General

Hatheway, A. W., and Stevens, M. M., 1998, Military geology should be upgraded as the U.S. Army stands down, *in* Underwood, J. R., Jr., and Guth, P. L., eds., Military Geology in War and Peace: Boulder, Colorado, Geological Society of America Reviews in Engineering Geology, v. XIII.

Staff College (CGSC) for majors, and, for a tiny fraction of them, the Army War College for lieutenant colonels.

Underlying this official doctrine of officer qualification training is the strong believe that each and every officer of a particular branch of the army is equally qualified to take any command or staff position and perform satisfactorily or at a higher level of competence. There are important professional specialist branches in the army providing Combat Service Support, such as the Medical, Nurse, Dental, and Veterinary Corps, Judge Advocate General Corps, and the Chaplains Corps. Officers sought and assigned to these branches are rated on their capabilities and performance in professions mirroring those of civilian life. In fact, the supply of these officers is nearly always short owing to competition from the private sector.

Geologists, on the other hand, have not been considered mission-essential to any particular branch of the army and are often preferentially assigned—without their consensual agreement—to the combat arms for the sole recognition of their capacity to interpret topographic maps and remote imagery, to deal with terrain factors, and to understand topographic survey techniques (especially valuable in the Field Artillery). Yet, no formal credit for this unique capacity is given to geologists, and relatively few graduate geologists are preferentially assigned to the Corps of Engineers or to Military Intelligence, the most logical branches to appreciate and make frequent and important use of their talents.

Military geology, where and when employed, has demonstrated, for the past two centuries, a generally outstanding string of battlefield accomplishments in support of both the offense and defense. As the U.S. Army becomes smaller and depends more and more on sophistication of military technologies to maintain, or even increase, its combat effectiveness, there is an even greater need for the integration of a new wave of military geologists.

Integration always has been a significant weakness in the application of technology. Where technology leaves the shelves and racks of weapons and other military hardware, people become the substance of technology. Here has been and will be the obstacle and possible shortfall of the employment of military geology for the army.

The army has experimented over the 20th century with the use of enlisted and warrant personnel to perform in subprofessional roles, especially in signal (communications) and intelligence fields. Under the aegis of Major General Adolphus W. Greeley, Congressional Medal of Honor holder and renowned arctic explorer, exemplary noncommissioned officers of the Signal Corps were assigned duties to conduct Alaskan and arctic explorations and to compile formal printed reports of their expeditions. Enlisted personnel were trained as intelligence specialists for both World Wars, and, with the advent of aerial photography, these troops also participated in tactical image interpretation. Nevertheless, the enlisted and warrant personnel operated mainly on detection and evaluation of obvious image evidence of the presence and movement of enemy troops, not on geologic details influencing tactical options and results.

During the Viet Nam War the U.S. Army employed college-level enlisted personnel in an advancement in image interpretation. Here, drafted or volunteer soldiers with some degree of formal academic training as geologists, geographers, foresters, and other allied earth scientists were turned into *terrain evaluation specialists*. Their products deal mainly with the location of construction materials and, to some degree, route location selection designed to avoid geologic problems usually associated with the presence of surface water and poor-quality soil, weak rock, and rock. The experience generally was successful; yet the work product was placed in the hands of commanders largely unappreciative of its inherent value.

COMBAT MULTIPLIERS IN THE NEW WORLD ORDER

With the 1989 fall of the Soviet Union and the disintegration of the Warsaw Pact, the United States, the North Atlantic Treaty Organization (NATO), and the former Warsaw Pact nations have found themselves engaging in an increasing tenor of constructive communication, negotiations, and emerging forms of cooperation against mutual threats. Mutual threats facing the former world-class adversaries now come in the form of insurrections and civil war among the Third World nations or such fragmented former entities as Yugoslavia.

World order now depends on the United Nations (U.N.) to act responsibly in the face of insurrection, civil war, and genocide. In some instances, due to the inability of the U.N. to take constructive action, the United States, the United Kingdom, and to a lesser degree, France, have seen fit to project their influence into an unusual array of regional geographic conditions at great distances from home. In most of these situations, the projection of a First-World military force, as an instrument of national interest, is performed at considerable risk to the military personnel involved. Such risk is enveloped in geologic considerations such as the condition of the airfields or landing zones to accept incoming flights to trafficability, cross-country mobility of wheeled and tracked vehicles, road and bridge construction, field sanitation and water supply, and terrain analysis for defense and offense.

HISTORIC PRECEDENCE

To the authors' knowledge, the greatest use of military geology over the long period in the 20th century has been Germany. The German military geological community was highly integrated into what we know today as engineering geology. The military reserve structure of Germany allowed for and encouraged the participation of applied geologists in the entire spectrum of preparation for and conduct of warfare.

Between World Wars I and II, military geologists were many of the leading German academic and State survey engineering geologists. A spate of their published works appears in the bibliography appearing in this volume. An oft-forgotten success was the military geographic series of formal printed volumes, which was made ready for the blitzkrieg actions that accompanied the unfortunate opening of World War II and that carried the Wehrmacht to its initial successes throughout Europe and the Mediterranean.

There is no question but that Britain and the United States were able to at least match the military geologic competence of Germany, but the ramp-up to locating and employing such geologists was not successful until early 1943. Equally delaying was the fact that the General Staffs of the two armies had not been sensitized to the capabilities and advantages of embracing military geology. Generally, it was Corps of Engineers officers at the highest level who managed to promote acceptance of military geologic assessments.

MILITARY GEOLOGY IN THE U.S. ARMY

Recognizing that military geology can always be employed to achieve some degree of enhanced success on the battlefield the problem has been, and will continue to be, that typical troop-oriented field commanders have little or no appreciation of the nature of geology or of the many ways in which its correct interpretation can be used to their significant advantage.

Technical specialists have traditionally been protected in the army through assignment to separate branches, commanded and led by officers of their own discipline. The traditional branches of nonmilitary professional disciplines are the Medical, Dental, Nurse, and Veterinary Corps, along with the Judge Advocate General branch, which includes the army's lawyers. The engineers construct and demolish, and they are the first troops to be committed in combat to support the infantry, in association with Artillery, Armor, and Aviation, in dealing blows directly to the enemy.

Virtually nothing has changed in terms of the genuine need for employable military geologic information in the hands of the field commander. Military geology in today's U.S. Army, however, is essentially nonexistent, except where promoted and utilized by stubborn professional officers who possess degrees in geology. Former Chairman of the U.S. Joint Chiefs of Staff, General Colin C. Powell, for example, is a geologist (B.S., City College of New York) who never served in such an assignment throughout his career. The army did not recognize this capability, and the general survived and prospered in the army as a troop leader, not as a geologist.

While there is an outstanding and critical need for rapid access to military geologic products in the army today, there is no such mechanism to deliver such products. The Military Intelligence branch has been transformed into an electronic collection and assessment role, and it remains that only the Corps of Engineers retains the terrain intelligence role.

Today, when the Army General Staff is directed to contingency or operational planning, there are only two sources of military geologic input. The traditional source is to request a product from the small group of military geologic specialists at the U.S. Geological Survey National Headquarters at Reston, Virginia. We sense that the U.S. Geological Survey is not geared to rapid production of operational-level military geological products.

The nontraditional source of military geology is embodied in the army's two largest engineer units, both lodged appropriately in the U.S. Army Reserve. These are the 412th and 416th Engineer Commands (ENCOMs) of Vicksburg, Mississippi, and Chicago, Illinois. The former has strong ties with the U.S. Army Corps of Engineers Waterways Experiment Station (WES) in the same city.

Of the two, the 416th ENCOM has a particularly responsive nature, having been carried to prominence by two consulting engineers from Kansas City, Missouri. The unique nature of the 416th ENCOM is that it will provide military geological input on extremely short notice to any requesting army combat command. Colonels Robert Van Zandt and Robert Bay (later major general and a president of the American Society of Civil Engineers), both now retired from the Army Reserve, were principals of the architectural/engineering and environmental firm of Black & Veatch, founded in Kansas City, Missouri, in 1895.

In the late 1970s, Colonels Van Zandt and Bay, in visiting the Pentagon as drilling reservists, became aware of the need not only for military geologists but also for civil engineers to handle "what-if" contingency planning as well as to provide rapid response and operational technical expertise in the event of post–Viet Nam military commitments around the world. The officers promised to deliver the expertise through a network of identified and screened professional geologists and engineers of the Army Reserve, both as paid-status reservists and as members of the unpaid Individual Ready Reserve (IRR). Van Zandt and Bay created their network, and the 416th ENCOM now has a strong reputation of being able to locate and deliver qualified expertise on short notice.

It should be recalled, as an example of high-level practical technical experience, that then-Colonel Lief J. Sverdrup (a young World War I artillery officer who was never sent to France) was recommissioned in 1942 from civilian life by the army and spent 39 uninterrupted months in the South Pacific during World War II in engineer technical support of all manner of military operations, particularly airfield location and design. During the war his small, struggling St. Louis consulting engineering firm was managed by two trusted partners. Major General Sverdrup, a graduate civil engineer, later served a full career in the Army Reserve, and his firm has prospered to become one of America's largest architectural/engineering concerns.

WHAT CAN BE DONE IN AND FOR TODAY'S ARMY

Providing military geology at the time, scale, and breadth sufficient to be included in tactical operations calls for three main ingredients:

(1) topographic maps of the scale appropriate to the unit's mission;

(2) recent aerial photography in stereo coverage, along with whatever satellite photographs that can be identified and furnished in a timely manner; and

(3) a competent military geologist on the Special Staff for units commanded by field-grade officers at Colonel or lower

ranks or at the General Staff for units commanded by general officers.

Topographic maps are usually either present or can be acquired in hours to days. The army has the combat capability to procure aerial photographs (sometimes with the aid of the supporting Tactical Air Force unit). Competent military geologists are the subject of this chapter. These officers should ideally be captains at the Battalion level; majors at Brigade; lieutenant colonels at Division, and colonels at higher headquarters. Lieutenants normally will be detailed to the introductory levels of tactical and troop-leading experience.

We propose that military geologists be commissioned in and assigned to the Corps of Engineers, the historic home for the greatest number of federal civilian geologists in the past, at least since 1929. This would be the ideal branch for their career development. The fact that most geologists are scientists, not engineers, does not present an obstacle to the way in which the Corps functions. The Corps has a constant shortfall of officers holding engineering degrees. Part of the engineer shortfall is that the army insists on sharing its incoming supply of commissioned graduate engineers across the force. Military education for Corps of Engineers officers has been standardized into an almost wholly empirical engineering content, so that most graduates of four-year academic institutions have the innate intellectual capacity to grasp and make use of the various rules of thumb, tables, charts, and diagrams that have been developed as solutions to combat engineering construction and demolition tasks. The senior author recalls once being assigned with a successful Corps Major who held a baccalaureate in music!

The proposed military geologist is therefore a fully qualified combat leader who functions within the Corps of Engineers. He or she can utilize topographic maps and aerial photographs in the combat operations zone and produce a transparent overlay showing various forms of combat restrictions and enhancements for the commander at the same scale (generally 1:50,000) used by the commander for all other purposes.

Contrary to the high cost and restricted employment of most military weapons and intelligence systems, this military geologic capability can be built, maintained, and employed at the most minimal of costs. The standard U.S. Army Photointerpretation Kit (FSN 6675-202-8542) is an olive drab fiberglass attaché case weighing about 10 kg and containing everything necessary for geologic field work except a Brunton compass and a rock hammer. Additionally, several key technical references should be offset printed and bound for issue, including the English translation of von Bulow's classic 1938 handbook, *Wehrgeologie*, along with one bound volume of reprints dealing with the nature and history of military geology.

Annual training (A.T.) for these military geology specialist Reserve officers would be on a two-and-one basis: One year a two-week refresher course/workshop is held as a tactical exercise with compatriots from the combat force structure, and two alternate years of two-week periods of annual training are devoted to work with the staff to which the military geologist is assigned. The one-in-three-year A.T. workshops should be held at a military installation in a geologic region for which field reconnaissance can be made, aerial observation is a possibility, and aerial photographs have been secured. This training should be placed under the directorship and direct guidance of a full colonel who is in the military geology specialty. Ideally, the director would be a university professor or manager of the engineering geologic group of a public- or private-sector organization.

Doctrine for the military geologic specialty should come from a Corps of Engineers activity at its relocated Engineer Center and School, at Fort Leonard Wood, Missouri, where senior Reserve military geologists in the grade of colonel are brought in for two-week A.T. periods (say, two individuals per year) to develop new and mission-relevant materials for succeeding workshops. The Appendix lists the current doctrinal publications on military geology. The U.S. Geological Survey and the National Imagery and Mapping Agency (formerly the Defense Mapping Agency) can be charged with providing background geologic materials, providing added realism to the workshop, and producing materials that may actually prove useful on a contingency basis against American force projection into overseas areas.

MAKING IT ALL WORK

Nothing is successful in the U.S. Army without command emphasis. The successful recipe calls for the ingredients of an operational plan, a schedule for implementation, a commitment of officer personnel on the *Tables of Organization and Equipment* (TOE), and a requirement for follow-up and evaluation. Furthermore, officers being service-school trained for command will be instructed in the need and reason for inclusion of military geologic input. We lost a window of opportunity during the time when General Colin Powell, a geologist, was chairman of the Joint Chiefs of Staff.

Military geologists would be assigned from the Army Reserve to the Regular Army for call to active duty in the event the unit is mobilized. Army National Guard combat units (independent brigades and divisions) would have military geologists assigned directly to them from within the National Guard. There should be a backup cadre of casualty-replacement military geologist officers available from the Individual Ready Reserve or from geologists who are commissioned officers in the "gray area" of retirement (retired on paper but unable to receive retirement pay until age 60, as stipulated by Congress). Although these officers do not receive retirement benefits, they are mandatorily subjected to instant recall and carry so-called hip-pocket orders directing them to report to predesigned duty stations in the event of an authorization by the President.

The concept presented herein would call for identification of a cadre of about 100 Reserve Office graduate geologists in the paid-drill units of the Army Reserve and the Army National Guard, and in the nonpaid Individual Ready Reserve of the Army Reserve.

Commanders would be taught the capabilities and advantages of military geology at each of the branch service schools and at the senior officer service schools such as the Command and General Staff College and the Army War College. This instruction could be conducted in a one- or two-hour block during which the instruction would emphasize historic case histories of campaigns and battles influenced by geologic conditions and by strong, graphic representations of military geologic products for use by the command and staff. Ideally, an additional two hours of field demonstration, such as the army's traditional "walk-through" form of instruction, could bring the message home.

SUMMARY

Significant combat multipliers can be achieved on a minimal basis by employment of otherwise-trained geologists with a military background. No special weapons systems or costly hardware are required—just intelligent and dedicated professional geologists who are experienced army officers. Such people can already be found in the Reserve components and be recruited for reassignments to such duty.

APPENDIX. U.S. ARMY DOCTRINE ON GEOLOGY

The most recent army manual on geology (U.S. Army, 1994) is an initial draft of a field manual. It replaces a field circular (U.S. Army, 1986) that was a photocopied reproduction made to respond to needed training and operational technical doctrine and fielded for limited periods of time against perceived needs. The field circular had replaced a technical manual (U.S. Army, 1971) that was the third edition of a work previously published in 1952 and 1967.

None of these documents matches the classic *Wehrgeologie* (von Bulow, 1938). The 1938 Leipzig edition by Dr. von Bulow (professor at the University of Rostock, director of the Geological Institute of Mecklenberg), in collaboration with Major Dr. Walter Kranz (retired; Society of German Engineers and state geologist of Wurtemberg) and Major Erich Sonne (graduate engineer, former inspector of fortifications, western front, World War I), with contributions by Prof. Dr. Otto Burre and Prof. Dr. Wilhelm Dienemann (District Geologists of Prussia), was translated February 15, 1943, as Engineer Research Office translation No. T-23, by the Office of Chief of Engineers, Intelligence Branch, New York City. It was edited from an in-house translation, managed by Dr. Kurt E. Lowe, associate geologist, with a typewritten manuscript and hand-colored photocopies of the maps. Lowe was then a refugee emigrant from Germany and teaching geology at the City College of New York; he was inducted into the U.S. Army Air Corps shortly after completion of the translation editing. Copy no. 10 of the translation is held at U.S. Geological Society National Library, Reston, Virginia, as 203.5 qB86We).

REFERENCES CITED

U.S. Army, 1971, Geology: Washington, D.C., Office of Chief of Engineers, TM-5-545, 8 July, 93 p.

U.S. Army, 1986, Geology: Ft. Belvoir, Virginia, U.S. Army Engineer School, April, Field Circular, 355 p.

U.S. Army, 1994, Geology: FM-5-445: Washington, D.C., Headquarters, Department of the Army, Initial Draft, December, var. p.

von Bulow, K., 1938, Wehrgeologie (Military Geology): Leipzig, Quelle and Meyer, 234 p. (See notes in the Appendix on its translation into English.)

Manuscript Accepted by the Society October 29, 1997

Printed in U.S.A.

Index

[Italic page numbers indicate major references]

U

V

W

Y

Printed in U.S.A.